Asymptotic Analysis of Random Walks

This is a companion book to *Asymptotic Analysis of Random Walks: Heavy-Tailed Distributions* by A. A. Borovkov and K. A. Borovkov. Its self-contained systematic exposition provides a highly useful resource for academic researchers and professionals interested in applications of probability in statistics, ruin theory, and queuing theory. The large deviation principle for random walks was first established by the author in 1967, under the restrictive condition that the distribution tails decay faster than exponentially. (A close assertion was proved by S. R. S. Varadhan in 1966, but only in a rather special case.) Since then, the principle has been treated in the literature only under this condition. Recently, the author, jointly with A. A. Mogul'skii, removed this restriction, finding a natural metric for which the large deviation principle for random walks holds without any conditions. This new version is presented in the book, as well as a new approach to studying large deviations in boundary crossing problems. Many results presented in the book, obtained by the author himself or jointly with co-authors, are appearing in a monograph for the first time.

A. A. Borovkov is Principal Research Fellow and a Russian Academy of Sciences Adviser at the Sobolev Institute of Mathematics, where he also served as the founding head of the Department of Probability Theory and Mathematical Statistics. He was the founding Chair in Probability Theory and Mathematical Statistics at the Novosibirsk State University in 1965. He has authored and co-authored several influential research monographs. In 1979 he was awarded (together with V. V. Sazonov and V. Statulevičius) the State Prize of the USSR for outstanding research results on asymptotic methods of probability theory, and was elected a full member of the Academy of Sciences of the USSR in 1990. He received the 2002 Russian Government Prize in Education, the 2003 Russian Academy of Sciences A. A. Markov Prize, and in 2015 the Russian Academy of Sciences A. A. Kolmogorov Prize for his joint work with A. A. Mogul'skii on the extended large deviation principle (2015).

Encyclopedia of Mathematics and Its Applications

This series is devoted to significant topics or themes that have wide application in mathematics or mathematical science and for which a detailed development of the abstract theory is less important than a thorough and concrete exploration of the implications and applications.

Books in the **Encyclopedia of Mathematics and Its Applications** cover their subjects comprehensively. Less important results may be summarized as exercises at the ends of chapters. For technicalities, readers can be referred to the bibliography, which is expected to be comprehensive. As a result, volumes are encyclopedic references or manageable guides to major subjects.

ENCYCLOPEDIA OF MATHEMATICS AND ITS APPLICATIONS

All the titles listed below can be obtained from good booksellers or from Cambridge University Press. For a complete series listing, visit www.cambridge.org/mathematics

126 O. Calin and D.-C. Chang *Sub-Riemannian Geometry*
127 M. Grabisch *et al. Aggregation Functions*
128 L. W. Beineke and R. J. Wilson (eds.) with J. L. Gross and T. W. Tucker *Topics in Topological Graph Theory*
129 J. Berstel, D. Perrin and C. Reutenauer *Codes and Automata*
130 T. G. Faticoni *Modules over Endomorphism Rings*
131 H. Morimoto *Stochastic Control and Mathematical Modeling*
132 G. Schmidt *Relational Mathematics*
133 P. Kornerup and D. W. Matula *Finite Precision Number Systems and Arithmetic*
134 Y. Crama and P. L. Hammer (eds.) *Boolean Models and Methods in Mathematics, Computer Science, and Engineering*
135 V. Berthé and M. Rigo (eds.) *Combinatorics, Automata and Number Theory*
136 A. Kristály, V. D. Rădulescu and C. Varga *Variational Principles in Mathematical Physics, Geometry, and Economics*
137 J. Berstel and C. Reutenauer *Noncommutative Rational Series with Applications*
138 B. Courcelle and J. Engelfriet *Graph Structure and Monadic Second-Order Logic*
139 M. Fiedler *Matrices and Graphs in Geometry*
140 N. Vakil *Real Analysis through Modern Infinitesimals*
141 R. B. Paris *Hadamard Expansions and Hyperasymptotic Evaluation*
142 Y. Crama and P. L. Hammer *Boolean Functions*
143 A. Arapostathis, V. S. Borkar and M. K. Ghosh *Ergodic Control of Diffusion Processes*
144 N. Caspard, B. Leclerc and B. Monjardet *Finite Ordered Sets*
145 D. Z. Arov and H. Dym *Bitangential Direct and Inverse Problems for Systems of Integral and Differential Equations*
146 G. Dassios *Ellipsoidal Harmonics*
147 L. W. Beineke and R. J. Wilson (eds.) with O. R. Oellermann *Topics in Structural Graph Theory*
148 L. Berlyand, A. G. Kolpakov and A. Novikov *Introduction to the Network Approximation Method for Materials Modeling*
149 M. Baake and U. Grimm *Aperiodic Order I: A Mathematical Invitation*
150 J. Borwein *et al. Lattice Sums Then and Now*
151 R. Schneider *Convex Bodies: The Brunn–Minkowski Theory (Second Edition)*
152 G. Da Prato and J. Zabczyk *Stochastic Equations in Infinite Dimensions (Second Edition)*
153 D. Hofmann, G. J. Seal and W. Tholen (eds.) *Monoidal Topology*
154 M. Cabrera García and Á. Rodríguez Palacios *Non-Associative Normed Algebras I: The Vidav–Palmer and Gelfand–Naimark Theorems*
155 C. F. Dunkl and Y. Xu *Orthogonal Polynomials of Several Variables (Second Edition)*
156 L. W. Beineke and R. J. Wilson (eds.) with B. Toft *Topics in Chromatic Graph Theory*
157 T. Mora *Solving Polynomial Equation Systems III: Algebraic Solving*
158 T. Mora *Solving Polynomial Equation Systems IV: Buchberger Theory and Beyond*
159 V. Berthé and M. Rigo (eds.) *Combinatorics, Words and Symbolic Dynamics*
160. B. Rubin *Introduction to Radon Transforms: With Elements of Fractional Calculus and Harmonic Analysis*
161 M. Ghergu and S. D. Taliaferro *Isolated Singularities in Partial Differential Inequalities*
162 G. Molica Bisci, V. D. Radulescu and R. Servadei *Variational Methods for Nonlocal Fractional Problems*
163 S. Wagon *The Banach–Tarski Paradox (Second Edition)*
164 K. Broughan *Equivalents of the Riemann Hypothesis I: Arithmetic Equivalents*
165 K. Broughan *Equivalents of the Riemann Hypothesis II: Analytic Equivalents*
166 M. Baake and U. Grimm (eds.) *Aperiodic Order II: Crystallography and Almost Periodicity*
167 M. Cabrera García and Á. Rodríguez Palacios *Non-Associative Normed Algebras II: Representation Theory and the Zel'manov Approach*
168 A. Yu. Khrennikov, S. V. Kozyrev and W. A. Zúñiga-Galindo *Ultrametric Pseudodifferential Equations and Applications*
169 S. R. Finch *Mathematical Constants II*
170 J. Krajíček *Proof Complexity*
171 D. Bulacu, S. Caenepeel, F. Panaite and F. Van Oystaeyen *Quasi-Hopf Algebras*
172 P. McMullen *Geometric Regular Polytopes*
173 M. Aguiar and S. Mahajan *Bimonoids for Hyperplane Arrangements*
174 M. Barski and J. Zabczyk *Mathematics of the Bond Market: A Lévy Processes Approach*
175 T. R. Bielecki, J. Jakubowski and M. Niewęgłowski *Fundamentals of the Theory of Structured Dependence between Stochastic Processes*
176 A. A. Borovkov *Asymptotic Analysis of Random Walks: Light-Tailed Distributions*

Asymptotic Analysis of Random Walks

Light-Tailed Distributions

A. A. BOROVKOV
Sobolev Institute of Mathematics, Novosibirsk

Translated by

V. V. ULYANOV
Lomonosov Moscow State University
and HSE University, Moscow

M. V. ZHITLUKHIN
Steklov Institute of Mathematics, Moscow

CAMBRIDGE
UNIVERSITY PRESS

CAMBRIDGE
UNIVERSITY PRESS

University Printing House, Cambridge CB2 8BS, United Kingdom

One Liberty Plaza, 20th Floor, New York, NY 10006, USA

477 Williamstown Road, Port Melbourne, VIC 3207, Australia

314–321, 3rd Floor, Plot 3, Splendor Forum, Jasola District Centre, New Delhi – 110025, India

79 Anson Road, #06–04/06, Singapore 079906

Cambridge University Press is part of the University of Cambridge.

It furthers the University's mission by disseminating knowledge in the pursuit of education, learning, and research at the highest international levels of excellence.

www.cambridge.org
Information on this title: www.cambridge.org/9781107074682
DOI: 10.1017/9781139871303

First published 2020

A catalogue record for this publication is available from the British Library.

Library of Congress Cataloging-in-Publication Data
Names: Borovkov, A. A. (Aleksandr Alekseevich), 1931– author.
Title: Asymptotic analysis of random walks : light-tailed distributions / A.A. Borovkov, Sobolev Institute of Mathematics, Novosibirsk ; translated by V.V. Ulyanov, Higher School of Economics, Mikhail Zhitlukhin, Steklov Institute of Mathematics, Moscow.
Description: Cambridge, United Kingdom ; New York, NY : Cambridge University Press, 2020. | Series: Encyclopedia of mathematics and its applications | Includes bibliographical references and index.
Identifiers: LCCN 2020022776 (print) | LCCN 2020022777 (ebook) | ISBN 9781107074682 (hardback) | ISBN 9781139871303 (epub)
Subjects: LCSH: Random walks (Mathematics) | Asymptotic expansions. | Asymptotic distribution (Probability theory)
Classification: LCC QA274.73 .B6813 2020 (print) | LCC QA274.73 (ebook) | DDC 519.2/82–dc23
LC record available at https://lccn.loc.gov/2020022776
LC ebook record available at https://lccn.loc.gov/2020022777

ISBN 978-1-107-07468-2 Hardback

Contents

Introduction

This book is devoted mainly to the study of the asymptotic behaviour of the probabilities of rare events (large deviations) for trajectories of random walks. By random walks we mean the sequential sums of independent random variables or vectors, and also processes with independent increments. It is assumed that those random variables or vectors (jumps of random walks or increments of random processes) have distributions which are 'rapidly decreasing at infinity'. The last term means distributions which satisfy Cramér's moment condition (see below).

The book, in some sense, continues the monograph [42], where more or less the same scope of problems was considered but it was assumed that the jumps of random walks have distributions which are 'slowly decreasing at infinity', i.e. do not satisfy Cramér's condition. Such a division of the objects of study according to the speed of decrease of the distributions of jumps arises because for rapidly and slowly decreasing distributions those objects form two classes of problems which essentially differ, both in the methods of study that are required and also in the nature of the results obtained.

Each of these two classes of problems requires its own approach, to be developed, and these approaches have little in common. So, the present monograph, being a necessary addition to the book [42], hardly intersects with the latter in its methods and results. In essence, this is the second volume (after [42]) of a monograph with the single title *Asymptotic Analysis of Random Walks*.

The asymptotic analysis of random walks for rapidly decreasing distributions and, in particular, the study of the probabilities of large deviations have become one of the main subjects in modern probability theory. This can be explained as follows.

- *A random walk* is a classical object of probability theory, which presents huge theoretical interest and is a mathematical model for many important applications in mathematical statistics (sequential analysis), insurance theory (risk theory), queuing theory and many other fields.
- *Asymptotic analysis and limit theorems* (under the unbounded growth of some parameters, for example the number of random terms in a sum) form the chief

method of research in probability theory. This is due to the nature of the main laws in probability theory (they have the form of limit theorems), as well as the fact that explicit formulas or numerical values for the characteristics under investigation in particular problems, generally, do not exist and one has to find approximations for them.

- *Probabilities of large deviations* present a considerable interest from the mathematical point of view as well as in many applied problems. Finding the probabilities of large deviations allows one, for example, to find the small error probabilities in statistical hypothesis testing (error probabilities should be small), the small probabilities of the bankruptcy of insurance companies (they should be small as well), the small probabilities of the overflow of bunkers in queuing systems and so on. The so-called 'rough' theorems about the probabilities of large deviations (i.e. about the asymptotics of the logarithms of those probabilities; see Chapters 4 and 5) have found application in a series of related fields such as statistical mechanics (see, for example, [83], [89], [178]).
- *Rapidly decreasing distributions* deserve attention because the first classical results about the probabilities of large deviations of sums of random variables were obtained for rapidly decreasing distributions (i.e. distributions satisfying Cramér's condition). In many problems in mathematical statistics (especially those related to the likelihood principle), the condition of rapid decrease turns out to be automatically satisfied (see, for example, sections 6.1 and 6.2 below). A rapid (in particular, exponential) decrease in distributions often arises in queuing theory problems (for example, the Poisson order flow is widely used), in risk theory problems and in other areas. Therefore, the study of problems with rapidly decreasing jump distributions undoubtedly presents both theoretical and applied interest. Let us add that almost all the commonly used distributions in theory and applications, such as the normal distribution, the Poisson distribution, the Γ-distribution, the distribution in the Bernoulli scheme, the uniform distribution, etc. are rapidly decreasing at infinity.

Let $\xi, \xi_1, \xi_2, \ldots$ be a sequence of independent identically distributed random variables or vectors. Put $S_0 = 0$ and

$$S_n := \sum_{k=1}^{n} \xi_k, \qquad n = 1, 2, \ldots$$

The sequence $\{S_n;\ n \geqslant 0\}$ is called a *random walk*. As has been noted, a random walk is a classical object of probability theory. Let us mention the following fundamental results related to random walks.

- *The strong law of large numbers*, on the convergence $S_n/n \underset{\text{a.s.}}{\to} \mathbf{E}\xi$ almost surely as $n \to \infty$.
- *The functional central limit theorem*, on the convergence in distribution of the process $\zeta_n(t)$, $t \in [0, 1]$, with values $(S_k - A_k)/B_n$ at the points $t = k/n$, $k = 0, 1, \ldots, n$, to a stable process, where A_k, B_n are appropriate normalising constants. For example, in the case $\mathbf{E}\xi = 0$, $\mathbf{E}\xi^2 = \sigma^2 < \infty$, the polygonal

line $\zeta_n(t)$ for $A_k = 0$, $B_n^2 = n\sigma^2$ converges in distribution to a standard Wiener process (the invariance principle; for details, see e.g. [11], [39] or the introduction in [42]).

- *The law of the iterated logarithm*, which establishes upper and lower bounds for the trajectories of $\{S_k\}$.

None of the above results describes the *asymptotic behaviour of the probabilities of large deviations of trajectories of* $\{S_k\}$. We mention the following main classes of problems.

(a) The study of the probabilities of large deviations of sums of random variables (or vectors); for example, the study of the asymptotics (for $\mathbf{E}\xi = 0$, $\mathbf{E}\xi^2 < \infty$ in the one-dimensional case) of probabilities

$$\mathbf{P}(S_n \geqslant x) \quad \text{for} \quad x \gg \sqrt{n}, \quad n \to \infty. \tag{0.0.1}$$

(b) The study of the probabilities of large deviations in boundary crossing problems. For example, in this class of problems belongs a problem concerning the asymptotics of the probabilities

$$\mathbf{P}\biggl(\max_{t\in[0,1]}\Bigl(\zeta_n(t) - \frac{x}{\sqrt{n}}\,g(t)\Bigr) \geqslant 0\biggr) \tag{0.0.2}$$

for an arbitrary function $g(t)$ on $[0, 1]$ in the case $x \gg \sqrt{n}$ as $n \to \infty$ (the process $\zeta_n(t)$ is defined above).

(c) The study of the more general problem about the asymptotics of the probabilities

$$\mathbf{P}\Bigl(\zeta_n(\cdot) \in \frac{x}{\sqrt{n}} B\Bigr), \qquad x \gg \sqrt{n}, \tag{0.0.3}$$

where B is an arbitrary measurable set in one or another space of the functions on $[0, 1]$.

This monograph is largely devoted to the study of problems concerning probabilities (0.0.1)–(0.0.3) and other closely related problems, under the assumption that Cramér's moment condition

$$\textbf{[C]} \qquad \psi(\lambda) := \mathbf{E}e^{\lambda\xi} < \infty$$

is satisfied *for some* $\lambda \neq 0$ (its statement here is given for a scalar ξ). This condition means (see subsection 1.1.1) that at least one of the 'tails' $\mathbf{P}(\xi \geqslant x)$ or $\mathbf{P}(\xi < -x)$ of the distribution of the variable ξ decreases as $x \to \infty$ faster than some exponent. In the monograph [42], it was assumed that condition **[C]** is not satisfied but the tails of the distribution of ξ behave in a sufficiently regular manner.

As we noted above, the first general results for problem (a) about the asymptotics of the probabilities of large deviations of sums of random variables go back to the paper of Cramér. Essential contributions to the development of this direction were also made in the papers of V.V. Petrov [149], R.R. Bahadur and R. Ranga Rao [5], C. Stone [175] and others. One should also mention the papers of B.V. Gnedenko [95], E.A. Rvacheva [163], C. Stone [174] and L.A. Shepp [166] on integro-local limit theorems for sums of random variables in the zone of normal deviations, which played an important role in extending such theorems to the zone of large deviations and forming the most adequate approach (in our opinion) to problems concerning large deviations for sums S_n. This approach is presented in Chapter 2 (see also the papers of A.A. Borovkov and A.A. Mogul'skii [56], [57], [58], [59]).

The first general results about the joint distribution of S_n and $\overline{S}_n := \max_{k\leqslant n} S_k$ (this is a particular case of the boundary crossing problem (b); see (0.0.2)) in the zone of large deviations were obtained by A. A. Borovkov in the papers [15] and [16], using analytical methods based on solving generalised Wiener–Hopf equations (in terms of Stieltjes-type integrals) for the generating function of the joint distribution of S_n and $\overline{S}_n$. The solution was obtained in the form of double transforms of the required joint distribution, expressed in terms of factorisation components of the function $1 - z\psi(\lambda)$. The double transforms, as functions of the variables λ and z, can be inverted if one knows the poles of those transforms as a function of the variable λ and applies modifications of the steepest descent methods in the variable z. Those results allowed author A.A. Borovkov in [17]–[19] to find a solution to a more general problem about the asymptotics of the probabilities (0.0.2). Later in the work by A. A. Borovkov and A. A. Mogul'skii [53], the asymptotics of the probability (0.0.2) were found for some cases using direct probability methods without the factorisation technique (see Chapter 3).

The first general results on the rough asymptotics in problem (c), i.e. on the logarithmic asymptotics of the probability of the general form (0.0.3) for arbitrary sets B in the one-dimensional case (these results constitute the large deviation principle) were obtained in the papers of S.R.S. Varadhan [177] (for a special case and $x \gg n$) and A.A. Borovkov [19] (for $x = O(n)$). In the paper of A.A. Mogul'skii [131], the results of the paper [19] were transferred to the multidimensional case. The large deviation principle was later extended to a number of other objects (see, for example, [83], [178], [179]). However, the large deviation principle in the papers [177], [19], [131] was established under a very restrictive version of Cramér's condition, that $\psi(\lambda) < \infty$ for all λ, and only for continuous trajectories $\zeta_n(t)$ with $A_k = 0$, $B_n = n$. In a more recent series of papers by A.A. Borovkov and A.A. Mogul'skii [60]–[68] (see also Chapter 4) substantial progress in the study of the large deviation principle was made: Cramér's condition, mentioned above in its strong form, was weakened or totally removed and the space of trajectories was extended up to the space of functions without discontinuities of the second kind. At the same time the form of the results changed, so it became necessary to introduce the notion of the 'extended large deviation principle' (see Chapter 4).

Let us mention that in the paper of A.A. Borovkov [19] the principle of *moderately* large deviations was established as well for $\zeta_n(t)$ as $x = o(n)$, $n \to \infty$. After that A.A. Mogul'skii [131] extended those results to the multidimensional case. In a recent paper [67] those results were strengthened (see Chapter 5 for details).

Also, note that a sizable literature has been devoted to the large deviation principle for a wide class of random processes, mainly Markov processes and processes arising in statistical mechanics (see, for example, [83], [89], [93], [178], [179]). Those publications have little in common with the present monograph in their methodology and the nature of the results obtained, so we will not touch upon them.

The above overview of the results does not in any way pretend to be complete. Our goal here is simply to identify the main milestones in the development of the limit theorems in probability theory that are presented in this monograph. A more detailed bibliography will be provided in the course of the presentation.

Let us now provide a brief description of the contents of the book.

This monograph contains main and supplemental sections. The presentation of the main sections is, generally, self-sufficient; they contain full proofs. The supplemental sections contain results which are close to the main results, but the presentation in those sections is concise and, typically, proofs are not given. We have included in these supplements results whose proofs are either too cumbersome or use methods beyond the scope of this monograph. In those cases, we supply references to the omitted proofs.

This book consists of six chapters. The first chapter contains preliminary results which will be used in what follows. In sections 1.1 and 1.2 we discuss Cramér's condition and state the properties of the *deviation function*, which plays an important role throughout the book. We obtain inequalities for the distributions of the random variables S_n and their maxima $\overline{S}_n$. In section 1.3 we establish exponential Chebyshev-type inequalities for sums of random vectors in terms of deviation functions. Section 1.5 is devoted to the integro-local limit theorems of Gnedenko and Stone.

In the second chapter we study the asymptotics of the distributions of the sums S_n of random variables and vectors (problem (a)). In section 2.1, we introduce the notion of the Cramér transform and establish the so-called reduction formula, which reduces a problem concerning integro-local theorems in the zone of large deviations to the same problem in the zone of normal deviations. In section 2.2, integro-local limit theorems in the so-called Cramér deviation zone are obtained. Section 2.3 contains supplements to the results of section 2.2, which are provided without proofs. They include local theorems for densities in the homogeneous case and integro-local limit theorems in the multidimensional case. In section 2.4, we obtain integro-local theorems on the border of the Cramér zone of deviations and in section 2.5 theorems that apply outside the Cramér zone. In the latter case, it is possible to obtain substantial results only for special classes of distributions which vary sufficiently regularly at infinity. Section 2.6 contains supplements to the results of sections 2.4 and 2.5 relating to the multidimensional case. In section 2.7, we present the large deviation principle for the sums S_n. The influence of one or several 'non-homogeneous' terms in the sum S_n on the form of the results obtained in sections 2.2–2.7 is studied in section 2.8. Section 2.9 contains additions concerning the large deviation principle for renewal functions, the probability that the sequence $\{S_n\}_{n=1}^{\infty}$ reaches a distant set and other related problems.

The third chapter is devoted to boundary crossing problems (problem (b); see (0.0.2)) for random walks. In section 3.1, we investigate the limiting behaviour of the conditional distribution of the jumps of a random walk when the end of a trajectory (the sum S_n) is fixed. This allows us to understand the probabilistic meaning of the Cramér transform. In section 3.2, conditional invariance principles and the law of the iterated logarithm are established (again, when the end of a trajectory is fixed). The problem of the crossing of an arbitrary boundary by a trajectory $\{S_k\}$ is considered in section 3.3. In section 3.4, we study the joint distribution of the first time that a random walk crosses a high level and the overshoot over this level.

The first passage time over a fixed level (in particular, the zero level) is studied in section 3.5. In section 3.6, the distribution of the first passage time over a curvilinear boundary for a class of asymptotically linear boundaries is considered.

In section 3.7, we consider the same problem as in section 3.6 but for arbitrary boundaries and normalised trajectories on the segment $[0,1]$. A generalisation of those results to the multidimensional case is provided without proof in section 3.8. In section 3.9, also without proof, we give an account of the analytical approach to the study of the joint distribution of the variables S_n and $\overline{S}_n$. The probability

$$u_{x,n}^{y} = \mathbf{P}(\overline{S}_{n-1} < x,\ S_n \geqslant x+y)$$

satisfies an integro-difference equation, so the generating function $u_x^y(z) = \sum_{n=1}^{\infty} z^n u_{x,n}^y$ satisfies the generalised (in terms of the Stieltjes integral) Wiener–Hopf integral transform on the half-line. For the Laplace transform $u^y(\lambda, z)$ of $u_x^y(z)$ (in x), we find an explicit form in terms of the so-called V-factorisation of the function $1-z\psi(\lambda)$. It turns out that this double transform can be asymptotically inverted, since it is possible to find an explicit form for the pole of the function $u^y(\lambda, z)$ in the plane λ. This allows us to find the asymptotics in x of the function $u_x^y(z)$, and then, using a modification of the steepest descent method, the asymptotics of $u_{x,n}^y$.

In the last section of Chapter 3 (section 3.10) we consider numerical methods for finding values of parameters in terms of which the investigated probabilities can be described.

Chapter 4 is devoted to the large deviation principles (l.d.p.) for trajectories $\{S_k\}_{k=1}^n$ (problem (c); see (0.0.3)). The first three sections are of a preliminary nature. In the first section, we introduce the notions of a local and an extended l.d.p. for random elements in an arbitrary metric space. We find conditions when a local l.d.p. implies an extended l.d.p. In the second section, we study the functional (i.e. the integral) of deviations, in terms of which all the main results of Chapters 4 and 5 are obtained. In section 4.3, we obtain exponential Chebyshev-type inequalities for trajectories of a random walk (extensions of the inequalities in section 1.3 to the case of trajectories); using these inequalities, upper bounds on the l.d.p. will be obtained. In section 4.4, we establish strong (compared to the already known) versions of the l.d.p. for continuous normalised trajectories $\zeta_n(t)$ of a random walk as $x \sim n$, $n \to \infty$. The strengthening consists in eliminating some conditions or substantially weakening them. In section 4.5, we consider an extended setting of the problem, which arises when the Cramér condition is not required to hold over the the whole axis or under simultaneous extension of the space where the trajectories are defined to the space $\mathbb{D}$ of functions without discontinuities of the second type. We introduce a metric in this space which is more relevant to the problems we consider than the Skorokhod metric, as it allows the convergence of continuous processes to discontinuous ones. In section 4.6, we establish the local l.d.p. in the space $\mathbb{D}$ and also the so-called *extended* l.d.p. In section 4.7, l.d.p.'s in the space of functions of bounded variations are presented without proof. Conditional l.d.p.'s in the space $\mathbb{D}$ of trajectories of random walks with the end that is localised (in some form) are considered in section 4.8. Some results obtained earlier in Chapter 4 are extended in section 4.9 to processes with independent increments. As a corollary, we obtain Sanov's theorem about large deviations of empirical distribution functions. In section 4.10, we briefly discuss approaches to how one can obtain l.d.p.'s for compound renewal processes and, in section 4.11, for sums of random variables defined on a finite Markov chain.

Chapter 5 is devoted to the moderately large deviation principle (m.l.d.p.) for trajectories of random walks $\zeta_n(t)$ and processes with independent increments, when $x = o(n)$, $x \gg \sqrt{n}$ as $n \to \infty$. Section 5.1 contains a presentation of the m.l.d.p. for sums S_n of random variables. The moderately large deviation principle for trajectories is formulated and proved in section 5.2. Similar results for processes with independent increments are established in section 5.3. As a corollary, we obtain a counterpart of Sanov's theorem for moderately large deviations of empirical distribution functions. The connection of the m.l.d.p. to the invariance principle is considered in section 5.4. The end of section 5.5 is devoted to conditional m.l.d.p. with a localised end.

Some applications of the results obtained in Chapters 2–5 to problems of mathematical statistics are provided in Chapter 6. The following problems are considered: the finding of parameters (the small probabilities of errors of the first and second types) for the most powerful test for two simple hypotheses (section 6.1); the finding of parameters for optimal tests in sequential analysis (also with small probabilities of errors; section 6.2).

In section 6.3, we construct asymptotically optimal non-parametric goodness of fit tests. Some asymptotic results concerning the testing of two complex parametric hypotheses are provided in section 6.4.

In section 6.5 we find asymptotics for the main characteristics in some change point problems.

Let us now mention the main distinguishing features of the book.

(1) The traditional circle of problems about limit theorems for sums S_n in the book is considerably extended. It includes the so-called boundary crossing problems related to the crossing of given boundaries by trajectories of random walks. To this circle belong, in particular, problems about probabilities of large deviations of the maxima $\overline{S}_n = \max_{k \leqslant n} S_k$ of sums of random variables, which are widely used in applications. For the first time, a systematic presentation of a unified approach to solve the above-mentioned problems is provided. In particular, we present a direct probabilistic approach to the study of boundary crossing problems for random walks.
(2) For the first time in the monographic literature, the so-called extended large deviation principle, which is valid under considerably wider assumptions than before and under a wider problem setup, is presented.
(3) For the first time in the monographic literature, the conditional l.d.p. and the moderately large deviation principle are presented.
(4) Results concerning the large deviation principle are extended to multidimensional random walks and in some cases to processes with independent increments.
(5) The book contains a considerable number of applications of the results we obtain to certain problems in mathematical statistics.

The author is grateful to A.A. Mogul'skii and A.I. Sakhanenko for useful remarks, and also to T.V. Belyaeva for the help with preparing the manuscript for printing.

1

Preliminary results

1.1 Deviation function and its properties in the one-dimensional case

1.1.1 Cramér's conditions

Let $\xi, \xi_1, \xi_2, \ldots$ be a sequence of independent identically distributed random variables,

$$S_0 = 0, \quad S_n = \sum_{k=1}^{n} \xi_k \quad \text{for } n \geqslant 1, \quad \overline{S}_n = \max_{0 \leqslant k \leqslant n} S_k.$$

An important role in describing the distribution asymptotics of the values S_n and $\overline{S}_n$, as well as the whole trajectory of $\{S_k\}_{k=0}^{n}$ for large n, is played by the so-called *deviation function* (rate function). The deviation function is most informative if at least one of Cramér's moment conditions is met:

[C$_\pm$] *There exists* $\lambda \gtrless 0$ *such that*

$$\psi(\lambda) := \int e^{\lambda t} \mathbf{P}(\xi \in dt) < \infty.$$

A condition **[C]** will be used to mark the fulfilment of at least one of these conditions:

[C] = **[C$_+$]** $\bigcup$ **[C$_-$]**.

We denote an intersection of the conditions **[C$_\pm$]** as

[C$_0$] = **[C$_+$]** $\bigcap$ **[C$_-$]**.

Condition **[C$_0$]** means, evidently, that

$$\psi(\lambda) < \infty \quad \text{for sufficiently small } |\lambda|.$$

If

$$\lambda_+ := \sup\big\{\lambda : \psi(\lambda) < \infty\big\}, \qquad \lambda_- = \inf\big\{\lambda : \psi(\lambda) < \infty\big\},$$

then correspondingly conditions **[C$_\pm$]**, **[C]**, **[C$_0$]** can be written in the forms

$$\lambda_\pm \gtrless 0, \quad \lambda_+ - \lambda_- > 0, \quad |\lambda_\pm| > 0.$$

These conditions, which are called Cramér's conditions, characterise the decay rate of the 'tails' $F_{\pm}(t)$ for the distribution of a random variable ξ. When the condition $[\mathbf{C}_+]$ is met, by virtue of Chebyshev's exponential inequality we have

$$F_+(t) := \mathbf{P}(\xi \geqslant t) \leqslant e^{-\lambda t}\psi(\lambda) \quad \text{for } \lambda \in (0, \lambda_+), \quad t > 0,$$

and, therefore, $F_+(t)$ decreases exponentially as $t \to \infty$. Conversely, if $F_+(t) < ce^{-\mu t}$ for some $c < \infty$, $\mu > 0$ and for all $t > 0$, then for $\lambda \in (0, \mu)$ we have

$$\int_{-\infty}^{0} e^{\lambda t}\mathbf{P}(\xi \in dt) \leqslant 1 - F_+(0),$$

$$\int_{0}^{\infty} e^{\lambda t}\mathbf{P}(\xi \in dt) = -\int_{0}^{\infty} e^{\lambda t} dF_+(t) = F_+(0) + \lambda \int_{0}^{\infty} e^{\lambda t} F_+(t)\, dt$$

$$\leqslant F_+(0) + c\lambda \int_{0}^{\infty} e^{(\lambda-\mu)t} dt = F_+(0) + \frac{c\lambda}{\mu - \lambda} < \infty,$$

$$\psi(\lambda) \leqslant 1 + \frac{c\lambda}{\mu - \lambda} < \infty.$$

There is a similar connection between the decay rate of $F_-(t) = \mathbf{P}(\xi \leqslant -t)$ as $t \to \infty$ and the finiteness of $\psi(\lambda)$ under condition $[\mathbf{C}_-]$.

It is clear that condition $[\mathbf{C}_0]$ implies the exponential decay of $F_+(t) + F_-(t) = \mathbf{P}\big(|\xi| \geqslant t\big)$, and vice versa.

Hereafter we also use the conditions

$$[\mathbf{C}_{\infty\pm}] = \{\lambda_{\pm} = \pm\infty\}$$

and the condition

$$[\mathbf{C}_{\infty}] = \big\{|\lambda_{\pm}| = \infty\big\} = [\mathbf{C}_{\infty+}] \bigcap [\mathbf{C}_{\infty-}].$$

It follows from the above that the condition $[\mathbf{C}_{\infty+}]$ ($[\mathbf{C}_{\infty-}]$) is equivalent to the fact that the tail $F_+(t)$ ($F_-(t)$) diminishes faster than any exponent, as t increases.

It is clear that, for instance, an exponential distribution meets condition $[\mathbf{C}_+] \bigcap [\mathbf{C}_{\infty-}]$ while a normal distribution meets condition $[\mathbf{C}_{\infty}]$.

Along with Cramér's conditions we will also assume that the random variable ξ is *not degenerate*, i.e. $\xi \not\equiv$ const. (or $\mathbf{D}\xi > 0$, which is the same).

The properties of the Laplace transform $\psi(\lambda)$ of a distribution of random variable ξ are set forth in various textbooks; see e.g. [39]. Let us mention the following three properties, which we are going to use further on.

(Ψ1) *The functions $\psi(\lambda)$ and $\ln \psi(\lambda)$ are strictly convex; the ratio $\dfrac{\psi'(\lambda)}{\psi(\lambda)}$ strictly increases on (λ_-, λ_+).*

The analyticity property of $\psi(\lambda)$ in a strip $\operatorname{Re}\lambda \in (\lambda_-, \lambda_+)$ can be supplemented with the following 'extended' continuity property on a segment $[\lambda_-, \lambda_+]$ (on the strip $\operatorname{Re}\lambda \in [\lambda_-, \lambda_+]$).

(Ψ2) *The function $\psi(\lambda)$ is continuous 'from within' a segment $[\lambda_-, \lambda_+]$; i.e. $\psi(\lambda_{\pm} \mp 0) = \psi(\lambda_{\pm})$ (the cases $\psi(\lambda_{\pm}) = \infty$ are not excluded).*

The continuity on the whole line might fail as, for instance, $\lambda_+ < \infty$, $\psi(\lambda_+) < \infty$, $\psi(\lambda_+ + 0) = \infty$, which is the case for the distribution of a random variable ξ with density $f(x) = cx^{-3}e^{-\lambda_+ + x}$ for $x \geqslant 1$, $c = \text{const}$.

(Ψ3) *If* $\mathbf{E}\,|\xi|^k < \infty$ *and the right-hand side of Cramér's condition* $[\mathbf{C}_+]$ *is met, then the function* ψ *is k times right-differentiable at the point* $\lambda = 0$,

$$\psi^{(k)}(0) = \mathbf{E}\,\xi^k =: a_k,$$

and, as $\lambda \downarrow 0$,

$$\psi_\xi(\lambda) = 1 + \sum_{j=1}^{k} \frac{\lambda^j}{j!} a_j + o(\lambda^k).$$

It also follows that the next representation takes place as $\lambda \downarrow 0$:

$$\ln \psi_\xi(\lambda) = \sum_{j=1}^{k} \frac{\gamma_j \lambda^j}{j!} + o(\lambda^k), \tag{1.1.1}$$

where the γ_j are so-called *semi-invariants* (or *cumulants*) of order j of a random variable ξ. It is not difficult to check that

$$\gamma_1 = a_1, \qquad \gamma_2 = a_2^0 = \sigma^2, \qquad \gamma_3 = a_3^0, \ \ldots, \tag{1.1.2}$$

where $a_k^0 = \mathbf{E}\,(\xi - a_1)^k$ is a central moment of kth order.

1.1.2 Deviation function

Under the condition $[\mathbf{C}]$, a pivotal role in describing the asymptotics of probabilities $\mathbf{P}(S_n \geqslant x)$ is played by the *deviation function.*

Definition 1.1.1. *The deviation function of a random variable* ξ is the function [1]

$$\Lambda(\alpha) := \sup_\lambda \big(\alpha\lambda - \ln \psi(\lambda)\big). \tag{1.1.3}$$

The meaning of the name will become clear later. In classical convex analysis the right-hand side of (1.1.3) is known as the *Legendre transform* of the function $A(\lambda) := \ln \psi(\lambda)$.

Consider a function $A(\alpha, \lambda) := \alpha\lambda - A(\lambda)$ presented under the sup sign in (1.1.3). The function $-A(\lambda)$ is strictly concave (see property (Ψ1)), so the function $A(\alpha, \lambda)$ is the same (note also that $A(\alpha, \lambda) = -\ln \psi_\alpha(\lambda)$, where $\psi_\alpha(\lambda) = e^{-\lambda\alpha}\psi(\lambda)$ is the Laplace transform of the distribution of the random variable $\xi - \alpha$ and, therefore, from the 'qualitative' point of view, $A(\alpha, \lambda)$ possesses all the properties of the function $-A(\lambda)$). It follows from what has been said that there always exists a *unique* point $\lambda = \lambda(\alpha)$ on the 'extended' real line $[-\infty, \infty]$),

[1] This function is often referred to as a rate function.

where the sup in (1.1.3) is attained. When α increases, the values of $A(\alpha,\lambda)$ for $\lambda > 0$ will increase (in proportion to λ), and for $\lambda < 0$ they will decrease. Therefore, a graph of $A(\alpha,\lambda)$ as a function of λ, will, roughly speaking, 'roll over' to the right as α increases. It means that the maximum point $\lambda(\alpha)$ will also shift to the right (or will stay still, if $\lambda(\alpha) = \lambda_+$).

Let us move on to an exact formulation. The following three sets play an important role in studying the properties of the function $\Lambda(\alpha)$:

$$\mathcal{A} = \{\lambda : A(\lambda) < \infty\} = \{\lambda : \psi(\lambda) < \infty\}, \qquad \mathcal{A}' = \{A'(\lambda) : \lambda \in \mathcal{A}\},$$

and the convex envelope S of the support of the distribution of ξ. It is clear that the values $\lambda_\pm$ are bounds for the set $\mathcal{A}$. The right and left bounds $\alpha_\pm$, $s_\pm$ for the sets $\mathcal{A}'$, S, are evidently given by

$$\begin{aligned}
\alpha_\pm &= A'(\lambda_\pm \mp 0) = \frac{\psi'(\lambda_\pm \mp 0)}{\psi(\lambda_\pm \mp 0)};\\
s_+ &= \sup\{t : \mathbf{P}(\xi \geqslant t) < 1\},\\
s_- &= \inf\{t : \mathbf{P}(\xi \leqslant t) > 0\},
\end{aligned}$$

where $A'(\lambda_+ - 0) = \lim_{\lambda \uparrow \lambda_+} A'(\lambda)$ and $A'(\lambda_- + 0)$ is defined analogously.

If $s_+ < \infty$ (the variable ξ is bounded above), then asymptotically the function $A(\lambda)$, as $\lambda \to \infty$, will increase linearly, so that $\Lambda(\alpha) = \infty$ for $\alpha > s_+$. In a similar way $\Lambda(\alpha) = \infty$ for $s_- > -\infty$, $\alpha < s_-$. Thus, we may confine ourselves to considering the properties of the function Λ on $[s_-, s_+]$.

The value of α_+ determines the angle at which a curve $A(\lambda) = \ln \psi(\lambda)$ meets a point $\big(\lambda_+, A(\lambda_+)\big)$. The value of α_- has an analogous meaning. If $\alpha \in [\alpha_-, \alpha_+]$ then the equation $A'_\lambda(\alpha,\lambda) = 0$ or, which is the same, the equation

$$\frac{\psi'(\lambda)}{\psi(\lambda)} = \alpha, \qquad (1.1.4)$$

always has a unique solution $\lambda(\alpha)$ on the segment $[\lambda_-, \lambda_+]$ (the values of $\lambda_\pm$ can be infinite). This solution $\lambda(\alpha)$, as the inverse function to the function $\psi'(\lambda)/\psi(\lambda)$ (see (1.1.4)), which is analytic and strictly increasing on (λ_-, λ_+), is also analytic and strictly increasing on (α_-, α_+):

$$\lambda(\alpha) \uparrow \lambda_+ \quad \text{as} \quad \alpha \uparrow \alpha_+; \qquad \lambda(\alpha) \downarrow \lambda_- \quad \text{as} \quad \alpha \downarrow \alpha_-. \qquad (1.1.5)$$

From the equations

$$\Lambda(\alpha) = \alpha\lambda(\alpha) - A\big(\lambda(\alpha)\big), \qquad \frac{\psi'(\lambda(\alpha))}{\psi(\lambda(\alpha))} = \alpha \qquad (1.1.6)$$

we obtain

$$\Lambda'(\alpha) = \lambda(\alpha) + \alpha\lambda'(\alpha) - \frac{\psi'(\lambda(\alpha))}{\psi(\lambda(\alpha))}\lambda'(\alpha) = \lambda(\alpha).$$

Taking into account that $\psi'(0)/\psi(0) = a_1 = \mathbf{E}\xi$, $0 \in [\lambda_-, \lambda_+]$, $a_1 \in [\alpha_-, \alpha_+]$, we get the following representation for the function Λ:

(Λ1) *If* $\alpha_0 \in [\alpha_-, \alpha_+]$, $\alpha \in [\alpha_-, \alpha_+]$, *then*

$$\Lambda(\alpha) = \Lambda(\alpha_0) + \int_{\alpha_0}^{\alpha} \lambda(v)dv. \tag{1.1.7}$$

Since $\lambda(a_1) = \Lambda(a_1) = 0$ (*which follows from* (1.1.4) *and* (1.1.6)), *in particular for* $\alpha_0 = a_1$ *we have*

$$\Lambda(\alpha) = \int_{a_1}^{\alpha} \lambda(v)dv. \tag{1.1.8}$$

The functions $\lambda(\alpha)$, $\Lambda(\alpha)$ *are analytic on* (α_-, α_+).

Now let us consider what is happening outside the segment $[\alpha_-, \alpha_+]$. For definiteness, let $\lambda_+ > 0$. We are going to study the behaviour of the functions $\lambda(\alpha)$, $\Lambda(\alpha)$ for $\alpha \geqslant \alpha_+$. Similar considerations can be made in the case $\lambda_- < 0$, $\alpha \leqslant \alpha_-$.

First let $\lambda_+ = \infty$, i.e. let the function $\ln \psi(\lambda)$ be analytic on the whole semiaxis $\lambda > 0$, so that the tail $F_+(t)$ decreases, as $t \to \infty$, faster than any exponent. We will assume, without loss of generality, that

$$s_+ > 0, \qquad s_- < 0. \tag{1.1.9}$$

This can always be achieved using a shift transformation of a random variable; further, we assume, without loss of generality, as in many limit theorems about the distribution of S_n, that $\mathbf{E}\,\xi = 0$, using the fact that the problem of examining the distribution of S_n is 'translation-invariant'. It can also be noted that $\Lambda_{\xi-a}(\alpha - a) = \Lambda_\xi(\alpha)$ (see property (Λ4) below) and that (1.1.9) always holds, if $\mathbf{E}\,\xi = 0$.

(Λ2) (i) *If* $\lambda_+ = \infty$ *then* $\alpha_+ = s_+$.

Hence, if $\lambda_+ = \infty$, $s_+ = \infty$, then we always have $\alpha_+ = \infty$ and for any $\alpha \geqslant \alpha_-$ both (1.1.7) and (1.1.8) hold.

(ii) *If* $s_+ < \infty$ *then* $\lambda_+ = \infty$, $\alpha_+ = s_+$,

$$\Lambda(\alpha_+) = -\ln \mathbf{P}(\xi = s_+), \qquad \Lambda(\alpha) = \infty \quad \textit{for} \quad \alpha > \alpha_+.$$

Similar statements are true for s_-, α_-, λ_-.

Proof. (i) First let $s_+ < \infty$. Then the asymptotics of $\psi(\lambda)$ and $\psi'(\lambda)$, as $\lambda \to \infty$, are determined by the corresponding integrals in the vicinity of s_+:

$$\psi(\lambda) \sim \mathbf{E}\,(e^{\lambda\xi};\, \xi > s_+ - \varepsilon), \qquad \psi'(\lambda) \sim \mathbf{E}\,(\xi e^{\lambda\xi};\, \xi > s_+ - \varepsilon)$$

as $\lambda \to \infty$ and for any fixed $\varepsilon > 0$. It follows that

$$\alpha_+ = \lim_{\lambda\to\infty} \frac{\psi'(\lambda)}{\psi(\lambda)} = \lim_{\lambda\to\infty} \frac{\mathbf{E}\,(e^{\lambda\xi};\, \xi > s_+ - \varepsilon)}{\mathbf{E}\,(\xi e^{\lambda\xi};\, \xi > s_+ - \varepsilon)} = s_+.$$

If $s_+ = \infty$, then $\ln \psi(\lambda)$ increases faster than any linear function as $\lambda \to \infty$, and, therefore, the derivative $\big(\ln \psi(\lambda)\big)'$ increases without limit, $\alpha_+ = \infty$.

(ii) The first two statements are evident. Let $p_+ = \mathbf{P}(\xi = s_+) > 0$. Then

$$\psi(\lambda) \sim p_+ e^{\lambda s_+},$$
$$\alpha\lambda - \ln \psi(\lambda) = \alpha\lambda - \ln p_+ - \lambda s_+ + o(1) = (\alpha - \alpha_+)\lambda - \ln p_+ + o(1)$$

as $\lambda \to \lambda_+ = \infty$. It follows from this and (1.1.6) that

$$\Lambda(\alpha) = \begin{cases} -\ln p_+ & \text{for} \quad \alpha = \alpha_+, \\ \infty & \text{for} \quad \alpha > \alpha_+. \end{cases}$$

If $p_+ = 0$, then from the relation $\psi(\lambda) = o(e^{\lambda s_+})$ as $\lambda \to \infty$, we obtain in a similar way that $\Lambda(\alpha_+) = \infty$. The property ($\Lambda$2) is proved. □

Now let $0 < \lambda_+ < \infty$. Then $s_+ = \infty$. If $\alpha_+ < \infty$, then it is necessary that $\psi(\lambda_+) < \infty$, $\psi(\lambda_+ + 0) = \infty$, $\psi'(\lambda_+) < \infty$. The left derivative is meant. If $\psi(\lambda_+) = \infty$, then $\ln \psi(\lambda_+) = \infty$ and $\big(\ln \psi(\lambda)\big)' \to \infty$ as $\lambda \uparrow \lambda_+$, $\alpha_+ = \infty$, which contradicts the assumption $\alpha_+ < \infty$. Since $\psi(\lambda) = \infty$ for $\lambda > \lambda_+$, it follows that $\lambda(\alpha)$, having reached λ_+ with increasing α, stops at this point, so that for $\alpha \geqslant \alpha_+$ we have

$$\lambda(\alpha) = \lambda_+, \quad \Lambda(\alpha) = \Lambda(\alpha_+) + \lambda_+(\alpha - \alpha_+) = \alpha\lambda_+ - A(\lambda_+). \tag{1.1.10}$$

Thus, for $\alpha \geqslant \alpha_+$ the function $\lambda(\alpha)$ is a constant, and $\Lambda(\alpha)$ grows linearly. Moreover, the relations (1.1.7), (1.1.8) remain valid.

If $\alpha_+ = \infty$, then $\alpha < \alpha_+$ for all finite $\alpha \geqslant \alpha_-$, and again we are dealing with the 'regular' situation considered before (see (1.1.7), (1.1.8)). Since $\lambda(\alpha)$ is non-decreasing, those relations imply the convexity of $\Lambda(\alpha)$.

In sum, we can formulate the next property.

(Λ3) *The functions $\lambda(\alpha)$, $\Lambda(\alpha)$ may have discontinuities only at the points $s_\pm$ in the case $\mathbf{P}(\xi = s_\pm) > 0$. These points separate the domain (s_-, s_+) of finiteness and continuity (in the extended sense) of the function Λ from the domain $\alpha \notin [s_-, s_+]$, where $\Lambda(\alpha) = \infty$. On $[s_-, s_+]$ the function Λ is convex. (If one defines convexity in the 'extended' sense, i.e. allowing infinite values, then Λ is convex on the whole line.) The function Λ is analytic on the interval $(\alpha_-, \alpha_+) \subset (s_-, s_+)$. If $\lambda_+ < \infty$, $\alpha_+ < \infty$, then the function $\Lambda(\alpha)$ is linear on (α_+, ∞) with a slope angle of λ_+; at the boundary point α_+ the continuity of the first derivatives persists. If $\lambda_+ = \infty$, then $\Lambda(\alpha) = \infty$ on (α_+, ∞). An analogous property is valid for the function $\Lambda(\alpha)$ on $(-\infty, \alpha_-)$.*

If $\lambda_- = 0$, then $\alpha_- = a_1$ and $\lambda(\alpha) = \Lambda(\alpha) = 0$ with $\alpha \leqslant a_1$.

In fact, since $\lambda(a_1) = 0$ and $\psi(\lambda) = \infty$ for $\lambda < \lambda_- = 0 = \lambda(a_1)$, with the decrease of α down to $\alpha_- = a_1$, the point $\lambda(\alpha)$, having reached 0, stops, and $\lambda(\alpha) = 0$ for $\alpha \leqslant \alpha_- = a_1$. It follows from this and from the first identity in (1.1.6) that $\Lambda(\alpha) = 0$ for $\alpha \leqslant a_1$.

If $\lambda_- = \lambda_+ = 0$ (the condition **[C]** is not satisfied), then $\lambda(\alpha) = \Lambda(\alpha) \equiv 0$ for all α. This is evident, since the value under the sup sign in (1.1.3) is equal to $-\infty$ for all $\lambda \neq 0$. In this case the limit theorems stated in the subsequent sections would not be informative.

By summarising the properties of Λ, we can conclude, in particular, that on the whole line the function Λ is

(a) *convex: for* $\alpha, \beta \in \mathbb{R}$, $p \in [0, 1]$

$$\Lambda\big(p\alpha + (1-p)\beta\big) \leqslant p\Lambda(\alpha) + (1-p)\Lambda(\beta); \tag{1.1.11}$$

(b) *lower semicontinuous*:

$$\varliminf_{\alpha \to \alpha_0} \Lambda(\alpha) \geqslant \Lambda(\alpha_0), \qquad \alpha_0 \in \mathbb{R}. \tag{1.1.12}$$

Properties (a), (b) are known as general properties of the Legendre transform of a convex lower-semicontinuous function $A(\lambda)$ (see e.g. [159]).

We will also need the following properties of function Λ.

(Λ4) *Under trivial conventions about notation, for independent random variables* ξ *and* η *we have*

$$\Lambda^{(\xi+\eta)}(\alpha) = \sup_\lambda \big(\alpha\lambda - A^{(\xi)}(\lambda) - A^{(\eta)}(\lambda)\big) = \inf_\gamma \big(\Lambda^{(\xi)}(\gamma) + \Lambda^{(\eta)}(\alpha - \gamma)\big),$$

$$\Lambda^{(c\xi+b)}(\alpha) = \sup_\lambda \big(\alpha\lambda - \lambda b - A^{(\xi)}(\lambda c)\big) = \Lambda^{(\xi)}\left(\frac{\alpha - b}{c}\right).$$

It is clear that the infimum $\inf_\gamma$ in the first relation is reached at a point γ, such that $\lambda^{(\xi)}(\gamma) = \lambda^{(\eta)}(\alpha - \gamma)$. If ξ and η are identically distributed, then $\gamma = \alpha/2$ and, therefore,

$$\Lambda^{(\xi+\eta)}(\alpha) = \Lambda^{(\xi)}\left(\frac{\alpha}{2}\right) + \Lambda^{(\eta)}\left(\frac{\alpha}{2}\right) = 2\Lambda^{(\xi)}\left(\frac{\alpha}{2}\right).$$

It is also evident that for all $n \geqslant 2$

$$\Lambda^{(S_n)}(\alpha) = \sup\big(\alpha\lambda - nA^{(\xi)}(\lambda)\big) = n\sup\left(\frac{\alpha\lambda}{n} - A^{(\xi)}(\lambda)\right) = n\Lambda^{(\xi)}\left(\frac{\alpha}{n}\right).$$

(Λ5) *The function* $\Lambda(\alpha)$ *attains its minimal value, which is equal to* 0, *at the point* $\alpha = \mathbf{E}\,\xi = a_1$. *For definiteness, let* $\alpha_+ > 0$. *If* $a_1 = 0$, $\mathbf{E}\,|\xi^k| < \infty$, *then*

$$\lambda(0) = \Lambda(0) = \Lambda'(0) = 0, \quad \Lambda''(0) = \frac{1}{\gamma_2}, \quad \Lambda'''(0) = -\frac{\gamma_3}{\gamma_2^2}, \ \ldots \tag{1.1.13}$$

(*in the case* $\alpha_- = 0$, *right derivatives are meant*). *As* $\alpha \downarrow 0$, *the next representation takes place:*

$$\Lambda(\alpha) = \sum_{j=2}^{k} \frac{\Lambda^{(j)}(0)}{j!}\alpha^j + o(\alpha^k). \tag{1.1.14}$$

The semi-invariants γ_j *are defined in* (1.1.1) and (1.1.2).

If the double-sided Cramér's condition $[\mathbf{C}_0]$ is met, then the expansion of $\Lambda(\alpha)$ into a series (1.1.14) for $k = \infty$ holds and is called a *Cramér's series*.

The proof of property (Λ5) should not cause any trouble, and we leave it to the reader.

(Λ6) *If condition $[\mathbf{C}_+]$ is met, then there exist constants $c_1 > 0$ and c_2 such that, for all α,*

$$\Lambda(\alpha) \geqslant c_1\alpha - c_2.$$

If $\lambda_+ = \infty$, then $\Lambda(\alpha) \gg \alpha$ as $\alpha \to \infty$, i.e. there exists a function $v_+(\alpha) \to \infty$, as $\alpha \to \infty$, such that $\Lambda(\alpha) \geqslant \alpha v_+(\alpha)$.

Proof. By virtue of (1.1.10), $\lambda(\alpha) \uparrow \lambda_+$ as $\alpha \uparrow \alpha_+$, so $\lambda(\alpha) = \lambda_+$ for $\alpha \geqslant \alpha_+$. Since the function Λ is convex, $\Lambda'(\alpha) = \lambda(\alpha) \uparrow$ as $\alpha \uparrow$; then, having taken a point $\alpha_0 > 0$, such that $\lambda(\alpha_0) > 0$, we obtain

$$\Lambda(\alpha) \geqslant \Lambda(\alpha_0) + \lambda(\alpha_0)(\alpha - \alpha_0) \quad \text{for all } \alpha.$$

If $\lambda_+ = \infty$, then $\lambda(\alpha) \uparrow \infty$ as $\alpha \uparrow \infty$ and

$$v_+(\alpha) := \frac{1}{\alpha}\,\Lambda(\alpha) = \frac{1}{\alpha}\left(\Lambda(0) + \int_0^\alpha \lambda(t)dt\right) \to \infty \qquad \text{as } \alpha \to \infty.$$

The property (Λ6) is proved. □

If follows from this property that under condition $[\mathbf{C}_0]$ there exist constants $c_1 > 0$, $c_2 > 0$, such that

$$\Lambda(\alpha) \geqslant c_1|\alpha| - c_2 \quad \text{for all } \alpha.$$

If $|\lambda_\pm| = \infty$, then $\Lambda(\alpha) \gg |\alpha|$ as $|\alpha| \to \infty$.

(Λ7) *An inversion formula holds: for $\lambda \in (\lambda_-, \lambda_+)$*

$$\ln \psi(\lambda) = \sup_\alpha \big(\alpha\lambda - \Lambda(\alpha)\big). \tag{1.1.15}$$

This means that when condition $[\mathbf{C}]$ is met, the deviation function uniquely determines the Laplace transform of $\psi(\lambda)$, and, hence, the distribution of a random variable ξ. The formula (1.1.15) also indicates that the iterated Legendre transform of the convex function $\ln \psi(\lambda)$ leads to the same original function.

Proof. Let us denote the right-hand side of (1.1.15) by $T(\lambda)$ and show that $T(\lambda) = \ln \psi(\lambda)$ for $\lambda \in (\lambda_-, \lambda_+)$. If, in order to find the sup in (1.1.15), we set the derivative with respect to α of the function under the sup sign equal to zero, then we obtain the equation

$$\lambda = \Lambda'(\alpha) = \lambda(\alpha). \tag{1.1.16}$$

Since $\lambda(\alpha)$ on (α_-, α_+) is the function inverse to $\big(\ln \psi(\lambda)\big)'$ (see (1.1.4)), then for $\lambda \in (\lambda_-, \lambda_+)$ the equation (1.1.16) has an evident solution

$$\alpha = a(\lambda) := \big(\ln \psi(\lambda)\big)'. \tag{1.1.17}$$

Taking into account that $\lambda(a(\lambda)) \equiv \lambda$, we obtain

$$T(\lambda) = \lambda a(\lambda) - \Lambda(a(\lambda)),$$
$$T'(\lambda) = a(\lambda) + \lambda a'(\lambda) - \lambda(a(\lambda))a'(\lambda) = a(\lambda).$$

Since $a(0) = a_1$ and $T(0) = -\Lambda(a_1) = 0$,

$$T(\lambda) = \int_0^\lambda a(u)du = \ln \psi(\lambda). \tag{1.1.18}$$

The statement is proved, as well as another inversion formula (the last equality in (1.1.18); this expresses $\ln \psi(\lambda)$ in terms of the integral of the function $a(\lambda)$, which is inverse to $\lambda(\alpha)$). □

By virtue of Chebyshev's exponential inequality, for all $n \geqslant 1$, $\lambda \geqslant 0$, $x \geqslant 0$, we have

$$\mathbf{P}(S_n \geqslant x) \leqslant e^{-\lambda x}\psi(\lambda) = \exp\{-\lambda x + n \ln \psi(\lambda)\}. \tag{1.1.19}$$

Since $\lambda(\alpha) \geqslant 0$ for $\alpha \geqslant a_1$, by setting $\alpha = x/n$ and by substituting $\lambda = \lambda(\alpha) \geqslant 0$ in (1.1.19) we obtain the property

(Λ8) *For all $n \geqslant 1$ and $\alpha = x/n \geqslant a_1$,*

$$\mathbf{P}(S_n \geqslant x) \leqslant e^{-n\Lambda(\alpha)}.$$

The next property can be named as an exponential modification of the Kolmogorov–Doob inequality.

(Λ9) **Theorem 1.1.2.** (i) *For all $n \geqslant 1$, $x \geqslant 0$ and $\lambda \geqslant 0$, one has*

$$\mathbf{P}(\overline{S}_n \geqslant x) \leqslant e^{-\lambda x} \max\{1, \psi^n(\lambda)\}. \tag{1.1.20}$$

(ii) *Let $\mathbf{E}\xi < 0$, $\lambda_1 := \max\{\lambda : \psi(\lambda) \leqslant 1\}$. Then, for all $n \geqslant 1$ and $x \geqslant 0$, one has*

$$\mathbf{P}(\overline{S}_n \geqslant x) \leqslant e^{-\lambda_1 x}. \tag{1.1.21}$$

If $\lambda_+ > \lambda_1$ then $\psi(\lambda_1) = 1$, $\Lambda(\alpha) \geqslant \lambda_1\alpha$ for all α and $\Lambda(\alpha_1) = \lambda_1\alpha_1$, where

$$\alpha_1 := \arg\{\lambda(\alpha) = \lambda_1\} = \frac{\psi'(\lambda_1)}{\psi(\lambda_1)}, \tag{1.1.22}$$

so that a line $y = \lambda_1\alpha$ is tangent to the convex function $y = \Lambda(\alpha)$ at the point $(\alpha_1, \lambda_1\alpha_1)$. In addition, along with (1.1.21), the next inequality holds for $\alpha := x/n$:

$$\mathbf{P}(\overline{S}_n \geqslant x) \leqslant e^{-n\Lambda_1(\alpha)}, \tag{1.1.23}$$

where

$$\Lambda_1(\alpha) = \begin{cases} \lambda_1\alpha & \textit{for} \quad \alpha \leqslant \alpha_1, \\ \Lambda(\alpha) & \textit{for} \quad \alpha > \alpha_1. \end{cases}$$

If $\alpha \leqslant \alpha_1$ *then the inequality* (1.1.23) *coincides with* (1.1.21)*; for* $\alpha > \alpha_1$ *it is stronger than* (1.1.21).

(iii) *Let* $\mathbf{E}\xi \geqslant 0$, $\alpha = x/n \geqslant \mathbf{E}\xi$. *Then, for all* $n \geqslant 1$,

$$\mathbf{P}(\overline{S}_n \geqslant x) \leqslant e^{-n\Lambda(\alpha)}. \tag{1.1.24}$$

Theorem 1.1.2 distinguishes three non-overlapping possibilities,

(a) $\mathbf{E}\xi < 0$, $\lambda_+ = \lambda_1$,
(b) $\mathbf{E}\xi < 0$, $\lambda_+ > \lambda_1$,
(c) $\mathbf{E}\xi \geqslant 0$,

for which $\mathbf{P}(\overline{S}_n \geqslant x)$ is bounded by the right-hand sides of inequalities (1.1.21), (1.1.23), (1.1.24) correspondingly. However, by accepting some natural conventions, one can express all three stated inequalities in the unique form of (1.1.23). Indeed, let us turn to the definition (1.1.22) of α_1. As already noted (see (1.1.4)), $\lambda(\alpha)$ is a solution of the equation $\psi'(\lambda)/\psi(\lambda) = \alpha$, which is unique for

$$\alpha \in [\alpha_-, \alpha_+], \qquad \alpha_+ := \lim_{\lambda \uparrow \lambda_+} \frac{\psi'(\lambda)}{\psi(\lambda)}, \qquad \alpha_- := \lim_{\lambda \downarrow \lambda_-} \frac{\psi'(\lambda)}{\psi(\lambda)}.$$

For $\alpha \geqslant \alpha_+$ the function $\lambda(\alpha)$ is defined as a constant λ_+ (see (1.1.10)). This means that for $\lambda_1 = \lambda_+$ the value of α_1 is not uniquely defined and may take any value from α_+ to ∞, so that, by setting $\alpha_1 = \max\{\alpha : \lambda(\alpha) = \lambda_1 = \lambda_+\} = \infty$, we turn the inequality (1.1.23) in the case $\lambda_1 = \lambda_+$ (i.e. in case (a)) into the inequality

$$\mathbf{P}(\overline{S}_n \geqslant x) \leqslant e^{-n\lambda_1\alpha} = e^{-\lambda_1 x},$$

i.e. into the inequality (1.1.21).

If $\mathbf{E}\xi \geqslant 0$ then $\lambda_1 = 0$. If $\lambda_+ = 0$ then $\lambda_+ = \lambda_1$ and we have the same situation as before, but now $\mathbf{P}(\overline{S}_n \geqslant x)$ allows only the trivial bound of 1. If $\lambda_+ > 0$ then $\alpha_1 = \mathbf{E}\xi$; for $\alpha < \alpha_1$ the bound (1.1.23) is trivial again, and for $\alpha > \alpha_1$ it coincides with (1.1.24).

Corollary 1.1.3. *If we set*

$$\alpha_1 := \max\{\alpha : \lambda(\alpha) = \lambda_1\} = \begin{cases} \dfrac{\psi'(\lambda_1)}{\psi(\lambda_1)}, & \text{for} \quad \lambda_+ > \lambda_1, \\ \infty, & \text{for} \quad \lambda_+ = \lambda_1, \end{cases}$$

then the inequality (1.1.23) *holds without any additional conditions on* $\mathbf{E}\xi$ *and* λ_+ *and comprises inequalities* (1.1.21), (1.1.24).

From the results of [16] (where the asymptotics of the distributions of the maximum of sequential sums of random variables are studied; see also Chapter 3) it is not difficult to extract the information that the exponents in the inequality (1.1.23) are asymptotically unimprovable:

$$\lim_{n\to\infty} \frac{1}{n} \ln \mathbf{P}(\overline{S}_n \geqslant x) = -\Lambda_1(\alpha) \quad \text{as} \quad \frac{x}{n} = \alpha;$$

the same can be deduced using the large deviation principle for trajectories of $\{S_k\}$ (see Chapter 4).

Proof of Theorem 1.1.2. (i) The random variable

$$\eta(x) := \inf\{k > 0 : S_k \geqslant x\}$$

is a stopping time. Thus, the event $\{\eta(t) = k\}$ and the random variable $S_n - S_k$ are independent and

$$\psi^n(\lambda) = \mathbf{E}e^{\lambda S_n} \geqslant \sum_{k=1}^{n} \mathbf{E}(e^{\lambda S_n};\ \eta(x) = k) \geqslant \sum_{k=1}^{n} \mathbf{E}\big(e^{\lambda(x+S_n-S_k)};\ \eta(x) = k\big)$$

$$= e^{\lambda x} \sum_{k=1}^{n} \psi^{n-k}(\lambda)\mathbf{P}\big(\eta(x) = k\big) \geqslant e^{\lambda x} \min\big\{1,\ \psi^n(\lambda)\big\}\mathbf{P}\big(\eta(x) \leqslant n\big).$$

Hence we obtain statement (i).

(ii) Inequality (1.1.21) directly follows from (1.1.20), if one takes $\lambda = \lambda_1$. Now let $\lambda_+ > \lambda_1$. Then, obviously, $\psi(\lambda_1) = 1$ and from the definition of the function $\Lambda(\alpha)$ it follows that

$$\Lambda(\alpha) \geqslant \lambda_1\alpha - \ln \psi\big(\lambda_1\big) = \lambda_1\alpha.$$

Furthermore,

$$\Lambda(\alpha_1) = \lambda_1\alpha_1 - \ln \psi\big(\lambda_1\big) = \lambda_1\alpha_1,$$

so that the curves $y = \lambda_1\alpha$ and $y = \Lambda(\alpha)$ are tangent to each other at the point $(\alpha_1, \lambda_1\alpha_1)$.

Next, it is clear that $\psi\big(\lambda(\alpha)\big) \geqslant 1$ for $\alpha \geqslant \alpha_1$. For $\alpha = x/n$, the optimal choice of λ in (1.1.20) would be $\lambda = \lambda(\alpha)$. For such an $\lambda(\alpha)$, i.e. $\alpha = x/n$, we obtain

$$\mathbf{P}\,(\overline{S}_n \geqslant x) \leqslant e^{-n\Lambda(\alpha)}.$$

Together with (1.1.21) this proves (1.1.23). It is also clear that $\Lambda(\alpha) > \lambda_1\alpha$ for $\alpha > \lambda_1$, which proves the last statement of (ii).

(iii) Since $\lambda(a_1) = 0$ and $\lambda(\alpha)$ is non-decreasing, $\lambda(\alpha) \geqslant 0$ for $\alpha \geqslant a_1$. In the case $a_1 \geqslant 0$ one has $\psi(\lambda) \geqslant 1$ for $\lambda \geqslant 0$. Thus $\psi\big(\lambda(\alpha)\big) \geqslant 1$ for $\alpha \geqslant a_1$. By substituting $\lambda = \lambda(\alpha)$ in (1.1.20) for $\alpha \geqslant a_1$, one obtains (1.1.24). Theorem 1.1.2 is proved. □

The probabilistic sense of the deviation function is clarified by the following statement. Let $(\alpha)_\varepsilon = (\alpha - \varepsilon, \alpha + \varepsilon)$ be an ε-neighbourhood of α. For any set $B \subset \mathbb{R}$ denote

$$\Lambda(B) = \inf_{\alpha \in B} \Lambda(\alpha).$$

Theorem 1.1.4. *For all* $\alpha \in \mathbb{R}$ *and* $\varepsilon > 0$,

$$\lim_{n\to\infty} \frac{1}{n} \ln \mathbf{P}\left(\frac{S_n}{n} \in (\alpha)_\varepsilon\right) = -\Lambda\big((\alpha)_\varepsilon\big), \qquad (1.1.25)$$

$$\lim_{\varepsilon\to 0}\lim_{n\to\infty}\frac{1}{n}\ln \mathbf{P}\Big(\frac{S_n}{n}\in(\alpha)_\varepsilon\Big) = -\Lambda(\alpha). \tag{1.1.26}$$

The statement (1.1.26) in Theorem 1.1.4, referred to as the *local large deviation principle*, allows another forms. It will be shown in section 4.1 that the statement (1.1.26) is equivalent to the following: *given any α, for any sequence ε_n that converges to zero sufficiently slowly, one has*

$$\Lambda(\alpha) = -\lim_{n\to\infty}\frac{1}{n}\ln \mathbf{P}\left(\frac{S_n}{n}\in(\alpha)_{\varepsilon_n}\right). \tag{1.1.27}$$

This relation can be expressed in the form

$$\ln \mathbf{P}\left(\frac{S_n}{n}\in(\alpha)_{\varepsilon_n}\right) = -n\Lambda(\alpha)+o(n) \quad \text{as } n\to\infty.$$

Thus the dependence on n and α of the left-hand side of this equality is asymptotically *factorisable*: it is presented as a product of the factors n and $\Lambda(\alpha)$, where the former depends only on n, while the latter depends only on α.

Theorem 1.1.4 will be proved in section 2.7. Large deviation principles will be discussed in detail in Chapters 4 and 5 in a more general setting. Note that for a sequence of sums S_n, the so-called *moderately large deviation principle* holds; this is formulated and proved in Chapter 5.

Now let us consider a few examples in which the values of $\lambda_\pm$ and $\alpha_\pm$ and the functions $\psi(\lambda)$, $\lambda(\alpha)$, $\Lambda(\alpha)$ can be calculated in an explicit form.

Example 1.1.5. If ξ is normally distributed with parameters (0, 1) then

$$\psi(\lambda)=e^{\lambda^2/2}, \qquad |\lambda_\pm|=|\alpha_\pm|=\infty \qquad \lambda(\alpha)=\alpha, \qquad \Lambda(\alpha)=\frac{\alpha^2}{2}.$$

Example 1.1.6. For the Bernoulli scheme with parameter $p\in(0,1)$ we have

$$\psi(\lambda)=pe^\lambda+q, \qquad |\lambda_\pm|=\infty, \quad \alpha_+=1, \quad \alpha_-=0, \quad a_1=\mathbf{E}\xi=p,$$

$$\lambda(\alpha)=\ln\frac{\alpha(1-p)}{p(1-\alpha)}, \quad \Lambda(\alpha)=\alpha\ln\frac{\alpha}{p}+(1-\alpha)\ln\frac{1-\alpha}{1-p} \quad \text{for } \alpha\in(0,1),$$

$$\Lambda(0)=-\ln(1-p), \quad \Lambda(1)=-\ln p, \quad \Lambda(\alpha)=\infty \quad \text{for } \alpha\notin[0,1].$$

Example 1.1.7. For an exponential distribution with rate β we have

$$\psi(\lambda)=\frac{\beta}{\beta-\lambda}, \quad \lambda_+=\beta, \quad \lambda_-=-\infty, \quad \alpha_+=\infty, \quad \alpha_-=0, \quad a_1=\frac{1}{\beta},$$

$$\lambda(\alpha)=\beta-\frac{1}{\alpha}, \quad \Lambda(\alpha)=\alpha\beta-1-\ln\alpha\beta \quad \text{for } \alpha>0.$$

Example 1.1.8. For a centred Poisson distribution with rate β we have

$$\psi(\lambda)=\exp\left\{\beta[e^\lambda-1-\lambda]\right\}, \quad |\lambda_\pm|=\infty, \quad \alpha_-=-\beta, \quad \alpha_+=\infty, \quad a_1=0,$$

$$\lambda(\alpha)=\ln\frac{\beta+\alpha}{\beta}, \quad \Lambda(\alpha)=(\alpha+\beta)\ln\frac{\alpha+\beta}{\beta}-\alpha \quad \text{for } \alpha>-\beta.$$

1.2 Deviation function and its properties in the multidimensional case

1.2.1 Cramér's conditions

Now let $\xi, \xi_1, \xi_2, \ldots$ be a sequence of independent identically distributed random *vectors* in a d-dimensional Euclidean space $\mathbb{R}^d$. For vectors $\alpha = (\alpha_{(1)}, \ldots, \alpha_{(d)})$ and $\beta = (\beta_{(1)}, \ldots, \beta_{(d)})$ from $\mathbb{R}^d$ we denote the dot product and the norm respectively by

$$\langle \alpha, \beta \rangle = \alpha_{(1)}\beta_{(1)} + \cdots + \alpha_{(d)}\beta_{(d)}, \quad |\alpha| = \langle \alpha, \alpha \rangle^{1/2}.$$

As before, let $S_0 = 0$, $S_n = \xi_1 + \cdots + \xi_n$ for $n \geqslant 1$. We assume that the distribution of vector ξ, as in section 1.1, is not degenerate, i.e. there is no $b \in \mathbb{R}^d$, $b \neq 0$, such that $\langle \xi, b \rangle = \text{const.}$ with unit probability.

The Laplace transform of the distribution of a random vector ξ is denoted, as before, by

$$\psi(\lambda) := \mathbf{E} e^{\langle \lambda, \xi \rangle}, \quad \lambda \in \mathbb{R}^d.$$

In the multidimensional case we use Cramér's conditions in the following form:

[C] *There exists $\lambda \in \mathbb{R}^d$ such that $\psi(\lambda) < \infty$ in some vicinity of λ.*
[C$_0$] *Condition* **[C]** *is met for $\lambda = 0$.*
[C$_\infty$] *Condition* **[C]** *is met for all $\lambda \in \mathbb{R}^d$.*

It is evident that **[C$_\infty$]** $\subset$ **[C$_0$]** $\subset$ **[C]**.

1.2.2 Deviation function

The *deviation function* of a random vector ξ, which is also known as the *Legendre transform* of the function

$$A(\lambda) := \ln \psi(\lambda)$$

or as the *conjugate function to* $A(\lambda)$, in convex analysis (cf. [159]), is defined by

$$\Lambda(\alpha) = \Lambda^{(\xi)}(\alpha) := \sup_{\lambda} \{\langle \alpha, \lambda \rangle - A(\lambda)\}, \quad \alpha \in \mathbb{R}^d. \tag{1.2.1}$$

Let us clarify the main properties of the deviation function.

($\overrightarrow{\Lambda}$1) *Young's inequality: For all α and λ from $\mathbb{R}^d$,*

$$A(\lambda) + \Lambda(\alpha) \geqslant \langle \alpha, \lambda \rangle. \tag{1.2.2}$$

This inequality follows immediately from the definition. By putting $\lambda = 0$ in (1.2.2), we get $\Lambda(\alpha) \geqslant 0$ for all $\alpha \in \mathbb{R}^d$.

An important role in the multidimensional case is played (just as before) by the sets

$$\mathcal{A} = \{\lambda : A(\lambda) < \infty\},$$

$$\mathcal{A}' = \{A'(\lambda), \lambda \in \mathcal{A}\}, \text{ where } A'(\lambda) = \left(\frac{\partial A(\lambda)}{\partial \lambda_{(1)}}, \ldots, \frac{\partial A(\lambda)}{\partial \lambda_{(d)}}\right) = \operatorname{grad} A(\lambda),$$

and the convex envelope S of the support of the distribution of ξ.

The following properties of the function $\Lambda(\alpha)$ hold.

($\overrightarrow{\Lambda}$2) *$\Lambda(\alpha) = \infty$ for $\alpha \notin S$; $\mathcal{A}' \subset S$. If the support S is bounded then $\mathcal{A} = \mathbb{R}^d$, $\mathcal{A}' = S$. For all $\alpha \in \mathbb{R}^d$ there exists a unique point $\lambda(\alpha) \in \mathbb{R}^d$ such that*

$$\Lambda(\alpha) = \langle \lambda(\alpha), \alpha \rangle - A\big(\lambda(\alpha)\big).$$

The functions $\lambda(\alpha)$ and $\Lambda(\alpha)$ are analytic on $\mathcal{A}'$,

$$\Lambda(a_1) = 0, \quad \lambda(a_1) = \operatorname{grad} \Lambda(\alpha)|_{\alpha=a_1} = 0, \quad \Lambda''(a_1) := \left\| \frac{\partial^2 \Lambda(\alpha)}{\partial \alpha_{(i)} \partial \alpha_{(j)}} \right\|_{\alpha=a_1} = M^{-1},$$

where $a_1 = \mathbf{E}\xi$ and M^{-1} is the inverse covariance matrix

$$M = \|\mathbf{E}\xi^0_{(i)}\xi^0_{(j)}\|, \qquad \xi^0_{(i)} = \xi_{(i)} - \mathbf{E}\xi_{(i)}, \quad i = 1, \ldots, d.$$

If the vector ξ has a normal distribution with mean a_1 and covariance matrix M then it is easy to verify that

$$\Lambda(\alpha) = \frac{1}{2}(\alpha - a_1)M^{-1}(\alpha - a_1)^T.$$

($\overrightarrow{\Lambda}$3) *The function $\Lambda(\alpha)$ is convex: for $\alpha, \beta \in \mathbb{R}^d$, $p \in [0, 1]$ one has* (1.1.11). *The sets $\mathcal{A}$ and $\Lambda_v = \{\alpha : \Lambda(\alpha) \leqslant v\}$ for all $v \geqslant 0$ are convex.*

($\overrightarrow{\Lambda}$4) *The function $\Lambda(\alpha)$ is lower semicontinuous everywhere*:

$$\lim_{\alpha \to \alpha_0} \Lambda(\alpha) \geqslant \Lambda(\alpha_0), \qquad \alpha_0 \in \mathbb{R}^d. \tag{1.2.3}$$

($\overrightarrow{\Lambda}$5)

$$\Lambda_\infty(\alpha) := \lim_{t \to \infty} \frac{\Lambda(\alpha t)}{t} = \sup_{\lambda \in \mathcal{A}} \langle \lambda, \alpha \rangle. \tag{1.2.4}$$

($\overrightarrow{\Lambda}$6) *An inversion formula holds: for all $\lambda \in \mathbb{R}^d$*

$$\ln \psi(\lambda) = \sup_\alpha \big(\langle \alpha, \lambda \rangle - \Lambda(\alpha)\big).$$

Thus, as in the one-dimensional case, the deviation function $\Lambda(\alpha)$ under **[C]** uniquely defines the distribution of ξ.

The next property can be named as a *consistency property*. Under natural conventions about notation one has

($\overrightarrow{\Lambda}$7)

$$\Lambda^{(\xi_{(1)})}(\alpha_{(1)}) = \inf_{\alpha_{(2)}, \ldots, \alpha_{(d)}} \Lambda(\alpha)$$

and, more generally, for all $b \in \mathbb{R}^d$,

$$\Lambda^{(\langle \xi, b \rangle)}(\beta) = \inf\big\{\Lambda(\alpha) : \langle \alpha, b \rangle = \beta\big\}.$$

Proofs of the properties ($\vec{\Lambda}$2)–($\vec{\Lambda}$7) can be found in [159], where general properties of the Legendre transform of convex lower-semicontinuous functions are studied. In the same book and in [48] the consistency property can be found in a more general form, i.e. in the form of the deviation function of a random vector $\xi H + b$, where H is a matrix of order (d, r).

Directly from the definition of the function $\Lambda(\alpha)$ we obtain the property

($\vec{\Lambda}$8) (1) Let $b \in \mathbb{R}^d$ and let H be an invertible square matrix of order d. Then

$$\Lambda^{(\xi H+b)}(\alpha) = \Lambda\big((\alpha - b)H^{-1}\big).$$

(2)
$$\Lambda^{(S_n)}(\alpha) = n\Lambda\left(\frac{\alpha}{n}\right).$$

The proof is the same as for property (Λ4) in the one-dimensional case.

Property ($\vec{\Lambda}$5) allows one to obtain the following property:

($\vec{\Lambda}$9) *If* **[C$_0$]** *is met then there exist constants* $c_1 > 0$ *and* c_2 *such that*

$$\Lambda(\alpha) \geqslant c_1|\alpha| - c_2.$$

If **[C$_\infty$]** *is met then*

$$\Lambda(\alpha) \gg |\alpha| \quad as\ |\alpha| \to \infty.$$

The first statement follows from ($\vec{\Lambda}$5) and the fact that under **[C$_0$]** the point $\lambda = 0$ is an inner point of $\mathcal{A}$ and, therefore, that the right-hand side of (1.2.4) is uniformly positive over α. Under **[C$_\infty$]** one has $\mathcal{A} = \mathbb{R}^d$, and the right-hand side of (1.2.4) is infinite.

The probabilistic sense of the deviation function remains the same (see Theorem 1.1.4):

Theorem 1.2.1. *The statements of Theorem* 1.1.4 *remain valid in the case of a* d*-dimensional random vector* ξ, $d > 1$.

Our commentaries on Theorem 1.1.4 also remain valid, including the statement of the equivalence of relations (1.1.25) and (1.1.27). Theorem 1.2.1 will be proved in section 2.7.

In the next section we will need a strengthening of the property ($\vec{\Lambda}$4). Denote by $\Lambda_{<\infty} = \{\alpha : \Lambda(\alpha) < \infty\}$ the set for which Λ is finite. The function $\Lambda(\alpha)$ is continuous in the interior $(\Lambda_{<\infty})$ of $\Lambda_{<\infty}$ (this property is inherent to any convex function on $\mathbb{R}^d$), and lower semicontinuous, if the boundary is included. It turns out that the latter property can be strengthened up to *lower continuity and continuity inside* $\Lambda_{<\infty}$ on the boundary $\partial\Lambda_{<\infty}$ (see Lemma 1.2.2 below). Hereafter such an extension will be required for *convex lower-semicontinuous* functions $J = J(y) \geqslant 0$ defined on an *arbitrary linear metric* space $(\mathbb{Y}, \rho)$. Hence we will formulate and prove Lemma 1.2.2 in the general case, for which the functions have the aforementioned properties in an arbitrary metric space $(\mathbb{Y}, \rho)$.

Let the function $J = J(y) : \mathbb{Y} \to [0, \infty]$ be convex and lower semicontinuous. Since the function J is convex, the set $J_{<\infty} := \{y \in \mathbb{Y} : J(y) < \infty\}$ and its closure $[J_{<\infty}]$ are convex. For any set $B \subset \mathbb{Y}$ denote

$$J(B) := \inf_{y \in B} J(y).$$

Lemma 1.2.2. *Let* $(\mathbb{Y}, \rho)$ *be an arbitrary linear metric space, and let a function* $J : \mathbb{Y} \to [0, \infty]$ *be lower semicontinuous and convex,* $y \in [J_{<\infty}]$, $y_0 \in J_{<\infty}$. *Then the following properties hold.*

(i) (*Lower continuity*)

$$\lim_{\varepsilon \to 0} J((y)_\varepsilon) = J(y). \tag{1.2.5}$$

(ii) (*Continuity inside* $J_{<\infty}$ *along rays*)

$$\lim_{p \uparrow 1} J((1-p)y_0 + py) = J(y). \tag{1.2.6}$$

For a set $B \subset \mathbb{R}^d$ we denote by (B) and $[B]$ the interior and the closure of B respectively.

Corollary 1.2.3. *Let* $B \subset \mathbb{R}^d$ *be a convex set and let the function* J *be lower semicontinuous and convex,* $J((B)) < \infty$. *Then*

$$J([B]) = J(B) = J((B)). \tag{1.2.7}$$

Proof of Lemma 1.2.2. (i) Property (1.2.5) is a consequence of the lower semicontinuity of J (cf. (1.2.3)) and the relation (1.2.6). Indeed, by virtue of the lower semicontinuity, we have on the one hand

$$\lim_{\varepsilon \to 0} J((y)_\varepsilon) \geqslant J(y).$$

On the other hand, choose $y_0 \in J_{<\infty}$ and let $y_p := (1-p)y_0 + py$, so that $y = y_1$. According to the properties of a linear metric space (see [112], p. 23) we have $y_p \to y_1$ as $p \to 1$. Hence for every $\varepsilon > 0$ there exists an arbitrarily large $p = p(\varepsilon) < 1$ such that $y_{p(\varepsilon)} \in (y)_\varepsilon$ and $p(\varepsilon) \to 1$ as $\varepsilon \to 0$. Therefore, by virtue of (1.2.6),

$$\underline{\lim}_{\varepsilon \to 0} J((y)_\varepsilon) \leqslant \lim_{\varepsilon \to 0} J(y_{p(\varepsilon)}) = J(y).$$

Property (1.2.5) is proved.

(ii) Let us prove property (1.2.6). If $J(y) = \infty$ then (1.2.6) follows from (1.2.3) for J. If $J(y) < \infty$ then consider the function

$$g(p) := J(y_p), \quad p \in [0, 1].$$

The function J is convex and therefore on the one hand

$$g(p) \leqslant pJ(y_1) + (1-p)J(y_0), \quad \overline{\lim}_{p \to 1} g(p) \leqslant J(y_1).$$

On the other hand, by virtue of (1.2.3) and the convergence $y_p \to y_1$ as $p \to 1$, we have

$$\varliminf_{p\to 0} g(p) \geqslant J(y_1).$$

This proves (1.2.6). Lemma 1.2.2 is proved. □

Proof of Corollary 1.2.3. (i) Note the following in advance. Since B and $J_{<\infty}$ are convex sets, the sets $[B]$, $[B] \cap J_{<\infty}$ are also convex. According to the condition, $J((B)) < \infty$ and, therefore, the set $(B) \cap J_{<\infty}$ is not empty. Let us take points $y_0 \in (B) \cap J_{<\infty}$, $y_1 \in [B] \cap J_{<\infty}$ and show that a half-open interval

$$[y_0, y_1) := \{y_p := py_1 + (1-p)y_0;\ p \in [0,1)\} \quad \text{is a subset of} \quad (B). \qquad (1.2.8)$$

Both the points y_0 and y_1 belong to the convex set $[B] \cap J_{<\infty}$. Hence, y_p also belongs to that set for all $p \in [0,1]$ and, in order to prove (1.2.8), we have to exclude the possibility $y_p \in \partial B$ for $p \in [0,1)$. But if $y_p \in \partial B$ then there exists a sequence of points $v \to y_p$ such that $v \notin [B]$. In this case,

$$w := \frac{v - py_1}{1-p} \to y_0 \quad \text{as} \quad v \to y_p.$$

Since $y_0 \in (B)$, $w \in (B)$ as v is close enough to y_p. Since $v = py_1 + (1-p)w$ and $y_1 \in [B]$, $v \in [B]$. The derived contradiction proves (1.2.8).

(ii) Now let us turn to the direct proof of (1.2.7). Statement (1.2.8) implies the inequality

$$J((B)) \leqslant J(py_1 + (1-p)y_0), \quad 0 \leqslant p < 1. \qquad (1.2.9)$$

Since y_0, y_1 belong to $J_{<\infty}$, by virtue of statement (ii) of Lemma 1.2.2, the right-hand side of (1.2.9) converges to $J(y_1)$ as $p \to 1$. Therefore,

$$J((B)) \leqslant J(y_1).$$

Hence, due to the arbitrariness of $y_1 \in [B] \cap J_{<\infty}$, we obtain the inequality

$$J((B)) \leqslant J([B]),$$

which, along with the evident relations

$$J((B)) \geqslant J(B) \geqslant J([B]),$$

proves (1.2.7). The corollary is proved. □

Remark 1.2.4. In Corollary 1.2.3, the condition $J((B)) < \infty$ cannot be replaced with the condition $J(B) < \infty$, as evidenced (in the case $\mathbb{Y} = \mathbb{R}^d$, $J = \Lambda$) by the following example.

Example 1.2.5. A random vector $\xi \in \mathbb{R}^2$ takes three values, $\beta^{(1)} = (0,2)$, $\beta^{(2)} = (1,1)$, $\beta^{(3)} = (1,-1)$ with the probabilities $p^{(1)} = 1/2$, $p^{(2)} = 1/4$, $p^{(3)} = 1/4$, correspondingly. By virtue of the local large deviation principle (1.1.27) (see Theorem 1.2.1), one has

$$\Lambda(\beta^{(i)}) = -\ln p^{(i)}, \quad i = 1, 2, 3.$$

Now consider a set $B = B_1 \cup \{\beta^{(2)}\}$, where B_1 is an open triangle with vertices $\beta^{(1)}$, $\beta^{(2)}$, $\beta^{(4)} = (1, 2)$. Since $\Lambda(\alpha) = \infty$, if a point α lies outside the closed triangle with vertices $\beta^{(1)}$, $\beta^{(2)}$, $\beta^{(3)}$ then $\Lambda(B) = -\ln p^{(2)} = \ln 4$. In addition, $\Lambda([B]) \leqslant \Lambda(\beta^{(1)}) = \ln 2$. Therefore, we have

$$\Lambda([B]) < \Lambda(B) < \infty.$$

The equalities (1.2.7) fail.

1.3 Chebyshev-type exponential inequalities for sums of random vectors

1.3.1 Basic inequalities for random vectors

Consider an arbitrary measurable set $B \subset \mathbb{R}^d$ and, as before, put

$$\Lambda(B) := \inf_{\alpha \in B} \Lambda(\alpha).$$

If the set B is empty, we put $\Lambda(B) = \infty$. For an arbitrary set $B \subset \mathbb{R}^d$ we will denote by (B) the interior of B, i.e. the collection of all points α that belong to B along with a neighbourhood, and by $[B]$ the closure of the set B. For a $v \geqslant 0$ consider the set

$$\Lambda_v := \{\alpha : \ \Lambda(\alpha) \leqslant v\},$$

of all points at which the value of the deviation function does not exceed v. We will denote the set on which the deviation function is finite by

$$\Lambda_{<\infty} := \bigcup_{v \geqslant 0} \Lambda_v = \{\alpha : \ \Lambda(\alpha) < \infty\}.$$

Note that in the one-dimensional case $d = 1$ Chebyshev's exponential inequality (see property (Λ8) in section 1.1) for convex sets B can be written in the form

$$\mathbf{P}(\xi \in B) \leqslant e^{-\Lambda(B)}. \tag{1.3.1}$$

Indeed, if $\Lambda(B) = 0$ then the inequality is trivial. If $\Lambda(B) > 0$ then either $b_+ := \inf\{\alpha : \ \alpha \in B\} > \mathbf{E}\xi$ or $b_- := \sup\{\alpha : \ \alpha \in B\} < \mathbf{E}\xi$. If $b_+ > \mathbf{E}\xi$, $b_+ \in B$, then $\Lambda(b_+) = \Lambda(B) > 0$ and the relation (1.3.1) follows from Chebyshev's inequality

$$\mathbf{P}(\xi \in B) \leqslant \mathbf{P}(\xi \geqslant b_+) \leqslant e^{-\Lambda(b_+)}. \tag{1.3.2}$$

If $\Lambda(b_+) < \Lambda(B)$ (which is possible only if $b_+ \notin B$, $\mathbf{P}(\xi \leqslant b_+) = 1$ and Λ has a discontinuity at point b_+) then $\Lambda(B) = \infty$, $\mathbf{P}(\xi \in B) = 0$ and inequality (1.3.1) holds as before. The case $b_- < \mathbf{E}\xi$ is considered analogously. It is clear that 'one-sided' sets B (i.e. those lying wholly on one side of $\mathbf{E}\xi$) can be embedded in convex sets with the same bounds $b_\pm$, and the inequality (1.3.1) holds for them as well.

We return to the general case $d \geqslant 1$. The main assertions of the present section are stated below.

Theorem 1.3.1. *If B is an arbitrary open convex set then*

$$\mathbf{P}(\xi \in B) \leqslant e^{-\Lambda(B)}. \tag{1.3.3}$$

Theorem 1.3.2. *For an arbitrary set B, one has*

$$\mathbf{P}(\xi \in B) \leqslant e^{-\Lambda([B^{con}])}, \tag{1.3.4}$$

where B^{con} is the convex envelope of B.

It follows from Theorems 1.3.1 and 1.3.2 that inequality (1.3.3) holds for both open convex and closed convex sets. Thus, selecting a class of convex sets and using the deviation function allows one to find a unified simple form (1.3.3) of the inequality, which it is still natural to call Chebyshev's inequality and which holds in spaces of any dimension, including the infinite-dimensional case (see section 4.3 below).

Theorems 1.3.1 and 1.3.2 are consequences of a more general assertion (Theorem 1.3.4) that employs broader conditions ensuring that inequalities of the form (1.3.3) hold true. Verifying these conditions in spaces of high dimension can be quite difficult. For this reason, we have stated simpler versions of Theorems 1.3.1, 1.3.2 as our main results. These theorems will suffice for the purposes of forthcoming sections and for some applications as well.

To state Theorem 1.3.4, we will need several concepts.

If $\Lambda(B) = 0$ then also $\Lambda([B]) = 0$, and all the assertions we have already stated before and will state hereafter would be trivial. Therefore, *it would be assumed throughout the section* that $\Lambda(B) > 0$. This means that we also assume that Cramér's condition **[C]** is met:

$\mathbf{E}e^{\langle\lambda,\xi\rangle} < \infty$ for λ taking values in a body (i.e. a compact set with non-empty interior).

In this case the set $\Lambda_{<\infty}$ will also be a body.

To simplify the preliminaries, we will assume for the present that the stronger condition $[\mathbf{C}_0]$ is met.

In this case one can assume without loss of generality that $\mathbf{E}\xi = 0$, 'shifting' the set B if necessary. Then $\Lambda(\alpha) \to \infty$ as $|\alpha| \to \infty$, and the sets Λ_v form a family of increasing sets (as v increases) that eventually fill the set $\Lambda_{<\infty}$, which coincides (up to its boundary) with the convex envelope S of the support of the distribution of ξ. As the boundary of Λ_v approaches that of $\Lambda_{<\infty}$ (as v increases), the growth of Λ_v in that direction slows down or stops altogether, so that the boundaries of the sets Λ_v and $\Lambda_{<\infty}$ may coincide in some locations. The boundaries $\partial\Lambda_v$ of the sets Λ_v are called *level surfaces*. One should note that the level surfaces corresponding to large enough v may have 'cavities'; i.e. for some directions e, the equation $\Lambda(te) = v$ can have no solutions.

Let us state the conditions for Theorem 1.3.4. First let $\Lambda(B) < \infty$, i.e. there exist points $\alpha \in B$ for which $\Lambda(\alpha) < \infty$. In this case, under the condition **[C$_0$]**, there exists a $v = v_B$ at which, by virtue of the continuity properties of the function Λ (noted above in Lemma 1.2.2) and the monotone growth of the compacts Λ_v, these compacts will touch the set $[B]$ for the first time. In other words, there exists a point α_B, at which the minimum

$$\min_{\alpha \in [B]} \Lambda(\alpha) = \Lambda([B]) = \Lambda(\alpha_B) =: v_B$$

is attained. It is clear that the point α_B lies on the boundaries of the sets Λ_{v_B} and $[B]$, and that the sets touch each other: $\alpha_B \in \Lambda_{v_B} \cap [B] \neq \emptyset$.

If only condition **[C]** is met then, in the case where the set B is unbounded, the sets Λ_{v_B} and $[B]$ may ‘touch’ each other ‘at infinity’.

If $\Lambda(B) = \infty$ then the set B does not intersect the convex set $\Lambda_{<\infty}$.

The symbol e will be used for unit vectors.

Definition 1.3.3. Let $\Lambda(B) < \infty$. In this case, the set B will be called Λ-*separable* if there exists a hyperplane $\Pi_{=} := \{\alpha : \langle \mathrm{e}, \alpha \rangle = b\}$ (going through the point α_B if the latter exists) that separates the sets B and $\Lambda_{v_B} = \{\alpha : \Lambda(\alpha) \leqslant v_B\}$ in the following sense:

$$B \subset \Pi_{>} := \{\alpha : \langle \mathrm{e}, \alpha \rangle > b\}, \quad \Lambda_{v_B} \subset \Pi_{\leqslant} := \{\alpha : \langle \mathrm{e}, \alpha \rangle \leqslant b\}. \tag{1.3.5}$$

If $\Lambda(B) = \infty$ then the set B is called Λ-*separable* if there exists a hyperplane $\Pi_{=}$ that separates the sets B and $\Lambda_{<\infty}$:

$$B \subset \Pi_{>}, \quad \Lambda_{<\infty} \subset \Pi_{\leqslant}.$$

Theorem 1.3.4. *If B is a Λ-separable set then*

$$\mathbf{P}(\xi \in B) \leqslant e^{-\Lambda([B])}. \tag{1.3.6}$$

Clearly, a convex set B with $\Lambda(B) < \infty$ cannot be Λ-separable.

Note that Λ-separability is essential for the assertion of Theorem 1.3.4 to hold (without that assumption, inequality (1.3.6) is, generally speaking, wrong), but this property is quite restrictive. What bounds can one obtain for $\mathbf{P}(\xi \in B)$ if the set B is not Λ-separable?

If $B \subset \cup B_k$ is a subset of the union of an at most countable collection of convex sets B_k then Theorem 1.3.2 implies that

$$\mathbf{P}(\xi \in B) \leqslant \sum e^{-\Lambda([B_k])}. \tag{1.3.7}$$

Theorems 1.3.2 and 1.3.4 do not provide any further meaningful bounds. Using a somewhat different approach, we can consider a random variable $\gamma := \Lambda(\xi)$ and an ‘iterate’ deviation function $\Lambda^{(\gamma)}$, i.e. the deviation function for a random variable γ which is equal to the value of the deviation function $\Lambda = \Lambda^{(\xi)}(\alpha)$ at the point ξ. Properties of the random variable γ and the deviation function $\Lambda^{(\gamma)}$ are studied in section 1.4.

Theorem 1.3.5. *If* $\Lambda(B) \geqslant \mathbf{E}\gamma$ *then*

$$\mathbf{P}(\xi \in B) \leqslant e^{-\Lambda^{(\gamma)}(\Lambda(B))}.$$

1.3.2 Proofs of Theorems 1.3.1, 1.3.2, 1.3.4 and 1.3.5

Proof of Theorem 1.3.4. Put $\beta = \langle \mathrm{e}, \xi \rangle$. The property $(\overrightarrow{\Lambda}7)$ can be written as

$$\Lambda^{(\beta)}(v) = \inf_{\langle \mathrm{e}, \alpha \rangle = v} \Lambda(\alpha). \tag{1.3.8}$$

(i) Let the set B be Λ-separable and $\Lambda(B) = \infty$. Then $\Pi_> \cap \Lambda_{<\infty} = \varnothing$ and so $\Lambda(\Pi_>) = \infty$. In other words, $\Lambda(\alpha) = \infty$ for all $\alpha \in \Pi_>$. It follows from this and (1.3.8) that $\Lambda^{(\beta)}(v) = \infty$ for all $v > b$ and therefore $\Lambda^{(\beta)}((b, \infty)) = \infty$.

Further, by virtue of (1.3.5), one has

$$\mathbf{P}(\xi \in B) \leqslant \mathbf{P}(\xi \in \Pi_>) = \mathbf{P}(\langle \mathrm{e}, \xi \rangle > b) = \mathbf{P}(\beta > b). \tag{1.3.9}$$

Since $\mathbf{P}(\beta > b) = \lim_{k\to\infty} \mathbf{P}(\beta \geqslant b + 1/k)$, Chebyshev's exponential inequality (1.3.2) yields

$$\mathbf{P}(\beta > b) \leqslant \lim_{k\to\infty} e^{-\Lambda^{(\beta)}(b+1/k)} = e^{-\Lambda^{(\beta)}(b+0)}, \tag{1.3.10}$$

where $\Lambda^{(\beta)}(b + 0) = \Lambda^{(\beta)}((b, \infty)) = \infty$. Therefore, $\mathbf{P}(\xi \in B) = 0$, and so inequality (1.3.6) is proved.

(ii) Now let $\Lambda(B) < \infty$. It follows from (1.3.8) that

$$\Lambda^{(\beta)}((b, \infty)) = \inf_{\Pi_>} \Lambda(\alpha) = \Lambda(\Pi_>).$$

Therefore, once again using inequalities (1.3.9), (1.3.10), we get

$$\mathbf{P}(\xi \in B) \leqslant e^{-\Lambda^{(\beta)}((b,\infty))} = e^{-\Lambda(\Pi_>)}.$$

But the sets Λ_{v_B} and $\Pi_>$ are disjoint. Hence $\Lambda(\alpha) > v_B$ for any $\alpha \in \Pi_>$. This means that $\Lambda(\Pi_>) \geqslant v_B$. Inequality (1.3.6) is established and Theorem 1.3 is proved. □

Note that if the set B is not Λ-separable then, generally speaking, the equality $\Lambda([B]) = \infty$ does not imply that $\mathbf{P}(\xi \in B) = 0$ (recall that the set $\Lambda_{<\infty}$ coincides up to its boundary with the convex envelope of the support of the distribution of ξ). This is demonstrated by the following example.

Example 1.3.6. Let Γ be a unit sphere in $\mathbb{R}^d$, $d \geqslant 2$, and let a random vector ξ have a uniform distribution on Γ. Also let B be the closure of the Γ exterior. Then $\Gamma \subset B$, $\Lambda([B]) = \infty$ and $\mathbf{P}(\xi \in B) = 1$.

Proof of Theorem 1.3.1. Since the sets Λ_v, $\Lambda_{<\infty}$ are convex, according to the Hahn–Banach theorem (see e.g. [110], p. 137) the open convex set B is Λ-separable and so by Theorem 1.3.4 one has (1.3.6) for that set. It remains to make use of Corollary 1.2.3, by virtue of which the right-hand side of (1.3.6) coincides with the right-hand side of (1.3.3). Theorem 1.3.1 is proved. □

Proof of Theorem 1.3.2. Since $B \subset B^{con}$, it suffices to verify that (1.3.4) holds for any convex set $B = B^{con}$.

The set $B_N := \{\alpha \in B\colon |\alpha| \leqslant N\}$ is convex, along with B. The ε-neighbourhood $(B_N)_\varepsilon$ of the set B_N is also convex. Therefore, by Theorem 1.3.1,

$$\mathbf{P}(\xi \in B) \leqslant \mathbf{P}(\xi \in (B_N)_\varepsilon) + P_N \leqslant e^{-\Lambda((B_N)_\varepsilon)} + P_N, \quad P_N := \mathbf{P}(|\xi| > N).$$

Now let $\varepsilon = \varepsilon_k \to 0$ as $k \to \infty$, even out the spaces on this line and let α_k be a sequence of points from $(B_N)_{\varepsilon_k} \subset [(B_N)_{\varepsilon_k}]$ such that $\Lambda(\alpha_k) \leqslant \Lambda((B_N)_{\varepsilon_k}) + 1/k$. Assuming without loss of generality that α_k converge to $\alpha_0 \in [B_N]$ as $k \to \infty$, we obtain from the lower semicontinuity of the function Λ that

$$\begin{aligned}\mathbf{P}(\xi \in B) \leqslant \varlimsup_{k\to\infty} e^{-\Lambda((B_N)_{\varepsilon_k})} + P_N \leqslant \varlimsup_{k\to\infty} e^{-\Lambda(\alpha_k)+1/k} \\ + P_N \leqslant e^{-\Lambda(\alpha_0)} + P_N \leqslant e^{-\Lambda([B_N])} + P_N. \end{aligned} \tag{1.3.11}$$

Since $\Lambda([B_N]) \downarrow \Lambda([B])$ as $N \uparrow \infty$, passing to the limit in (1.3.11) as $N \uparrow \infty$ we obtain (1.3.6). Theorem 1.3.2 is proved. □

Proof of Theorem 1.3.5. Since $B \subset \{\alpha : \Lambda(\alpha) \geqslant \Lambda(B)\}$, one has, for $\Lambda(B) \geqslant \mathbf{E}\gamma$, that

$$\mathbf{P}(\xi \in B) \leqslant \mathbf{P}(\Lambda(\xi) \geqslant \Lambda(B)) = \mathbf{P}(\gamma \geqslant \Lambda(B)) \leqslant e^{-\Lambda^{(\gamma)}(\Lambda(B))}.$$

The theorem is proved. □

1.3.3 Inequalities for sums of random vectors

As before, set

$$S_n := \sum_{i=1}^{n} \xi_i \quad \text{for } n \geqslant 1.$$

Put

$$\zeta_n := \frac{S_n}{n}, \quad A^{(\zeta_n)}(\lambda) := \ln \mathbf{E}e^{\langle\lambda,\zeta_n\rangle} = nA\left(\frac{\lambda}{n}\right).$$

Then the deviation function $\Lambda^{(\zeta_n)}(\alpha)$ for ζ_n, by virtue of property $(\overrightarrow{\Lambda}8)$ is equal to

$$\Lambda^{(\zeta_n)}(\alpha) = n\Lambda(\alpha). \tag{1.3.12}$$

Thus, the following assertion immediately follows from Theorems 1.3.1 and 1.3.2.

Corollary 1.3.7. (i) *For an arbitrary open convex set B one has the inequality*

$$\mathbf{P}(\zeta_n \in B) \leqslant e^{-n\Lambda(B)}. \tag{1.3.13}$$

(ii) *For an arbitrary measurable set B, one has the inequality*

$$\mathbf{P}(\zeta_n \in B) \leqslant e^{-n\Lambda([B^{\mathrm{con}}])},$$

where B^{con} is the convex envelope of B.

From the results contained in this monograph (see below), it follows that the bounds in Corollary 1.3.7(i) are 'exponentially' unimprovable. Such *exact* bounds for the probability $\mathbf{P}(\zeta_n \in B)$ for arbitrary sets B cannot be found. However, the following assertion holds true. As before, we let $\gamma = \Lambda(\xi)$.

Corollary 1.3.8. *If $\Lambda(B) \geqslant \mathbf{E}\gamma$ then*

$$\mathbf{P}(\zeta_n \in B) \leqslant e^{-n\Lambda^{(\gamma)}(\Lambda(B))}. \tag{1.3.14}$$

Inequality (1.3.14) is not 'exponentially' unimprovable owing to the 'losses' in the first inequality in (1.3.15) (see below). This is also indicated by the inequality $\Lambda^{(\gamma)}(v) < v - \mathbf{E}\gamma$ (see (1.4.8) below), which holds for random variables ξ that are unbounded from above. However, for large v one has $\Lambda^{(\gamma)}(v) \sim v$, and so inequalities (1.3.13), (1.3.14), in a certain sense, converge (i.e. $\Lambda^{(\gamma)}(\Lambda(B)) \sim \Lambda(B)$ for large $\Lambda(B)$).

Proof of Corollary 1.3.8. By virtue of the convexity of the deviation function $\Lambda(\alpha)$, one has

$$n\Lambda(\zeta_n) \leqslant \sum_{k=1}^{n} \gamma_k, \quad \gamma_k := \Lambda(\xi_k).$$

Since $B \subset \{\alpha : \Lambda(\alpha) \geqslant \Lambda(B)\}$, we have the implications

$$\{\zeta_n \in B\} \subset \{\Lambda(\zeta_n) \geqslant \Lambda(B)\} \subset \left\{\sum\nolimits_{k=1}^{n} \gamma_k \geqslant n\Lambda(B)\right\}.$$

From this relation, in the case where $\Lambda(B) \geqslant \mathbf{E}\gamma$, using Chebyshev's exponential inequality we obtain

$$\mathbf{P}(\zeta_n \in B) \leqslant \mathbf{P}\left(\sum\nolimits_{k=1}^{n} \gamma_k \geqslant n\Lambda(B)\right) \leqslant e^{-n\Lambda^{(\gamma)}(\Lambda(B))}. \tag{1.3.15}$$

Corollary 1.3.8 is proved. □

1.3.4 Appendix1. Strengthening of exponential inequalities for non-convex sets

As was noted before, the inequality (1.3.14) in Corollary 1.3.8 is not 'exponentially' unimprovable. In [20] a strengthening of that inequality was obtained where the 'exponential' component has the form $e^{-n\Lambda(B)}$, the same as for convex sets B but with a multiplier which is exponential with respect to n. Let $B_u = \{\alpha : \Lambda(\alpha) \geqslant u\}$, so that $\Lambda(B_u) \equiv u$, and let ∂B_u be a boundary of the set B_u, V_u be a 'volume' of ∂B_u (the area for $d = 3$, the length for $d = 2$). The set B_u is the complement to a convex set and is the 'least convenient' for obtaining upper bounds for the probability $\mathbf{P}(\zeta_n \in B_u)$.

Theorem 1.3.9. [20] *Let the condition* $[\mathbf{C}_0]$ *be met. Then for every* $\varepsilon > 0$ *and for all sufficiently large* un *one has*

$$\mathbf{P}(\zeta_n \in B_u) \leqslant (1+\varepsilon)V_u \left(\frac{e\lambda_u n}{4r_u}\right)^{d-1/2} e^{-un}, \tag{1.3.16}$$

where $r_u = \sup_{\alpha\in\partial B_u} |\alpha|$, $\lambda_u = \sup_{\alpha\in\partial B_u} |\lambda(\alpha)|$,

If $u \to 0$, $un \geqslant 1$, *as* $n \to \infty$ *then for all sufficiently large* n *the inequality* (1.3.16) *can be written in the form*

$$\mathbf{P}(\zeta_n \in B_u) \leqslant (1+\varepsilon)\frac{2\sqrt{\pi}}{\Gamma(d/2)}\left(\frac{\pi e}{2}\right)^{(d-1)/2}\left(\frac{\mu_1}{\mu_d}\right)^{3(d-1)/2}(un)^{(d-1)/2}e^{-un},$$

where μ_1^2 *and* μ_d^2 *are respectively the largest and the smallest eigenvalues of the matrix*

$$M^{-1} = \left\| \frac{\partial^2 \Lambda(\alpha)}{\partial\alpha_{(i)}\partial\alpha_{(j)}} \right\|_{\alpha=\mathbf{E}\xi},$$

which is the inverse to the matrix $M = \|\mathbf{E}\xi^0_{(i)}\xi^0_{(j)}\|$, *where* $\xi^0_{(i)} = \xi_{(i)} - \mathbf{E}\xi_{(i)}$, *of central second moments.*

It is clear that for any $B \subset \mathbb{R}^d$ the probability $\mathbf{P}(\zeta_n \in B)$ is bounded by the right-hand side of (1.3.16) with $u = \Lambda(B)$. For small u the boundary ∂B_u is close to the ellipse $\alpha M^{-1}\alpha^T = u$.

Theorem 1.3.9 is proved in [20] by means of majorisation of the left-hand side of (1.3.16) with $\mathbf{P}(\zeta_n \in \widehat{B}_u)$, where $\widehat{B}_u$ is a complement to the polyhedron $\partial\widehat{B}_u$ located between ∂B_u and $\partial B_{u-\Delta}$, $\Delta > 0$. The boundary $\partial\widehat{B}_u$ has, as n grows, a growing number of faces (in order to make $\Delta \to 0$), and the probability $\mathbf{P}(\zeta_n \in \widehat{B}_u)$ is bounded by the sum of the probabilities of reaching the corresponding subspaces (see (1.3.7)), which allows simple exponential bounds.

Note that the inequalities in Theorem 1.3.9, in contrast with the exact inequalities met in previous sections, are *asymptotic*.

From the integro-local theorem 2.3.2, obtained below in section 2.3 for sums S_n (with some additional conditions), it is not difficult to find by means of integration the exact asymptotic behaviour of the probability $\mathbf{P}(\zeta_n \in B_u)$ for the 'least convenient' sets B_u. It has the form

$$c_u n^{d/2-1} e^{-un}.$$

Comparison with (1.3.16) shows that the error of inequality (1.3.16) is of order $\sqrt{n}$ as $n \to \infty$ – the same as for the exact inequalities in Theorems 1.3.1–1.3.4.

1.4 Properties of the random variable $\gamma = \Lambda(\xi)$ and its deviation function

1.4.1 Invariance of γ under linear transformations of ξ

By virtue of property ($\overrightarrow{\Lambda}8$) from subsection 1.2.2, for a fixed vector $b \in \mathbb{R}^d$ and square non-singular matrix H the following relations hold true:

$$\Lambda^{(\xi+b)}(\alpha) = \Lambda^{(\xi)}(\alpha - b), \quad \Lambda^{(\xi H)}(\alpha) = \Lambda^{(\xi)}(\alpha H^{-1}),$$

where H^{-1} is the inverse to the matrix H. Hence

$$\Lambda^{(\xi+b)}(\xi + b) = \Lambda^{(\xi)}(\xi + b - b) = \gamma,$$

i.e. the value of γ does not change under the shift ξ. Similarly, we get

$$\Lambda^{(\xi H)}(\xi H) = \Lambda^{(\xi)}(\xi H H^{-1}) = \gamma,$$

i.e. the value of γ does not change under 'rotation and contraction' of the vector ξ. Thus, a linear transformation of the vector ξ does not affect the value of γ.

1.4.2 Properties of the random variable γ and the function $\Lambda^{(\gamma)}$ in the one-dimensional case, $d = 1$

Let $s_\pm$ be the boundaries of the set $\Lambda_{<\infty}$ or, equivalently, the boundaries of the convex envelope S of the support for the distribution of ξ, $\Lambda_\pm = \Lambda(s_\pm)$, $\Lambda^* = \max\{\Lambda_+, \Lambda_-\}$, $\Lambda_* = \min\{\Lambda_+, \Lambda_-\}$.

Theorem 1.4.1. (i) *The distribution of a random variable $\gamma = \Lambda(\xi)$ satisfies the inequalities*

$$\mathbf{P}(\gamma \geqslant v) \leqslant \begin{cases} 2e^{-v}, & \text{if } v \leqslant \Lambda_*, \\ e^{-v}, & \text{if } v \in (\Lambda_*, \Lambda^*], \\ 0, & \text{if } v > \Lambda^*. \end{cases} \tag{1.4.1}$$

(ii) *The value $\mathbf{E}\gamma$ satisfies the inequalities*

$$\mathbf{E}\gamma \leqslant 2 - e^{-\Lambda_*} - e^{-\Lambda^*} \leqslant 2.$$

(iii) *The following dichotomy holds: either $\Lambda^* = \infty$ and then*

$$\Lambda^{(\gamma)}(v) \sim v \quad as \quad v \to \infty$$

(or, equivalently, $\lambda_+^{(\gamma)} := \sup\{t : \mathbf{E}e^{t\gamma} < \infty\} = 1$), or $\Lambda^* < \infty$, and then, in the case where $\Lambda^* = \Lambda_+$, one has $s_+ < \infty$, $\gamma \leqslant \Lambda_+$ (and hence, $\lambda_+^{(\gamma)} = \infty$), $\mathbf{P}(\gamma = \Lambda_+) = \mathbf{P}(\xi = s_+) = e^{-\Lambda_+}$. Similar relations hold true in the case where $\Lambda^* = \Lambda_-$.

Proof. (i) Consider the equation

$$\Lambda(\alpha) = v > 0.$$

If $v \leqslant \Lambda_*$ then, by virtue of the convexity of the function $\Lambda(\alpha)$ and its continuity inside $[s_-, s_+]$, there exist two solutions $\alpha_\pm(v)$ of that equation; $\alpha_+(v) > \mathbf{E}\xi$ and $\alpha_-(v) < \mathbf{E}\xi$. If $v \in (\Lambda_*, \Lambda^*]$ then there exists only one solution: $\alpha_+(v)$ (if $\Lambda_* = \Lambda_-$) or $\alpha_-(v)$ (if $\Lambda_* = \Lambda_+$). In that case, we will introduce the second 'solution', which does not exist, by setting it equal to $\mp\infty$. If $v > \Lambda^*$ then there are no solutions, and we set $\alpha_\pm(v) = \pm\infty$. It follows that $\{\Lambda(\xi) \geqslant v\}$ is the union of two disjoint events $\{\xi \leqslant \alpha_-(v)\}$, $\{\xi \geqslant \alpha_+(v)\}$ (these events may be empty). Therefore, for $\gamma = \Lambda(\xi)$ we obtain

$$\mathbf{P}(\gamma \geqslant v) = \mathbf{P}(\xi \leqslant \alpha_-(v)) + \mathbf{P}(\xi \geqslant \alpha_+(v)),$$

so that, by Chebyshev's inequality,

$$\mathbf{P}(\gamma \geqslant v) \leqslant e^{-\Lambda(\alpha_-(v))} + e^{-\Lambda(\alpha_+(v))} \leqslant \begin{cases} 2e^{-v}, & \text{if } v \leqslant \Lambda_*, \\ e^{-v}, & \text{if } v \in (\Lambda_*, \Lambda^*], \\ 0, & \text{if } v > \Lambda^*. \end{cases}$$

(ii) The second assertion of the theorem follows from the first and the equality

$$\mathbf{E}\gamma = \int_0^\infty \mathbf{P}(\gamma \geqslant v)dv.$$

(iii) Let $\Lambda^* = \infty$. It follows from (i) that $\lambda_+^{(\gamma)} \geqslant 1$. We have to prove the converse inequality. Put $\theta_n := \frac{1}{n}\sum_{k=1}^n \gamma_k$, where the γ_k are independent copies of γ. By inequality (1.3.1),

$$\overline{\lim_{n\to\infty}} \frac{1}{n} \ln \mathbf{P}(\theta_n \geqslant v) \leqslant -\Lambda^{(\gamma)}(v). \tag{1.4.2}$$

We will obtain a lower bound for the left-hand side of this inequality. By virtue of the convexity of the function Λ, for $\zeta_n := \frac{1}{n}\sum_{k=1}^n \xi_k$ one has

$$\Lambda(\zeta_n) \leqslant \frac{1}{n}\sum_{k=1}^n \Lambda(\xi_k) = \theta_n, \tag{1.4.3}$$

and therefore

$$\mathbf{P}(\theta_n \geqslant v) \geqslant \mathbf{P}(\Lambda(\zeta_n) \geqslant v) \geqslant \mathbf{P}(\zeta_n \geqslant \alpha_+(v)).$$

For the sake of definiteness, let $\Lambda_+ = \infty$. Then the function $\Lambda(\alpha)$ continuously increases from 0 to ∞ on $(\mathbf{E}\xi, \infty)$ and therefore $\Lambda(\alpha_+(v) + 0) = \Lambda(\alpha_+(v)) = v$. By the large deviation principle for ζ_n (see Theorem 1.1.4) it is not hard to obtain (see also Theorem 2.2.2)

$$\overline{\lim_{n\to\infty}} \frac{1}{n} \ln \mathbf{P}(\theta_n \geqslant v) \geqslant \underline{\lim_{n\to\infty}} \frac{1}{n} \ln \mathbf{P}(\theta_n \geqslant v)$$

$$\geqslant \underline{\lim_{n\to\infty}} \frac{1}{n} \ln \mathbf{P}(\zeta_n \geqslant \alpha_+(v)) \geqslant -\Lambda(\alpha_+(v) + 0) = -v. \tag{1.4.4}$$

It follows from (1.4.2) and (1.4.4) that

$$\Lambda^{(\gamma)}(v) \leqslant v. \tag{1.4.5}$$

Further, by the property (Λ1) (see (1.1.8)), the deviation function $\Lambda^{(\gamma)}(v)$ for the random variable γ (and for any other random variable as well) allows a representation of the form

$$\Lambda^{(\gamma)}(v) = \int_{\mathbf{E}\gamma}^{v} \lambda^{(\gamma)}(u)du, \tag{1.4.6}$$

where $\lambda^{(\gamma)}(u)$ is the value of λ at which the supremum is attained in the definition $\Lambda^{(\gamma)}(u) = \sup_\lambda\{\lambda u - \ln \mathbf{E}e^{\lambda\gamma}\}$. Moreover, $\lambda^{(\gamma)}(u) \uparrow \lambda_+^{(\gamma)} := \sup\{\lambda : \mathbf{E}e^{\lambda\gamma} < \infty\}$ as $u \uparrow \infty$. From that it follows that there exists the limit

$$\lim_{v\to\infty} \frac{\Lambda^{(\gamma)}(v)}{v} = \lambda_+^{(\gamma)}. \tag{1.4.7}$$

Now the required inequality $\lambda_+^{(\gamma)} \leqslant 1$ follows from (1.4.5) and (1.4.7), which proves that $\lambda_+^{(\gamma)} = 1$.

If $\Lambda^* = \Lambda_+ < \infty$ then the assertion of the theorem follows in an obvious way from the equality

$$\Lambda_+ = -\ln \mathbf{P}(\xi = s_+)$$

(see (Λ2ii)). Theorem 1.4.1 is proved. □

Relations (1.4.7) for $\lambda_+^{(\gamma)} = 1$ mean that, under broad assumptions,

$$-\ln \mathbf{P}(\gamma \geqslant v) \sim v \quad \text{as} \quad v \to \infty$$

(see also subsection 1.4.3). This is, in a certain sense, an analogue of the relation

$$\mathbf{P}\big(F(\xi) < t\big) = \mathbf{P}\big(\xi < F^{(-1)}(t)\big) = F\big(F^{(-1)}(t)\big) \equiv t,$$

which is true under some assumptions on the distribution function $F(t) = \mathbf{P}(\xi < t)$. Also observe that, by virtue of the relation (1.4.6) and the equality $\lambda_+^{(\gamma)} = 1$, one has the inequality

$$\Lambda^{(\gamma)}(v) < v - \mathbf{E}\gamma \quad \text{for} \quad v > \mathbf{E}\gamma. \tag{1.4.8}$$

Example 1.4.2. Suppose that ξ has a normal distribution. By virtue of subsection 1.4.1, one can assume that $\mathbf{E}\xi = 0$, $\mathbf{E}\xi^2 = 1$. Then $\Lambda(\alpha) = \alpha^2/2$ (see Example 1.1.5), so that $\gamma = \xi^2/2$, $\mathbf{E}\gamma = 1/2$,

$$\mathbf{P}(\gamma \geqslant v) = 2\mathbf{P}(\xi \geqslant \sqrt{2v}) = \frac{2}{\sqrt{2\pi}}\int_{\sqrt{2v}}^{\infty} e^{-u^2/2}du \sim \frac{1}{\sqrt{\pi v}}e^{-v} \quad \text{as} \quad v \to \infty.$$

Further,

$$\mathbf{E}e^{\lambda\gamma} = \mathbf{E}e^{\lambda\xi^2/2} = \frac{1}{\sqrt{2\pi}}\int_{-\infty}^{\infty} e^{\lambda x^2/2 - x^2/2}dx = \frac{1}{\sqrt{1-\lambda}} \quad \text{for } \lambda \leqslant 1.$$

Therefore the equation for the point $\lambda^{(\gamma)}(v)$ takes the form of a partial derivative with respect to λ

$$\left(\lambda v + \frac{1}{2}\ln(1-\lambda)\right)'_\lambda = v - \frac{1}{2(1-\lambda)} = 0$$

and it has the unique solution $\lambda^{(\gamma)}(v) = 1 - 1/(2v)$. Hence we find that

$$\Lambda^{(\gamma)}(v) = \int_{\mathbf{E}\gamma}^{v} \left(1 - \frac{1}{2u}\right) du = v - \frac{1}{2}(\ln v + \ln 2 + 1).$$

It is also clear that, in this example, one has $s_\pm = \pm\infty$, $\Lambda_\pm = \infty$.

Example 1.4.3. Suppose that ξ follows the exponential distribution. By virtue of subsection 1.4.1, one can assume

$$\mathbf{P}(\xi \geqslant t) = e^{-t} \quad \text{for} \quad t \geqslant 0.$$

Then (see Example 1.1.7)

$$\Lambda(\alpha) = \alpha - 1 - \ln\alpha \quad \text{for} \quad \alpha \geqslant 0,$$

so that $\gamma = \xi - 1 - \ln\xi$. The equation $\Lambda(\alpha) = v$ has the solution $\alpha_+(v) = v + \ln v + 1 + O(\ln v/v)$ as $v \to \infty$. Therefore,

$$\begin{aligned}\mathbf{P}(\gamma \geqslant v) &= \mathbf{P}\left(\xi \geqslant v + \ln v + 1 + O\left(\frac{\ln v}{v}\right)\right)\\ &= e^{-(v+\ln v+1+O(\ln v/v))} = \frac{1}{v}e^{-v-1}\left(1 + O\left(\frac{\ln v}{v}\right)\right) \quad \text{as} \quad v \to \infty.\end{aligned}$$

In this example, $s_- = 0$, $s_+ = \infty$, $\Lambda_\pm = \infty$.

Example 1.4.4. For a Bernoulli random variable ξ ($\mathbf{P}(\xi = 1) = p = 1 - \mathbf{P}(\xi = 0)$) one has (see Example 1.1.6)

$$\gamma = \Lambda(\xi) = \xi\ln\frac{\xi}{p} + (1-\xi)\ln\frac{1-\xi}{1-p}, \quad \mathbf{E}\gamma = -p\ln p - (1-p)\ln(1-p) \leqslant \ln 2.$$

Hence

$$\gamma = \begin{cases} -\ln p & \text{with probability } p,\\ -\ln(1-p) & \text{with probability } 1-p.\end{cases}$$

If $p = 1/2$ then the variable γ degenerates to the constant $\ln 2$. For $p < 1/2$, the variable γ is given by an affine transformation of ξ:

$$\gamma = \frac{\xi + a}{b - a} \quad \text{with} \quad a = \ln(1-p), \quad b = \ln(1-p) - \ln p,$$

so that

$$\Lambda^{(\gamma)}(v) = \Lambda(v(b-a) - a).$$

The case $p > 1/2$ is dealt with in a similar way. In this example we have $s_- = 0$, $s_+ = 1$, $\Lambda_- = -\ln(1-p)$, $\Lambda_+ = -\ln p$.

1.4.3 Properties of the random variable γ and the function $\Lambda^{(\gamma)}$ in the case $d > 1$

In the multidimensional case, $d > 1$, analysis of the properties of the random variable γ turns out to be more complicated. We will restrict ourselves to considering vectors ξ whose distributions behave at infinity in a sufficiently regular way.

Definition 1.4.5. We will say that *a one-dimensional random variable ξ belongs to the class $\mathcal{L}$* if

$$\ln \mathbf{P}(\xi \geqslant t) \sim -\Lambda^{(\xi)}(t) \quad \text{as } t \to \infty. \tag{1.4.9}$$

The class $\mathcal{L}$ is rather broad and contains the classes $\mathcal{ER}$, $\mathcal{ES}$, $\mathcal{SE}$ of random variables ξ for which one has, respectively, the relations

$$\mathbf{P}(\xi \geqslant t) = e^{-\lambda_+ t}t^{\alpha}L(t), \quad \mathbf{P}(\xi \geqslant t) = e^{-\lambda_+ t \pm t^{\beta}L(t)}, \quad \mathbf{P}(\xi \geqslant t) = e^{-t^{\nu}L(t)},$$

where $\lambda_+ := \sup\{\lambda : \mathbf{E}e^{\lambda\xi} < \infty\} > 0$, $\alpha \in (-\infty, \infty)$, $\beta \in (0, 1)$, $\nu > 2$ and $L(t)$ is a function that is slowly varying at infinity.

One can suggest the following hypothetical criterion: for a random variable ξ to belong to the class $\mathcal{L}$ *it is necessary and sufficient that there exists a convex function $H(t)$ such that $-\ln \mathbf{P}(\xi \geqslant t) \sim H(t)$ as $t \to \infty$. If such a function exists then $H(t) \sim \Lambda(t)$ as $t \to \infty$.*

We will assume in what follows that $\Lambda_{<\infty} = \mathbb{R}^d$. Then the sets Λ_v grow unboundedly as $v \to \infty$. Set $B_v := \{\alpha : \Lambda(\alpha) \geqslant v\}$ and $\Lambda_{v-} := \{\alpha : \Lambda(\alpha) < v\}$ and put $\Pi(\mathrm{e}, b) := \{\alpha : \langle \mathrm{e}, \alpha\rangle \geqslant b\}$. We will need the following condition.

$[\mathbf{\Pi}]$ *For any $\varepsilon > 0$ there exists a finite number R_ε of half-spaces $\Pi(\mathrm{e}_{k,v}, b_{k,v})$, $k = 1, \ldots, R_\varepsilon$, such that for any sufficiently large v one has*

(1) $B_v \subset U_v := \cup_k \Pi(\mathrm{e}_{k,v}, b_{k,v})$,
(2) $\Lambda_{v(1-\varepsilon)-} \cap U_v = \varnothing$ (one can assume that the $\Pi(\mathrm{e}_{k,v}, b_{k,v})$ touch the set $\Lambda_{v(1-\varepsilon)}$).

The condition $[\mathbf{\Pi}]$ means that the sets $\Lambda_{v(1-\varepsilon)}$ and B_v can be separated by a polyhedron with R_ε faces. This condition is met, for instance, by the class of random vectors ξ for which condition $[\mathbf{C_0}]$ is satisfied and the set $\mathcal{A} := \{\lambda : A(\lambda) < \infty\}$ is bounded in $\mathbb{R}^d$. Indeed, for such vectors the deviation function Λ is asymptotically linear in any direction e:

$$\Lambda(t\mathrm{e}) \sim t\Lambda_\infty(\mathrm{e}) \quad \text{as} \quad t \to \infty,$$

where $\Lambda_\infty(\mathrm{e}) := \sup_{\lambda \in \mathcal{A}} \langle \lambda, \mathrm{e}\rangle < \infty$ (see (1.2.4)). Hence the level-v surfaces $\partial\Lambda_v$ of the function $\Lambda(\alpha)$ are 'asymptotically concentric', and, when contracted with the factor $1/v$, converge to the surface $\Gamma_1 := \{\alpha : \Lambda_\infty(\alpha) = 1\}$ as $v \to \infty$, while the level-$v(1-\varepsilon)$ surface $\partial\Lambda_{v(1-\varepsilon)}$ approaches the surface $\Gamma_{1-\varepsilon} := \{\alpha : \Lambda_\infty(\alpha) = (1-\varepsilon)\}$. Since the condition $[\mathbf{C_0}]$ is met, one has $\inf_{\mathrm{e}} \Lambda_\infty(\mathrm{e}) > 0$ and the surfaces Γ_1 and $\Gamma_{1-\varepsilon}$ do not touch each other. Moreover, they are the boundaries of convex sets. Hence it is clear that there is a separating polyhedron between Γ_1 and $\Gamma_{1-\varepsilon}$, and so the condition $[\mathbf{\Pi}]$ is satisfied.

The boundedness of the set $\mathcal{A}$ is not essential for the condition $[\mathbf{\Pi}]$. Let, for instance, $\Lambda(\alpha)$ grow as a quadratic function as $|\alpha| \to \infty$, up to an additive term $o(|\alpha|^2)$ (in this case, $\mathcal{A} = \mathbb{R}^d$). Then, after an appropriate linear transformation, we will obtain the function $\sum |\alpha_{(i)}|^2 + o(|\alpha|^2)$, whose level-$v$ surfaces approach, after contracting them with the factor $\sqrt{v}$, the sphere $\sum |\alpha_{(i)}|^2 = 1$, while the level-$v(1-\varepsilon)$ surfaces approach the sphere $\sum |\alpha_{(i)}|^2 = 1-\varepsilon$. As linear transformations map polyhedra into polyhedra, the condition $[\mathbf{\Pi}]$ is obviously met in this case. Taking into account Remark 1.4.7 (see below), one can also consider other types of convergence of $\Lambda(\alpha)$ to ∞ as $|\alpha| \to \infty$.

Theorem 1.4.6. *Let $\Lambda_{<\infty} = \mathbb{R}^d$ and the condition $[\mathbf{\Pi}]$ be met. Then*

(i) $\lambda_+^{(\gamma)} = 1, \quad \Lambda^{(\gamma)}(v) = v + o(v)$ *as* $v \to \infty$.

(ii) *If, in addition, there is a unit vector* e *such that* $\langle \mathrm{e}, \xi \rangle \in \mathcal{L}$ *then* $\gamma \in \mathcal{L}$ *and*

$$-\ln \mathbf{P}(\gamma \geqslant v) \sim v$$

as $v \to \infty$.

Proof. (i) *The upper bound.* By Theorem 1.3.2,

$$\mathbf{P}(\gamma \geqslant v) = \mathbf{P}(\xi \in B_v) \leqslant \mathbf{P}\Big(\xi \in \bigcup\nolimits_{1 \leqslant k \leqslant R_\varepsilon} \Pi(\mathrm{e}_{k,v}, b_{k,v})\Big)$$

$$\leqslant \sum_{k=1}^{R_\varepsilon} \mathbf{P}(\xi \in \Pi(\mathrm{e}_{k,v}, b_{k,v})) \leqslant \sum_{k=1}^{R_\varepsilon} e^{-\Lambda(\Pi(\mathrm{e}_{k,v}, b_{k,v}))} \leqslant R_\varepsilon e^{-v(1-\varepsilon)}.$$

Therefore,

$$\ln \mathbf{P}(\gamma \geqslant v) \leqslant -(1-\varepsilon)v + \ln R_\varepsilon.$$

Since $\varepsilon > 0$ is an arbitrary positive number, one has

$$\overline{\lim_{v\to\infty}} \frac{1}{v} \ln \mathbf{P}(\gamma \geqslant v) \leqslant -1, \quad \lambda_+^{(\gamma)} \geqslant 1. \qquad (1.4.10)$$

The lower bound. We will make use of relations (1.4.2) and (1.4.3), which imply that

$$-\Lambda^{(\gamma)}(v) \geqslant \overline{\lim_{n\to\infty}} \frac{1}{n} \ln \mathbf{P}(\Lambda(\zeta_n) \geqslant v) = \overline{\lim_{n\to\infty}} \frac{1}{n} \ln \mathbf{P}(\zeta_n \in B_v).$$

Since $\Lambda_{<\infty} = \mathbb{R}^d$ the function $\Lambda(\alpha)$ is continuous everywhere, and hence $\Lambda((B_v)) = \Lambda(B_v) = v$. Therefore, by the large deviation principle proved below in section 2.2 (see Theorem 2.2.2), one has for the scaled sums $\zeta_n = \frac{1}{n}\sum_{k=1}^n \xi_k$,

$$-\Lambda^{(\gamma)}(v) \geqslant \overline{\lim_{n\to\infty}} \frac{1}{n} \ln \mathbf{P}(\zeta_n \in B_v) \geqslant \underline{\lim_{n\to\infty}} \frac{1}{n} \ln \mathbf{P}(\zeta_n \in B_v) \geqslant -\Lambda((B_v)) = -v.$$

We have obtained $\Lambda^{(\gamma)}(v) \leqslant v$, $\lambda_+^{(\gamma)} \leqslant 1$. Together with (1.4.10), this proves the first assertion of the theorem.

(ii) It follows from (1.4.10) that

$$\mathbf{P}(\gamma \geqslant v) \leqslant e^{-v+o(v)} \quad \text{as} \quad v \to \infty.$$

So, to prove the second assertion, it suffices to establish the converse inequality. For any $v > 0$ and any given e, one has

$$\mathbf{P}(\gamma \geqslant v) = \mathbf{P}(\xi \in B_v) \geqslant \mathbf{P}(\xi \in \Pi(\mathrm{e}, b_v)) = \mathbf{P}(\langle \mathrm{e}, \xi \rangle \geqslant b_v),$$

where b_v is chosen so that the half-space $\Pi(\mathrm{e}, b_v)$ touches ∂B_v. Then, as in section 1.3, we need to verify that $\Lambda^{(\langle \mathrm{e},\xi \rangle)}(b_v) = v$. Clearly $b_v \to \infty$ as $v \to \infty$. Choose an e such that $\langle \mathrm{e}, \xi \rangle \in \mathcal{L}$. Then

$$\ln \mathbf{P}(\gamma \geqslant v) \geqslant \ln \mathbf{P}(\langle \mathrm{e}, \xi \rangle \geqslant b_v) \sim -\Lambda^{(\langle \mathrm{e},\xi \rangle)}(b_v) = -v.$$

as $v \to \infty$. This means that

$$\mathbf{P}(\gamma \geqslant v) \geqslant e^{-v+o(v)} \quad \text{as} \quad v \to \infty.$$

The theorem is proved. □

Remark 1.4.7. (i) It can be seen from the proof of Theorem 1.4.6 that its assertion will remain true in the case where the number R_ε of half-spaces in condition [**Π**] grows along with v, but in a such way that $\ln R_\varepsilon = o(v)$ as $v \to \infty$.

(ii) If $\Lambda_{<\infty}$ does not coincide with $\mathbb{R}^d$, $d \geqslant 2$, then it may happen that the random variable $\gamma = \Lambda(\xi)$ does not satisfy Cramér's condition [$\mathbf{C}_0$], which is demonstrated by Example 1.3.6 presented in section 1.3. Recall that, in Example 1.3.6, the random vector ξ has a uniform distribution on the unit sphere Γ in $\mathbb{R}^d$, $d \geqslant 2$. Since for any $\alpha \in \Gamma$ one has $\Lambda(\alpha) = \infty$, we see that $\gamma = \Lambda(\xi) = \infty$ with probability 1, and thus Cramér's condition fails for γ.

Example 1.4.8. Assume that vector $\xi = (\xi_{(1)}, \dots, \xi_{(d)})$ has a normal distribution with zero mean and identity covariance matrix. Then it is easy to see that the deviation function has the form $\Lambda(\alpha) = \frac{1}{2}\sum_{i=1}^{d} \alpha_{(i)}^2$, and hence the random variable

$$2\gamma = 2\Lambda(\xi) = \xi_{(1)}^2 + \dots + \xi_{(d)}^2$$

has the distribution χ^2 with d degrees of freedom, so that

$$\mathbf{P}(\gamma \geqslant v) = \frac{1}{\Gamma(d/2)} \int_v^\infty u^{d/2-1} e^{-u} du \sim \frac{1}{\Gamma(d/2)} v^{d/2-1} e^{-v} \quad \text{as} \quad v \to \infty.$$

The conditions of Theorem 1.4.6 are clearly satisfied here. In the case $d = 2$, γ has the exponential distribution $\mathbf{P}(\gamma \geqslant v) = e^{-v}$, $v \geqslant 0$.

Observe also that, if the components $\xi_{(i)}$ of the vector ξ are jointly independent then $\Lambda(\alpha) = \sum_{i=1}^{d} \Lambda^{(\xi_{(i)})}(\alpha_i)$, and hence the study of the properties of γ and $\Lambda^{(\gamma)}(v)$ is broadly reduced to the one-dimensional case since Theorem 1.4.1 can be applied to independent summands $\Lambda^{(\xi_{(i)})}(\xi_{(i)})$. In particular, the inequality $\mathbf{E}\gamma \leqslant 2d$ is always true.

1.5 The integro-local theorems of Stone and Shepp and Gnedenko's local theorem

1.5.1 On integro-local theorems

In probability theory local and integral limit theorems are usually discerned while studying the limit distributions of sums of random variables standardised in a proper way.

If ξ has a *lattice* distribution then, when studying sums of random variables, such a distribution, without loss of generality, might be considered *arithmetic*, i.e. one might assume that ξ is integer-valued, so that the greatest common divisor between differences of possible values of ξ would be equal to 1. In this case *local* theorems would deal with the asymptotics of the probabilities $\mathbf{P}(S_n = k)$ as $n \to \infty$, $k = k(n)$. *Integral* theorems deal with probabilities $\mathbf{P}(S_n \geqslant k)$; typically, they can be deduced from local theorems easily enough but the converse does not hold.

In the *non-lattice* case, *local* theorems examine the behaviour of the density of the distribution of S_n, and they assume (along with some other conditions; see e.g. [39]) the existence of the density of the distribution of ξ. *Integral* theorems, which examine the asymptotics of $\mathbf{P}(S_n \geqslant x)$ as $n \to \infty$, $x = x(n)$, are proved under significantly broader conditions. These theorems can be obtained from local theorems (by integration), but not vice versa.

In the non-lattice case, so-called *integro-local* theorems, which study the asymptotics of the probability $\mathbf{P}\big(S_n \in \Delta[x)\big)$ that S_n will hit a half-open interval

$$\Delta[x) := [x, x + \Delta) \tag{1.5.1}$$

with fixed or diminishing (as n increases) length Δ, are also of interest. Here in the notation $\Delta[x)$ for a half-open interval $[x, x+\Delta)$, the symbol Δ, which also indicates the length of the interval, is inseparable from the symbol $[x)$, so that the 'dual' usage of the symbol Δ does not lead to misunderstanding. Statements describing the asymptotic behaviour of the probabilities $\mathbf{P}\big(S_n \in \Delta[x)\big)$ are called *integro-local theorems*, in order to distinguish them from local and integral theorems. In the arithmetic case, local and integro-local theorems coincide.

Integro-local theorems have a number of advantages, which (in contrast with local and integral theorems) make them a principal object of examination later on. Here are the aforementioned advantages.

(1) Integro-local theorems give, roughly speaking, the same information on the distribution of S_n as local theorems do (the asymptotics for probability that S_n will hit a small half-open interval $\Delta[x)$), but they are proved under the same general conditions as integral theorems.
(2) Integro-local theorems imply without any difficulty, the corresponding integral theorems (by means of summation). This fact has a singular importance in the multidimensional case, where not only in proofs but also in formulations of integral theorems for $\mathbf{P}(S_n \in B)$ difficulties appear caused by the form of the sets B. Integro-local theorems completely lack these difficulties, that have low connection to the nature of the limit distribution of S_n.

(3) In contrast with local and integral theorems, integro-local theorems allow a unified natural form for all kinds of deviations lying in the Cramér's area, i.e. when $S_n/n \in \mathcal{A}'$, in other words, for deviations that are normal, moderately large or large in current terminology (for the latter, possibly up to some limit defined by the border $\partial\mathcal{A}'$ of the set $\mathcal{A}'$). In addition, integro-local theorems turn out to be uniform with respect to the magnitude of the deviations.

Thus, integro-local theorems describe in a unified form and under general conditions the asymptotics of the distribution of S_n, allowing one to deduce both 'local' and integral theorems.

1.5.2 The theorem of Stone and Shepp and Gnedenko's theorem

In what follows, the integro-local limit theorem of Stone and Shepp plays an important role in the area of normal deviations; it was established in [166] and [174].

Theorem 1.5.1. *Let ξ be a non-lattice random variable, and let a convergence of the distributions of the normalised sums $(S_n - A_n)/B_n$ to the stable law $\mathbf{\Phi}$ hold for proper normalising constants A_n, B_n. Then, as $n \to \infty$,*

$$\mathbf{P}\left(S_n \in \Delta[x)\right) = \frac{\Delta}{B_n}\phi\left(\frac{x - A_n}{B_n}\right) + o(B_n^{-1}), \tag{1.5.2}$$

where ϕ is the density of the distribution $\mathbf{\Phi}$. The relations (1.5.2) *are uniform over x and over $\Delta \in [\Delta_1, \Delta_2]$, where $\Delta_1 > 0$, $\Delta_2 < \infty$ are fixed.*

Proof of Theorem 1.5.1. is given under a simplifying assumption, that the characteristic function (ch.f.) $\varphi(t) = \mathbf{E}e^{it\xi} = \psi(it)$ meets Cramér's condition

$$\limsup_{|t|\to\infty} \left|\varphi(t)\right| < 1. \tag{1.5.3}$$

Moreover, we will assume that $\mathbf{E}\xi = 0$ and $\sigma^2 = \mathbf{D}\xi < \infty$, i.e. $A_n = 0$, $B_n = \sigma\sqrt{n}$, and, therefore, the convergence of the distributions of $S_n/(\sigma\sqrt{n})$ to the normal law $\mathbf{\Phi} = \mathbf{\Phi}_{0,1}$ with parameters $(0, 1)$ holds. In this case the relation (1.5.2) takes the form

$$\mathbf{P}\left(S_n \in [x)\right) = \frac{\Delta}{\sigma\sqrt{2\pi n}} e^{-x^2/(2\sigma^2 n)} + o\left(\frac{1}{\sqrt{n}}\right).$$

The case of convergence to an arbitrary stable law does not cause any trouble (see, for instance, [174] and [39], Appendix 7).

From the properties of the ch. f. it follows that condition (1.5.3) is always met, if the distribution of the sum S_m for some $m \geqslant 1$ has a positive absolutely continuous component.

In the general case the proof of Theorem 1.5.1 is rather more complex; a detailed exposition can be found in the textbook [39].

In order to be able to use the inversion formula, we take advantage of the 'smoothing method' and consider along with S_n the sums

$$Z_n = S_n + \eta_\delta, \tag{1.5.4}$$

where η_δ has a uniform distribution on $[-\delta, 0]$. Since the ch. f. $\varphi_{\eta_\delta}(t)$ of η_δ, given by

$$\varphi_{\eta_\delta}(t) = \frac{1 - e^{-it\delta}}{it\delta}, \tag{1.5.5}$$

has the property that the function $\varphi_{\eta_\delta}(t)/t$ is integrable at infinity, then for increments of the cumulative distribution function $G_n(x)$ of the random variable Z_n (its ch. f., divided by t, is integrable as well) we can use the following formula (see e.g. (7.2.8) in [39]):

$$\begin{aligned} G_n(x+\Delta) - G_n(x) = \mathbf{P}\big(Z_n \in \Delta[x)\big) &= \frac{1}{2\pi} \int e^{-itx} \frac{1 - e^{-it\Delta}}{it} \varphi^n(t)\varphi_{\eta_\delta}(t)dt \\ &= \frac{\Delta}{2\pi} \int e^{-itx} \varphi^n(t)\widehat{\varphi}(t)dt, \end{aligned} \tag{1.5.6}$$

where $\widehat{\varphi}(t) = \varphi_{\eta_\delta}(t)\varphi_{\eta_\Delta}(t)$ is the ch. f. of the sum of the independent random variables η_δ and η_Δ. We have obtained that the difference $G_n(x+\Delta) - G_n(x)$, up to a multiplier Δ, is nothing other than the density at point x of the random variable $S_n + \eta_\delta + \eta_\Delta$.

Now split the integral on the right-hand side of (1.5.6) into two integrals: one over the interval $|t| < \beta$ for some $\beta < 1$ and the other over its complement. Set $x = v\sqrt{n}$ and consider first

$$I_1 := \int_{|t|<\beta} e^{-itv\sqrt{n}} \varphi^n(t)\widehat{\varphi}(t)dt = \frac{1}{\sqrt{n}} \int_{|u|<\beta\sqrt{n}} e^{-iuv} \varphi^n\left(\frac{u}{\sqrt{n}}\right) \widehat{\varphi}\left(\frac{u}{\sqrt{n}}\right) du.$$

Assuming, without loss of generality, that $\sigma = 1$, we obtain

$$\begin{aligned} 1 - \varphi(t) &= \frac{t^2}{2} + o(t^2), \\ \ln\varphi(t) &= \ln\Big[1 - \big(1 - \varphi(t)\big)\Big] = -\frac{t^2}{2} + o(t^2) \quad \text{as } t \to 0. \end{aligned} \tag{1.5.7}$$

Therefore

$$n \ln\varphi\left(\frac{u}{\sqrt{n}}\right) = -\frac{u^2}{2} + h_n(u), \tag{1.5.8}$$

where $h_n(u) \to 0$ for any fixed u as $n \to \infty$. Furthermore, for small enough β, in the interval $|u| < \beta\sqrt{n}$ one has

$$\big|h_n(u)\big| \leqslant \frac{u^2}{6},$$

so that the right-hand side of (1.5.8) is not larger than $-u^2/3$. Now I_1 can be expressed in the form

$$I_1 = \frac{1}{\sqrt{n}} \int_{|u|<\beta\sqrt{n}} e^{-iuv-u^2/2+h_n(u)} \widehat{\varphi}\left(\frac{u}{\sqrt{n}}\right) du, \tag{1.5.9}$$

where $|\widehat{\varphi}(u/\sqrt{n})| \leqslant 1$ and $\widehat{\varphi}(u/\sqrt{n}) \to 1$ for any fixed u as $n \to \infty$. Hence, by virtue of the above and the dominated convergence theorem, we have

$$\sqrt{n}\, I_1 \to \int e^{-iuv-u^2/2} du \tag{1.5.10}$$

uniformly over v, since the integral on the right-hand side of (1.5.9) is uniformly continuous over v. But the integral in the right-hand side of (1.5.10) is simply (up to factor $1/2\pi$) the result of applying the inversion formula to the ch. f. of the normal distribution, so that

$$\lim_{n\to\infty} \sqrt{n}\, I_1 = \sqrt{2\pi}\, e^{-v^2/2}. \tag{1.5.11}$$

It remains to consider the integral

$$I_2 := \int_{|t|\geqslant\beta} e^{-itv\sqrt{n}} \varphi^n(t) \widehat{\varphi}(t) dt.$$

Owing to the condition (1.5.3) and the fact that distribution of ξ is non-lattice, we have

$$q := \sup_{|t|\geqslant\beta} \left|\varphi(t)\right| < 1 \tag{1.5.12}$$

and, therefore,

$$\begin{gathered} |I_2| \leqslant q^n \int_{|t|\geqslant\beta} \left|\widehat{\varphi}(t)\right| dt \leqslant q^n c(\Delta, \delta), \\ \lim_{n\to\infty} \sqrt{n}\, I_2 = 0 \end{gathered} \tag{1.5.13}$$

uniformly over v, where $c(\Delta,\delta)$ depends only on Δ and δ. We have determined that, for $x = v\sqrt{n}$, $n \to \infty$, uniformly over v,

$$\begin{gathered} I_1 + I_2 = \sqrt{\frac{2\pi}{n}}\, e^{-v^2/2} + o\left(\frac{1}{\sqrt{n}}\right), \\ \mathbf{P}\big(Z_n \in \Delta[x)\big) = \frac{\Delta}{\sqrt{2\pi n}}\, e^{-x^2/2n} + o\left(\frac{1}{\sqrt{n}}\right) \end{gathered} \tag{1.5.14}$$

(see (1.5.6)). It means that the representation (1.5.14) holds uniformly over all x.

Next, by virtue of (1.5.4),

$$\big\{Z_n \in [x, x+\Delta-\delta)\big\} \subset \big\{S_n \in \Delta[x)\big\} \subset \big\{Z_n \in [x-\delta, x+\Delta)\big\} \tag{1.5.15}$$

and, therefore, in particular,

$$\mathbf{P}\big(S_n \in \Delta[x)\big) \leqslant \frac{\Delta+\delta}{\sqrt{2\pi n}}\, e^{-(x-\delta)^2/2n} + o\left(\frac{1}{\sqrt{n}}\right) = \frac{\Delta+\delta}{\sqrt{2\pi n}}\, e^{-x^2/2n} + o\left(\frac{1}{\sqrt{n}}\right).$$

By virtue of (1.5.15), the converse inequality holds. Since δ is arbitrary, this is possible only if

$$\mathbf{P}\big(S_n \in \Delta[x)\big) = \frac{\Delta}{\sqrt{2\pi n}} e^{-x^2/2n} + o\left(\frac{1}{\sqrt{n}}\right). \tag{1.5.16}$$

Theorem 1.5.1 is proved. □

A statement analogous to Theorem 1.5.1 holds in the arithmetic case. The following local theorem of Gnedenko holds (see [96], § 50).

Theorem 1.5.2. *Let ξ be an arithmetic random variable and let the conditions of Theorem 1.5.1 on the convergence of $(S_n - A_n)/B_n$ to a stable law $\mathbf{\Phi}$ be met. Then*

$$\mathbf{P}\,(S_n = x) = \frac{1}{B_n}\,\phi\left(\frac{x - A_n}{B_n}\right) + o(B_n^{-1}) \tag{1.5.17}$$

uniformly over all integers x.

The proof of Theorem 1.5.2 is similar to the proof of Theorem 1.5.1 but substantially simpler, so we omit it.

1.5.3 Uniform versions of Theorems 1.5.1, and 1.5.2

In what follows, in a number of cases we will require integral and local theorems for sums of random variables for which the distribution $\mathbf{F} = \mathbf{F}^{(\lambda)}$ depends on some parameter $\lambda > 0$. That parameter can, in turn, depend on n. Thus, we will deal with a *series scheme* in some specific form. We will need *uniform versions* of Theorems 1.5.1 and 1.5.2 over the parameter range $\lambda \in [\lambda_0, \lambda_1]$ for fixed $\lambda_0 < \lambda_1$. Here we consider only the case of convergence to the normal law $\mathbf{\Phi}_{0,1}$ with parameters (0, 1)

Denote

$$a^{(\lambda)} = \mathbf{E}\,\xi^{(\lambda)}, \qquad (\sigma^{(\lambda)})^2 = \mathbf{D}\xi^{(\lambda)}, \qquad \varphi^{(\lambda)}(t) = \mathbf{E}\,e^{it\xi^{(\lambda)}},$$

where $\xi^{(\lambda)}$ has distribution $\mathbf{F}^{(\lambda)}$. The following statement is an analogue of Theorem 1.5.1.

Theorem 1.5.3. *Let the distribution $\mathbf{F}^{(\lambda)}$ possess the following properties for $0 < \sigma_1 < \sigma^{(\lambda)} < \sigma_2 < \infty$, where σ_1, σ_2 do not depend on λ:*

(i)

$$\varphi^{(\lambda)}(t) - 1 - ia^{(\lambda)}t + \frac{t^2 a_2^{(\lambda)}}{2} = o(t^2), \qquad a_2^{(\lambda)} := \mathbf{E}\,(\xi^{(\lambda)})^2, \tag{1.5.18}$$

where the remainder term $o(t^2)$ is uniform over $\lambda \in [\lambda_0, \lambda_1]$ as $t \to 0$, i.e. there exist $t_0 > 0$ and a function $\varepsilon(t) \to 0$ as $t \to 0$, independent of λ, such that, for all $|t| \leqslant t_0$, $\lambda \in [\lambda_0, \lambda_1]$, the left-hand side of (1.5.18) *does not exceed $\varepsilon(t)t^2$ in absolute value;*

(ii) *for any fixed $0 < \theta_1 < \theta_2 < \infty$,*

$$q^{(\lambda)} := \sup_{\theta_1 \leqslant |t| \leqslant \theta_2} \left|\varphi^{(\lambda)}(t)\right| \leqslant q < 1, \tag{1.5.19}$$

where q does not depend on λ.

Then, for all fixed $\Delta > 0$,

$$\mathbf{P}\big(S_n^{(\lambda)} - na^{(\lambda)} \in \Delta[x)\big) = \frac{\Delta}{\sigma^{(\lambda)}\sqrt{n}}\,\phi\left(\frac{x}{\sigma^{(\lambda)}\sqrt{n}}\right) + o\left(\frac{1}{\sqrt{n}}\right), \tag{1.5.20}$$

where the remainder term $o(1/\sqrt{n})$ *is uniform over* x, $\lambda \in [\lambda_0, \lambda_1]$ *and* ϕ *is a normal distribution density with parameters* $(0, 1)$.

Proof. Looking at the proof of Theorem 1.5.1 in its general form (see [39], § 7), it is easy to verify that, in order to preserve the uniformity in all intermediate statements of the proof, it is sufficient to ensure:

(i) the uniformity over λ of the remainder $o(t^2)$ as $t \to 0$ in the relation (1.5.7) for the expansion of the ch. f. of the random variable $\xi = (\xi^{(\lambda)} - a^{(\lambda)})/\sigma^{(\lambda)}$;

(ii) the uniformity in

$$\sup_{\beta \leqslant t \leqslant \theta} \big|\varphi(t)\big| < 1$$

for the same ch. f. for any fixed $\theta > \beta$. It is easy to see that conditions (i), (ii) of Theorem 1.5.3 guarantee the desired uniformity. See [39], § 7 for details. □

An analogue of Theorem 1.5.2 is the following.

Theorem 1.5.4. *Let arithmetic distributions* $\mathbf{F}^{(\lambda)}$ *meet the conditions of Theorem* 1.5.3 *for* $\theta_2 = \pi$. *Then, uniformly over* x *and* $\lambda \in [\lambda_0, \lambda_1]$, *the representation* (1.5.17) *is valid, assuming that on its right-hand side*

$$A_n = a^{(\lambda)} n, \qquad B_n = \sigma^{(\lambda)}\sqrt{n}.$$

The proof of Theorem 1.5.4 is quite similar to the proof of Theorem 1.5.3, but simpler.

Uniform versions of integro-local theorems are given extensively in [54].

1.5.4 On multidimensional integro-local theorems

In the *multidimensional case* $\xi \in \mathbb{R}^d$, $d > 1$, statements of the aforementioned theorems are completely retained after trivial changes are made: A_n should be read as a centering vector and B_n as a normalising matrix. The common structural conditions in the multidimensional case look as follows. The non-degenerate *random vector* ξ is called *non-lattice* (i.e. it has a non-lattice distribution) if, for any unit vector $\mathrm{e} \in \mathbb{R}^d$, the random variable $\langle \mathrm{e}, \xi \rangle$ is non-lattice, i.e. the following condition is met: for any $\mathrm{e} \in \mathbb{R}^d$, $|\mathrm{e}| = 1$, $t \neq 0$,

$$\big|\varphi(\mathrm{e}t)\big| < 1.$$

A random vector ξ is *arithmetic*, if for all $t \in \mathbb{Z}^d$,

$$\varphi(2\pi t) = 1$$

and, for all $t \notin \mathbb{Z}^d$,

$$\big|\varphi(2\pi t)\big| < 1.$$

The structural conditions are described in detail in subsection 2.3.2.

In the non-lattice case the multiplier Δ on the right-hand side of (1.5.2) should be replaced by Δ^d. A multidimensional version of Theorem 1.5.1 was obtained in [174]. A multidimensional version of the local theorem 1.5.2 was established in [163].

For instance, in the case $\mathbf{E}\xi = 0$, $\mathbf{E}|\xi|^2 < \infty$, the multidimensional analogue of Theorem 1.5.1 has the following form.

Denote by $\Delta[x) = \{y = (y_{(1)}, \dots, y_{(d)}) : x_{(i)} \leqslant y_{(i)} < x_{(i)} + \Delta;\ 1 \leqslant i \leqslant d\}$ a half-open cube in $\mathbb{R}^d$ with a vertex at the point $x = (x_{(1)}, \dots, x_{(d)})$ and with side length equal to $\Delta > 0$.

Theorem 1.5.5. *Let ξ be a non-lattice vector, let $\mathbf{E}\xi = 0$ and let σ^2 be a positive definite matrix of second moments of the vector ξ. Then, as $n \to \infty$,*

$$\mathbf{P}\big(S_n \in \Delta[x)\big) = \frac{\Delta^d}{(2\pi n)^{d/2}|\sigma|} \exp\left\{-\frac{x\sigma^{-2}x^T}{2n}\right\} + o(n^{-d/2}), \qquad (1.5.21)$$

where σ^{-2} is the matrix inverse to σ^2. The relations (1.5.21) *are uniform over x and over $\Delta \in [\Delta_1, \Delta_2]$, where $\Delta_1 > 0$, $\Delta_2 > 0$ are fixed.*

For a detailed treatment of multidimensional integro-local theorems in the range of x values including the large deviation zone, see subsection 2.3.2.

2

Approximation of distributions of sums of random variables

2.1 The Cramér transform. The reduction formula

2.1.1 The one-dimensional case

In the following presentation an important role will be played by the so-called *Cramér transform* of the distribution $\mathbf{F}$ of a random variable ξ. Let, as before (see Section 1.1.1),

$$\psi(\lambda) = \mathbf{E}e^{\lambda\xi}, \qquad A(\lambda) = \ln \psi(\lambda).$$

Definition 2.1.1. Suppose that the condition **[C]** holds (see Section 1.1.1). The Cramér transform at a point λ of a distribution $\mathbf{F}$ is the distribution[1]

$$\mathbf{F}^{(\lambda)}(dt) := \frac{e^{\lambda t}\mathbf{F}(dt)}{\psi(\lambda)}. \tag{2.1.1}$$

Clearly, the distributions $\mathbf{F}$ and $\mathbf{F}^{(\lambda)}$ are mutually absolutely continuous with the density

$$\frac{\mathbf{F}^{(\lambda)}(dt)}{\mathbf{F}(dt)} = \frac{e^{\lambda t}}{\psi(\lambda)}.$$

A random variable with the distribution $\mathbf{F}^{(\lambda)}$ will be denoted by $\xi^{(\lambda)}$.

[1] In the literature, the transform (2.1.1) is also called the *Esscher transform*. However, a systematic use of this transform for the study of the probabilities of large deviations was first undertaken by Cramér.

When studying the probabilities of large deviations of sums of random variables using inversion formulas in the same way as for normal deviations, we necessarily arrive at the *steepest descent method*, which consists in moving the integration contour so that it passes through the so-called *saddle point*, where the power of the exponent of the integrand, when moving along the imaginary axis, attains its minimum (then at this point the maximum along the real axis is attained; hence the name 'saddle point'). The Cramér transform essentially does the same, performing this movement of the integration contour before passing to the inversion formula and reducing a problem about large deviations to a problem about normal deviations, where the inversion formula is no longer needed if one uses limit theorems in the zone of normal deviations. This is precisely the path that we will follow in this chapter.

The Laplace transform of the distribution $\mathbf{F}^{(\lambda)}$ is obviously equal to

$$\mathbf{E}\, e^{\mu\xi^{(\lambda)}} = \frac{\psi(\lambda+\mu)}{\psi(\lambda)}. \tag{2.1.2}$$

Clearly

$$\mathbf{E}\,\xi^{(\lambda)} = \frac{\psi'(\lambda)}{\psi(\lambda)} = \big(\ln\psi(\lambda)\big)' = A'(\lambda)$$

$$\mathbf{E}\,\big(\xi^{(\lambda)}\big)^2 = \frac{\psi''(\lambda)}{\psi(\lambda)}, \qquad \mathbf{D}\xi^{(\lambda)} = \frac{\psi''(\lambda)}{\psi(\lambda)} - \left(\frac{\psi'(\lambda)}{\psi(\lambda}\right)^2$$
$$= \big(\ln\psi(\lambda)\big)'' = A''(\lambda). \tag{2.1.3}$$

Denote

$$S_n^{(\lambda)} := \sum_{i=1}^{n} \xi_i^{(\lambda)},$$

where the $\xi_i^{(\lambda)}$ are independent copies of $\xi^{(\lambda)}$. The distribution $\mathbf{F}^{(\lambda)}$ of the random variable $\xi^{(\lambda)}$ will be called the *Cramér transform of* $\mathbf{F}$ *with parameter* λ. The random variables $\xi^{(\lambda)}$ will also be called Cramér transforms, but of the initial random variable ξ. The connection between the distributions of S_n and $S_n^{(\lambda)}$ is established in the following theorem.

Theorem 2.1.2. *For any measurable set B we have*

$$\mathbf{P}\big(S_n \in x + B\big) = e^{-\lambda x + nA(\lambda)} \int_B e^{-\lambda z}\mathbf{P}\big(S_n^{(\lambda)} - x \in dz\big). \tag{2.1.4}$$

Assuming in (2.1.4) $x = \alpha n$, $\lambda = \lambda(\alpha)$, we obtain the relation

$$\mathbf{P}(S_n \in x + B) = e^{-n\Lambda(\alpha)} \int_B e^{-\lambda(\alpha)z}\mathbf{P}\,(S_n^{(\lambda(\alpha))} - \alpha n \in dz), \tag{2.1.5}$$

which will be called the (?) *reduction formula*.

If $\alpha = x/n \in (\alpha_-, \alpha_+)$ (see section 1.1) then $\mathbf{E}\xi^{(\lambda(\alpha))} = A'\big(\lambda(\alpha)\big) \equiv \alpha$ (see (1.1.6)), $\mathbf{E}S_n^{(\lambda(\alpha))} - \alpha n = 0$ and, for $x > n\mathbf{E}\xi$, the reduction formula allows one to reduce the problem about the *probabilities of large deviations* for S_n to a problem about the distribution of $S_n^{(\lambda(\alpha))}$ in the zone of *normal deviations*.

The reduction formula will be useful in 'boundary' cases, when $\alpha = x/n \notin (\alpha_-, \alpha_+)$. If, for instance, $\lambda_+ < \infty$, $\alpha \geqslant \alpha_+$, then $\lambda(\alpha) = \lambda_+$ and (2.1.5) implies

$$\mathbf{P}\,(S_n \in x + B) = e^{-n\Lambda(\alpha_+)} \int_B e^{-\lambda_+ z}\mathbf{P}\,(S_n^{(\lambda_+)} - \alpha n \in dz).$$

Here, $\alpha n > \mathbf{E}S_n^{(\lambda_+)}$ and we have reduced a problem about large deviations for S_n to a problem that also concerns large deviations for $S_n^{(\lambda_+)}$ but with the important difference that now $\xi^{(\lambda_+)}$ *does not satisfy Cramér's condition* $[\mathbf{C}_+]$. Here we will find useful results about the large deviations of sums of random variables with distributions regularly varying at infinity (see § 2.5).

Proof of Theorem 2.1.2. The relation (2.1.4) is essentially an identity and its proof is almost obvious. On the one hand the Laplace transform of the distribution of the sum $S_n^{(\lambda)}$ is equal to (see (2.1.2))

$$\mathbf{E}\, e^{\mu S_n^{(\lambda)}} = \left[\frac{\psi(\mu+\lambda)}{\psi(\lambda)}\right]^n. \tag{2.1.6}$$

On the other hand, consider the Cramér transform $(S_n)^{(\lambda)}$ of S_n at the point λ. Applying formula (2.1.2) to the distribution of S_n, we obtain

$$\mathbf{E}\, e^{\mu (S_n)^{(\lambda)}} = \frac{\psi^n(\mu+\lambda)}{\psi^n(\lambda)}. \tag{2.1.7}$$

Since the right-hand sides of (2.1.6) and (2.1.7) coincide, *the Cramér transform of S_n at the point λ coincides in distribution with the sum $S_n^{(\lambda)}$ of the transforms of $\xi_i^{(\lambda)}$*. In other words,

$$\frac{\mathbf{P}(S_n \in dv)e^{\lambda v}}{\psi^n(\lambda)} = \mathbf{P}(S_n^{(\lambda)} \in dv), \tag{2.1.8}$$

or, equivalently,

$$\mathbf{P}\,(S_n \in dv) = e^{-\lambda v + n \ln \psi(\lambda)} \mathbf{P}\,(S_n^{(\lambda)} \in dv).$$

Integrating this equality over the set $x + B$ and making the change $v = z + x$, we obtain (2.1.4). The theorem is proved. □

Let us mention the property of the stochastic monotonicity of Cramér transforms. We say that a random variable η_1 does not exceed η_2 stochastically ($\eta_1 \underset{st}{\leqslant} \eta_2$) if for all t we have

$$\mathbf{P}\,(\eta_1 \geqslant t) \leqslant \mathbf{P}\,(\eta_2 \geqslant t).$$

In a similar way, one defines the inequalities

$$\eta_1 \underset{st}{<} \eta_2, \qquad \eta_1 \underset{st}{\geqslant} \eta_2, \qquad \eta_1 \underset{st}{>} \eta_2.$$

The above-mentioned stochastic monotonicity property consists in the following: *the sequence of transforms $\xi^{(\lambda)}$ stochastically does not increase in λ*:

$$\xi^{(\lambda_1)} \underset{st}{<} \xi^{(\lambda_2)} \quad \text{for} \quad \lambda_1 < \lambda_2.$$

Indeed, without loss of generality, one may assume that $\lambda_1 = 0$, $\lambda_2 = \lambda > 0$. The distribution of $\xi^{(\lambda)}$ is obtained from the distribution of ξ by multiplying the probabilities $\mathbf{P}\,(\xi \in dt)$ by the factor $e^{\lambda t}/\psi(\lambda)$, which increases with t.

More formally, let

$$t_\lambda := \frac{1}{\lambda} \ln \psi(\lambda).$$

This is the unique solution of the equation $e^{\lambda t} = \psi(\lambda)$. Then, for any $t \geqslant t_\lambda$,

$$\mathbf{P}(\xi^{(\lambda)} \geqslant t) = \int_t^\infty \mathbf{P}(\xi^{(\lambda)} \in du) = \int_t^\infty \frac{e^{\lambda t}}{\psi(t)} \mathbf{P}(\xi \in du) \geqslant \mathbf{P}(\xi \geqslant t),$$

and, for any $t < t_\lambda$,

$$\mathbf{P}(\xi^{(\lambda)} < t) < \mathbf{P}(\xi < t).$$

Since $\lambda(\alpha)$ is a monotone function of α, *the sequence (in α) of the random variables ${}^\alpha\xi := \xi^{(\lambda(\alpha))}$ forms a stochastically increasing sequence.*

Note also that some properties of the Cramér transform allow one to characterize the normal distribution (see Theorem 2.1.3 in the next section).

2.1.2 The multidimensional case

In the multidimensional case the Cramér transform $\mathbf{F}^{(\lambda)}$ of a distribution $\mathbf{F}$ at a point $\lambda \in \mathbb{R}^d$ is defined by

$$\mathbf{F}^{(\lambda)}(dt) := \frac{e^{\langle \lambda, t\rangle}\mathbf{F}(dt)}{\psi(\lambda)}.$$

Denote by $\xi^{(\lambda)}$ a random vector with the distribution $\mathbf{F}^{(\lambda)}$. Similarly to the one-dimensional case we find

$$\begin{aligned} \mathbf{E}e^{\langle \mu, \xi^{(\lambda)}\rangle} &= \frac{\psi(\lambda+\mu)}{\psi(\lambda)} \\ \mathbf{E}\xi^{(\lambda)} &= \operatorname{grad}\big(\ln\psi(\lambda)\big) = \operatorname{grad} A(\lambda) \end{aligned} \tag{2.1.9}$$

For brevity, the *gradient* of a function $g = g(\lambda) : \mathbb{R}^d \to \mathbb{R}^1$ will be denoted by $g'(\lambda) := \operatorname{grad} g(\lambda) \in \mathbb{R}^d$. In particular, $A'(\lambda) := \operatorname{grad} A(\lambda) = \varphi'(\lambda)/\varphi(\lambda)$. Along with the set $\mathcal{A} := \{\lambda : A(\lambda) < \infty\}$, consider the set $\mathcal{A}' := \{\alpha = A'(\lambda) \in \mathbb{R}^d : \lambda \in \mathcal{A}\}$. If the condition **[C]** is satisfied then these sets are non-empty and have non-empty interiors $(\mathcal{A})$, $(\mathcal{A}')$. The deviation function $\Lambda(\alpha)$ is analytic in the region $(\mathcal{A}')$, and for $\alpha \in (\mathcal{A}')$ the vector $\lambda(\alpha) = \Lambda'(\alpha)$ is the point where the supremum in the definition (1.2.1) is attained, i.e. $\Lambda(\alpha) = \langle\lambda(\alpha), \alpha\rangle - A\big(\lambda(\alpha)\big)$ for $\alpha \in (\mathcal{A}')$. The matrix of second derivatives $\Lambda''(\alpha)$ is positive definite in the region $(\mathcal{A}')$. Its inverse matrix coincides with the matrix $A''(\lambda(\alpha))$:

$$\big(\Lambda''(\alpha)\big)^{-1} := \sigma^2(\alpha) = A''\big(\lambda(\alpha)\big).$$

Moreover, the matrix $\sigma^2(\alpha)$ is the covariance matrix of the random vector $\xi^{(\lambda(\alpha))}$.

Denote

$$S_n^{(\lambda)} := \sum_{i=1}^n \xi_i^{(\lambda)},$$

where the $\xi_i^{(\lambda)}$ are independent copies of $\xi^{(\lambda)}$. The distribution $\mathbf{F}^{(\lambda)}$ of the random vector $\xi^{(\lambda)}$ is called the *Cramér transform of* $\mathbf{F}$ *with parameter* λ. The random vector $\xi^{(\lambda)}$ is also known as a Cramér transform, but of the initial random vector ξ.

The connection between the distributions of S_n and $S_n^{(\lambda)}$ is established in the following theorem.

Theorem 2.1.3. *For any measurable $B \subseteq \mathbb{R}^d$ we have*

$$\mathbf{P}\left(S_n \in x+B\right) = e^{-\langle \lambda, x\rangle + nA(\lambda)} \int_B e^{-\langle \lambda, z\rangle} \mathbf{P}\left(S_n^{(\lambda)} - x \in dz\right). \tag{2.1.10}$$

Assuming in (2.1.4) that $x = \alpha n$, $\lambda = \lambda(\alpha)$, we obtain the relation

$$\mathbf{P}\left(S_n \in x+B\right) = e^{-n\Lambda(\alpha)} \int_B e^{-\langle \lambda(\alpha), z\rangle} \mathbf{P}\left(S_n^{(\lambda(\alpha))} - \alpha n \in dz\right), \tag{2.1.11}$$

the reduction formula.

If $\alpha = x/n \in \mathcal{A}'$ then $\mathbf{E}\xi^{(\lambda(\alpha))} = A'\big(\lambda(\alpha)\big) \equiv \alpha$ (see (2.1.9)), $\mathbf{E}S_n^{(\lambda(\alpha))} - \alpha n = 0$ and, for $x \neq n\mathbf{E}\xi$, the reduction formula allows us to reduce a problem about the *probabilities of large deviations* for S_n to a problem about the distribution of $S_n^{(\lambda(\alpha))}$ in the zone of *normal deviations*.

The reduction formula will be also useful in 'boundary' cases, when $\alpha \in \partial\mathcal{A}'$.

The proof of Theorem 2.1.3 repeats the proof of Theorem 2.1.2 apart from obvious changes.

Note that some properties of the Cramér transform allow us to characterize the normal distribution: it turns out to be the unique distribution with covariance matrix invariant under the Cramér transform. In [57] the following result is proved (see Theorems 6.1 and 6.2 in [57]).

Theorem 2.1.4. *Suppose that Cramér's condition holds for the distribution* $\mathbf{F}$ *of a non-degenerate random vector $\xi \in \mathbb{R}^d$, i.e. the set $\mathcal{A}$ where the Laplace transform $\psi(\lambda)$ is finite has non-empty interior $(\mathcal{A})$. Then the following three statements are equivalent:*

(i) *the covariance matrix*

$$\mathbf{E}(\xi^{(\lambda)} - \mathbf{E}\xi^{(\lambda)})^T(\xi^{(\lambda)} - \mathbf{E}\xi^{(\lambda)})$$

does not depend on λ in a neighbourhood of some point $\lambda_0 \in (\mathcal{A})$;

(ii) *the distribution of the random vector $\xi^{(\lambda)} - \mathbf{E}\xi^{(\lambda)}$ does not depend on λ in a neighbourhood of some point $\lambda_0 \in (\mathcal{A})$;*

(iii) *the distribution* $\mathbf{F}$ *is normal.*

If one of the statements (i), (ii), (iii) is true then $\mathcal{A} = \mathbb{R}^d$ and, for all $\lambda \in \mathbb{R}^d$, the distribution of the random vector $\xi^{(\lambda)} - \mathbf{E}\xi^{(\lambda)}$ does not depend on λ (and is normal with zero mean).

2.2 Limit theorems for sums of random variables in the Cramér deviation zone. The asymptotic density

In this section we will use an approach the foundations of which were laid out by H. Cramér in the paper [80].

Denote for brevity

$$^{\alpha}\xi = \xi^{(\lambda(\alpha))}, \qquad {}^{\alpha}S_n = S_n^{(\lambda(\alpha))}.$$

We will use the reduction formula (2.1.5) for the probability of hitting a half-interval $x + B = \Delta[x) = [x, x + \Delta)$. On the right-hand side of this formula for $\alpha = x/n$, we have the probability $\mathbf{P}({}^{\alpha}S_n - \alpha n \in dz)$, where ${}^{\alpha}S_n = {}^{\alpha}\xi_1 + \cdots + {}^{\alpha}\xi_n$ and the distribution of the Cramér transform ${}^{\alpha}\xi = \xi^{(\lambda(\alpha))}$ does not depend on the parameter $\alpha = x/n \in (\alpha_-, \alpha_+)$. This means that in the case when $\alpha = x/n \neq$ const. as $n \to \infty$, the parameter α will depend on n and we will be dealing with a *scheme of series*. It means that we should use not Theorems 1.5.1, 1.5.2 but their modifications 1.5.3, 1.5.4, established for a scheme of series for $\lambda = \lambda(\alpha)$.

Under the assumption that Cramér's condition **[C]** holds ($\lambda_- < \lambda_+$), we will find the asymptotics of the probabilities $\mathbf{P}\big(S_n \in \Delta[x)\big)$ for normalised deviations $\alpha = x/n$ from the *Cramér* (or *regular*) *zone*, i.e. from the region $\alpha \in (\alpha_-, \alpha_+)$ where the deviation function $\Lambda(\alpha)$ is analytic. Note that for $\lambda_- < \lambda_+$ the interval (α_-, α_+) is not empty.

In some cases the length of the interval Δ will depend on n. In those cases, instead of Δ we will write Δ_n so that $\Delta_n[x) = [x, x + \Delta_n)$. It is also clear that, for $\alpha \in (\alpha_-, \alpha_+)$,

$$\mathbf{E}({}^{\alpha}\xi) = \alpha, \quad \mathbf{D}({}^{\alpha}\xi) =: \sigma^2(\alpha) = \frac{\psi''(\lambda(\alpha))}{\psi(\lambda(\alpha))} - \alpha^2 = A''\big(\lambda(\alpha)\big) \tag{2.2.1}$$

(see (2.1.2) and the definition of ${}^{\alpha}\xi$).

Theorem 2.2.1. *Let ξ be a non-lattice random variable. If $\Delta_n \to 0$ sufficiently slowly as $n \to \infty$, then*

$$\mathbf{P}\big(S_n \in \Delta_n[x)\big) = \frac{\Delta_n}{\sigma(\alpha)\sqrt{2\pi n}} e^{-n\Lambda(\alpha)}\big(1 + o(1)\big), \tag{2.2.2}$$

where $\alpha = x/n$, the remainder term $o(1)$ is uniform in $\alpha \in \mathcal{A}^* := [\alpha_*, \alpha^*]$ for any fixed $\alpha_* < \alpha^*$ from the interval (α_-, α_+).

Proof. The proof is based on Theorems 2.1.2 and 1.5.3. Since the conditions of Theorem 2.1.2 are satisfied, according to (2.1.5) we have

$$\mathbf{P}\big(S_n \in \Delta_n[x)\big) = e^{-n\Lambda(\alpha)} \int_0^{\Delta_n} e^{-\lambda(\alpha)z}\mathbf{P}({}^{\alpha}S_n - \alpha n \in dz).$$

Since $|\lambda(\alpha)| \leqslant \max\Big(|\lambda(\alpha_*)|, |\lambda(\alpha^*)|\Big) < \infty$, $\Delta_n \to 0$, then $e^{-\lambda(\alpha)z} \to 1$ uniformly in $z \in \Delta_n[0)$ and hence, as $n \to \infty$,

$$\mathbf{P}\big(S_n \in \Delta_n[x)\big) = e^{-n\Lambda(\alpha)}\mathbf{P}\big(S_n^{(\alpha)} - \alpha n \in \Delta_n[0)\big)\big(1 + o(1)\big) \tag{2.2.3}$$

uniformly in $\alpha \in [\alpha_*, \alpha^*]$.

Let us now show that Theorem 1.5.3 can be applied to the random variables ${}^{\alpha}\xi = \xi^{(\lambda(\alpha))}$. Since $A''(\lambda) > 0$ in (α_-, α_+), then, in view of (2.1.3), $\sigma(\alpha) = \sigma^{(\lambda(\alpha))}$ is uniformly separated from 0 and ∞ for $\alpha \in [\alpha_*, \alpha^*]$ (as also in all the subsequent theorems of this section). Therefore it remains to check the conditions (i), (ii) of Theorem 1.5.3 for $\lambda = \lambda(\alpha) \in [\lambda_*, \lambda^*]$, $\lambda_* = \lambda(\alpha_*) > \lambda_-$, $\lambda^* := \lambda(\alpha^*) < \lambda_+$ and $\varphi^{(\lambda)}(t) = \psi(\lambda + it)/\psi(\lambda)$ (see (2.1.2)). We have

$$\psi(\lambda + it) = \psi(\lambda) + it\psi'(\lambda) - \frac{t^2}{2}\psi''(\lambda) + o(t^2)$$

as $t \to 0$, where the remainder term will be uniform in λ if the function $\psi''(\lambda + iu)$ is equicontinuous in u with respect to λ. This continuity is easily proved in the same way as the uniform continuity of a characteristic function (see e.g. property 4 in § 7.1, [39]). This proves that condition (i) of Theorem 1.5.3 is satisfied for $a^{(\lambda)} = \psi'(\lambda)/\psi(\lambda)$, $a_2^{(\lambda)} = \psi''(\lambda)/\psi(\lambda)$.

Let us now check the condition (ii) of Theorem 1.5.3. Suppose the contrary: that there exists a sequence $\lambda_k \in \left[\lambda_*, \lambda^*\right]$ such that

$$q^{(\lambda_k)} := \sup_{\theta_1 \leqslant |t| \leqslant \theta_2} \frac{|\psi(\lambda_k + it)|}{\psi(\lambda_k)} \to 1$$

as $k \to \infty$. Since ψ is uniformly continuous, in the region under consideration one can find points $t_k \in [\theta_1, \theta_2]$ such that, as $k \to \infty$,

$$\frac{\psi(\lambda_k + it_k)}{\psi(\lambda_k)} \to 1.$$

Since the region $\lambda \in \left[\lambda_*, \lambda^*\right]$, $|t| \in [\theta_1, \theta_2]$, is compact, there exists a subsequence $(\lambda_{k'}, t_{k'}) \to (\lambda_0, t_0)$ as $k' \to \infty$. Using again the continuity of ψ we obtain

$$\frac{|\psi(\lambda_0 + it_0)|}{\psi(\lambda_0)} = 1, \tag{2.2.4}$$

which contradicts the assumption that $\xi^{(\lambda_0)}$ is non-lattice. The property (ii) is proved.

Thus, we can apply Theorem 1.5.3 to the probability on the right-hand side of (2.2.3). Since $\mathbf{E}\,({}^{\alpha}\xi) = \alpha$ and $\mathbf{E}\,({}^{\alpha}\xi)^2 = \psi''(\lambda(\alpha))/\psi(\lambda(\alpha))$, this yields

$$\begin{aligned} \mathbf{P}\big(S_n \in \Delta_n[x)\big) &= e^{-n\Lambda(\alpha)}\left(\frac{\Delta_n}{\sigma(\alpha)\sqrt{n}}\,\phi(0) + o\left(\frac{1}{\sqrt{n}}\right)\right) \\ &= \frac{\Delta_n}{\sigma(\alpha)\sqrt{2\pi n}}\,e^{-n\Lambda(\alpha)}\big(1 + o(1)\big) \end{aligned} \tag{2.2.5}$$

uniformly in $\alpha \in [\alpha_*, \alpha^*]$ (or in $x \in [\alpha_* n, \alpha^* n]$), where the values

$$\sigma^2(\alpha) = \mathbf{E}\,({}^{\alpha}\xi - \alpha)^2 = \frac{\psi''(\lambda(\alpha))}{\psi(\lambda(\alpha))} - \alpha^2$$

are uniformly bounded away from 0 and ∞. The theorem is proved. □

Remark 2.2.2. It is not difficult to see that in the case $\psi(\lambda_-) < \infty$, $\psi''(\lambda_-) < \infty$ ($\psi(\lambda_+) < \infty$, $\psi''(\lambda_+) < \infty$) the region of uniformity $\alpha \in [\alpha_*, \alpha^*]$ can be enlarged to $\alpha \in [\alpha_-, \alpha^*]$ ($\alpha \in [\alpha_*, \alpha_+]$). Here, by $\psi''(\lambda_\pm)$ we denote the left and right derivatives respectively. In part this will also follow from the results of section 2.4, where we obtain integro-local theorems on the boundary of the Cramér zone.

From Theorem 2.2.1 we can now obtain integro-local theorems for fixed or growing Δ and also integral theorems.

If $\alpha \in [\alpha_*, \alpha^*]$ and $|\alpha - \mathbf{E}\xi| = O(1/\sqrt{n})$ (this is possible when $\mathbf{E}\xi \in [\alpha_*, \alpha^*]$), one is dealing with the normal deviation region, when the Stone–Shepp theorem can be applied. Since this case has been well studied, we will not consider it here but will assume that

$$|\alpha - \mathbf{E}\xi| > \frac{N}{\sqrt{n}},$$

where $N = N(n) \to \infty$ as $n \to \infty$. If we want to obtain integro-local theorems in a unified form for fixed and growing Δ, then in the case $\alpha - \mathbf{E}\xi > N/\sqrt{n}$ it is natural to consider the probabilities of the events $\{S_n \in \Delta[x)\}$, and in the case $\alpha - \mathbf{E}\xi < -N/\sqrt{n}$ the probabilities of events $\{S_n \in (x - \Delta, x]\}$. Since these two possibilities are symmetric, we need only consider the former. Thus, we will assume that

$$\alpha \in [\alpha_*, \alpha^*], \quad \alpha > \mathbf{E}\xi + \frac{N}{\sqrt{n}}, \quad N = N(n) \to \infty \quad \text{as} \quad n \to \infty. \tag{2.2.6}$$

Theorem 2.2.3. *Let ξ be non-lattice and let (2.2.6) hold. Then, for any $\Delta \geqslant \Delta_0 > 0$ and $n \to \infty$, we have*

$$\mathbf{P}\big(S_n \in \Delta[x)\big) = \frac{e^{-n\Lambda(\alpha)}}{\sigma(\alpha)\lambda(\alpha)\sqrt{2\pi n}} (1 - e^{-\lambda(\alpha)\Delta})\big(1 + o(1)\big), \tag{2.2.7}$$

where the remainder $o(1)$ is uniform in

$$\alpha = \frac{x}{n} \in \mathcal{A}^+ := \left[\max\left(\alpha_*, \mathbf{E}\xi + \frac{N(n)}{\sqrt{n}}\right), \alpha^*\right]$$

and in $\Delta \geqslant \Delta_0$, for any fixed $\alpha_* < \alpha^*$ from (α_-, α_+).

In particular, for $\Delta = \Delta(n) > \sqrt{n}$,

$$\mathbf{P}\big(S_n \in \Delta[x)\big) = \frac{e^{-n\Lambda(\alpha)}}{\sigma(\alpha)\lambda(\alpha)\sqrt{2\pi n}} \big(1 + o(1)\big). \tag{2.2.8}$$

Clearly, the probabilities $\mathbf{P}\big(S_n \in \Delta[x - \Delta)\big)$ will have the same form for

$$\alpha = \frac{x}{n} \in \mathcal{A}^- := \left[\alpha_*, \min\left(\alpha^*, \mathbf{E}\xi - \frac{N(n)}{\sqrt{n}}\right)\right].$$

In particular, the probability $\mathbf{P}(S_n < x)$ for $\alpha \in \mathcal{A}^-$ will be determined by the right-hand side of (2.2.8).

Proof of Theorem 2.2.3. Suppose first that $\Delta = o(n)$ as $n \to \infty$. Divide the half-interval $\Delta[x)$ into half-intervals $\Delta_n[x + k\Delta_n)$, $k = 0, \ldots, \Delta/\Delta_n - 1$, where

$\Delta_n \to 0$, assuming for simplicity that $R = \Delta/\Delta_n$ is integer. Due to Theorem 2.1.2, as $\Delta_n \to 0$ we have

$$\begin{aligned}\mathbf{P}(S_n \in \Delta_n[x + k\Delta_n)) &= \mathbf{P}(S_n \in [x, x + (k+1)\Delta_n)) - \mathbf{P}(S_n \in [x, x + k\Delta_n)) \\ &= e^{-n\Lambda(\alpha)} \int_{k\Delta_n}^{(k+1)\Delta_n} e^{-\lambda(\alpha)z} \mathbf{P}({}^{\alpha}S_n - \alpha n \in dz) \\ &= e^{-n\Lambda(\alpha) - \lambda(\alpha)k\Delta_n} \mathbf{P}\Big({}^{\alpha}S_n - \alpha n \in \Delta_n\big[k\Delta_n\big)\Big)\big(1 + o(1)\big)\end{aligned} \tag{2.2.9}$$

uniformly in $\alpha \in \mathcal{A}^+$. Here, similarly to (2.2.5) and according to Theorem 1.5.3,

$$\mathbf{P}\Big({}^{\alpha}S_n - \alpha n \in \Delta_n\big[k\Delta_n\big)\Big) = \frac{\Delta_n}{\sigma(\alpha)\sqrt{n}} \phi\left(\frac{k\Delta_n}{\sigma(\alpha)\sqrt{n}}\right) + o\left(\frac{1}{\sqrt{n}}\right) \tag{2.2.10}$$

uniformly in k and α. Since

$$\mathbf{P}(S_n \in \Delta[x)) = \sum_{k=0}^{R-1} \mathbf{P}(S_n \in \Delta_n[x + k\Delta)),$$

inserting into the right-hand side of this equality the values (2.2.9), (2.2.10), we obtain

$$\begin{aligned}\mathbf{P}(S_n \in \Delta[x)) &= \frac{e^{-n\Lambda(\alpha)}}{\sigma(\alpha)\sqrt{n}} \sum_{k=0}^{R-1} \Delta_n e^{-\lambda(\alpha)k\Delta_n} \left(\phi\Big(\frac{k\Delta_n}{\sigma(\alpha)\sqrt{n}}\Big) + o(1)\right) \\ &= \frac{e^{-n\Lambda(\alpha)}}{\sigma(\alpha)\sqrt{n}} \int_0^{\Delta - \Delta_n} e^{-\lambda(\alpha)z} \left(\phi\Big(\frac{z}{\sigma(\alpha)\sqrt{n}}\Big) + o(1)\right) dz.\end{aligned} \tag{2.2.11}$$

Setting $\lambda(\alpha)z = u$, the right-hand side can be written in the form

$$\frac{e^{-n\lambda(\alpha)}}{\sigma(\alpha)\lambda(\alpha)\sqrt{n}} \int_0^{(\Delta - \Delta_n)\lambda(\alpha)} e^{-u} \left(\phi\Big(\frac{u}{\sigma(\alpha)\lambda(\alpha)\sqrt{n}}\Big) + o(1)\right) du, \tag{2.2.12}$$

where the remainder $o(1)$ is uniform in $\alpha \in \mathcal{A}^+$, $\Delta \geqslant \Delta_0$, $\Delta = o(n)$ and the u from the region of integration. Since for small $\alpha - \mathbf{E}\xi$ we have $\lambda(\alpha) \sim (\alpha - \mathbf{E}\xi)/\sigma^2$ (see (1.1.7) and (1.1.13)), then for $\alpha - \mathbf{E}\xi \geqslant N(n)/\sqrt{n}$ we have

$$\lambda(\alpha) > \frac{N(n)}{\sigma^2\sqrt{n}} \big(1 + o(1)\big),$$

$$\sigma(\alpha)\lambda(\alpha)\sqrt{n} > \frac{\sigma(\alpha)N(n)}{\sigma^2} \to \infty.$$

Hence, for any fixed u,

$$\phi\left(\frac{u}{\sigma(\alpha)\lambda(\alpha)\sqrt{n}}\right) \to \phi(0) = \frac{1}{\sqrt{2\pi}}.$$

Moreover, $\phi(v) \leqslant 1/\sqrt{2\pi}$ for all v. Therefore, by (2.2.11) and (2.2.12),

$$\mathbf{P}\big(S_n \in \Delta[x)\big) = \frac{e^{-n\Lambda(\alpha)}}{\sigma(\alpha)\lambda(\alpha)\sqrt{2\pi n}} \int_0^{\lambda(\alpha)\Delta} e^{-u}du\big(1+o(1)\big)$$
$$= \frac{e^{-n\Lambda(\alpha)}}{\sigma(\alpha)\lambda(\alpha)\sqrt{2\pi n}}\big(1-e^{-\lambda(\alpha)\Delta}\big)\big(1+o(1)\big)$$

uniformly in $\alpha \in \mathcal{A}^+$ and $\Delta \geqslant \Delta_0$, $\Delta = o(n)$, which implies (2.2.7).

If $\Delta > \sqrt{n}$ then $\lambda(\alpha)\Delta \geqslant N(n)/\sigma^2 \to \infty$ and the relation (2.2.8) follows immediately from (2.2.7).

If the condition $\Delta = o(n)$ is not satisfied, we use the relation

$$\mathbf{P}\left(S_n \in \Delta[x)\right) = \mathbf{P}\left(S_n \in [x, x+n^{2/3})\right) + \mathbf{P}\left(S_n \in [x+n^{2/3}, x+\Delta)\right).$$

The first term on the right-hand side is equal to the right-hand side of (2.2.8); the second term can be bounded as follows:

$$\mathbf{P}\left(S_n \geqslant x+n^{2/3}\right) \leqslant e^{-n\Lambda(\alpha+n^{-1/3})} \leqslant e^{-n\Lambda(\alpha)-\lambda(\alpha)n^{2/3}}.$$

Since $\lambda(\alpha)n^{2/3} \geqslant N(n)n^{1/6}/\sigma^2$, the second term is negligible compared with the right-hand side of (2.2.8). The theorem is proved. □

Here, Remark 2.2.2 about the extension of the region of uniformity in α, if $\psi''(\alpha_-) < \infty$, is also valid (in the definition of the region $\mathcal{A}^+$ the value α_* can be replaced with α_-).

Suppose, for the sake of definiteness, that $\lambda_+ > 0$, $\psi''(0) = \mathbf{E}\xi^2 < \infty$ and also $\mathbf{E}|\xi|^k < \infty$ for some integer $k \geqslant 2$. Then $\mathbf{E}\xi$ exists and, without loss of generality, one can assume that $\mathbf{E}\xi = 0$ and, hence, $\alpha_- \leqslant 0$, $\alpha_+ > 0$. Then the following statement about moderately large deviations is true; it does not exclude the equality $\alpha_- = 0$.

Corollary 2.2.4. *Under the above assumptions for $x \geqslant N(n)\sqrt{n}$, $N(n) \to \infty$, $n\alpha^k = x^k/n^{k-1} \leqslant c =$ const., $\Delta > \sqrt{n}$, $n \to \infty$, the representation (2.2.8) is true if $\sigma(\alpha)$ is replaced with σ and the exponent $n\Lambda(\alpha)$ with*

$$n\Lambda(\alpha) = n\sum_{j=2}^{k} \frac{\Lambda^{(j)}(0)}{j!}\alpha^j + o(n\alpha^k), \qquad (2.2.13)$$

where the $\Lambda^{(j)}(0)$ were given in (1.1.13).

In particular, if $\mathbf{E}|\xi|^3 < \infty$ $(k=3)$ and $x = o(n^{2/3})$ then

$$\mathbf{P}(S_n \geqslant x) \sim \frac{\sigma\sqrt{n}}{x\sqrt{2\pi}}e^{-x^2/(2n\sigma^2)} \sim \Phi\left(-\frac{x}{\sigma\sqrt{n}}\right). \qquad (2.2.14)$$

Here, in the last relation we have used the symmetry of the normal law, i.e. $1-\Phi(t) = \Phi(-t)$. Formula (2.2.14) shows that in the case $\lambda_+ > 0$, $\mathbf{E}\,|\xi|^3 < \infty$, the asymptotic equivalence

$$\mathbf{P}(S_n \geqslant x) \sim \Phi\left(-\frac{x}{\sigma\sqrt{n}}\right)$$

is also preserved outside the zone of normal deviations up to values $x = o(n^{2/3})$. If $\mathbf{E}\,\xi^3 = 0$ and $\mathbf{E}\,\xi^4 < \infty$, this equivalence is preserved up to values $x = o(x^{3/4})$. For larger x the equivalence in general does not hold.

Proof of Corollary 2.2.4. The first relation in (2.2.14) follows from Theorem 2.2.3, (2.2.13) and Remark 2.2.2. The second follows from the asymptotic equivalence

$$\int_x^\infty e^{-u^2/2}du \sim \frac{e^{-x^2/2}}{x},$$

which can be easily established by using, for example, the l'Hôpital rule. □

Limit theorems about the distribution of S_n for arithmetic ξ are analogous to Theorems 2.2.1 and 2.2.3.

Theorem 2.2.5. *Suppose that the distribution of ξ is arithmetic. Then, for integer x,*

$$\mathbf{P}(S_n = x) = \frac{e^{-n\Lambda(\alpha)}}{\sigma(\alpha)\sqrt{2\pi n}}\big(1 + o(1)\big),$$

where the remainder $o(1)$ is uniform in $\alpha = x/n \in [\alpha_*, \alpha^*]$ for any fixed $\alpha_* < \alpha^*$ from (α_-, α_+).

Proof. The proof of the theorem does not much differ from the proof of Theorem 2.2.1. According to (2.1.8),

$$\mathbf{P}(S_n = x) = e^{-\lambda(\alpha)x}\psi^{-n}\big(\lambda(\alpha)\big)\mathbf{P}({}^{\alpha}S_n = x) = e^{-n\Lambda(\alpha)}\mathbf{P}({}^{\alpha}S_n = x),$$

where $\mathbf{E}\,{}^{\alpha}\xi = \alpha$ for $\alpha \in (\alpha_-, \alpha_+)$. To compute $\mathbf{P}({}^{\alpha}S_n = x)$, one should use Theorem 1.5.4. Verification of the conditions (i), (ii) of Theorem 1.5.3, which are assumed in Theorem 1.5.4, can be achieved in the same way as in the proof of Theorem 2.2.1, the only difference being that the relation (2.2.4) for $t^* \in [\theta_1, \pi]$ will contradict the assumption that the distribution of ξ is arithmetic. Since $\mathbf{E}\,{}^{\alpha}\xi = \alpha$, by Theorem 1.5.4 we have

$$\mathbf{P}({}^{\alpha}S_n = x) = \frac{1}{\sigma(\alpha)\sqrt{2\pi n}}\big(1 + o(1)\big)$$

uniformly in $\alpha = x/n \in [\alpha_*, \alpha^*]$. The theorem is proved. □

From Theorem 2.2.5, it is not difficult to obtain counterparts of Theorem 2.2.3 and Corollary 2.2.4.

The results of this section can be also found in [39], [56] and [57].

Denote

$$f_{(n)}(x) = \frac{e^{-n\Lambda(\alpha)}}{\sigma(\alpha)\sqrt{2\pi n}}. \tag{2.2.15}$$

The block of factors on the right-hand side of (2.2.15) appears in all the statements of this section. In Theorem 2.2.1 we have, for $\alpha \in [\alpha_*, \alpha^*] \subset (\alpha_-, \alpha_+)$ $\Delta_n \to 0$ sufficiently slowly as $n \to \infty$,

$$\frac{\mathbf{P}\left(S_n \in \Delta_n[x)\right)}{\Delta_n} \sim f_{(n)}(x). \tag{2.2.16}$$

In Theorem 2.2.5, for arithmetic ξ we have

$$\mathbf{P}(S_n = x) \sim f_{(n)}(x). \tag{2.2.17}$$

We will also see in Theorem 2.3.1 (see section 2.3) that in the case when the density of the distribution of ξ exists, the density $f_n(x)$ of the distribution of the sum S_n has the property

$$f_n(x) \sim f_{(n)}(x) \tag{2.2.18}$$

for $\alpha \in [\alpha_*, \alpha^*]$, $n \to \infty$.

Thus, the block of factors $f_{(n)}(x)$ defined in (2.2.15) is the main part of the asymptotics of the distribution density of the sum S_n (with respect to the counting measure in (2.2.17) and the Lebesgue measure in (2.2.18)) as $n \to \infty$. In the general non-lattice case it determines the asymptotics

$$\frac{\mathbf{P}(S_n \in \Delta_n[x))}{\Delta_n}$$

as $\Delta_n \to 0$ sufficiently slowly, i.e. it is 'almost' the density of the distribution of the sum S_n. Therefore, the function $f_{(n)}(x)$ can be naturally called *the asymptotic density*. Since ξ is either non-lattice or arithmetic, these form the exhaustive alternatives in the study of the distribution of the sums S_n, we have quite a general fact: *for* $\alpha \in \mathcal{A}^* = [\alpha_*, \alpha^*]$, $n \to \infty$, *there always exists an asymptotic density* $f_{(n)}(x)$, i.e. a function which satisfies the properties (2.2.16)–(2.2.18).

Integral theorems about the asymptotics of $\mathbf{P}(S_n \geqslant x)$ in the Cramér deviation zone can be also found in [102], [151] and [165] (in a somewhat different form).

In the next section we provide without proof a number of statements (accompanied by references), which supplement the main results provided above.

2.3 Supplement to section 2.2

2.3.1 Local theorems in the one-dimensional case

To simplify the presentation, we will formulate the theorem for densities assuming that the following condition is satisfied.

[D] *The condition* **[C$_+$]** *is satisfied and the distribution* **F** *has a bounded density* $f(x)$ *which has the following properties:*

$$f(x) = e^{-\lambda_+ x + o(x)} \quad \textit{as } x \to \infty \text{ if } \lambda_+ < \infty; \tag{2.3.1}$$

$$f(x) \leqslant c e^{-\lambda x} \quad \textit{for any fixed } \lambda > 0,\ c = c(\lambda), \text{ if } \lambda_+ = \infty. \tag{2.3.2}$$

Since for $\lambda_+ > 0$ inequalities of the form (2.3.1), (2.3.2) are always satisfied, which follows from the exponential Chebyshev inequality for the right tails $F_+(x) = \int_x^\infty f(u)du$, the condition **[D]** is not too restrictive compared with the condition $\lambda_+ > 0$. It just eliminates sharp 'peaks' of $f(x)$ as $x \to \infty$.

Denote by $f_n(x)$ the density of the sum S_n.

If $\mathbf{E}\xi \in (\alpha_-, \alpha_+)$ or $\mathbf{E}\xi = \alpha_-$, $\psi''(\alpha_-) < \infty$, we put $\alpha_* = \mathbf{E}\xi$. In other cases, we will assume, as before, that α_* is a number from (α_-, α_+).

Theorem 2.3.1. *Suppose that $\lambda_+ > 0$ and the condition* **[D]** *holds. Then*

$$f_n(x) = \frac{e^{-n\Lambda(\alpha)}}{\sigma(\alpha)\sqrt{2\pi n}}\big(1 + o(1)\big),$$

where $o(1)$ is uniform in $\alpha \in [\alpha_*, \alpha^*]$ for any fixed $\alpha^* \in (\alpha_*, \alpha_+)$.

The proof of the theorem can be found in [39] or [56]. The theorem confirms the asymptotic equivalence relation (2.2.18). The case when the condition **[C₋]** holds can be considered in a similar way.

2.3.2 Integro-local theorems in the multidimensional case. The Cramér deviation zone

The characteristic function of a random vector ξ will be denoted by

$$\varphi(t) = \psi(it) = \mathbf{E}e^{i\langle t,\xi\rangle}, \qquad t \in \mathbb{R}^d.$$

We will use the following conditions on the structure of the distribution $\mathbf{F}$ of a non-degenerate random vector ξ. Such a random vector ξ is called *arithmetic* (it has an arithmetic distribution) if $\mathbf{P}\{\xi \in \mathbb{Z}^d\} = 1$ and, for some $y_0 \in \mathbb{Z}^d$ such that $\mathbf{P}\{\xi = y_0\} > 0$, the additive group generated by the set $\{y : \mathbf{P}\{\xi = y_0 + y\} > 0\}$ coincides with $\mathbb{Z}^d$. By Lemma 21.6 in [9], p. 235, a non-degenerate vector ξ is arithmetic if and only if the following condition is satisfied.

[Z] (*Condition of arithmeticity*) *For any $t \in \mathbb{Z}^d$,*

$$\varphi(2\pi t) = 1,$$

and, for any $t \notin \mathbb{Z}^d$,

$$\big|\varphi(2\pi t)\big| < 1.$$

As has been already noted, a non-degenerate *scalar* random variable ξ is said to be *lattice with lattice step h* (having a lattice distribution), if for some $c \in \mathbb{R}^1$ and $h > 0$ the random variable $\xi' = c + h^{-1}\xi$ is arithmetic. If this property is not satisfied then the random variable ξ is called *non-lattice* (it has a non-lattice distribution). According to what was said above, *a scalar random variable ξ is non-lattice if and only if for all $t \neq 0$ we have $\big|\varphi(t)\big| < 1$.*

We say that a non-degenerate *random vector* ξ is *non-lattice* (has a non-lattice distribution) if, for any unit vector $\mathrm{e} \in \mathbb{R}^d$, the random variable $\langle \mathrm{e}, \xi \rangle$ is non-lattice, i.e. the following condition is satisfied.

[R] (*Condition of non-latticeness*) *For any* $\mathrm{e} \in \mathbb{R}^d$, $|\mathrm{e}| = 1$, $t \neq 0$,

$$\big|\varphi(\mathrm{e}t)\big| < 1.$$

In the known sense, the conditions **[R]** and **[Z]** exhaust all the possibilities (see [175]), since using a non-degenerate linear transformation one can always transform a vector ξ into a random vector $\zeta = (\zeta_{(1)}, \ldots, \zeta_{(d)})$ such that this vector ζ is arithmetic or non-lattice or, for some $m \in \{1, \ldots, d-1\}$, its first m coordinates $(\zeta_{(1)}, \ldots, \zeta_{(m)})$ form an arithmetic subvector and the other $d-m$ coordinates $(\zeta_{(m+1)}, \ldots, \zeta_{(d)})$ form a non-lattice vector. In what follows, in order to simplify the presentation, we will consider the non-lattice and arithmetic cases separately. The intermediate case $0 < m < d$ can be considered in the same way (see e.g. [175]).

Along with **[R]**, we will also need a stronger condition, which depends on a parameter $\kappa \in (0, \infty]$ (Cramér's condition on a characteristic function):

[R$^\kappa$]

$$\limsup_{|t| \to \infty} \big|\varphi(t)\big| \leqslant e^{-d\kappa}.$$

Note that the constant $\kappa > 0$, which appears in the condition **[R$^\kappa$]** and characterizes the 'strength' of this condition, plays an important role in integro-local theorems (see Theorem 2.3.2 below). If a distribution **F** is absolutely continuous then it satisfies the condition **[R$^\infty$]**.

Henceforth we will assume that *Cramér's moment condition* **[C]** is also satisfied, unless otherwise stated.

Denote by $\Delta[x) = \{y = (y_{(1)}, \ldots, y_{(d)}) : x_{(i)} \leqslant y_{(i)} < y_{(i)} + \Delta,\ 1 \leqslant i \leqslant d\}$ a half-open cube in the space $\mathbb{R}^d$ with vertex at a point $x = (x_{(1)}, \ldots, x_{(d)})$ and with sides of length $\Delta > 0$. The term *integro-local theorems* for the sums S_n (see also [51], [52], [56], [57]) refers to statements about the asymptotic behaviour as $n \to \infty$ of the probability

$$\mathbf{P}\big(S_n \in \Delta[x)\big), \tag{2.3.3}$$

where the vector $x = x(n)$ and the positive number $\Delta = \Delta(n)$ generally depend on n. Let $a = \mathbf{E}\xi$. Since $\mathbf{E}S_n = an$, the quantity $s = |x - an|$ in (2.3.3) characterises the deviation of the sum S_n from the 'most probable' value an. If $|x - an| = O(\sqrt{n})$ then, in the case when the second moment $\mathbf{E}|\xi|^2$ is finite, such deviations are called *normal deviations*; if $s = |x - an|'' \sqrt{n}$, $s = o(n)$, they are often called *moderately large deviations* of the sums S_n. If $s = |x - an|$ is of order cn, i.e. $\liminf_{n \to \infty} |x - an|/n > 0$, $\limsup_{n \to \infty} |x - an|/n < \infty$, then such deviations form the 'usual' large deviations (or simply *large deviations*). If in addition the point

$\alpha = x/n$ belongs to the region of analyticity of the function $\Lambda(\alpha)$, then we will speak about *regular* large deviations, or about deviations in the Cramér zone.

Deviations $s = |x - an|$ for which $s/n \to \infty$ will be called *'superlarge deviations'* of sums of random vectors.

If the asymptotics (2.3.3) in an integro-local theorem is known then it is possible to estimate the probabilities in the corresponding integral theorem; for example, we can estimate the probabilities $\mathbf{P}(S_n \in rB)$, where the real number $r = r(n)$ depends on n, and the measurable set $B \subseteq \mathbb{R}^d$ is fixed. In order to do this, one has to approximate the set rB 'from above' and 'from below' (if it has a sufficiently smooth boundary) by a union of non-intersecting cubes $\Delta(x)$ and use the asymptotics of the probabilities (2.3.3). Obviously, the statement of an integro-local theorem will become stronger when the parameter Δ tends to 0 faster. The smaller is Δ, the more exact is the approximation of the set rB by a union of cubes. Suppose that only the condition **[R]** is satisfied; then an integro-local theorem in the zone of large deviations will be valid for any fixed parameter (Delta) (and, hence, for any parameter tending to zero sufficiently slowly). If the condition $[\mathbf{R}^\kappa]$ holds then the parameter Δ can decrease exponentially fast, i.e. as the function $e^{-n\kappa_1}$ for any $\kappa_1 < \kappa$ (see Theorem 2.3.2 below).

It turns out that if the condition **[C]** is satisfied then integro-local theorems for sums of random vectors in the zone of *normal, moderately large* and *regularly large deviations* can be formulated in a unified form.

Recall that the matrix of second derivatives $\Lambda''(\alpha)$ is positive definite in the region $(\mathcal{A}')$. Its inverse coincides with the matrix $A''(\lambda(\alpha))$:

$$(\Lambda''(\alpha))^{-1} =: \sigma^2(\alpha) = A''(\lambda(\alpha)).$$

The matrix $\sigma^2(\alpha)$ is the covariance matrix of the random vector ${}^{\alpha}\xi$ with distribution

$${}^{\alpha}\mathbf{F}(B) := \mathbf{P}({}^{\alpha}\xi \in B) = \frac{\mathbf{E}(e^{\langle\lambda(\alpha),\xi\rangle};\ \xi \in B)}{\psi(\lambda(\alpha))}, \quad \alpha \in (\mathcal{A}'), \tag{2.3.4}$$

so the Laplace transform of ${}^{\alpha}\mathbf{F}$ has the form

$$\mathbf{E}e^{\langle\lambda,{}^{\alpha}\xi\rangle} = \frac{\psi(\lambda(\alpha)+\lambda)}{\psi(\lambda(\alpha))}, \quad \alpha \in (\mathcal{A}').$$

Here, the distribution ${}^{\alpha}\mathbf{F}$ is the so-called *Cramér transform* at a point $\lambda(\alpha)$ of the distribution $\mathbf{F}$ of the vector ξ. Obviously, the distributions $\mathbf{F}$ and ${}^{\alpha}\mathbf{F}$ are mutually absolutely continuous. In particular, if the distribution $\mathbf{F}$ of the random vector ξ has a density $f(x)$ then the distribution ${}^{\alpha}\mathbf{F}$ has a density which is given by

$${}^{\alpha}f(x) = \frac{e^{\langle\lambda(\alpha),x\rangle}}{\psi(\lambda(\alpha))} f(x), \quad \alpha \in (\mathcal{A}'), \quad x \in \mathbb{R}^d.$$

Along with the structural conditions **[Z]**, **[R]**, $[\mathbf{R}^\kappa]$ already introduced, we will need the condition

[$\mathbf{R}_{\text{den}}$] *The distribution* $\mathbf{F}$ *is absolutely continuous and, for any compact set* $\mathcal{A}^* \subseteq (\mathcal{A}')$, *the density* ${}^{\alpha}f(x)$ *of the distribution* ${}^{\alpha}\mathbf{F}$ *for* $\alpha \in \mathcal{A}^*$ *is uniformly bounded:*

$$\sup_{\substack{\alpha\in\mathcal{A}^*\\ x\in\mathbb{R}^d}} {}^{\alpha}f(x) < \infty.$$

We will now consider the properties of the Cramér transform ${}^{\alpha}\mathbf{F}$. It is important to note that at any point α from the region $(\mathcal{A}')$ the equality $\mathbf{E}{}^{\alpha}\xi = \alpha$ holds. Moreover, for all $\alpha \in (\mathcal{A}')$ the matrix $\sigma^2(\alpha)$ is positive definite. Therefore, a unique *positive definite* matrix $\sigma(\alpha)$ is defined (the positive definite root of $\sigma^2(\alpha)$, the square of which is equal to $\sigma^2(\alpha)$). Denote the determinant of the matrix $\sigma(\alpha)$ by $|\sigma(\alpha)| = \sqrt{|\sigma^2(\alpha)|}$, where $\alpha \in (\mathcal{A}')$. The value $|\sigma(\alpha)|$ is equal to the volume of the transformed cube

$$\Delta[0)\sigma(\alpha) := \{x = y\sigma(\alpha) :\ y \in \Delta[0)\},$$

for $\Delta = 1$. Define the function

$$c(\alpha, \Delta) = \Delta^{-d} \int_{\Delta[0)} e^{-\langle\lambda(\alpha),u\rangle} du = \prod_{i=1}^{d} \frac{1 - e^{-\Delta\lambda_i(\alpha)}}{\Delta\lambda_i(\alpha)}.$$

It is easy to see that $c(\alpha, \Delta) \to 1$ for $\Delta \to 0$ uniformly in $\alpha \in \mathcal{A}^*$, where $\mathcal{A}^*$ is an arbitrary compact set which lies in the region $(\mathcal{A}')$ ($\mathcal{A}^* = [\alpha_*, \alpha^*]$ in the one-dimensional case).

Theorem 2.3.2. *Suppose that Cramér's condition* **[C]** *holds.*

(i) *If the condition* **[R]** *is satisfied then*

$$\mathbf{P}(S_n \in \Delta[x)) = \frac{\Delta^d}{(2\pi n)^{d/2}|\sigma(\alpha)|} e^{-n\Lambda(\alpha)}(c(\alpha, \Delta) + \varepsilon_n), \tag{2.3.5}$$

where $\alpha = x/n$, *the remainder term* $\varepsilon_n = \varepsilon_n(x, \Delta)$ *tends to* 0 *uniformly in* $\Delta \in [\Delta_1, \Delta_2]$ *and* $\alpha = x/n \in \mathcal{A}^*$ *for any fixed* $0 < \Delta_1 \leqslant \Delta_2 < \infty$ *and any fixed compact set* $\mathcal{A}^*$ *lying in the region* $(\mathcal{A}')$*:*

$$\lim_{n\to\infty} \sup_{\substack{\alpha\in\mathcal{A}^*\\ \Delta_1\leqslant\Delta\leqslant\Delta_2}} |\varepsilon_n(x, \Delta)| = 0. \tag{2.3.6}$$

(ii) *If instead of the condition* **[R]** *the stronger condition* **[R$^{\kappa}$]** *holds then in* (2.3.5) *the remainder term* $\varepsilon_n = \varepsilon_n(x, \Delta)$ *satisfies* (2.3.6), *where* Δ_1 *is replaced by* $e^{-n\kappa_1}$ *for any fixed* $\kappa_1 < \kappa$.

(iii) *If instead of the condition* **[R]** *the condition* **[$\mathbf{R}_{\text{den}}$]** *holds then, for the density* $f_n(x)$ *of the distribution of the sum* S_n *and for any compact set* $\mathcal{A}^*$ *lying in the region* $(\mathcal{A}')$, *we have the equality*

$$f_n(x) = \frac{1}{(2\pi n)^{d/2}|\sigma(\alpha)|} e^{-n\Lambda(\alpha)}(1 + \varepsilon_n(x)), \tag{2.3.7}$$

where the remainder term $\varepsilon_n(x)$ *satisfies the relation*

$$\lim_{n\to\infty} \sup_{\substack{\alpha\in\mathcal{A}^* \\ x\in\mathbb{R}^d}} |\varepsilon_n(x)| = 0.$$

(iv) *If instead of the condition* **[R]** *the condition* **[Z]** *holds, then for* $x \in \mathbb{Z}^d$,

$$\mathbf{P}(S_n = x) = \frac{1}{(2\pi n)^{d/2}|\sigma(\alpha)|} e^{-n\Lambda(\alpha)}(1 + \varepsilon_n(x)), \tag{2.3.8}$$

where for any compact set $\mathcal{A}^*$ *lying in the region* $(\mathcal{A}')$

$$\lim_{n\to\infty} \sup_{\substack{\alpha\in\mathcal{A}^* \\ x\in\mathbb{Z}^d}} |\varepsilon_n(x)| = 0.$$

The first statement of Theorem 2.3.2 is true for any fixed Δ_1, Δ_2 and, hence, it is also true for $\Delta = \Delta_n$, which tends to zero as $n \to \infty$ sufficiently slowly. Since the function $c(\alpha, \Delta)$ in (2.3.5) converges to 1 as $\Delta \to 0$ uniformly in $\alpha \in \mathcal{A}^*$, the claim (2.3.5) can be simplified and expressed in a form close to (2.3.7) and (2.3.8).

Corollary 2.3.3. *Suppose that the conditions* **[R]**, **[C]** *are satisfied, that* $\alpha = x/n \in \mathcal{A}^*$ *where* $\mathcal{A}^*$ *is a fixed compact lying in* $(\mathcal{A}')$ *and that* $\Delta = \Delta_n$ *tends to* 0 *as* $n \to \infty$ *sufficiently slowly. Then for* $n \to \infty$ *we have the uniform (in* $\alpha \in \mathcal{A}^*$*) relation*

$$\mathbf{P}\big(S_n \in \Delta[x)\big) = \frac{\Delta^d}{(2\pi n)^{d/2}|\sigma(\alpha)|} e^{-n\Lambda(\alpha)}(1 + o(1)). \tag{2.3.9}$$

The statements of parts (i), (ii), (iv) of Theorem 2.3.2 were essentially obtained in the work of Stone [175] (Theorem 3). However, in that theorem, instead of the condition **[C]** Stone assumed the following somewhat stronger condition: $\psi(\lambda) < \infty$ *in some neighbourhood of the origin* (i.e. the condition **[C$_0$]**). Moreover, for $\mathcal{A}^*$ Stone considered only sets $\{\alpha : |\alpha| \leqslant c\}$ for sufficiently small $c > 0$. It is not difficult to see that the proof given in [175] is preserved without essential changes under the condition **[C]**. Note also that in Theorem 3 from [175], Stone considers a more general situation than in Theorem 2.3.2, in which some coordinates of the random vector $\xi = (\xi_{(1)}, \ldots, \xi_{(d)})$ satisfy the condition **[R]** and the other coordinates satisfy the condition **[Z]**. As we have already noted, in Theorem 2.3.2 for simplicity we considered only the non-lattice and arithmetic cases. The statement of part (iii) of Theorem 2.3.2 is proved in [56].

So, from Theorem 2.3.2 one can see that in the case $\alpha = x/n \in \mathcal{A}'$ all the three types of deviations mentioned above (normal, moderately large and large) can be described in a completely unified way using the analytical representations (2.3.7)—(2.3.9) (this is one advantage of integro-local theorems). In the next section we will see that under some additional conditions (close to the necessary conditions), the asymptotic representations (2.3.7)–(2.3.9) under the conditions of the parts (ii), (iii), (iv) of Theorem *2.3.2* also hold also as $|\alpha| \to \infty$, $\alpha \in \mathcal{A}'$ (i.e. in the zone of superlarge deviations). If $\alpha \notin \mathcal{A}'$, the asymptotics of the probabilities $\mathbf{P}(S_n \in \Delta[x))$ for large and superlarge deviations will be different.

2.3.3 Large and superlarge deviations in the multidimensional case. General theorems

In this section we will consider deviations $x = x(n) \in \mathbb{R}^d$ such that $\alpha = x/n$ belongs 'from some moment of time', for n large enough, to the truncated (from below) cone

$$\underline{K} = \underline{K}(\mathrm{e}, \delta, N) := \{\alpha = \mathrm{e}'t : \ |\mathrm{e} - \mathrm{e}'| \leqslant \delta, \ t \geqslant N\}, \tag{2.3.10}$$

for some e, $\delta > 0, N < \infty$, where e, e$'$ are unit vectors in $\mathbb{R}^d$. *We will assume that the set $\underline{K}$ is included in $(\mathcal{A}')$ and will not exclude the case $|\alpha| \to \infty$ as $n \to \infty$.*

In order to generalise Theorem 2.3.2 to the case of superlarge deviations we will need additional conditions. Denote by ${}^{\alpha}\zeta$ the normalized random variable

$${}^{\alpha}\zeta = ({}^{\alpha}\xi - \alpha)\sigma^{-1}(\alpha),$$

so that in our case

$$\mathbf{E}({}^{\alpha}\zeta) = 0, \qquad \mathbf{E}({}^{\alpha}\zeta)^T({}^{\alpha}\zeta) = E,$$

where E is the identity matrix; the superscript T stands for transposition.

The first condition is as follows.

[UI$_{\underline{K}}$] (*The condition of uniform integrability in $\alpha \in \underline{K}$ of the square of the norm of ${}^{\alpha}\zeta$*)

$$\lim_{T\to\infty} \sup_{\alpha\in\underline{K}} \mathbf{E}\big(|{}^{\alpha}\zeta|^2;\ |{}^{\alpha}\zeta| > T\big) = 0.$$

The meaning of the condition **[UI$_{\underline{K}}$]** is that it allows us to use a normal approximation for the distribution of sums ${}^{\alpha}S_n$ in the reduction formula (2.1.10). If one considers a scheme of series $\xi_{1,n}, \ldots, \xi_{n,n}$, where $\xi_{k,n} =_d {}^{\alpha}\zeta$ and the vector $\alpha \in \underline{K}$ depends on n, then **[UI$_{\underline{K}}$]** implies the validity of the Lindeberg condition for $\{\xi_{k,n}\}$, and therefore the central limit theorem holds for the sum

$${}^{\alpha}S_n^0 := \xi_{1,n} + \cdots + \xi_{n,n} := {}^{\alpha}\zeta_1 + \cdots + {}^{\alpha}\zeta_n$$

(the ${}^{\alpha}\zeta_j$ are independent copies of ${}^{\alpha}\zeta$). It is well known that the condition **[UI$_{\underline{K}}$]** is satisfied if, for example, the uniform (in $\alpha \in \underline{K}$) Lyapunov condition holds:

$$\sup_{\alpha\in\underline{K}} \mathbf{E}|{}^{\alpha}\zeta|^{2+\delta} < \infty, \qquad \delta > 0.$$

The second condition that we will need is *Cramér's condition that is uniform in $\alpha \in \underline{K}$ for a characteristic function* (the parameter $\varkappa = \varkappa_{\underline{K}}$ generally depends on $\underline{K}$):

$[\mathbf{R}^{\kappa}_{\underline{K}}]$

$$\limsup_{|t|\to\infty} \sup_{\alpha\in\underline{K}} |\mathbf{E}e^{i\langle t, {}^{\alpha}\zeta\rangle}| \leqslant e^{-d\kappa}.$$

The condition $[\mathbf{R}^{\kappa}_{\underline{K}}]$ holds for some $\kappa > 0$ if the following condition is satisfied (see [54]).

$[\mathbf{R}^{*}_{\underline{K}}]$ *There exists a cube* $\Delta[v)$ *with side* $\Delta > 0$ *and a vertex at a point* $v \in \mathbb{R}^d$ *(which may depend on* α*) such that the distribution of the vector* ${}^{\alpha}\zeta$ *has an absolutely continuous component and its density* ${}^{\alpha}f(y)$ *in this cube is uniformly separated from zero:*

$${}^{\alpha}f(y) \geqslant p > 0 \quad \textit{for} \quad y \in \Delta[v), \quad \alpha \in \Omega.$$

The meaning of the condition $[\mathbf{R}^{\kappa}_{\underline{K}}]$ is that it allows us to use statement (ii) in Theorem 2.3.2 for the distribution of ${}^{\alpha}S_n$ in the reduction formula.

Note also that the condition $[\mathbf{UI}_{\underline{K}}]$ will not hold if there are large lacunas in the distribution of ξ. The condition $[\mathbf{R}^{\kappa}_{\underline{K}}]$ will not hold if the distribution of ξ 'at infinity', in some sense, approaches a lattice distribution (then ${}^{\alpha}\xi$ approaches a lattice distribution as $\alpha \to \infty$).

In the case when the distribution $\mathbf{F}$ has a density, the following condition for the absolute continuity of the distribution of ${}^{\alpha}\zeta$ in an appropriate half-space is an analogue of the condition $[\mathbf{R}^{*}_{\underline{K}}]$.

$[\mathbf{R}^{\mathrm{den}}_{\underline{K}}]$ *For some* $\mathrm{e} \in \mathbb{R}^d$, $|\mathrm{e}| = 1$, $c > 0$ *(the vector* e *and the number* c *may depend on* α*) the distribution of the vector* ${}^{\alpha}\zeta$ *is absolutely continuous in the half-space* $\Pi = \{y : \langle \mathrm{e}, y\rangle \geqslant -c\}$ *with density* ${}^{\alpha}g(y)$, $y \in \Pi$, *such that*

$$\sup_{\alpha\in\underline{K}} \sup_{y\in\Pi} {}^{\alpha}g(y) < \infty.$$

Moreover, for some $\Delta > 0$, $h > 0$ *and* $v \in \mathbb{R}^d$ *(the vector* v *may depend on* α*),*

$${}^{\alpha}g(y) \geqslant h$$

for $y \in \Delta[v) \subseteq \Pi$, $\alpha \in \underline{K}$.

The condition $[\mathbf{UI}_{\underline{K}}]$ in the arithmetic case remains without any change, and an analogue of the condition $[\mathbf{R}^{*}_{\underline{K}}]$ is the following condition for the uniform (in $\alpha \in \underline{K}$) arithmeticity of the distribution of ${}^{\alpha}\xi$:

$[\mathbf{Z}_{\underline{K}}]$ *Given that the condition* $[\mathbf{Z}]$ *holds, for some* $p > 0$, $\Delta > 0$ *and* $v \in \mathbb{R}^d$, *for a cube* $\Delta_{\alpha}[v)$ *with vertex at a point* v *(which may depend on* α*) and with side* $\Delta_{(\alpha)} = \Delta|\sigma(\alpha)|^{1/d}$, *the following inequality holds:*

$$\mathbf{P}({}^{\alpha}\xi = y) \geqslant \frac{p}{|\sigma(\alpha)|} \quad \textit{for} \quad y \in \mathbb{Z}^d \cap \Delta_{(\alpha)}[v), \quad \alpha \in \underline{K}.$$

Recall that $|\sigma(\alpha)|$ is the volume of the parallelepiped $\Delta[0)\sigma(\alpha)$, for $\Delta = 1$, which is obtained if every vector from the unit cube with vertex at the origin is multiplied by the matrix $\sigma(\alpha)$. We also have $|\sigma(\alpha)| = \prod_{i=1}^{d} \sigma_i(\alpha)$, where $0 < \sigma_1(\alpha) \leqslant \cdots \leqslant \sigma_d(\alpha)$ are the eigenvalues of the matrix $\sigma(\alpha)$.

Theorem 2.3.4. *Suppose for some fixed truncated cone $\underline{K} \subseteq (\mathcal{A}')$, as defined in (2.3.10), the condition* $[\mathbf{UI}_{\underline{K}}]$ *holds. Then*

(i) *If the condition* $[\mathbf{R}_{\underline{K}}^{\kappa}]$ *holds, we have*

$$\mathbf{P}\big(S_n \in \Delta[x)\big) = \frac{\Delta^d}{(2\pi n)^{d/2}|\sigma(\alpha)|} e^{-n\Lambda(\alpha)}(1+\varepsilon_n), \tag{2.3.11}$$

where $\varepsilon_n = \varepsilon_n(\Delta, \alpha)$ for any fixed $\kappa_1 < \kappa$ and the sequence $\delta(n) = o(1)$ satisfies the relation

$$\lim_{n\to\infty} \sup_{\substack{\alpha\in\underline{K} \\ \Delta_1 \leqslant \Delta \leqslant \Delta_2}} \varepsilon_n(\Delta, \alpha)| = 0, \tag{2.3.12}$$

where $\Delta_1 = \sigma_d(\alpha)e^{-n\kappa_1}$, $\Delta_2 = \delta(n)\min\{|\lambda(\alpha)|^{-1}$ and $\sqrt{n}\sigma_1(\alpha)\} = o(1)$.

(ii) *If the condition* $[\mathbf{R}_{\underline{K}}^{\mathrm{den}}]$ *holds then the distribution of S_n in the set $n\underline{K}$ is absolutely continuous with density $f_n(x)$ and, for $\alpha = x/n \in \underline{K}$, we have*

$$f_n(x) = \frac{1}{(2\pi n)^{d/2}|\sigma(\alpha)|} e^{-n\Lambda(\alpha)}(1+\varepsilon_n), \tag{2.3.13}$$

where, for $\varepsilon_n = \varepsilon_n(\alpha)$,

$$\lim_{n\to\infty} \sup_{\alpha\in\underline{K}} |\varepsilon_n(\alpha)| = 0.$$

(iii) *If the condition* $[\mathbf{Z}_{\underline{K}}]$ *holds and also $\sqrt{n}\sigma_1(\alpha) \to \infty$ as $|\alpha| \to \infty$, $\alpha \in \underline{K}$, then, for $x \in \mathbb{Z}^d$,*

$$\mathbf{P}(S_n = x) = \frac{1}{(2\pi n)^{d/2}|\sigma(\alpha)|} e^{-n\Lambda(\alpha)}(1+\varepsilon_n), \tag{2.3.14}$$

where for $\varepsilon_n = \varepsilon_n(\alpha)$ we have

$$\lim_{n\to\infty} \sup_{\substack{x\in\mathbb{Z}^d \\ \alpha\in\underline{K}}} |\varepsilon_n(\alpha)| = 0. \tag{2.3.15}$$

Remark 2.3.5. The statement of part (i) of Theorem 2.3.4 for fixed $\Delta > 0$ can be presented in the form

$$\mathbf{P}\big(S_n \in \Delta[x)\big) = \frac{1}{(2\pi n)^{d/2}|\sigma(\alpha)|} \int_{\Delta[0)} e^{-n\Lambda(\alpha+u)} du (1+\varepsilon_n),$$

where $\varepsilon_n = \varepsilon_n(\Delta, \alpha)$ satisfies the relation $\lim_{n\to\infty} \sup_{\alpha\in\underline{K}} |\varepsilon_n(\Delta, \alpha)| = 0$. Note also that the inequality $\Delta_1 \leqslant \Delta_2$ under the supremum in (2.3.12) implies the inequality

$$\sigma_d(\alpha)e^{-n\kappa_1} \leqslant \delta(n)\min\big\{|\lambda(\alpha)|^{-1},\ \sqrt{n}\sigma_1(\alpha)\big\}.$$

This inequality gives the following upper bound on the order of the deviation $|\alpha| = |x/n|$ in part (i) of Theorem 2.3.4, for which it turns out to be possible to obtain the integro-local theorem in the universal form (2.3.11) (cf. Theorem 2.3.2):

$$\sigma_d(\alpha)\left(|\lambda(\alpha)| + \frac{1}{\sqrt{n}\sigma_1(\alpha)}\right) \leqslant e^{n\kappa_1}. \tag{2.3.16}$$

In the examples which will be considered in the next section, the left-hand side of (2.3.16) does not grow faster than some power of $|\alpha|$, so in those examples the bound (2.3.16) will allow us to study deviations $\alpha = x/n$ of order $|\alpha| \leqslant e^{nc}$ for some $c > 0$.

Theorem 2.3.4 is proved in [56].

2.3.4 Large and superlarge deviations for three classes of one-dimensional distributions

We will introduce three classes of distributions $\mathbf{F}(\cdot) = \mathbf{P}(\xi \in \cdot)$ of random variables ξ with rapidly decreasing 'regularly varying' right tails $F_+(t) := \mathbf{P}(\xi \geqslant t)$. For convenience, we will associate these classes with the known classes of regularly varying and semi-exponential distributions (see [42]). Everywhere below, by $l = l(t)$ we denote a function regularly varying as $t \to \infty$:

$$l(t) = t^\beta L(t), \quad t \geqslant 0, \tag{2.3.17}$$

where $L(t)$ is a slowly varying function (s.v.f.) as $t \to \infty$. Wherever it becomes necessary, the exponent β of the function l will be denoted by $\beta(l)$.

Denote by $\mathcal{R}$ the class of *distributions regularly varying at infinity*, i.e. such that

$$\mathbf{P}(\xi \geqslant t) = F_+(t) = l(t), \quad t \geqslant 0, \tag{2.3.18}$$

where the function $l(t)$ is defined in (2.3.17) and $\beta = \beta(l) < 0$.

By $\mathcal{S}e$, denote the class of *semi-exponential distributions*, i.e. distributions such that

$$F_+(t) = e^{-l(t)}, \quad t \geqslant 0, \tag{2.3.19}$$

where $\beta(l) \in (0, 1]$,

$$0 < l(t) = o(t), \tag{2.3.20}$$

and also the following regularity condition is satisfied: for $t \to \infty$, $s = o(t)$,

$$l(t+s) - l(t) = s\frac{\beta l(t)}{t}(1 + o(1)) + o(1). \tag{2.3.21}$$

In other words,

$$l(t+s) - l(t) \sim s\frac{\beta l(t)}{t}, \quad \text{if } \liminf_{t\to\infty} \frac{sl(t)}{t} > 0,$$

and

$$l(t+s) - l(t) = o(1), \quad \text{if} \quad \lim_{t\to\infty} \frac{sl(t)}{t} = 0.$$

The following condition is sufficient for the validity of (2.3.21).

[D$_1$] *The function $L(t)$ for some $t_0 < \infty$ and all $t \geqslant t_0 > 0$ is differentiable, and*

$$L'(t) = o\left(\frac{L(t)}{t}\right)$$

as $t \to \infty$.

Indeed, by **[D$_1$]**,

$$l'(t) = \frac{\beta l(t)}{t}(1 + o(1)); \tag{2.3.22}$$

so, using (2.3.22) and the equality $l(t+s) - l(t) = \int_t^{t+s} l'(u)du$, we obtain the relation (2.3.21) without the last correction term $o(1)$.

If ξ is a random variable with an arithmetic distribution then the arguments t and s in (2.3.18)–(2.3.21) *take on only integer values.*

We will call distributions that are regularly varying and semi-exponential *regular*. Let us introduce now the classes of exponentially decreasing distributions associated with the classes $\mathcal{R}$ and $\mathcal{S}e$:

(1) *the class $\mathcal{ER}$ of distributions such that*

$$F_+(t) = e^{-\lambda_+ t} l(t), \quad t \geqslant 0, \tag{2.3.23}$$

where $\lambda_+ \in (0, \infty)$ but in contrast with (2.3.18) *the parameter $\beta = \beta(l)$ may take on non-negative values as well;*

(2) *the class $\mathcal{ES}e$ of distributions such that*

$$F_+(t) = e^{-\lambda_+ t + l(t)}, \quad t \geqslant 0, \tag{2.3.24}$$

where $\lambda_+ \in (0, \infty)$ and the function $l \in \mathcal{R}$ satisfies (2.3.21), *$\beta = \beta(l) \in [0, 1]$, $l(t) = o(t)$, but in contrast with* (2.3.19) *and* (2.3.20) *the function $l(t)$ may be both negative or positive.*

Finally, we consider the class of *superexponentially* decreasing distributions (i.e. the class of distributions with right tail decreasing faster than any exponent):

(3) *the class $\mathcal{SE}$ of superexponential distributions of the form*

$$F_+(t) = e^{-l(t)}, \quad t \geqslant 0, \tag{2.3.25}$$

where $\beta = \beta(l) > 1$, so that the following property holds:

$$l(t) \gg t \quad as \quad t \to \infty,$$

which, in an obvious sense, supplements (2.3.20), (2.3.23).

In the relations (2.3.23)–(2.3.25), for arithmetic distributions we use what was previously agreed: *these relations for the mentioned distributions hold only for integer t.*

Distributions from the classes $\mathcal{ER}$, $\mathcal{ES}e$ will be called *regularly exponentially decreasing* distributions; distributions from the class $\mathcal{SE}$ will be called *regularly superexponentially decreasing* (or simply *superexponential*) distributions.

It is clear that a distribution with a 'purely exponentially' decreasing tail $ce^{-\lambda_+ t}$ for $t \geqslant t_0 > 0$ and an arithmetic distribution with a tail $F_+(k) = ce^{-\lambda_+ k}$ for $k \geqslant k_0 > 0$ belong to the class $\mathcal{ER}$, and a distribution with a tail of the form ce^{-vt^2} for $t \geqslant t_0 > 0$ and the normal distribution belong to the class $\mathcal{SE}$.

Consider now the smoothness conditions which will be used for the classes of distributions we have introduced. These conditions will be given in terms of the functions $l(t) = t^\beta L(t)$, which define the classes $\mathcal{ER}$, $\mathcal{ES}e$, $\mathcal{SE}$.

The condition (2.3.21) and the stronger condition $[\mathbf{D}_1]$ have been already formulated above (the condition (2.3.21) appears in the definition of the class $\mathcal{S}e$). The following condition is a broad analogue of the condition on the existence of a smooth second derivative of the function $l = l(t)$:

$[\mathbf{D}_2]$ *The function $L(t)$ satisfies the condition $[\mathbf{D}_1]$ and, as $t \to \infty$, $s = o(t)$,*

$$l(t+s) - l(t) = sl'(t) + s^2\frac{\beta(\beta-1)l(t)}{2t^2}(1+o(1)) + o(1). \qquad (2.3.26)$$

If the second derivative $l''(t) = \beta(\beta-1)\frac{l(t)}{t^2}(1+o(1))$ exists as $t \to \infty$, the relation (2.3.26) will be satisfied; then the last term $o(1)$ in (2.3.26) will be absent:

$$l(t+s) - l(t) = \int_t^{t+s}\left[l'(t) + \int_t^v l''(u)du\right]dv = sl'(t) + s^2\frac{\beta(\beta-1)l(t)}{2t^2}(1+o(1)).$$

In the paper [57], limit theorems for the Cramér transforms of the above-mentioned distribution as $\lambda \to \infty$ were found. This allows one to check the validity of the conditions of Theorem 2.3.4 for the classes of distributions under consideration.

Note that in the one-dimensional case the truncated cone $\underline{K}$ (see (2.3.10)) turns into the half-axis $\{t : t \geqslant c > 0\}$. Because of this, in the notation for the conditions which have the subscript $\underline{K}$ in the multidimensional case, in the one-dimensional case this superscript will be replaced by the subscript $>c$. Thus instead of $[\mathbf{UI}_{\underline{K}}]$, $[\mathbf{R}^\kappa_{\underline{K}}]$, $[\mathbf{R}^{\text{den}}_{\underline{K}}]$, $[\mathbf{Z}_{\underline{K}}]$ we will write $[\mathbf{UI}_{>c}]$, $[\mathbf{R}^\kappa_{>c}]$, $[\mathbf{R}^{\text{den}}_{>c}]$, $[\mathbf{Z}_{>c}]$, respectively.

Recall that the function $\sigma(\alpha)$ is the square root of the variance of the random variable ${}^\alpha\xi = \xi$ and the function $\lambda(\alpha)$ satisfies the equation

$$\frac{\psi'(\lambda)}{\psi(\lambda)} = \mathbf{E}\xi^{(\lambda)} = \alpha, \qquad (2.3.27)$$

In Theorems 2.3.6–2.3.9, given below, we choose a fixed number $\alpha_* > \alpha_-$ as a lower bound for deviations $\alpha = x/n$, where α_- is the left boundary of the region of analyticity of the deviation function $\Lambda(\alpha)$. This means that in these theorems we consider not only superlarge deviations, for which $\alpha = x/n \to \infty$, but also large deviations $\alpha \in (\alpha_*, \infty)$, if $\alpha_* > \mathbf{E}\xi$, and even normal deviations $\alpha \to \mathbf{E}\xi$, if $\mathbf{E}\xi > \alpha_*$. However, the statements about large and normal deviations remain valid under wider conditions (see Theorem 2.3.2).

First consider the case $\mathbf{F} \in \mathcal{ER}$, $\beta \in (-2, -1)$. In [57] it is shown that, as $\alpha \to \infty$,

$$\lambda(\alpha) = \lambda_+ - \alpha^{-1/(\beta+2)} L_1(\alpha), \quad \sigma(\alpha) = \alpha^{(\beta+3)/(2\beta+4)} L_2(\alpha)$$
$$\Lambda(\alpha) = \lambda_+ \alpha + c + o(1),$$

where L_1, L_2 are positive s.v.fs. as $\alpha \to \infty$, $c =$ const.

Theorem 2.3.6. *Let $\mathbf{F} \in \mathcal{ER}$, $\beta \in (-2, -1)$. Then*

(i) *If the distributions $\mathbf{F}$, $\mathbf{F}^{(\lambda_+)}$ satisfy the condition $[\mathbf{R}^\kappa]$ for some $\kappa > 0$ (for which it is sufficient that a non-zero absolutely continuous component of the distribution $\mathbf{F}$ exists) then, for any fixed numbers $\delta_1 > 0$, $\delta_2 > 0$, $\alpha_* > \alpha_-$ and any sequence $\delta(n) \downarrow 0$ we have*

$$\mathbf{P}\big(S_n \in \Delta[x)\big) = \frac{\Delta}{\sqrt{2\pi n}\,\sigma(\alpha)} e^{-n\Lambda(\alpha)}(1 + \varepsilon_n), \tag{2.3.28}$$

where $\alpha = x/n$, and the remainder $\varepsilon_n = \varepsilon_n(x, \Delta)$ satisfies

$$\lim_{n\to\infty} \sup_{\substack{\Delta_1 \leqslant \Delta \leqslant \Delta_2 \\ \alpha_* \leqslant \alpha \leqslant n^{(\beta+2)/|\beta+1|-\delta_1}}} |\varepsilon_n(x, \Delta)| = 0,$$

for $\Delta_1 = e^{-n/m(\alpha)}$, $m(\alpha) = \alpha^{|\beta+1|/(\beta+2)+\delta_2}$, $\Delta_2 = \delta(n)$.

(ii) *If the condition $[\mathbf{Z}]$ holds then, for any fixed numbers $\alpha_* > \alpha_-$, $\delta_1 > 0$, we have the representation*

$$\mathbf{P}(S_n = x) = \frac{1}{\sqrt{2\pi n}\,\sigma(\alpha)} e^{-n\Lambda(\alpha)}(1 + \varepsilon_n), \quad x \in \mathbb{Z}, \tag{2.3.29}$$

where $\alpha = x/n$, and the remainder $\varepsilon_n = \varepsilon_n(x)$ satisfies the relation

$$\lim_{n\to\infty} \sup_{\substack{x \in \mathbb{Z} \\ \alpha_* \leqslant \alpha \leqslant n^{(\beta+2)/|\beta+1|-\delta_1}}} |\varepsilon_n(x)| = 0.$$

Consider now the deviations $\alpha = x/n > n^{(\beta+2)/|\beta+1|-\delta_1}$ (or $x > n^{|\beta+1|^{-1}-\delta_1}$), which are not considered in Theorem 2.3.6. Since we have $\psi(\lambda_+) < \infty$ for $\mathbf{F} \in \mathcal{ER}$ when $\beta \in (-2, -1)$, the random variable $\xi^{(\lambda_+)}$ is well defined and has a regularly varying right tail of the distribution:

$$V_+(t) := \mathbf{P}(\xi^{(\lambda_+)} \geqslant t) = \frac{\lambda_+}{|\beta+1|\psi(\lambda_+)} t^{\beta+1} L(t)(1 + o(1)) \quad \text{as} \quad t \geqslant 0.$$

Therefore, according to well-known theorems (see e.g. Theorem 2 on p. 646 in [92]), the distribution $\mathbf{F}^{(\lambda_+)}$ belongs the domain of attraction of the stable law $\mathbf{\Phi}$

with exponent $\beta_* = |\beta + 1|$ concentrated on the right semi-axis. This means that for a function $B_t = V_+^{(-1)}(1/t)$ regularly varying with exponent $1/\beta^*$, where $V_+^{(-1)}(u)$ is the inverse function of $V_+(t)$, we have the weak convergence

$$\mathbf{P}\left(\frac{S_n^{(\lambda_+)}}{B_n} \in \cdot\right) \Longrightarrow \mathbf{\Phi}(\cdot) \quad \text{as} \quad n \to \infty.$$

It is known that the limiting distribution $\mathbf{\Phi}$ is absolutely continuous and that its density $\phi = \phi(t)$ is continuous, positive for all $t > 0$ and equal to 0 for all non-positive t (see e.g. [92], p. 652).

Now put

$$\Lambda_+(\alpha) := \lambda_+\alpha - \ln \psi(\lambda_+).$$

Theorem 2.3.7. *Let* $\mathbf{F} \in \mathcal{ER}$, $\beta(l) \in (-2, -1)$ *and* c, C *be any fixed numbers such that* $0 < c \leqslant C < \infty$.

(i) *If the condition* **[R]** *holds then, for some sequence* $\Delta_1 = \Delta_1(n) = o(1)$ *and any sequence* $\Delta_2 = \Delta_2(n) = o(1)$, $\Delta_2 \geqslant \Delta_1$ *as* $n \to \infty$, *we have the relation*

$$\mathbf{P}\big(S_n \in \Delta[x)\big) = \frac{\Delta}{B_n} e^{-n\Lambda_+(\alpha)} \phi\left(\frac{x}{B_n}\right)(1 + \varepsilon_n), \tag{2.3.30}$$

where $\alpha = x/n$, *the remainder* $\varepsilon_n = \varepsilon_n(x, \Delta)$ *tends to* 0 *uniformly in* $\Delta \in [\Delta_1, \Delta_2]$ *and* $\alpha = x/n \in [cB_n/n,\ CB_n/n]$ *i.e.*,

$$\lim_{n\to\infty} \sup_{\substack{x\in[cB_n,\ CB_n] \\ \Delta_1\leqslant\Delta\leqslant\Delta_2}} \big|\varepsilon_n(x, \Delta)\big| = 0. \tag{2.3.31}$$

(ii) *If instead of the condition* **[R]** *for the distribution* $\mathbf{F}^{(\lambda_+)}$ *the stronger condition* $[\mathbf{R}^\kappa]$ *holds then in the relation* (2.3.30) *the remainder* $\varepsilon_n = \varepsilon_n(x, \Delta)$ *satisfies the relation* (2.3.31), *where* $\Delta_1 = e^{-n\kappa_1}$ *for any fixed* $\kappa_1 < \kappa$ *and* $\Delta_2 = o(1)$, $\Delta_2 \geqslant \Delta_1$.

(iii) *If instead of the condition* **[R]** *the condition* **[Z]** *holds then, for* $x \in \mathbb{Z}$, $\alpha = x/n \in [cB_n/n,\ CB_n/n]$, *we have*

$$\mathbf{P}(S_n = x) = \frac{1}{B_n} e^{-n\Lambda_+(\alpha)} \phi\left(\frac{x}{B_n}\right)(1 + o(1)).$$

Consider now the case $\mathbf{F} \in \mathcal{ER}$, $\beta > -1$. In [57] it was shown that in this case

$$\lambda(\alpha) = \lambda_+ - \frac{1}{\alpha(\beta + 1)}(1 + o(1)),$$

$$\sigma(\alpha) \sim \frac{\alpha}{\sqrt{\beta + 1}}, \quad \Lambda(\alpha) = \lambda_+\alpha - \frac{1}{\beta + 1} \ln \alpha(1 + o(1)) \tag{2.3.32}$$

as $\alpha \to \infty$.

Theorem 2.3.8. *Let* $\mathbf{F} \in \mathcal{ER}$, $\beta > -1$. *Then*

(i) *If the distribution* $\mathbf{F}$ *has a non-zero absolutely continuous component then, for any fixed numbers* $\delta_1 > 0$, $\delta_2 > 0$, $\alpha_* > \alpha_-$ *and any sequence* $\delta(n) \downarrow 0$, *we have the representation* (2.3.28), *where the remainder* $\varepsilon_n = \varepsilon_n(x, \Delta)$ *satisfies the relation*

$$\lim_{n\to\infty} \sup_{\substack{\Delta_1 \leqslant \Delta \leqslant \Delta_2 \\ \alpha_* \leqslant \alpha \leqslant n^{1/(\beta+1)-\delta_1}}} \left|\varepsilon_n(x, \Delta)\right| = 0,$$

for $\Delta_1 = e^{-n/m(\alpha)}$, $m(\alpha) = \alpha^{1+\beta+\delta_2}$, $\Delta_2 = \delta(n)$.

(ii) *If the function* L *in* (2.3.17) *satisfies the condition* $[\mathbf{D}_1]$, *then, for any fixed* $\alpha_* > \max\{0, \alpha_-\}$ *and all sufficiently large* n, *the distribution of the sum* S_n *is absolutely continuous in the region* $x \geqslant n\alpha_*$ *with density*

$$f_n(x) = \frac{1}{\sqrt{2\pi n}\,\sigma(\alpha)} e^{-n\Lambda(\alpha)}(1 + \varepsilon_n), \tag{2.3.33}$$

where $\alpha = x/n$ *and the remainder* $\varepsilon_n = \varepsilon_n(x)$ *satisfies the relation*

$$\lim_{n\to\infty} \sup_{\alpha \geqslant \alpha_*} \left|\varepsilon_n(x)\right| = 0.$$

(iii) *If the condition* $[\mathbf{Z}]$ *holds then, for any fixed* $\alpha_* > \alpha_-$, *the probability* $\mathbf{P}(S_n = x)$ *can be represented in the form* (2.3.29), *where* $\alpha = x/n$ *and the remainder* $\varepsilon_n = \varepsilon_n(x)$ *satisfies the relation*

$$\lim_{n\to\infty} \sup_{\substack{x \in \mathbb{Z} \\ \alpha \geqslant \alpha_*}} |\varepsilon_n(x)| = 0.$$

Statement (ii) of Theorem 2.3.8 agrees with the results of [145], where for an absolutely continuous distribution $\mathbf{F} \in \mathcal{ER}$, $\beta > -1$, an asymptotic expansion for the density $f_n(x)$ of the distribution of the sum S_n was obtained.

Consider now the case $\mathbf{F} \in \mathcal{E}e$. In [57] it was shown that, as $\alpha \to \infty$,

$$\lambda(\alpha) = \lambda_+ - \alpha^{\beta-1}L_1(\alpha), \qquad \sigma(\alpha) = \alpha^{1-\beta/2}L_2(\alpha),$$
$$\Lambda(\alpha) = \lambda_+\alpha - \frac{1}{\beta}\alpha^{\beta}L_1(\alpha)(1 + o(1)),$$

where L_1, L_2 are s.v.f. as $\alpha \to \infty$.

In the case $\mathbf{F} \in \mathcal{SE}$, it was established that

$$\lambda(\alpha) = \alpha^{\beta-1}L_1(\alpha), \quad \sigma(\alpha) = \alpha^{1-\beta/2}L_2(\alpha), \quad \Lambda(\alpha) = \alpha^{\beta}L_3(\alpha), \quad \alpha \to \infty,$$

where L_1, L_2, L_3 are s.v.fs. as $\alpha \to \infty$.

Theorem 2.3.9. *Suppose that one of the following two conditions is satisfied:*

(1) $\mathbf{F} \in \mathcal{ES}e$ *and the function* l *is negative* (see (2.3.24)) *and satisfies the condition* $[\mathbf{D}_2]$;

(2) $\mathbf{F} \in \mathcal{SE}$, *the function* l *satisfies the condition* $[\mathbf{D}_2]$, $\beta(l) > 1$ *in the non-lattice case and* $\beta \in (1, 2)$ *in the arithmetic case.*

Then the following statements are true.

(i) *If the condition* **[R]** *holds then, for any fixed* $\alpha_* > \max\{0, \alpha_-\}$ *and all sufficiently large* n*, the distribution of the sum* S_n *is absolutely continuous in the region* $x \geqslant n\alpha_*$ *with density* $f_n(x)$ *which admits the representation* (2.3.33), *where* $\alpha = x/n$ *and the remainder* $\varepsilon_n = \varepsilon_n(x)$ *satisfies the relation*

$$\lim_{n\to\infty} \sup_{\alpha \geqslant \alpha_*} \left|\varepsilon_n(x)\right| = 0.$$

(ii) *If the condition* **[Z]** *holds then, for any fixed* $\alpha_* > \alpha_-$ *and all* $\alpha = x/n \geqslant \alpha_*$, *the probability* $\mathbf{P}(S_n = x)$ *admits the representation* (2.3.29), *where the remainder* $\varepsilon_n = \varepsilon_n(x)$ *satisfies the relation*

$$\lim_{n\to\infty} \sup_{\substack{x\in\mathbb{Z} \\ \alpha \geqslant \alpha_*}} \left|\varepsilon_n(x)\right| = 0.$$

Theorem 2.3.9 does not include the case of an arithmetic distribution $\mathbf{F} \in \mathcal{SE}$ for $\beta > 2$. In that case there is no universal form of the result (2.3.29) in the region of superlarge deviations, as indicated by the results in [140] (see also [141], [137]). In Theorems 2–4 of the paper [140], for an arithmetic distribution $\mathbf{F} \in \mathcal{SE}$ in the particular case $l(n) = n^{\beta}$, $2 < \beta < 3$, local theorems in the zone of superlarge deviations were obtained.

Theorems 2.3.4–2.3.9 were proved in [57].

2.4 Integro-local theorems on the boundary of the Cramér zone

2.4.1 Introduction

For the sake of specificity, in this section we will assume that $\lambda_+ > 0$. If $\alpha_+ = \infty$ then the theorems of sections 2.2.1–2.2.3 describe the probability of large deviations for any $\alpha = x/n$. If $\alpha_+ < \infty$, it is not always possible to find the asymptotics of large deviations of S_n for normalised deviations $\alpha = x/n$ located in a neighbourhood of the point α_+ by the methods of section 2.2.

In this section we will consider the case $\alpha_+ < \infty$. If also $\lambda_+ = \infty$ then by the property (Λ2i) we have $\alpha_+ = s_+ = \sup\{t : F_+(t) > 0\}$ and hence the random variables ξ_k are bounded from above by the value α_+, $\mathbf{P}(S_n \geqslant x) = 0$

for $\alpha = x/n > \alpha_+$. This case will not be considered in what follows. So, we will study the case

$$\lambda_+ < \infty, \qquad \alpha_+ < \infty.$$

In this and the next sections we will consider only integro-local theorems in the non-lattice case for $\Delta = \Delta_n \to 0$, since local theorems, as we saw in the previous sections, are simpler than integro-local ones. Integral theorems, as in section 2.2, can be easily obtained from integro-local theorems.

2.4.2 Probabilities of large deviations of S_n located in a $o(n)$-neighbourhood of the point $\alpha_+ n$; the case $\psi''(\lambda_+) < \infty$

In this subsection we will study the asymptotics of $\mathbf{P}(S_n \in \Delta[x))$, $x = \alpha n$, when α is located in a neighbourhood of the point $\alpha_+ < \infty$ and it holds that $\psi''(\lambda_+) < \infty$ (the left derivative $\psi''(\lambda_+ - 0)$ distributions $\mathbf{F}$ for which $\lambda_+ < \infty$, $\alpha_+ < \infty$ and $\psi''(\lambda_+) < \infty$ will be illustrated later in Lemma 2.4.3). When the above conditions hold, the Cramér transform $\mathbf{F}^{(\lambda_+)}$ at the point λ_+ is well defined and a random variable ${}^{\alpha_+}\xi$ with the distribution $\mathbf{F}^{(\lambda_+)}$ has mean α_+ and finite variance (see (2.2.1)):

$$\mathbf{E}\,{}^{\alpha_+}\xi = \frac{\psi'(\lambda_+)}{\psi(\lambda_+)} = \alpha_+, \qquad \mathbf{D}\,{}^{\alpha_+}\xi = \sigma^2(\alpha_+) = \frac{\psi''(\lambda_+)}{\psi(\lambda_+)} - \alpha_+^2. \qquad (2.4.1)$$

Theorem 2.4.1. *Let ξ be a non-lattice random variable and let $\lambda_+ > 0$, $\psi''(\lambda_+) < \infty$ and $y = x - \alpha_+ n = o(n)$. If $\Delta_n \to 0$ sufficiently slowly as $n \to \infty$, we have*

$$\mathbf{P}(S_n \in \Delta_n[x)) = \frac{\Delta_n}{\sigma(\alpha_+)\sqrt{2\pi n}}\, e^{-n\Lambda(\alpha_+) - \lambda_+ y}\left(\exp\left\{-\frac{y^2}{2\sigma^2(\alpha_+)n}\right\} + o(1)\right), \qquad (2.4.2)$$

where $\alpha = x/n$, $\sigma^2(\alpha_+) = \psi''(\lambda_+)/\psi(\lambda_+) - \alpha_+^2$ and the remainder $o(1)$ is uniform in y.

Clearly, under suitable conditions a similar statement is true for $x = \alpha_- n + y$, $y = o(n)$.

Proof. As in the proof of Theorem 2.2.1, let us use the Cramér transform but now at the fixed point λ_+, so that the scheme of series in the analysis of ${}^{\alpha_+}S_n$ will not appear here. In this case we have the following analogue of Theorem 2.1.2 (see (2.1.5)). □

Theorem 2.1.1A. *Let $\lambda_+ \in (0, \infty)$, $\alpha_+ < \infty$, $y = x - n\alpha_+$. Then for $x = n\alpha$ and any fixed $\Delta > 0$ we have*

$$\mathbf{P}(S_n \in \Delta[x)) = e^{-n\Lambda(\alpha_+) - \lambda_+ y} \int_0^\Delta e^{-\lambda_+ z}\mathbf{P}(S_n^{(\lambda_+)} - \alpha n \in dz). \qquad (2.4.3)$$

Proof. The proof of Theorem 2.1.1A repeats the proof of Theorem 2.1.2, only difference being that now, as has been already noted, the Cramér transform is applied at the fixed point λ_+, which does not depend on $\alpha = x/n$. In this case, due to (2.1.8),

$$\mathbf{P}(S_n \in dv) = e^{-\lambda_+ v + n \ln \psi(\lambda_+)} \mathbf{P}(S_n^{(\lambda_+)} \in dv)$$
$$= e^{-n\Lambda(\alpha_+) + \lambda_+(\alpha_+ n - v)} \mathbf{P}(S_n^{(\lambda_+)} \in dv).$$

Integrating this equality from x to $x + \Delta$, changing the variable to $v = x + z$ ($x = n\alpha$) and observing that $\alpha_+ n - v = -y - z$, we obtain (2.4.3). □

Let us return to the proof of Theorem 2.4.1. Assuming that $\Delta = \Delta_n \to 0$ for $n \to \infty$, by Theorem 2.1.1A we obtain

$$\mathbf{P}\big(S_n \in \Delta_n[x)\big) = e^{-n\Lambda(\alpha_+) - \lambda_+ y} \mathbf{P}\big(S_n^{(\lambda_+)} - \alpha_+ n \in \Delta_n[y)\big)\big(1 + o(1)\big). \quad (2.4.4)$$

By (2.4.1) we can apply Theorem 1.5.1 to the probability on the right-hand side of (2.4.4). This implies that when $\Delta_n \to 0$ sufficiently slowly we have

$$\mathbf{P}\big(S_n^{(\lambda_+)} - \alpha_+ n \in \Delta_n[y)\big) = \frac{\Delta_n}{\sigma(\alpha_+)\sqrt{n}} \phi\left(\frac{y}{\sigma(\alpha_+)\sqrt{n}}\right) + o\left(\frac{1}{\sqrt{n}}\right)$$
$$= \frac{\Delta_n}{\sigma(\alpha_+)\sqrt{2\pi n}} \exp\left\{-\frac{y^2}{2\sigma^2(\alpha_+)n}\right\} + o\left(\frac{1}{\sqrt{n}}\right)$$

uniformly in y. Together with (2.4.4), this proves Theorem 2.4.1. □

As usual, the integro-local theorem implies the integral theorem.

Corollary 2.4.2. *Suppose that the conditions of Theorem* 2.4.1 *are met* (except the condition on Δ_n). *Then*

$$\mathbf{P}(S_n \geqslant x) = \frac{1}{\lambda_+ \sigma(\alpha_+)\sqrt{2\pi n}} e^{-n\Lambda(\alpha_+) - \lambda_+ y} \left[\exp\left\{-\frac{y^2}{2\sigma^2(\alpha_+)n}\right\} + o(1)\right],$$

where the remainder is uniform in y.

Similarly, if $\lambda_- < 0$, $\psi''(\lambda_-) < \infty$, $x = \alpha_- n + y$, $y = o(n)$, *then*

$$\mathbf{P}(S_n < x) = \frac{1}{|\lambda_-|\sigma(\alpha_-)\sqrt{2\pi n}} e^{-n\Lambda(\alpha_-) - \lambda_- y} \left[\exp\left\{-\frac{y^2}{2\sigma^2(\alpha_-)n}\right\} + o(1)\right].$$

2.4.3 Probability of large deviations of S_n in a $o(n)$-neighbourhood of the point $\alpha_+ n$ for distributions F from the class $\mathcal{ER}$ in the case $\psi''(\lambda_+) = \infty$

In the study of the asymptotics of $\mathbf{P}(S_n \geqslant \alpha n)$ (or $\mathbf{P}\big(S_n \in \Delta[\alpha n)\big)$) in the case when $\psi''(\lambda_+) = \infty$ and α is in a neighbourhood of the point $\alpha_+ < \infty$, we have to make additional assumptions about the distribution **F** as when the convergence to stable laws is studied.

As before, *we will say that a distribution* **F** (*or a random variable* ξ) *belongs to the class* $\mathcal{ER}$ *if in the representation* $F_+(t) = e^{-\lambda_+ t} l(t)$ *the function* l *is regularly varying* (*this will be written as* $l \in \mathcal{R}$; see (2.3.18), (2.3.23)).

The following result clarifies to which distributions from $\mathcal{ER}$ the cases $\alpha_+ = \infty$, $\alpha_+ < \infty$, $\psi''(\lambda_+) = \infty$, $\psi''(\lambda_+) < \infty$ correspond.

Lemma 2.4.3. *Suppose that* $\mathbf{F} \in \mathcal{ER}$. *Then, in order for* α_+ *to be finite, it is necessary and sufficient that*

$$\int_1^\infty tl(t)dt < \infty.$$

For the finiteness of $\psi''(\lambda_+)$ *it is necessary and sufficient that*

$$\int_1^\infty t^2 l(t)dt < \infty.$$

The lemma means that $\alpha_+ < \infty$ if $\beta > 2$ in the representation $l(t) = t^{-\beta}L(t)$, L is an s.v.f. and $\alpha_+ = \infty$ if $\beta < 2$. For $\beta = 2$ the finiteness of α_+ is equivalent to the finiteness of $\int_1^\infty t^{-1}L(t)dt$. The question whether $\psi''(\lambda_+)$ is finite is similar to this question.

Proof of Lemma 2.4.3. We first prove the claim about α_+. Since $\alpha_+ = \psi'(\lambda_+)/\psi(\lambda_+)$, we need to estimate the values $\psi'(\lambda_+)$ and $\psi(\lambda_+)$. The finiteness of $\psi'(\lambda_+)$ is equivalent to the finiteness of

$$-\int_1^\infty te^{\lambda_+ t}dF_+(t) = \int_1^\infty t\big[\lambda_+ l(t)dt - dl(t)\big], \tag{2.4.5}$$

where, for $l(t) = o(1/t)$,

$$-\int_1^\infty t\,dl(t) = l(1) + \int_1^\infty l(t)\,dt.$$

Hence, the finiteness of the integral on the left-hand side of (2.4.5) is equivalent to the finiteness of the sum

$$\lambda_+ \int_1^\infty tl(t)dt + \int_1^\infty l(t)dt$$

or, equivalently, the finiteness of the integral $\int_1^\infty tl(t)dt$. In a similar way we can see that the finiteness of $\psi(\lambda_+)$ is equivalent to the finiteness of $\int_1^\infty l(t)dt$. This implies the claim of the lemma in the case $\int_1^\infty l(t)dt < \infty$, where $l(t) = o(1/t)$. If $\int_1^\infty l(t)dt = \infty$ then $\psi(\lambda_+) = \infty$, $\ln\psi(\lambda) \to \infty$ as $\lambda \uparrow \lambda_+$ and so $\alpha_+ = \lim_{\lambda\uparrow\lambda_+} \big(\ln\psi(\lambda)\big)' = \infty$.

The claim concerning $\psi''(\lambda_+)$ can be proved in an absolutely similar way. □

This lemma implies the following.

(a) If $\beta < 2$ or $\beta = 2$, $\int_1^\infty t^{-1}L(t) = \infty$, then $\alpha_+ = \infty$ and one can apply the theorems of the previous section when studying $\mathbf{P}(S_n \geqslant x)$.

(b) If $\beta > 3$ or $\beta = 3$, $\int_1^\infty t^{-1}L(t)dt < \infty$, then $\alpha_+ < \infty$, $\psi''(\lambda_+) < \infty$ and one can apply Theorem 2.4.1.

It remains to consider the case

(c) $\beta \in [2, 3]$, where for $\beta = 2$ the integral $\int_1^\infty t^{-1}L(t)dt$ is finite and for $\beta = 3$ it is infinite.

Let

$$l_+(t) = \frac{\lambda_+ t\, l(t)}{\beta\psi(\lambda_+)}, \qquad B_n = l_+^{(-1)}\left(\frac{1}{n}\right),$$

where $l_+^{(-1)}(1/n)$ is the value at the point $1/n$ of the function inverse to l_+.

Theorem 2.4.4. *Let ξ be a non-lattice random variable, $\mathbf{F} \in \mathcal{ER}$ and the condition (c) be satisfied. If $\Delta_n \to 0$ sufficiently slowly as $n \to \infty$ then, for $y = x - \alpha_+ n = o(n)$,*

$$\mathbf{P}\big(S_n \in \Delta_n[x)\big) = \frac{\Delta_n e^{-n\Lambda(\alpha_+) - \lambda_+ y}}{B_n}\left(\phi^{(\beta-1,1)}\left(\frac{y}{B_n}\right) + o(1)\right), \tag{2.4.6}$$

where $\phi^{(\beta-1,1)}$ is the density of the stable law $\mathbf{\Phi}_{(\beta-1,1)}$ with parameters $\beta - 1$, 1 and the remainder $o(1)$ is uniform in y.

It will become clear from the proof that it is essentially impossible to study the probabilities of large deviations in the case $\alpha_+ < \infty$, $\psi''(\lambda_+) = \infty$ outside of the class $\mathcal{ER}$, just as it is impossible to find theorems about the limit distribution of S_n in the case $\mathbf{D}\xi = \infty$ without the condition that the tails of $\mathbf{F}$ vary regularly.

Proof of Theorem 2.4.4. The condition (c) implies that $\alpha_+ = \mathbf{E}^{\alpha_+}\xi < \infty$, $\mathbf{D}^{\alpha_+}\xi = \infty$. Apply Theorem 2.1.1A. If $\Delta_n \to 0$ slowly enough, we obtain, as in the proof of Theorem 2.4.1, relation (2.4.4). But now, in contrast with Theorem 2.4.1, in order to compute the probability on the right-hand side of (2.4.4) we have to use integro-local theorem 8.8.3 in [39] on convergence to a stable law. In our case, by the properties of regularly varying functions,

$$\begin{aligned}\mathbf{P}\big({}^{\alpha_+}\xi \geqslant t\big) &= -\frac{1}{\psi(\lambda_+)}\int_t^\infty e^{\lambda_+ u} dF_+(u) = \frac{1}{\psi(\lambda_+)}\int_t^\infty \big(\lambda_+ l(u)du - dl(u)\big)\\ &= \frac{\lambda_+}{\beta\psi(\lambda_+)}\, t^{-\beta+1} L_+(t) \sim l_+(t),\end{aligned} \tag{2.4.7}$$

where $L_+(t) \sim L(t)$ is a regularly varying function. Moreover, the left tails of the distribution ${}^{\alpha_+}\mathbf{F}$ decrease at least exponentially fast. In view of the results in § 8.8 in [39], this means that for $B_n = l_+^{(-1)}(1/n)$ we have convergence of the distributions of $(S_n^{(\lambda_+)} - \alpha_+ n)/B_n$ to the stable law $\mathbf{\Phi}_{\beta-1,1}$ with parameters $\beta - 1 \in [1,2]$ and 1. It remains to use the representation (2.4.4) and Theorem 8.8.3 in [39], according to which for $\Delta_n \to 0$ sufficiently slowly we have

$$\mathbf{P}\big(S_n^{(\lambda_+)} - \alpha_+ n \in \Delta_n[y)\big) = \frac{\Delta_n}{B_n}\phi^{(\beta-1,1)}\left(\frac{y}{B_n}\right) + o\left(\frac{1}{B_n}\right)$$

uniformly in y. □

From Theorems 2.4.1 and 2.4.4, one can easily obtain integral theorems about the asymptotics of $\mathbf{P}\,(S_n \geqslant x)$. These asymptotics will be determined by the right-hand sides of (2.4.2) and (2.4.6), when the factor $(1 - e^{\lambda_+})^{-1}$ is added and the factor Δ_n removed.

The results of this section can be also found in [39] and [42].

2.4.4 The multidimensional case. Large deviations near the boundary of the Cramér zone when the vector $A'(\lambda_+)$ and the matrix $A''(\lambda_+)$ exist

Let $\alpha_+ = A'(\lambda_+) \in \partial \mathcal{A}'$, where $\lambda_+ = \lambda(\alpha_+)$ is the point from $\partial \mathcal{A}$ corresponding to α_+. Under the conditions of this section, when the covariance matrix $\sigma^2(\alpha_+)$ of the random vector $\xi^{(\lambda_+)}$ exists the following result holds.

Theorem 2.4.5. *Suppose that the matrix $A''(\lambda_+)$ exists. Then, in the non-lattice case* **[R]**, *for $y := x - n\alpha_+$, $|y| = o(n)$ and $0 < \Delta = \Delta_n \to 0$ sufficiently slowly as $n \to \infty$, we have*

$$\mathbf{P}\big(S_n \in \Delta[x)\big) = \frac{\Delta^d}{(2\pi n)^{d/2}|\sigma(\alpha_+)|} e^{-n\Lambda(\alpha_+) - \langle \lambda_+, y\rangle} [e^{-y\Lambda''(\alpha_+)y^T/2n} + o(1)].$$

In the arithmetic case **[Z]**, *for integer x and $n \to \infty$,*

$$\mathbf{P}(S_n = x) = \frac{1}{(2\pi n)^{d/2}|\sigma(\alpha_+)|} e^{-n\Lambda(\alpha_+) - \langle \lambda_+, y\rangle} [e^{-y\Lambda''(\alpha_+)y^T/2n} + o(1)].$$

The remainders $o(1)$ in these relations are uniform in y, $|y| = o(n)$.

In the one-dimensional case, $d = 1$, Theorem 2.4.5 coincides with Theorem 2.4.1. In the multidimensional case the proof essentially remains the same, but in the case **[R]** it is based on the multidimensional integro-local Stone theorem (see [175] and [174]), and on Gnedenko's theorem in the case **[Z]** (see subsection 1.5.2, §§ 49, 50 of the monograph [96], [163], Theorem 4.2.1 in [102] and Theorem 8.4.1 in [12]).

Clearly, from Theorem 2.4.5 one can easily obtain integro-local theorems for the probabilities that S_n hits the cube $\Delta[x)$ for $|x - \alpha_+ n| = O(\sqrt{n})$ and *any fixed* $\Delta > 0$ (see the derivation of Theorem 2.2.3 from Theorem 2.2.1).

2.5 Integro-local theorems outside the Cramér zone

2.5.1 The one-dimensional case. The classes of distributions $\mathcal{ER}$ with parameter $\beta < -3$

In this section we assume that $\alpha_+ < \infty$ and $\alpha = x/n > \alpha_+$ (the cases $\alpha_+ = \infty$ and $\alpha \leqslant \alpha_+ + o(1)$ as $n \to \infty$ were considered in the previous sections). The definition of the classes $\mathcal{ER}$ and $\mathcal{ES}e$ was given in subsection 2.3.4.

So, suppose the parameter β of the distribution from $\mathcal{ER}$ is less than -3 (see (2.3.23), (2.3.17)). In this case $\psi''(\lambda_+) < \infty$ and the variance

$$b^2 := \mathbf{D}\xi^{(\lambda_+)} = \frac{\psi''(\lambda_+)}{\psi(\lambda_+)} - \left(\frac{\psi'(\lambda_+)}{\psi(\lambda_+)}\right)^2$$

is finite. For such a β, Theorem 2.4.5 (regarding the zone of deviations close to the boundary $\partial\mathcal{A}'$) can be supplemented by the following result. Let $\mathbf{I}(t)$ denote the indicator of the set $[0,\infty)$:

$$\mathbf{I}(t) = \begin{cases} 1, & \text{if } \ t \geqslant 0, \\ 0, & \text{if } \ t < 0. \end{cases}$$

Theorem 2.5.1. *Let* $\mathbf{F} \in \mathcal{ER}$ *for* $\beta < -3$. *Then, for any fixed* $C < \infty$ *in the region* $y := x - n\alpha_+ \geqslant -C\sqrt{n}$, *the following is true.*

(i) *In the non-lattice case* **[R]**, *for any* $\Delta = \Delta_n > 0$ *tending to* 0 *sufficiently slowly as* $n \to \infty$, *we have*

$$\mathbf{P}\left(S_n \in \Delta[x)\right) \sim \Delta e^{-n\Lambda(\alpha_+)-\lambda_+ y}\left[\frac{1}{b\sqrt{2\pi n}}\, e^{-y^2/(2b^2 n)} + n\frac{\lambda_+}{\psi(\lambda_+)}\, l(y)\mathbf{I}(y-\sqrt{n})\right]. \tag{2.5.1}$$

(ii) *In the arithmetic case* **[Z]**, *for* $x \in \mathbb{Z}$ *we have*

$$\mathbf{P}\left(S_n = x\right) \sim e^{-n\Lambda(\alpha_+)-\lambda_+ y}\left[\frac{1}{b\sqrt{2\pi n}}\, e^{-y^2/(2b^2 n)} + n\frac{1-e^{-\lambda_+}}{\psi(\lambda_+)}\, l(y)\mathbf{I}(y-\sqrt{n})\right].$$

Corollary 2.5.2. *Suppose that the conditions of Theorem 2.5.1 are satisfied. Then, in the non-lattice case* **[R]**, *for any* $\Delta \geqslant \Delta_n > 0$, *where* Δ_n *tends to* 0 *sufficiently slowly as* $n \to \infty$, *we have*

$$\mathbf{P}\left(S_n \in \Delta[x)\right)$$
$$\sim (1-e^{-\lambda_+\Delta})e^{-n\Lambda(\alpha_+)-\lambda_+ y}\left[\frac{1}{\lambda_+ b\sqrt{2\pi n}} e^{-y^2/(2b^2 n)} + \frac{n}{\psi(\lambda_+)} l(y)\mathbf{I}(y-\sqrt{n})\right].$$

In particular, if $\Delta = \infty$ *then*

$$\mathbf{P}\left(S_n \geqslant x\right) \sim e^{-n\Lambda(\alpha_+)-\lambda_+ y}\left[\frac{1}{\lambda_+ b\sqrt{2\pi n}} e^{-y^2/(2b^2 n)} + \frac{n}{\psi(\lambda_+)} l(y)\mathbf{I}(y-\sqrt{n})\right].$$

In the arithmetic case **[Z]**, *for integer* x *we have*

$$\mathbf{P}\left(S_n \geqslant x\right)$$
$$\sim e^{-n\Lambda(\alpha_+)-\lambda_+ y}\left[\frac{1}{(1-e^{-\lambda_+})b\sqrt{2\pi n}} e^{-y^2/(2b^2 n)} + \frac{n}{\psi(\lambda_+)} l(y)\mathbf{I}(y-\sqrt{n})\right].$$

Corollary 2.5.2 almost obviously follows from Theorem 2.5.1 (see the derivation of Corollary 6.1.1 from Theorem 6.1.4 and Corollary 6.2.2 from Theorem 6.2.2 in [42]).

Remark 2.5.3. As is the case for many other statements, one can see from Theorem 2.5.1 and Corollary 2.5.2 that under the conditions of these statements the asymptotics of $\mathbf{P}\left(S_n \in \Delta[x)\right)$ can be obtained from the asymptotics of $\mathbf{P}\,(S_n \geqslant x)$, i.e. an integro-local theorem can be obtained from a local one.

Proof of Theorem 2.5.1. Consider the Cramér transform of the distribution $\mathbf{F}$ at the point λ_+:

$$\mathbf{F}^{(\lambda_+)}(dv) := \mathbf{P}(\xi^{(\lambda_+)} \in dv) = \frac{e^{\lambda_+ v}\mathbf{P}(\xi \in dv)}{\psi(\lambda_+)}.$$

The corresponding Laplace transform is

$$\mathbf{E}e^{\lambda\xi^{(\lambda_+)}} = \frac{\psi(\lambda_+ + \lambda)}{\psi(\lambda_+)}, \quad \lambda \in \mathbb{R}^1.$$

Considering the same transform for the distribution of S_n, we obtain

$$\mathbf{P}\,(S_n \in dv) = \psi^n(\lambda_+)e^{-\lambda_+ v}\mathbf{P}(S_n^{(\lambda_+)} \in dv), \tag{2.5.2}$$

where, as we have already seen, $S_n^{(\lambda_+)} \stackrel{d}{=} \xi_1^{(\lambda_+)} + \cdots + \xi_n^{(\lambda_+)}$, i.e. one can treat $S_n^{(\lambda_+)}$ as a sum of independent copies of the random variable $\xi^{(\lambda_+)}$. The first two factors on the right-hand side of (2.5.2) can be written as

$$e^{-n(\lambda_+\alpha_+ - \ln\psi(\lambda_+)) - \lambda_+(v - n\alpha_+)} = \exp\left\{e^{-n\Lambda(\alpha_+) - \lambda_+(v - n\alpha_+)}\right\},$$

and, therefore,

$$\mathbf{P}\left(S_n \in dv\right) = \exp\left\{e^{-n\Lambda(\alpha_+) - \lambda_+(v - n\alpha_+)}\,\mathbf{P}\left(S_n^{(\lambda_+)} \in dv\right)\right\}. \tag{2.5.3}$$

In the case **[R]**, consider an arbitrary sequence $\Delta_n \to 0$ as $n \to \infty$. Then, integrating (2.5.3) in the region $v \in \Delta_n[x) = [x, x + \Delta_n)$, we obtain

$$\begin{aligned}\mathbf{P}\left(S_n \in \Delta_n[x)\right) &= e^{-n\Lambda(\alpha_+)} \int_{\Delta_n[x)} e^{-\lambda_+(v - n\alpha_+)}\mathbf{P}\left(S_n^{(\lambda_+)} \in dv\right) \\ &\sim e^{-n\Lambda(\alpha_+) - \lambda_+(x - n\alpha_+)}\mathbf{P}\left(S_n^{(\lambda_+)} \in \Delta_n[x)\right).\end{aligned} \tag{2.5.4}$$

In the arithmetic case **[Z]**, for integer x the relation (2.5.4) becomes the following:

$$\mathbf{P}\left(S_n = x\right) = e^{-n\Lambda(\alpha_+) - \lambda_+(x - n\alpha_+)}\mathbf{P}\left(S_n^{(\lambda_+)} = x\right). \tag{2.5.5}$$

To compute the right-hand sides in (2.5.4) (2.5.5) we will need integro-local theorems for the sums $S_n^{(\lambda_+)} = \xi_1^{(\lambda_+)} + \cdots + \xi_n^{(\lambda_+)}$. In the non-lattice case **[R]**, the tail $F_+^{(\lambda_+)}(t) := \mathbf{P}\,(\xi^{(\lambda_+)} \geqslant t)$ has the form

$$\begin{aligned}F_+^{(\lambda_+)}(t) &= \frac{\mathbf{E}(e^{\lambda_+\xi};\ \xi \geqslant t)}{\psi(\lambda_+)} \\ &\frac{-\int_t^\infty e^{\lambda_+ u}d(e^{-\lambda_+ u}l(u))}{\psi(\lambda_+)} = \frac{l(t)}{\psi(\lambda_+)} + \frac{\lambda_+\int_t^\infty l(u)du}{\psi(\lambda_+)},\end{aligned}$$

where for $\beta < -1$ and $t \to \infty$

$$\int_t^\infty l(u)du \sim \frac{1}{|\beta + 1|}t^{\beta+1}L(t)$$

(see e.g. Theorem 1.1.3 in [42]). Thus, in the non-lattice case, the function $F_+^{(\lambda_+)}(t)$ has the form

$$F_+^{(\lambda_+)}(t) = t^\gamma L_+(t), \tag{2.5.6}$$

where $\gamma := \beta + 1 < -2$ and $L_+(t) \sim \dfrac{\lambda_+}{\psi(\lambda_+)|\gamma|} L(t)$ is an s.v.f.

In the arithmetic case **[Z]**, for integer t we have

$$\begin{aligned}\mathbf{P}(\xi^{(\lambda_+)} = t) = \frac{e^{\lambda_+ t}\mathbf{P}(\xi = t)}{\psi(\lambda_+)} &= \frac{e^{\lambda_+ t}[\mathbf{P}(\xi \geqslant t) - \mathbf{P}(\xi \geqslant t+1)]}{\psi(\lambda_+)} \\ &= \frac{l(t) - e^{-\lambda_+} l(t+1)}{\psi(\lambda_+)} \sim \frac{1 - e^{-\lambda_+}}{\psi(\lambda_+)} l(t). \end{aligned} \tag{2.5.7}$$

Therefore, in this case,

$$F_+^{(\lambda_+)}(t) := \mathbf{P}(\xi^{(\lambda_+)} \geqslant t) = \sum_{k \geqslant t} \frac{l(k) - e^{-\lambda_+} l(k+1)}{\psi(\lambda_+)} = t^\gamma L_+(t), \tag{2.5.8}$$

where $\gamma := \beta + 1 < -2$ and $L_+(t) \sim \dfrac{1 - e^{-\lambda_+}}{\psi(\lambda_+)|\gamma|} L(t)$ is an s.v.f.

By (2.5.6) and (2.5.8), in the case $\mathbf{F} \in \mathcal{ER}$, $\beta < -3$, the distribution of $\xi^{(\lambda_+)}$ belongs to the class $\mathcal{R}$ with exponent $\gamma = \beta + 1 < -2$, and the left tail of the distribution of the random variable $\xi^{(\lambda_+)}$ decreases exponentially fast. Hence, $b^2 = \mathbf{D}\xi^{(\lambda_+)} < \infty$.

Moreover, the representation (2.5.6) implies that *in the non-lattice case for any fixed $\Delta > 0$ as $t \to \infty$,*

$$\mathbf{P}\big(\xi^{(\lambda_+)} \in \Delta[t)\big) = \frac{\lambda_+}{\psi(\lambda_+)} \Delta l(t)\big(1 + o(1)\big). \tag{2.5.9}$$

For an arithmetic distribution $\mathbf{F} \in \mathcal{ER}$, the analogue of (2.5.9) is the relation (see (2.5.7))

$$\mathbf{P}(\xi^{(\lambda_+)} = t) = \frac{1 - e^{-\lambda_+}}{\psi(\lambda_+)} l(t)(1 + o(1)),$$

as $t \to \infty$, $t \in \mathbb{Z}$.

When these relations hold, an *integro-local theorem on the 'whole half-line' holds* (see Theorems 2.1 and 2.2 in [133]) which implies the following integro-local lemma for the sums $S_n^{(\lambda_+)} - n\alpha_+$.

Lemma 2.5.4. *Suppose that* $\mathbf{F} \in \mathcal{ER}$ *for* $\beta < -3$. *Then, in the non-lattice case* **[R]**, *for any fixed* $\Delta > 0$ *and* $C < \infty$ *in the region* $y := x - n\alpha_+ \geqslant -C\sqrt{n}$, *we have*

$$\mathbf{P}\big(S_n^{(\lambda_+)} - n\alpha_+ \in \Delta[y)\big) \sim \Delta\left[\frac{1}{b\sqrt{2\pi n}} e^{-y^2/(2b^2 n)} + n\frac{\lambda_+}{\psi(\lambda_+)} l(y)\mathbf{I}(y - \sqrt{n})\right]. \tag{2.5.10}$$

In the arithmetic case **[Z]**, *formula* (2.5.10) *remains valid for integer x and* $\Delta = 1$ *if one replaces the factor* λ_+ *on the right-hand side of* (2.5.10) *with* $(1 - e^{-\lambda_+})$.

Let us return to the proof of Theorem 2.5.1. In the case **[Z]**, the claim follows directly from (2.5.5) and Lemma 2.5.4. In the case **[R]**, observe that since Lemma 2.5.4 is true for any fixed $\Delta > 0$, then it will remain true for a sequence $\Delta = \Delta_n$ converging to 0 as $n \to \infty$ sufficiently slowly. Applying this to (2.5.4), we obtain (2.5.1). Theorem 2.5.1 is proved. □

2.5.2 The class $\mathcal{ER}$ with parameter $\beta \in (-3, -2)$

If $\beta \in (-3, -2)$ in (2.3.23) and (2.3.17), then $\psi''(\lambda_+) = \infty$, $\alpha_+ < \infty$ (recall that $\alpha_+ = \infty$ when $\beta > -2$). In this case we do not have theorems for 'the whole half-line' (cf. Theorems 2.1, 2.2 in [133]). However we have integro-local theorems for the distribution of $S_n^{(\lambda_+)} - n\alpha_+$ in the region $y = x - n\alpha_+ > n^{1/\gamma}$ for any fixed $\gamma < -\beta$. Namely, in view of Theorem 3.7.1 in [42] and relation (2.5.9), the next lemma follows.

Lemma 2.5.5. *Let* $\mathbf{F} \in \mathcal{ER}$ *for* $\beta \in (-3, -2)$. *Then, in the non-lattice case* **[R]**, *for any fixed* $\Delta > 0$ *in the region* $y = x - n\alpha_+ > n^{1/\gamma}$ *and any fixed* $\gamma < \beta$, *one has*

$$\mathbf{P}\left(S_n^{(\lambda_+)} - \alpha_+ n \in \Delta[y)\right) = \Delta n \frac{\lambda_+}{\psi(\lambda_+)} l(y)\left(1 + o(1)\right) \quad as \quad N \to \infty,$$

where the remainder $o(1)$ *is uniform in y, n and* $\Delta \in [\Delta_1, \Delta_2]$ *such that*

$$y \geqslant \max\{N, n^{1/\gamma}\}, \quad \Delta_1 \geqslant y^{-1}, \quad \Delta_2 \leqslant y\varepsilon_N$$

for any fixed sequence $\varepsilon_N \to 0$ *as* $N \to \infty$.

A similar claim is true in the arithmetic case.

Using Lemma 2.5.5, by the same arguments as those used in the previous section, we obtain the following supplement to Theorem 2.4.4.

Theorem 2.5.6. *Let* $\mathbf{F} \in \mathcal{ER}$ *for* $\beta \in (-3, -2)$. *Then, in the region* $y = x - \alpha_+ n \geqslant Nn^{1/\gamma}$, *for any fixed* $\gamma < -\beta$ *and any sequence* $N \to \infty$, *the following claims are true.*

(i) *In the non-lattice case* **[R]**,

$$\mathbf{P}\left(S_n \in \Delta[x)\right) = \frac{\Delta n \lambda_+}{\psi(\lambda_+)} l(y) e^{-n\Lambda(\alpha_+) - \lambda_+ y}\left(1 + o(1)\right) \quad as \quad N \to \infty,$$

where the remainder $o(1)$ *is uniform in x, n and* $\Delta \in [\Delta_1, \Delta_2]$ *such that*

$$y \geqslant max(N, n^{1/\gamma}), \quad \Delta_1 \geqslant y^{-1}, \quad \Delta_2 \leqslant y\varepsilon_N$$

for any sequence $\varepsilon_N \to 0$ *as* $N \to \infty$.

(ii) *In the arithmetic case* **[Z]**, *for* $x \in \mathbb{Z}$,

$$\mathbf{P}(S_n = x) = \frac{n(1 - e^{-\lambda_+})}{\psi(\lambda_+)}\, l(y) e^{-n\Lambda(\alpha_+) - \lambda_+ y}\big(1 + o(1)\big) \quad as \quad N \to \infty,$$

where the remainder $o(1)$ *is uniform in* $y \geqslant \max(N, n^{1/\gamma})$.

Note that there is no condition $n \to \infty$ in Theorem 2.5.6, so it holds also for $y \to \infty$ and any fixed n.

Clearly, Theorems 2.5.1 and 2.5.6, along with the zone of the 'usual' large deviations $x \sim cn$, $n \to \infty$, also cover the zone of very large deviations $x \gg n$, $n \to \infty$.

From Theorem 5.1 in [58] one can easily obtain an analogue of Theorem 2.5.1 for the class of distributions $\mathbf{F} \in \mathcal{ES}e''$ for $\beta \in (0, 1)$.

2.6 Supplement to section 2.5. The multidimensional case. The class of distributions $\mathcal{ER}$

In the multidimensional case we confine our analysis to the class of distributions $\mathcal{ER}$ (an analogue of the class $\mathcal{ER}$ in the one-dimensional case). We will consider a case which is alternative (in the known sense) to Theorem 2.4.5 when the deviation of the point $\alpha = x/n$ from the set $\mathcal{A}'$ is much greater than $\sqrt{\ln n/n}$.

In order to formulate the main results, we will need some preliminary facts. First we will find out at which point the Cramér transform should be applied when we consider deviations $x = \alpha n$, $\alpha \notin \mathcal{A}'$, and we will find the form of the deviation function for such an α.

To simplify the exposition, we will impose several 'excessive' restrictions on the distribution $\mathbf{F}$. A clarification about the possibility of the weakening or the complete removal of these restrictions can be found in [59]. One of these restrictions is the following condition.

[A] *The set* $\mathcal{A}$, *where the function* $\psi(\lambda) = \mathbf{E}e^{\langle\lambda,\xi\rangle}$ *(or* $A(\lambda) = \ln\psi(\lambda)$*) is finite, is bounded, and the vector-valued function*

$$\psi'(\lambda) = \mathbf{E}e^{\langle\lambda,\xi\rangle}\xi$$

(or $A'(\lambda) = \psi'(\lambda)/\psi(\lambda)$*) is uniformly bounded and continuous in* $\mathcal{A}$.

Note that the set $\mathcal{A}$ is always convex, and under the condition **[A]** it is closed as well; the set $\mathcal{A}' := \{\alpha = A'(\lambda) : \lambda \in \mathcal{A}\}$ is also closed and bounded.

In fact, if one considers deviations x only in some given direction then the condition **[A]** is not required in its full generality; only the assumption that the intersection of $\mathcal{A}$ with some cone is closed and bounded is needed.

Recall that by $\lambda(\alpha)$ we denote the point from $\mathcal{A}$ where the following is attained:

$$\sup_{\lambda \in \mathcal{A}}\{\langle\lambda, \alpha\rangle - A(\lambda)\} = \langle\lambda(\alpha), \alpha\rangle - A(\lambda(\alpha)) = \Lambda(\alpha). \qquad (2.6.1)$$

Since the maximum of a continuous function on a compact set is attained at some point of this set, the point $\lambda(\alpha)$ exists and is unique owing to the strong convexity of the function $A(\lambda)$. For $\alpha \in \mathcal{A}'$, the value of $\lambda(\alpha)$ is determined as the solution of the equation

$$\alpha = \alpha(\lambda) := A'(\lambda). \tag{2.6.2}$$

This solution (i.e. the inverse vector function of $A'(\lambda)$) is uniquely defined and has the form (see section 1.2)

$$\lambda(\alpha) = \Lambda'(\alpha), \tag{2.6.3}$$

so that $A'(\Lambda'(\alpha)) \equiv \alpha$. Also, the one-to-one mappings (2.6.2), (2.6.3) map the interior $(\mathcal{A})$ (the region of analyticity of the function $A(\lambda)$) to the interior $(\mathcal{A}')$ (the region of analyticity of the function $\Lambda(\alpha)$) and vice versa. This implies that if **[A]** holds then the functions (2.6.2) and (2.6.3) also provide the same one-to-one correspondence between the boundaries $\partial\mathcal{A}$ and $\partial\mathcal{A}'$. Recall that by λ_+ and α_+ we denote points of the boundaries $\partial\mathcal{A}$ and $\partial\mathcal{A}'$ respectively, and let

$$\alpha_+(\lambda_+) := A'(\lambda_+), \quad \lambda_+(\alpha_+) := \Lambda'(\alpha_+) \tag{2.6.4}$$

(where the derivatives are 'interior').

Now we describe the behaviour of the deviation function $\Lambda(\alpha)$ outside the set $\mathcal{A}'$.

Lemma 2.6.1. *Suppose that the condition* **[A]** *holds. Then*

(i) *For any $\alpha \notin \mathcal{A}'$, there exists a unique point $\lambda_+[\alpha] \in \partial\mathcal{A}$ (or a point $\alpha_+[\alpha] \in \partial\mathcal{A}'$). By* (2.6.2)–(2.6.4), *the values $\lambda_+[\alpha]$ and $\alpha_+[\alpha]$ are uniquely determine each other according to the equalities*

$$\alpha_+[\alpha] = \alpha_+(\lambda_+[\alpha]), \quad \lambda_+[\alpha] = \lambda_+(\alpha_+[\alpha])), \tag{2.6.5}$$

which are such that

$$\Lambda(\alpha) = \Lambda(\alpha_+[\alpha]) + \langle \lambda_+[\alpha], \alpha - \alpha_+[\alpha] \rangle. \tag{2.6.6}$$

(ii) *For $\alpha \notin \mathcal{A}'$ and $\lambda_+ \in \partial\mathcal{A}$, in the obvious inequality*

$$\sup_{\lambda \in \mathcal{A}} \langle \lambda, \alpha - A'(\lambda_+) \rangle \geqslant \langle \lambda_+, \alpha - A'(\lambda_+) \rangle$$

one has the case of equality,

$$\sup_{\lambda \in \mathcal{A}} \langle \lambda, \alpha - A'(\lambda_+) \rangle = \langle \lambda_+, \alpha - A'(\lambda_+) \rangle, \tag{2.6.7}$$

if and only if $\lambda_+ = \lambda_+[\alpha]$.

The relations (2.6.5) establish a connection between $\alpha_+[\alpha]$ and $\lambda_+[\alpha]$, while the relation (2.6.7) allows us to construct an algorithm to find these values (see also (2.6.13) below). The relation (2.6.7) means that the point $\lambda_+[\alpha]$ lying on the boundary $\partial\mathcal{A}$ is 'extreme in the set $\mathcal{A}$ in the direction $\mathrm{e}[\alpha] := \mathrm{e}(\alpha - \alpha_+[\alpha])$', where $\mathrm{e}(v) := v/|v|$. In other words, the hyperplane orthogonal to $\mathrm{e}[\alpha]$ is tangent to $\partial\mathcal{A}$ at the point $\lambda_+[\alpha]$. If the boundary $\partial\mathcal{A}$ of the convex set $\mathcal{A}$ at the point $\lambda_+[\alpha]$ has a unit normal e (directed outwards from the set $\mathcal{A}$), then $\mathrm{e} = \mathrm{e}[\alpha]$.

It follows from the above that for the points

$$\alpha_c := \alpha + c(\alpha - \alpha_+[\alpha]) \notin \mathcal{A}'$$

which constitute the ray $\mathcal{L}_\alpha := \{\alpha_c : c > -1\}$, one has

$$\begin{aligned} \alpha_+[\alpha_c] = \alpha_+[\alpha], \qquad \lambda_+[\alpha_c] &= \Lambda'(\alpha_+[\alpha_c]) = \Lambda'(\alpha_+[\alpha]) = \alpha_+[\alpha], \\ \mathrm{e}[\alpha_c] &= \mathrm{e}(\alpha_c - \alpha_+[\alpha]) = \mathrm{e}(\alpha + c(\alpha - \alpha_+[\alpha]) - \alpha_+[\alpha]) \\ &= \mathrm{e}(\alpha(1+c) - \alpha_+[\alpha](1+c)) = \mathrm{e}[\alpha], \end{aligned}$$

so that for the points α_c on this ray the direction $\mathrm{e}[\alpha_c]$ and the points $\alpha_+[\alpha_c]$, $\lambda_+[\alpha_c]$ remain the same. Also, the two rays $\mathcal{L}_{\alpha'}$, $\mathcal{L}_{\alpha''}$, constructed by the two points $\alpha', \alpha'' \notin \mathcal{A}'$ as specified above, either coincide or do not intersect.

The deviation function $\Lambda(\alpha)$ outside the region $\mathcal{A}'$ behaves linearly along each ray $\mathcal{L}_\alpha$ (see (2.6.6); as noted above, the values $\alpha_+[\alpha_c]$ and $\lambda_+[\alpha_c]$ for the points α_c on the ray $\mathcal{L}_\alpha$ remain the same).

Remark 2.6.2. If, in addition to the condition **[A]** one requires that the set $\mathcal{A}$ is strictly convex and that the boundary $\partial\mathcal{A}$ has a normal at each point (i.e. there are no 'kink' points; these additional conditions are not satisfied, for example, in the case when the coordinates of the vector ξ are independent), then the second statement of Lemma 2.6.1 can be supplemented by the following claim.

(iiA) *For each $\alpha \notin \mathcal{A}'$, the value of* $\sup_{\lambda\in\mathcal{A}}\langle\lambda, \alpha - \alpha_+[\alpha]\rangle$ *is attained at a unique point $\lambda = \lambda_+[\alpha]$ such that*

$$\sup_{\lambda\in\mathcal{A}}\langle\lambda, \alpha - \alpha_+[\alpha]\rangle = \langle\lambda_+[\alpha], \alpha - \alpha_+[\alpha]\rangle. \tag{2.6.8}$$

Lemma 2.6.1 and the statement (iiA) allow us to establish a one-to-one correspondence between each of the sets $\partial\mathcal{A}$, $\partial\mathcal{A}'$ and the field of rays $\mathcal{L}_\alpha$. Indeed, each ray $\mathcal{L}_\alpha$ is determined by its point of origin $\alpha_+[\alpha]$ and direction $\mathrm{e}[\alpha]$. However (see also (iiA)), the direction $\mathrm{e}[\alpha]$ uniquely determines the point $\lambda_+[\alpha]$ (the unique extreme point of the set $\mathcal{A}$ in this direction) and hence the point $\alpha_+[\alpha]$. Thus, each ray $\mathcal{L}_\alpha$ is completely characterized by a direction $\mathrm{e}[\alpha]$ which is a point on the unit sphere

$$\mathrm{E} = \{\mathrm{e} : |\mathrm{e}| = 1\}.$$

Further, as we have already observed, to each direction $\mathrm{e} \in \mathrm{E}$ corresponds the unique 'extreme point in this direction', $\lambda_+ \in \partial\mathcal{A}$, of the set $\mathcal{A}$. This proves the one-to-one correspondence between $\partial\mathcal{A}$ and E. The same can be said about the one-to-one correspondence between $\partial\mathcal{A}'$ and E.

Statement (ii) of Lemma 2.6.1 also implies that the random vector $\xi^{(\lambda_+[\alpha])}$ 'does not satisfy Cramér's condition in the direction e$[\alpha]$' in the following sense: *the Laplace transform of the distribution* $\xi^{(\lambda_+[\alpha])}$*, which is*

$$\frac{\psi(\lambda+\lambda_+[\alpha])}{\psi(\lambda_+[\alpha])},$$

equals ∞ *at any point* λ *such that* $\langle\lambda, \mathrm{e}[\alpha]\rangle > 0$.

Now we return to the study of the asymptotics of the probability $\mathbf{P}\big(S_n \in \Delta[x)\big)$. Consider deviations $x = n\alpha$, $\alpha \notin \mathcal{A}'$. Apply the Cramér transform, in the same way as above, to the vectors ξ and S_n, and choose the point $\lambda_+ = \lambda_+[\alpha]$ as the parameter λ. Then, as before, for $\Delta = \Delta_n > 0$ converging to zero sufficiently slowly as $n \to \infty$, we obtain

$$\mathbf{P}\big(S_n \in \Delta[x)\big) \sim e^{-n\Lambda(\alpha)}\mathbf{P}\big(S_n^{(\lambda_+)} - n\alpha_+[\alpha] \in \Delta[y)\big), \tag{2.6.9}$$

where $y := x - n\alpha_+[\alpha] = n(\alpha - \alpha_+[\alpha])$ and

$$S_n^{(\lambda_+)} \underset{d}{=} \xi_1^{(\lambda_+)} + \cdots + \xi_n^{(\lambda_+)}$$

can be considered as the sum of independent copies of the random vector $\xi^{(\lambda_+)}$ with distribution not satisfying Cramér's condition in the direction $\alpha - \alpha_+[\alpha]$.

In order to find the asymptotics of the right-hand side of (2.6.9), we will need to impose additional conditions on the regularity of variation of the distribution $\mathbf{F}$ 'in the direction $\alpha - \alpha_+[\alpha]$'. Here again, in order to simplify the exposition, we introduce a simplifying assumption that can be weakened. Namely, we will assume that the following condition is satisfied.

$[\mathcal{ER}]_1$ *For all t,* $|t| \geqslant N$*, and sufficiently large N, the distribution* $\mathbf{F}$ *has density* $f(t)$ *given by*

$$f(t) = g\big(\mathrm{e}(t)\big)l\big(|t|\big)e^{-h(t)}, \qquad \mathrm{e}(t) := \frac{t}{|t|}, \tag{2.6.10}$$

where the function $g(\mathrm{e})$ *is positive and continuous on* E, $l(v) := v^{\beta}L(v)$ *and the function* $L(v)$ *is an s.v.f. as* $v \to \infty$*; the convex function* $h(t)$ *can be represented as*

$$h(t) = \rho(\mathrm{e}(t))|t|,$$

where the function $\rho(\mathrm{e})$ *is positive and continuous on* E.

The class of such distributions will be denoted by $\mathcal{ER}$ (in [181], the class of absolutely continuous distributions, which is close to $\mathcal{ER}$, is called the class of gamma-like distributions; see Chapter 3.1 in [181]).

To study the asymptotics of the probability on the right-hand side of (2.6.9), we will need the following additional conditions on the functions $l(v)$ and $h(t)$ which determine the class $\mathcal{ER}$.

$[\mathcal{ER}]_2$ *The condition* $[\mathcal{ER}]_1$ *is satisfied, the function* $l(v)$ *satisfies the condition*

$$\int_1^{\infty} v^{(d-1)/2+1}l(v)dv < \infty \tag{2.6.11}$$

and the convex function $h(t)$ for all $t \neq 0$ is twice continuously differentiable; the symmetric non-negative definite matrix

$$h''(\mathrm{e}) := h''(t)|_{t=\mathrm{e}}$$

for any $\mathrm{e} \in \mathrm{E}$ *has rank* $d-1$.

The last condition, on the rank of the matrix $h''(\mathrm{e})$, is equivalent to the section of the cone $z = h(t)$ by the hyperplane $z = 1$ (i.e. the set $\{t : h(t) \leqslant 1\}$) being strictly convex and its boundary having a non-zero finite curvature at each point. For the validity of the inequality (2.6.11), it is enough to have $\beta < -(d+1)/2 - 1$.

Lemma 2.6.3. *Suppose that the condition* $[\mathcal{ER}]_2$ *holds. Then:*

(i) *For any* $t \neq 0$,

$$h(t) = \langle h'(t), t\rangle, \quad h'(t) = h'(\mathrm{e}(t)), \quad h''(t) = \frac{1}{|t|} h''(\mathrm{e}(t)), \qquad (2.6.12)$$

where $h'(\mathrm{e}) := h'(t)|_{t=\mathrm{e}}$, $h''(\mathrm{e}) := h''(t)|_{t=\mathrm{e}}$.

(ii) *The condition* **[A]** *holds,*

$$\partial\mathcal{A} = \{\lambda_+ = h'(\mathrm{e}) : \mathrm{e} \in \mathrm{E}\}$$

and, for any point $\alpha \notin \mathcal{A}'$, *the pair of points* $\lambda_+ = \lambda_+[\alpha]$, $\alpha_+ = \alpha_+[\alpha]$ *(see Lemma* 2.6.1*) is the unique solution in* $\partial\mathcal{A} \times \partial\mathcal{A}'$ *of the system of equations*

$$h'(\alpha - \alpha_+) = \lambda_+, \qquad A'(\lambda_+) = \alpha_+, \qquad (2.6.13)$$

where the vector $\alpha - \alpha_+[\alpha]$ *is the normal to the surface* $\partial\mathcal{A}$ *directed outwards* $\mathcal{A}$ *at the point* $\lambda_+[\alpha]$.

To formulate a theorem, we need the following notation. We denote by P an operator projecting $\mathbb{R}^d$ to $\mathbb{R}^{d-1}$ which is of the following form: for $t = (t_1, \ldots, t_d)$,

$$\mathrm{P}t = (t_2, \ldots, t_d),$$

so that $t = (t_1, \mathrm{P}t)$. For any direction $\mathrm{e} \in \mathrm{E}$, we denote by M_e the orthogonal matrix composed of the row eigenvectors of the matrix $h''(\mathrm{e})$, so that $\mathrm{e}M_\mathrm{e} = \mathrm{e}$ and the matrix

$$M_\mathrm{e} h''(\mathrm{e}) M_\mathrm{e}^T = \left\| \sum_{i=1}^{d} \delta_{i,j} b_i(\mathrm{e}) \right\| \quad (\text{where } \delta_{i,j} = 1 \text{ for } i = j, \ \delta_{i,j} = 0 \text{ for } i \neq j)$$

is a diagonal matrix with diagonal $b(\mathrm{e}) = (b_1(\mathrm{e}), \ldots, b_d(\mathrm{e}))$, where $b_1(\mathrm{e}) = 0$, $\min_{2 \leqslant i \leqslant d} b_i(\mathrm{e}) > 0$. Denote by $B^2(\mathrm{e})$ the covariance matrix (of order $d-1$) of the random vector $\mathrm{P}\zeta(\mathrm{e})$, where

$$\zeta(\mathrm{e}) := \xi^{(\lambda_+^\mathrm{e})} M_\mathrm{e}^T.$$

Further, denote by $T^2(\mathrm{e})$ the following diagonal matrix of order $d-1$:

$$T^2(\mathrm{e}) := \left\| \sum_{i=1}^{d-1} \delta_{i,j} b_{i+1}(\mathrm{e}) \right\|.$$

Now denote by $|G|$ the determinant of a square matrix G; for example

$$|T^2(\mathrm{e})| = b_2(\mathrm{e}) \times \cdots \times b_d(\mathrm{e}).$$

For $v > 0$, set

$$R_{\mathrm{e}}(v) := \frac{|B^2(\mathrm{e})|^{1/2}}{|B^2(\mathrm{e}) + vT^2(\mathrm{e})|^{1/2}}.$$

Theorem 2.6.4. *Suppose that the condition* $[\mathcal{ER}]_2$ *is satisfied for* $\beta < -(d+1)/2 - 2$. *For* $\alpha = x/n \notin \mathcal{A}'$, *construct the points* $\lambda_+ = \lambda_+[\alpha] \in \partial\mathcal{A}$, $\alpha_+ = \alpha_+[\alpha] \in \partial\mathcal{A}'$ *(see Lemmas* 2.6.1, 2.6.3*). Let*

$$y := x - n\alpha_+[\alpha] = n(\alpha - \alpha_+[\alpha]).$$

Suppose that, for any fixed $\kappa > 0$, *two fixed functions* $\Delta_1 = \Delta_1(y) = |y|^{-\kappa}$ *are given, and* $\Delta_2 = \Delta_2(y) \geqslant \Delta_1(y)$, $\Delta_2 = o(1)$ *as* $|y| \to \infty$. *Then, uniformly in* $\Delta \in [\Delta_1, \Delta_2]$ *and* y *such that* $|y| \gg \sqrt{n \ln n}$ *as* $n \to \infty$, *the following asymptotic relation is true:*

$$\mathbf{P}\big(S_n \in \Delta[x)\big) = \frac{\Delta^d}{\psi(\lambda_+)} nR_{\mathrm{e}}\left(\frac{n}{|y|}\right) e^{-n\Lambda(\alpha_+) - \langle\lambda_+, y\rangle} g(\mathrm{e})\, l\big(|y|\big)\big(1 + o(1)\big), \tag{2.6.14}$$

where for brevity we write $\mathrm{e} = \mathrm{e}(y) := y/|y|$, $\lambda_+ = \lambda_+[\alpha]$, $\alpha_+ = \alpha_+[\alpha]$. *Moreover, in the region* $|y| \gg n$, *the relation* (2.6.14) *remains true for* $\beta < -(d+1)/2 - 1$. *If* $|y| \gg n$ *then* $R_{\mathrm{e}}(n/|y|)$ *is replaced by* 1; *if* $|y| = o(n)$, *we replace it by* $(|y|/n)^{(d-1)/2} |B(\mathrm{e})|\, |T(\mathrm{e})|^{-1}$.

The uniformness in Theorem 2.6.4 is understood in the following natural sense: *for any fixed sequence* N_n, *which is increasing and unbounded as* $n \to \infty$, *the remainder* $o(1)$ *in* (2.6.14) *can be replaced by a function* $\varepsilon = \varepsilon(n, \Delta, y)$ *that satisfies the relation*

$$\lim_{n\to\infty} \sup_{\substack{\Delta \in [\Delta_1, \Delta_2] \\ |y| \geqslant N_n \sqrt{n \ln n}}} \big|\varepsilon(n, \Delta, y)\big| = 0.$$

If a distribution $\mathbf{F}$ from the class $\mathcal{ER}$ is absolutely continuous then the statement of Theorem 2.6.4 simplifies somewhat:

Theorem 2.6.5. *Suppose that the conditions of Theorem* 2.6.4 *are satisfied and also that the distribution* $\mathbf{F}$ *is absolutely continuous. Then the density* $f_n(t)$ *of the sum* S_n *satisfies the relation*

$$f_n(x) = \frac{1}{\psi(\lambda_+)} nR_{\mathrm{e}}\left(\frac{n}{|y|}\right) e^{-n\Lambda(\alpha_+) - \langle\lambda_+, y\rangle} g(\mathrm{e}) l(|y|)(1 + o(1)) \tag{2.6.15}$$

under the notation and conditions of Theorem 2.6.4.

Let us now provide an analogue of Theorem 2.6.4 in the arithmetic case by considering the class of arithmetic distributions $\mathbf{F}$ that satisfy, for all $t \in \mathbb{Z}^d$, the condition (cf. (2.6.10))

$$\mathbf{P}(\xi = t) = g\big(\mathrm{e}(t)\big)\, l\big(|t|\big) e^{-h(t)}, \qquad |t| \to \infty, \tag{2.6.16}$$

where the functions $g(\mathrm{e})$, $l(v) = v^\beta L(v)$, $h(t)$ *satisfy the same conditions as those in* $[\mathcal{ER}]_2$.

Theorem 2.6.6. *Suppose that an arithmetic distribution* $\mathbf{F}$ *satisfies the condition* (2.6.16) *for* $\beta < -(d+1)/2 - 2$. *Then, uniformly in* $y := x - n\alpha_+[\alpha]$ *such that* $x = y + n\alpha_+[\alpha] \in \mathbb{Z}^d$, $|y| \gg \sqrt{n \ln n}$ *as* $n \to \infty$, *the relation*

$$\mathbf{P}\big(S_n = x\big) = \frac{1}{\psi(\lambda_+)} nR_{\mathrm{e}}\left(\frac{n}{|y|}\right) e^{-n\Lambda(\alpha_+) - \langle \lambda_+, y\rangle} g(\mathrm{e})\, l\big(|y|\big)\big(1 + o(1)\big)$$

holds under the notation and conditions of Theorem 2.6.4.

Theorems 2.6.4–2.6.6 and Lemmas 2.6.1, 2.6.3 are proved in [59].

The statement of Theorem 2.6.5 is illustrated in [59] with an example in which the distribution $\mathbf{F}$ of the random vector $\xi = (\xi_{(1)}, \ldots, \xi_{(d)}) \in \mathbb{R}^d$ is absolutely continuous and has the uniform spherical density

$$f(t) = e^{-\rho|t|} l(|t|), \qquad t \in \mathbb{R}^d,$$

where $l(v) := v^\beta L(v)$, $L(v)$ is an s.v.f. as $v \to \infty$, $\rho > 0$ (this corresponds to $\rho(\mathrm{e}) = \rho = \text{const.}$, $g(\mathrm{e}) = 1$ in the representation (2.6.10)) and the condition $\beta < -(d+1)/2 - 2$ is satisfied. Then, obviously, all the conditions of Theorem 2.6.5 are satisfied and the statement (2.6.15) of this theorem holds. An explicit form for the functions on the right-hand side of (2.6.15) can be found (see [59]).

2.7 Large deviation principles

2.7.1 A local large deviation principle (l.l.d.p.) in the one-dimensional case

In this section we prove Theorem 1.1.4. In order to do this we will need upper and lower bounds for the probabilities under consideration. The lower bound is formulated as the following separate result.

Lemma 2.7.1. *For any* $\alpha > 0$, $\varepsilon > 0$,

$$\varliminf_{n\to\infty} \frac{1}{n} \ln \mathbf{P}\left(\frac{S_n}{n} \in (\alpha)_\varepsilon\right) \geqslant -\Lambda(\alpha). \tag{2.7.1}$$

Proof. The method of obtaining the lower bound depends on the location of the parameter $\alpha = x/n$. We will consider the following three possibilities.

(i) $\alpha \in (\alpha_-, \alpha_+)$. In this case the lower bound (and the upper bound as well) can be easily obtained from Theorem 2.3.2, on the exact asymptotics of $\mathbf{P}\big(S_n \in n(\alpha - \varepsilon, \alpha + \varepsilon)\big)$. However, to achieve a unified exposition, we will use another approach, which seems to be preferable in the multidimensional case (see below).

Let α belong to the Cramér deviation region (α_-, α_+). Consider the reduction formula (2.1.5). Denote, as before,

$$^{\alpha}\xi = \xi^{(\lambda(\alpha))}, \qquad ^{\alpha}S_n = S_n^{(\lambda(\alpha))}.$$

Then $\mathbf{E}\,(^{\alpha}\xi) = \alpha$ and hence, for $B_n = n(-\varepsilon, \varepsilon)$, by the law of large numbers we have

$$\mathbf{P}\left(^{\alpha}S_n - \alpha n \in B_n\right) \to 1 \quad \text{as} \quad n \to \infty. \tag{2.7.2}$$

For z (member) B_n the function $e^{-\lambda(\alpha)z}$ in the integral in (2.1.5) assumes values from $(e^{-n\varepsilon\lambda(\alpha)}, e^{n\varepsilon\lambda(\alpha)})$. Therefore, for $x = \alpha n$, from (2.1.5) we obtain that, for all sufficiently large n,

$$\mathbf{P}\big(S_n \in x + B_n\big) \geqslant \frac{1}{2}\, e^{-n\Lambda(\alpha) - |\lambda(\alpha)|\varepsilon n},$$

and so

$$\varliminf_{n\to\infty} \frac{1}{n} \ln \mathbf{P}\left(\frac{S_n}{n} \in (\alpha)_\varepsilon\right) \geqslant -\Lambda(\alpha) - \big|\lambda(\alpha)\big|\varepsilon.$$

On the left-hand side of this inequality we have an increasing function of ε, and on the right-hand side a decreasing function. So, for all $\varepsilon > 0$,

$$\varliminf_{n\to\infty} \frac{1}{n} \ln \mathbf{P}\left(\frac{S_n}{n} \in (\alpha)_\varepsilon\right) \geqslant -\Lambda(\alpha).$$

The inequality (2.7.1) in the case $\alpha \in (\alpha_-, \alpha_+)$ is proved.

Now consider the case $\alpha \notin (\alpha_-, \alpha_+)$. Suppose, for example, $\alpha \geqslant \alpha_+$. There are two possibilities: either $\alpha < s_+$ (this is always so if $s_+ = \infty$) or $\alpha \geqslant s_+$ (for $s_+ < \infty$). First, suppose that

(ii) $\alpha < s_+$, $\alpha > \alpha_-$. Introduce the 'truncated' random variables $^{(N)}\xi$ with distribution

$$\mathbf{P}(^{(N)}\xi \in B) = \frac{\mathbf{P}(\xi \in B;\ |\xi| < N)}{\mathbf{P}(|\xi| < N)} = \mathbf{P}\big(\xi \in B \big|\ |\xi| < N\big).$$

We will furnish all the notation corresponding to $^{(N)}\xi$ with the upper left index (N). Obviously, for any λ,

$$\mathbf{E}\left(e^{\lambda\xi};\ |\xi| < N\right) \uparrow \psi(\lambda), \qquad \mathbf{P}\big(|\xi| < N\big) \uparrow 1$$

as $N \uparrow \infty$, so that

$$^{(N)}\psi(\lambda) = \frac{\mathbf{E}\,(e^{\lambda\xi};\ |\xi| < N)}{\mathbf{P}(|\xi| < N)} \to \psi(\lambda), \quad ^{(N)}A(\lambda) \to A(\lambda), \quad ^{(N)}\alpha_- \to \alpha_-. \tag{2.7.3}$$

The functions $^{(N)}\Lambda(\alpha)$ and $\Lambda(\alpha)$ are, respectively, the upper bounds of the concave functions $\alpha\lambda - {}^{(N)}A(\lambda)$ and $\alpha\lambda - A(\lambda)$. Therefore, for any α we also have the convergence $^{(N)}\Lambda(\alpha) \to \Lambda(\alpha)$ as $N \to \infty$.

Then,

$$\mathbf{P}\left(\frac{S_n}{n} \in (\alpha)_\varepsilon\right) \geqslant \mathbf{P}\left(\frac{S_n}{n} \in (\alpha)_\varepsilon;\ |\xi_j| < N,\ j = 1,\dots,n\right)$$
$$= \mathbf{P}^n\big(|\xi| < N\big)\mathbf{P}\left(\frac{^{(N)}S_n}{n} \in (\alpha)_\varepsilon\right). \tag{2.7.4}$$

Since $\alpha < s_+ < \infty$ and ${}^{(N)}\alpha_+ = \min(s_+, N)$, for sufficiently large N we have $\alpha \in \big({}^{(N)}\alpha_-, {}^{(N)}\alpha_+\big)$. Hence, the first part of the proof of the theorem can be applied, which gives as $n \to \infty$

$$\frac{1}{n}\ln \mathbf{P}\left(\frac{^{(N)}S_n}{n} \in (\alpha)_\varepsilon\right) \geqslant -{}^{(N)}\Lambda(\alpha) + o(1).$$

Therefore, from (2.7.4) we obtain

$$\frac{1}{n}\ln \mathbf{P}\left(\frac{S_n}{n} \in (\alpha)_\varepsilon\right) \geqslant -{}^{(N)}\Lambda(\alpha) + o(1) + \ln \mathbf{P}\big(|\xi| < N\big).$$

The right-hand side of the last inequality can be made arbitrarily close to $-\Lambda(\alpha)$ by an appropriate choice of N and n. Since the left-hand side of this inequality does not depend on N, we have

$$\varliminf_{n\to\infty} \frac{1}{n}\ln \mathbf{P}\left(\frac{S_n}{n} \in (\alpha)_\varepsilon\right) \geqslant -\Lambda(\alpha). \tag{2.7.5}$$

Finally, consider the case

(iii) $s_+ < \infty$, $\alpha \geqslant s_+$. Since $s_+ < \infty$, then by the property (Λ2ii) (see section 1.1) we have $\alpha_+ = s_+$, $\lambda_+ = \infty$. If $\alpha > \alpha_+ = s_+$, then $\Lambda(\alpha) = \infty$ and the bound (2.7.1) becomes trivial. If $\alpha = \alpha_+$, $\Lambda(\alpha_+) = \infty$ then we get the same result.

It remains to consider the case $\alpha = \alpha_+$, $\Lambda(\alpha_+) < \infty$. We will use the continuity of the function $\Lambda(\alpha)$ inside (α_-, α_+). Suppose that the sequences α_k, ε_k are such that $\alpha_k \uparrow \alpha = \alpha_+$, $\alpha_k + \varepsilon_k < \alpha$ as $k \to \infty$. For sufficiently large k and sufficiently small ε_k we have $(\alpha_k)_{\varepsilon_k} \in (\alpha)_\varepsilon$ and hence, by part (i),

$$\varliminf_{n\to\infty} \frac{1}{n}\ln \mathbf{P}\left(\frac{S_n}{n} \in (\alpha)_\varepsilon\right) \geqslant \varliminf_{n\to\infty} \frac{1}{n}\ln \mathbf{P}\left(\frac{S_n}{n} \in (\alpha_k)_{\varepsilon_k}\right) \geqslant -\Lambda(\alpha_k).$$

But $\Lambda(\alpha_k) \to \Lambda(\alpha)$ as $k \to \infty$ and $\Lambda(\alpha_k)$ can be made arbitrarily close to $\Lambda(\alpha)$. This is possible only if the required inequality (2.7.1) is true.

The case $\alpha \leqslant \alpha_-$ can be considered in the same way. Lemma 2.7.1 is proved. □

Proof of Theorem 1.1.4. The upper bound. We need to prove that, for any $\alpha \in \mathbb{R}$, $\varepsilon > 0$ (see (1.1.25)),

$$\varlimsup_{n\to\infty} \frac{1}{n}\ln \mathbf{P}\left(\frac{S_n}{n} \in (\alpha)_\varepsilon\right) \leqslant -\Lambda\big((\alpha)_\varepsilon\big). \tag{2.7.6}$$

Since $(\alpha)_\varepsilon$ is an open convex set, by Corollary 1.3.7 we have

$$\mathbf{P}\left(\frac{S_n}{n} \in (\alpha)_\varepsilon\right) \leqslant e^{-n\Lambda((\alpha)_\varepsilon)},$$

$$\frac{1}{n}\ln \mathbf{P}\left(\frac{S_n}{n} \in (\alpha)_\varepsilon\right) \leqslant -\Lambda\big((\alpha)_\varepsilon\big).$$

This implies (2.7.6).

The lower bound. For any β and δ such that $(\beta)_\delta \subset (\alpha)_\varepsilon$ we have

$$\mathbf{P}\left(\frac{S_n}{n} \in (\alpha)_\varepsilon\right) \geqslant \mathbf{P}\left(\frac{S_n}{n} \in (\beta)_\delta\right).$$

So, by Lemma 2.7.1,

$$\varliminf_{n\to\infty} \frac{1}{n} \ln \mathbf{P}\left(\frac{S_n}{n} \in (\alpha)_\varepsilon\right) \geqslant -\Lambda(\beta).$$

Since one can choose any point from $(\alpha)_\varepsilon$ to be β, we have

$$\varliminf_{n\to\infty} \frac{1}{n} \ln \mathbf{P}\left(\frac{S_n}{n} \in (\alpha)_\varepsilon\right) \geqslant -\Lambda\big((\alpha)_\varepsilon\big). \tag{2.7.7}$$

Comparing (2.7.6) and (2.7.7), we obtain (1.1.25). The second statement (1.1.26) of Theorem 1.1.4 follows from the first, and the properties of the function Λ (see Lemma 1.2.2). Theorem 1.1.4 is proved. □

As already noted, Theorem 1.1.4 remains true without any additional assumptions on the distribution $\mathbf{F}$. If one assumes that ξ is a non-lattice or arithmetic random variable, then in these cases Theorem 1.1.2 can be strengthened and one can obtain estimates for considerably smaller neighbourhoods of a point α. Namely, the following result is true.

Theorem 2.7.2. *Suppose that ξ is a non-lattice random variable. Then there exists a sequence $\Delta_{1,n} \to 0$ as $n \to \infty$ such that, for any sequence $\Delta_{2,n} = o(n)$ of all $\Delta = \Delta(n)$ from the segment $[\Delta_{1,n}, \Delta_{2,n}]$ and all α, we have*

$$\lim_{n\to\infty} \frac{1}{n} \ln \mathbf{P}\left(\frac{S_n}{n} \in (\alpha)_{\Delta/n}\right) = -\Lambda(\alpha). \tag{2.7.8}$$

If ξ is an arithmetic random variable then, for any α,

$$\lim_{n\to\infty} \frac{1}{n} \mathbf{P}\big(S_n = [\alpha n]\big) = -\Lambda(\alpha).$$

Proof. The proof of Theorem 2.7.2 differs from the proof of Theorem 1.1.2, generally, only by arguments needed in the case $\alpha \in (\alpha_-, \alpha_+)$. In this case, by Theorem 2.3.2 for $\Delta_{1,n} \to 0$ sufficiently slowly as $n \to \infty$ and $\Delta_{2,n} = O(1)$ for non-lattice ξ, we have

$$\mathbf{P}\big(S_n \in n(\alpha)_{\Delta/n}\big) \sim \frac{c(\alpha, \Delta)}{\sqrt{n}} e^{-n\Lambda(\alpha)} \tag{2.7.9}$$

as $n \to \infty$, where the function $c(\alpha, \Delta)$ does not depend on n and is separated from 0 and ∞. From this, we get

$$\lim_{n\to\infty} \frac{1}{n} \ln \mathbf{P}\left(\frac{S_n}{n} \in (\alpha)_{\Delta/n}\right) = -\Lambda(\alpha). \tag{2.7.10}$$

The rest of the proof for $\Delta_{2,n} = O(1)$ remains the same, since for $\alpha \geqslant \alpha_+$, $s_+ = \infty$, the truncations ${}^{(N)}\xi$ of the non-lattice random variables for sufficiently large N are also non-lattice and one can apply the relations (2.7.10) to them. Hence, as in the proof of Theorem 1.1.4, we obtain the lower bound

$$\varliminf_{n\to\infty} \frac{1}{n} \ln \mathbf{P}\left(\frac{S_n}{n} \in (\alpha)_{\Delta/n}\right) \geqslant -\Lambda(\alpha). \tag{2.7.11}$$

In the case $s_+ < \infty$, $\alpha = s_+$ and $\Lambda(s_+) < \infty$, we need to use the relations

$$\mathbf{P}\big(S_n \in n(s_+)_{\Delta/n}\big) \geqslant \mathbf{P}(S_n = ns_+) = \big[\mathbf{P}(\xi = s_+)\big]^h,$$
$$\Lambda(s_+) = -\ln \mathbf{P}(\xi = s_+).$$

If $\alpha = s_+$ and $\Lambda(s_+) = \infty$, then inequality (2.7.11) is trivial, as in the case $\alpha > s_+$. The arguments relating to the point s_- are similar. Obviously, inequality (2.7.11) remains valid also for $\Delta_{2,n} \to \infty$ as $n \to \infty$. The lower bound is proved.

Now we obtain the upper bound. By Theorem 1.1.4, for any $\varepsilon > 0$,

$$\varlimsup_{n\to\infty} \frac{1}{n} \ln \mathbf{P}\left(\frac{S_n}{n} \in (\alpha)_{\Delta/n}\right) \leqslant \varlimsup_{n\to\infty} \frac{1}{n} \ln \mathbf{P}\left(\frac{S_n}{n} \in (\alpha)_\varepsilon\right) = -\Lambda\big((\alpha)_\varepsilon\big). \tag{2.7.12}$$

Since the function Λ is lower semicontinuous, we have

$$\varliminf_{\varepsilon\to 0} \Lambda\big((\alpha)_\varepsilon\big) \geqslant \Lambda(\alpha).$$

This and (2.7.12) imply that

$$\varlimsup_{n\to\infty} \frac{1}{n} \ln \mathbf{P}\left(\frac{S_n}{n} \in (\alpha)_{\Delta/n}\right) \leqslant -\Lambda(\alpha).$$

Together with (2.7.11), this proves (2.7.8).

In the arithmetic case the proof is similar. In that case $s_\pm$ are integer and, for $\alpha = s_\pm$, $|s_\pm| < \infty$,

$$[\alpha n] = s_\pm n,$$
$$\ln \mathbf{P}\big(S_n = [\alpha n]\big) = n \ln \mathbf{P}(\xi = s_\pm) = -n\Lambda(\alpha).$$

The theorem is proved. □

2.7.2 A large deviation principle in the one-dimensional case

From Theorem 2.1.2 one can easily obtain a corollary about the asymptotics of the probability that S_n/n hits an arbitrary Borel set. Denote by (B) and $[B]$, respectively, the interior and the closure of B ((B) is the union of all the points belonging to B together with some intervals) and let, as before,

$$\Lambda(B) = \inf_{\alpha \in B} \Lambda(\alpha).$$

Theorem 2.7.3. *For any Borel set B the following inequalities hold:*

$$\overline{\lim_{n\to\infty}} \frac{1}{n} \ln \mathbf{P}\left(\frac{S_n}{n} \in B\right) \leqslant -\Lambda\big([B]\big), \tag{2.7.13}$$

$$\underline{\lim_{n\to\infty}} \frac{1}{n} \ln \mathbf{P}\left(\frac{S_n}{n} \in B\right) \geqslant -\Lambda\big((B)\big). \tag{2.7.14}$$

If $\Lambda\big((B)\big) = \Lambda\big([B]\big)$ *then there exists*

$$\lim_{n\to\infty} \frac{1}{n} \ln \mathbf{P}\left(\frac{S_n}{n} \in B\right) = -\Lambda(B). \tag{2.7.15}$$

This results is an 'integral' or 'usual' *large deviation principle*. It belongs to the class of so-called 'rough' ('logarithmic') limit theorems, which describe the asymptotics of $\ln \mathbf{P}(S_n/n \in B)$. Usually, it is impossible to get from such a theorem the asymptotics of the probability $\mathbf{P}(S_n/n \in B)$ (in the equality $\mathbf{P}(S_n/n \in B) = \exp\{-n\Lambda(B) + o(n)\}$ the term $o(n)$ can grow in absolute value).

Proof. Without loss of generality, we can assume that $B \subset [s_-, s_+]$ (outside this region $\Lambda(\alpha) = \infty$).

First we prove (2.7.14). Let $\alpha_{(B)}$ be a number such that $\Lambda\big((B)\big) \equiv \inf_{\alpha\in(B)} \Lambda(\alpha) = \Lambda(\alpha_{(B)})$ (recall that $\Lambda(\alpha)$ is continuous on $[s_-, s_+]$). Then there exist sequences of points α_k and intervals $(\alpha_k)_{\delta_k} = (\alpha_k - \delta_k, \alpha_k + \delta_k)$, where $\delta_k \to 0$, which belong to (B) and converge to the point $\alpha_{(B)}$, such that

$$\Lambda\big((B)\big) = \inf_k \Lambda\big((\alpha_k - \delta_k, \alpha_k + \delta_k)\big).$$

Here, obviously,

$$\inf_k \Lambda\big((\alpha_k - \delta_k, \alpha_k + \delta_k)\big) = \inf_k \Lambda(\alpha_k)$$

and for given $\varepsilon > 0$ there exists $k = K$ such that $\Lambda(\alpha_K) < \Lambda\big((B)\big) + \varepsilon$. By Theorem 1.1.4, as $n \to \infty$ we have

$$\frac{1}{n} \mathbf{P}\left(\frac{S_n}{n} \in B\right) \geqslant \frac{1}{n} \ln \mathbf{P}\left(\frac{S_n}{n} \in (B)\right) \geqslant \frac{1}{n} \ln \mathbf{P}\left(\frac{S_n}{n} \in (\alpha_K)_{\delta_k}\right)$$
$$= -\Lambda\big((\alpha_K)_{\delta_k}\big) + o(1) \geqslant -\Lambda(\alpha_K) + o(1) \geqslant -\Lambda\big((B)\big) - \varepsilon + o(1).$$

Since the left-hand side of this inequality does not depend on ε, this proves inequality (2.7.14).

Let us prove inequality (2.7.13). Denote by $\alpha_{[B]}$ the point where $\inf_{\alpha\in[B]} \Lambda(\alpha) = \Lambda(\alpha_{[B]})$ is attained (this point always belongs to $[B]$ since $[B]$ is closed). If $\Lambda(\alpha_{[B]}) = 0$ then the inequality is obvious. Suppose now that $\Lambda(\alpha_{[B]}) > 0$.

Owing to the convexity of Λ, the equation $\Lambda(\alpha) = \Lambda(\alpha_{[B]})$ can have a second solution $\alpha'_{[B]}$. Suppose it exists and, for the sake of definiteness, $\alpha'_{[B]} < \alpha_{[B]}$. The relation $\Lambda([B]) = \Lambda(\alpha_{[B]})$ means that the set $[B]$ is concentrated outside $(\alpha'_{[B]}, \alpha_{[B]})$ and

$$\mathbf{P}\left(\frac{S_n}{n} \in B\right) \leqslant \mathbf{P}\left(\frac{S_n}{n} \in [B]\right) \leqslant \mathbf{P}\left(\frac{S_n}{n} \leqslant \alpha'_{[B]}\right) + \mathbf{P}\left(\frac{S_n}{n} \geqslant \alpha_{[B]}\right). \quad (2.7.16)$$

Moreover, in this case $\mathbf{E}\xi \in (\alpha'_{[B]}, \alpha_{[B]})$ and each probability on the right-hand side of (2.7.16) can be bounded by the value $e^{-n\Lambda(\alpha_{[B]})}$, according to the exponential Chebyshev inequality (see (Λ8)). This implies (2.7.13).

If the second solution $\alpha'_{[B]}$ does not exist then one term on the right-hand side of (2.7.16) equals zero, and we get the same result.

The second statement of Theorem (2.7.15) is obvious. The theorem is proved. □

Note that the statements of Theorems 1.1.4 and 2.7.3 and their corollaries are 'universal' in the sense that they do not contain any conditions on the distribution $\mathbf{F}$.

Below, in section 5.1, we will also establish *moderately* large deviation principles for S_n/x, where $x = o(n)$, $x \gg \sqrt{n}$ as $n \to \infty$.

2.7.3 The multidimensional case

The purpose of this section is to prove the l.l.d.p. (Theorem 1.2.1) for sums S_n of random vectors $\xi_1, \ldots, \xi_n$ and the 'usual' (or 'integral') l.d.p. (an analogue of Theorem 2.7.2).

Proof of Theorem 1.2.1. As in the one-dimensional case, we need to prove inequalities (2.7.6) and (2.7.1) in the multidimensional case, when $(\alpha)_\varepsilon$ is an open ball in $\mathbb{R}^d$ with radius ε and center α.

The lower bounds, as in the one-dimensional case, immediately follow from Corollary 1.3.7.

The upper bounds can also be obtained in a way completely similar to that for the one-dimensional case.

(i) First suppose $\alpha \in (\mathcal{A}')$. Assuming, as before, that

$$^{\alpha}\xi = \xi^{\lambda(\alpha))}, \qquad ^{\alpha}S_n = S_n^{(\lambda(\alpha))},$$

we obtain

$$\mathbf{E}(^{\alpha}\xi) = \alpha$$

(see section 2.1.2). By the law of large numbers,

$$\mathbf{P}\left(\frac{^{\alpha}S_n}{n} \in (\alpha)_\varepsilon\right) \to 1$$

as $n \to \infty$. Apply the reduction formula (2.1.11) for $x = \alpha n$, $B = n(0)_\varepsilon$. In this formula,

$$e^{-\langle \lambda(\alpha), z\rangle} \geqslant e^{-|\lambda(\alpha)|\varepsilon n}.$$

Therefore, for all sufficiently large n,

$$\mathbf{P}\left(\frac{{}^{\alpha}S_n}{n} \in (\alpha)_\varepsilon\right) \geqslant \frac{1}{2} e^{-n\Lambda(\alpha)-|\lambda(\alpha)|\varepsilon n}.$$

This, as in the one-dimensional case, implies that

$$\varliminf_{n\to\infty} \frac{1}{n} \ln \mathbf{P}\left(\frac{S_n}{n} \in (\alpha)_\varepsilon\right) \geqslant -\Lambda(\alpha).$$

Now suppose that $\alpha \notin (\mathcal{A}')$ and that S is, as before, the convex envelope of the support of the distribution $\mathbf{F}$. Consider the case

(ii) $\alpha \in (\mathrm{S})$. Define the 'truncated' random variables ${}^{(N)}\xi$ with distribution

$$\mathbf{P}\left({}^{(N)}\xi \in B\right) = \mathbf{P}\big(\xi \in B \big| \, |\xi| < N\big)$$

and, as in the one-dimensional case, furnish all the notation corresponding to ${}^{(N)}\xi$ with the left superscript (N). Then we have the relations (2.7.3),

$$ {}^{(N)}\Lambda(\alpha) \to \Lambda(\alpha) \quad \text{as} \quad N \to \infty. \tag{2.7.17}$$

Further, similarly to (2.7.4), we obtain

$$\mathbf{P}\left(\frac{S_n}{n} \in (\alpha)_\varepsilon\right) \geqslant \mathbf{P}^n\big(|\xi| < N\big)\mathbf{P}\left(\frac{{}^{(N)}S_n}{n} \in (\alpha)_\varepsilon\right).$$

But $\left({}^{(N)}\mathcal{A}'\right) = (\mathrm{S}) \bigcap \mathrm{U_N}$, where $U_N = \big\{\alpha : |\alpha| < N\big\}$ and, since $\alpha \in (\mathrm{S})$, for all sufficiently large N we have $\alpha \in \left({}^{(N)}\mathcal{A}'\right)$. Hence, it is possible to apply the first part of the proof, which gives for $n \to \infty$

$$\frac{1}{n} \ln \mathbf{P}\left(\frac{{}^{(N)}S_n}{n} \in (\alpha)_\varepsilon\right) \geqslant -{}^{(N)}\Lambda(\alpha) + o(1),$$

$$\frac{1}{n} \ln \mathbf{P}\left(\frac{S_n}{n} \in (\alpha)_\varepsilon\right) \geqslant -{}^{(N)}\Lambda(\alpha) + o(1) + \ln \mathbf{P}\big(|\xi| < N\big).$$

Using the arbitrariness of N and the convergence (2.7.17), we obtain the required bound.

Now consider the case

(iii) $\alpha \in \partial\mathrm{S}$, $\Lambda(\alpha) < \infty$. Here we use the inequality

$$\mathbf{P}\left(\frac{S_n}{n} \in (\alpha)_\varepsilon\right) \geqslant \mathbf{P}\left(\frac{S_n}{n} \in (\alpha')_\delta\right)$$

for appropriate $\delta < \varepsilon$ and $\alpha' \in (\mathrm{S}) \cap (\alpha)_\varepsilon$, repeat the arguments from part (ii) and use the continuity of $\Lambda(\alpha')$ inside (S) along rays with regard to points $\alpha' \in (\mathrm{S}) \cap (\alpha)_\varepsilon$ (see Lemma 1.2.2(ii)).

It remains to consider the case

(iv) $\alpha \notin S$ or $\alpha \in \partial S$, $\Lambda(\alpha) = \infty$. Here, the estimate (2.7.1) is trivial, since in both cases $\Lambda(\alpha) = \infty$.

Theorem 1.2.1 is proved. □

In a completely similar way, one can also prove a multidimensional counterpart of Theorem 2.7.2 for non-lattice and arithmetic ξ.

Now we establish the 'usual' l.d.p.

Theorem 2.7.4. *If the condition* **[C₀]** *is satisfied then, for any Borel set* $B \subset \mathbb{R}^d$, *the inequalities* (2.7.13), (2.7.14) *hold.*

See e.g. [46], [47], [1], [6], [13], regarding an l.d.p. for sums of random elements in spaces of a more general nature. An l.d.p. for so-called asymptotically homogeneous Markov chains was obtained in [43].

Proof. Theorem 4.1.1 (see section 4.1 below) states the following. Suppose that there exists a family of compacts $K_v \in \mathbb{R}^d$, $v > 0$, such that for any $N > 0$ there exists $v = v(N)$, for which

$$\mathbf{P}\left(\frac{S_n}{n} \notin K_v\right) \leqslant e^{-nN}. \tag{2.7.18}$$

Then the l.l.d.p. implies the l.d.p. (i.e. the inequalities (2.7.13), (2.7.14)). Thus, it is enough to verify that the condition $[\mathbf{C}_0]$ implies (2.7.18).

First, observe that, together with ξ, the absolute values $|\xi_{(i)}|$ of all the coordinates $\xi_{(i)}$ of the vector ξ also satisfy the condition $[\mathbf{C}_0]$. Since $|\xi| \leqslant \sqrt{d} \max |\xi_{(i)}|$, we have

$$\mathbf{P}\left(|\xi| > t\right) \leqslant \sum_{i=1}^{d} \mathbf{P}\left(|\xi_{(i)}| > \frac{t}{\sqrt{d}}\right)$$

and $|\xi|$ also satisfies the condition $[\mathbf{C}_0]$. Hence, by the property (Λ6) (see section 1.1) there exist constants c_1, c_2 such that

$$\Lambda^{(|\xi|)}(\alpha) \geqslant c_1|\alpha| - c_2. \tag{2.7.19}$$

This inequality allows us to prove (2.7.18). Indeed, take the balls $\{\alpha : |\alpha| \leqslant v\}$ as the K_v. Then

$$\mathbf{P}\left(\frac{S_n}{n} \notin K_v\right) = \mathbf{P}\left(|S_n| > vn\right) \leqslant \mathbf{P}\left(\sum_{j=1}^{n} |\xi_j| > vn\right) \leqslant e^{-n\Lambda^{(|\xi|)}(v)}.$$

Clearly, owing to (2.7.19), for all sufficiently large v the inequality $\Lambda^{(|\xi|)}(v) \geqslant N$ holds. The theorem is proved. □

In view of Theorem 2.7.4, one can ask a question about the importance of the condition $[\mathbf{C}_0]$. In the one-dimensional case, $d = 1$, this condition is not necessary for the l.d.p. to hold; see Theorem 2.7.3. It was shown in [136] that the condition

[C₀] is not necessary in the two-dimensional case $d = 2$ either. However, for $d = 3$, in [136], [84], [170], an example was constructed of a closed set $B \subset \mathbb{R}^3$ and a distribution $\mathbf{F}$ in $\mathbb{R}^3$ which satisfy only the condition **[C]** (the condition **[$\mathbf{C_0}$]** is not satisfied), for which the inequality (2.7.13) does not hold. Thus, at least when $d = 3$, the condition **[$\mathbf{C_0}$]** cannot be essentially relaxed for (2.7.13) and it cannot be relaxed up to the condition **[C]**. The above-mentioned example was obtained in [136] by a slight modification of the results of the papers [84], [170].

2.8 Limit theorems for sums of random variables with non-homogeneous terms

In certain problems (for example, in the study of compound renewal processes; see section 4.10) it becomes necessary to investigate the asymptotics of the distributions of sums of random vectors which contain one or several non-homogeneous terms. These terms, without loss of generality, can be grouped, and one can consider distributions of sums

$$S_{0,n} := \xi_0 + S_n, \qquad n \geqslant 0,$$

where S_n retains its previous meaning, $S_0 = 0$ and the distribution of the vector ξ_0, which does not depend on S_n, differs from the distribution of ξ.

2.8.1 Integro-local theorems

First we will establish integro-local theorems in the zone of normal deviations. We consider only the case when ξ is a non-lattice vector,

$$\mathbf{E}\xi = 0,$$

and there exists a non-degenerate matrix of second moments σ^2. As before, let $\Delta[x)$ be the half-open cube with vertex at the point x and side Δ.

Theorem 2.8.1. *Suppose that the above conditions are satisfied and ξ_0 is an arbitrary random vector independent of $\{\xi_k\}$ and such that $|\xi_0| = o_p(\sqrt{n})$ (i.e. there exists a sequence $\varepsilon(n) \to 0$ as $n \to \infty$ such that $\mathbf{P}\big(|\xi_0| > \varepsilon(n)\sqrt{n}\big) \to 0$ as $n \to \infty$). Then*

$$\mathbf{P}\big(S_{0,n} \in \Delta[x)\big) = \frac{\Delta^d}{(2\pi n)^{d/2}|\sigma|} e^{-x\sigma^{-2}x^T/2} + o(n^{-d/2}) \qquad (2.8.1)$$

uniformly in x and in $\Delta \in [\Delta_1, \Delta_2]$, where $\Delta_1 < \Delta_2$ are fixed; σ^{-2} is the inverse matrix of σ^2.

Clearly, the statement (2.8.1) remains true if $\Delta = \Delta_n \to 0$ sufficiently slowly as $n \to \infty$, and this new statement is equivalent to (2.8.1).

In the arithmetic case, under the same assumptions the following claim is true: $\mathbf{P}(\xi_0 + S_n = x)$ for integer x is equal to the right-hand side of (2.8.1) with $\Delta = 1$; however, additionally we need to assume that ξ_0 is arithmetic.

The statement of Theorem 2.8.1 shows that if the $o(n)$ terms in the sum S_n have zero mean, bounded variance and a distribution different from the distribution of ξ, then the asymptotics of $\mathbf{P}(S_n \in \Delta[x))$ remains the same as in the case of identically distributed terms.

Proof of Theorem 2.8.1. Let $N = \varepsilon(n)\sqrt{n}$. Then

$$\mathbf{P}(S_{0,n} \in \Delta[x)) = \mathbf{P}(|\xi_0| < N, S_{0,n} \in \Delta[x)) + \mathbf{P}(|\xi_0| \geqslant N, S_{0,n} \in \Delta[x)). \tag{2.8.2}$$

Since, for all x and y

$$\mathbf{P}(S_n \in \Delta[x-y)) < \frac{c}{n^{d/2}},$$

where c does not depend on x and y, the second term on the right-hand side of (2.8.2) is not greater than

$$\frac{c}{n^{d/2}} \mathbf{P}(|\xi_0| > N) = o(n^{-d/2}). \tag{2.8.3}$$

Consider the first term in the right-hand side of (2.8.2). For $|y| < N$ and $|x| = O(\sqrt{n})$, by Theorem 1.5.1 we have

$$\mathbf{P}(S_n \in \Delta[x-y)) = \frac{1}{(2\pi n)^{d/2}|\sigma|} \exp\left\{-\frac{1}{2n}(x-y)\sigma^{-2}(x-y)^T\right\} + o(n^{-d/2}),$$

where

$$|x\sigma^{-2}y^T| = o(n), \qquad |y\sigma^{-2}x^T| = o(n), \qquad |y|^2 = o(n).$$

Therefore, uniformly in $|x| = O(\sqrt{n})$, $|y| < N$,

$$\mathbf{P}(S_n \in \Delta[x-y)) = \frac{1}{(2\pi n)^d|\sigma|} e^{-(x\sigma^{-2}x^T)/2n}\left(1 + o(1)\right) + o(n^{-d/2});$$

$$\begin{aligned}\mathbf{P}(|\xi_0| < N, S_{0,n} \in \Delta[x)) &= \int_{|y|<N} \mathbf{P}(\xi_0 \in dy)\mathbf{P}(S_n \in \Delta[x-y)) \\ &= \frac{1}{(2\pi n)^{d/2}|\sigma|} e^{-(x\sigma^{-2}x^T)/2n}\left(1 + o(1)\right) + o(n^{-d/2}).\end{aligned} \tag{2.8.4}$$

If $|x| \gg \sqrt{n}$ then in (2.8.4) one has

$$\mathbf{P}(S_n \in \Delta[x-y)) = o(n^{-d/2}).$$

Together with (2.8.3), this proves (2.8.1). The theorem is proved. □

Remark 2.8.2. Here, the uniform versions of Theorem 2.8.1 are also true (the analogues of Theorems 1.5.3 and 1.5.4) in the case when the distribution of the variable ξ depends on some parameter λ, but the dependence is not too 'essential'.

Now we establish an integro-local theorem for $S_{0,n}$ in the Cramér deviation zone. Denote

$$\psi_0(\lambda) := \mathbf{E}e^{<\lambda,\xi_0>}$$

and let $\xi_0^{(\lambda)}$ be the Cramér transform of ξ_0 at a point $\lambda \in \mathcal{A}$. Assume, as before, that $\alpha = x/n$.

Theorem 2.8.3. *Suppose that the conditions of part* (i) *of Theorem* 2.3.2 *are satisfied and that*

$$\psi_0\big(\lambda(\alpha)\big) < \infty, \qquad \big|\xi_0^{(\lambda(\alpha))}\big| = o_p\big(\sqrt{n}\big) \tag{2.8.5}$$

(see Theorem 2.8.1). *Then, with the notation of Theorem* 2.3.2,

$$\mathbf{P}\big(S_{0,n} \in \Delta[x)\big) = \frac{\psi_0\big(\lambda(\alpha)\big)\Delta^d}{(2\pi n)^{d/2}|\sigma(\alpha)|}\, e^{-n\Lambda(\alpha)}\big(c(\alpha,\Delta) + \varepsilon_n\big), \tag{2.8.6}$$

where the properties of the remainder ε_n are specified in Theorem 2.3.2.

If $\Delta = \Delta_n \to 0$ slowly enough as $n \to \infty$, we can replace the function $c(\alpha,\Delta)$ in (2.8.6) *by* 1. *The statement thus obtained is equivalent to* (2.8.6).

It is clear that the conditions (2.8.5) of the theorem are always satisfied if ξ_0 is a random vector with a fixed (independent of n) distribution and $\lambda(\alpha) \in \mathcal{A}_0^*$, where $\mathcal{A}_0^*$ is a compact from $(\mathcal{A}_0)$, $\mathcal{A}_0 := \{\lambda : \psi_0(\lambda) < \infty\}$.

Proof of Theorem 2.8.3. Repeating the arguments from the proof of the reduction formulas (2.1.4), (2.1.5) under the present conditions, we obtain

$$\begin{aligned}\mathbf{P}(S_{0,n} \in x + B)& \\ &= \psi_0(\lambda)e^{-\langle\lambda,x\rangle + nA(\lambda)} \int_B e^{-\langle\lambda,z\rangle}\mathbf{P}(\xi_0^{(\lambda)} + S_n^{(\lambda)} - x \in dz),\end{aligned}$$

$$\begin{aligned}\mathbf{P}\big(S_{0,n} \in \Delta[x)\big)& \\ &= \psi_0\big(\lambda(\alpha)\big)e^{-n\Lambda(\alpha)} \int_{\Delta[0)} e^{-\langle\lambda(\alpha),z\rangle}\mathbf{P}(\xi_0^{(\lambda(\alpha))} + S_n^{(\lambda)(\alpha))} - \alpha n \in dz).\end{aligned}$$

From here, for $\Delta = \Delta_n \to 0$ sufficiently slowly, by Theorem 2.8.1 we find that for fixed $\alpha \in (\mathcal{A})$

$$\mathbf{P}\big(S_{0,n} \in \Delta_n[x)\big) = \psi_0\big(\lambda(\alpha)\big)\frac{\Delta_n^d}{(2\pi n)^{d/2}|\sigma(\alpha)|}(1 + \varepsilon_n),$$

where the remainder $\varepsilon_n \to 0$ as $n \to \infty$ and has the required property of uniformity. This also implies the statement (2.8.6) (by the same reasoning as that by which Theorem 2.2.3 was obtained from Theorem 2.2.1).

If $\alpha \in \mathcal{A}^*$ depends on n then one should use the uniform version of Theorem 2.8.1, which was mentioned above in Remark 2.8.2. The theorem is proved. □

Remark 2.8.4. As was shown in [175], the integro-local theorem 2.2.3 remains true in the case when some coordinates of the vector ξ have a non-lattice distribution and others are arithmetic. Then, one can take a rectangular open parallelepiped with sides of length $\Delta_{(1)}, \ldots, \Delta_{(d)}$ as $\Delta[x)$, where the $\Delta_{(k)}$ are arbitrary fixed numbers if $\xi_{(k)}$ is non-lattice, and the $\Delta_{(k)} \geqslant 1$ are arbitrary integer numbers if $\xi_{(k)}$ is arithmetic. The factor Δ^d on the right-hand side of the representation of form (2.8.6) should be replaced by $\prod_{k=1}^{d} \Delta_{(k)}$. By this reasoning, the statement (2.8.6) of Theorem 2.8.3 remains true for vectors ξ with 'mixed' coordinates. In that case, for the arithmetic coordinates of $\xi_{(k)}$ the corresponding coordinates of the vector ξ_0 should be integer-valued.

2.8.2 Inequalities in the one-dimensional case

Let us begin with the main exponential inequalities of Chebyshev type in the one-dimensional case. For all $x \geqslant 0$, $\lambda \geqslant 0$ we have

$$\mathbf{P}(S_{0,n} \geqslant x) \leqslant e^{-\lambda x} \psi_0(\lambda) \psi^n(\lambda). \tag{2.8.7}$$

Therefore, if $\lambda(\alpha) \geqslant 0$ (this is always the case for $\alpha = x/n \geqslant \mathbf{E}\xi$) then

$$\mathbf{P}(S_{0,n} \geqslant x) \leqslant \psi_0\big(\lambda(\alpha)\big) e^{-n\Lambda(\alpha)}. \tag{2.8.8}$$

Remark 2.8.5. Regarding the statement of Theorem 2.8.1 (see (2.8.6)) and inequality (2.8.8), let us provide a 'typical' example, which shows that if the condition $\psi_0\big(\lambda(\alpha)\big) < \infty$ is not satisfied then the exponential parts of the probabilities under consideration will be different from $e^{-n\Lambda(\alpha)}$ in a large class of cases. Suppose that

$$\mathbf{P}(\xi_0 > t) = e^{-t}, \qquad \mathbf{P}(\xi > t) = e^{-2t}, \qquad t > 0.$$

Then $\mathbf{E}\xi = 1/2$,

$$\psi_0(\lambda) = \frac{1}{1-\lambda} \quad \text{for} \quad \lambda \leqslant 1 \qquad \psi(\lambda) = \frac{2}{2-\lambda} \quad \text{for} \quad \lambda \leqslant 2$$

$$\lambda(\alpha) = 2 - \frac{1}{\alpha}, \qquad \Lambda(\alpha) = 2\alpha - 1 - \ln 2\alpha \quad \text{for} \quad \alpha > 0$$

(see Example 1.1.7). For deviations $x = \alpha n$ for $\alpha \in (1/2, 1)$ we have $\lambda(\alpha) \in (0, 1)$, $\psi_0\big(\lambda(\alpha)\big) < \infty$ and the relations (2.8.6), (2.8.8) hold. For $\alpha > 1$ (in this case $\psi_0\big(\lambda(\alpha)\big) = \infty$), let us find the exponential part of the probability $\mathbf{P}(S_{0,n} > \alpha n)$. By considering the convolution of the distributions of ξ_0 and S_n and taking into account that $\mathbf{E}S_n = n/2$, it is not difficult to see that the exponential part of $\mathbf{P}(\xi_0 + S_n > \alpha n)$ is determined by the value

$$\exp\Big\{-n \min_{t \in (0, \alpha - 1/2)} \big(t + 2(\alpha - t) - 1 - \ln 2(\alpha - t)\big)\Big\}, \tag{2.8.9}$$

and this value is determined by

$$2\alpha - 1 - \max_{t \in (0, \alpha - 1/2)} g(t), \qquad g(t) = t + \ln 2(\alpha - t).$$

We have

$$g'(t) = 1 - \frac{1}{\alpha - t},$$

so $g'(0) = 1 - 1/\alpha > 0$ and the maximum of $g(t)$ is attained at the point $t_0 = \alpha - 1 \subset [0, \alpha - 1/2)$ and is equal to

$$g(t_0) = \alpha - 1 + \ln 2;$$

thus the value (2.8.9) is equal to $e^{-n(\alpha - \ln 2)}$, where $\alpha - \ln 2 < \Lambda(\alpha)$ for $\alpha > 1$.

It is not difficult to obtain a similar result for the probability $\mathbf{P}\big(S_{0,n} \in \Delta[x)\big)$ for $x = \alpha n$.

The following analogue of Theorem 1.1.2 is also true. Let

$$\overline{S}_{0,n} = \max_{0 \leqslant k \leqslant n} S_{0,k}.$$

Theorem 2.8.6. (i) *For all* $n \geqslant 0$, $x \geqslant 0$, $\lambda \geqslant 0$,

$$\mathbf{P}(\overline{S}_{0,n} \geqslant x) \leqslant \psi_0(\lambda) e^{-\lambda x} \max\big(1, \psi^n(\lambda)\big). \tag{2.8.10}$$

(ii) *For* $\mathbf{P}(\overline{S}_{0,n} \geqslant x)$, *statements* (i), (ii) *of Theorem* 1.1.2 (inequalities (1.1.21) and (1.1.24)) *are true if on the right-hand side of* (1.1.21) *one adds the factor* $\psi_0(\lambda_1)$ *and on the right-hand side of* (1.1.24) *the factor* $\psi_0\big(\lambda(\alpha)\big)$.

(iii) *Also true is the inequality* (an analogue of Corollary 1.1.3)

$$\mathbf{P}(\overline{S}_{0,n} \geqslant x) \leqslant \psi_0\big(\lambda(\alpha_1)\big) e^{-n\Lambda_1(\alpha)},$$

where α_1 *and* $\Lambda_1(\alpha)$ *are defined in Theorem* 1.1.2 *and Corollary* 1.1.3.

Proof. (i) Let us begin with the main inequality (2.8.10). Let

$$\eta_0(x) = \min\{k \geqslant 0 : S_{0,k} \geqslant x\}.$$

Since $S_{0,n} - S_{0,k} = S_n - S_k$ for $k \leqslant n$, using reasoning similar to that in the proof of Theorem 1.1.2, (i), we find

$$\begin{aligned}
\psi_0(\lambda)\psi^n(\lambda) = \mathbf{E} e^{\lambda S_{0,n}} &\geqslant \sum_{k=0}^{n} \mathbf{E}\big[e^{\lambda S_{0,n}};\ \eta_0(x) = k\big] \\
&\geqslant \sum_{k=0}^{n} \mathbf{E}\big(e^{(x + S_{0,n} - S_{0,k})\lambda};\ \eta_0(x) = k\big) \\
&= e^{\lambda x} \sum_{k=0}^{n} \psi^{n-k}(\lambda) \mathbf{P}\big(\eta(x) = k\big) \\
&\geqslant e^{\lambda x} \min\big(1, \psi^n(\lambda)\big) \mathbf{P}\big(\eta_0(x) \leqslant n\big).
\end{aligned}$$

This implies (2.8.10).

Statements (ii), (iii) follow from (2.8.10) if again one uses (with obvious changes) the same argument as that used in the proofs of Theorem 1.1.2 and Corollary 1.1.3. The theorem is proved. □

2.8.3 Inequalities in the multidimensional case

We will now consider the multidimensional case. Let $B \subset \mathbb{R}^d$ be an open convex set which does not contain the point $\mathbf{E}\xi$, let $v_B = \Lambda(B) = \inf_{\alpha \in B} \Lambda(\alpha) < \infty$ and let

$$\Pi_= := \{\alpha : \langle e, \alpha \rangle = b\}, \qquad |e| = 1,$$

be the hyperplane separating the sets B and $\Lambda_{v_B} := \{\alpha : \Lambda(\alpha) \leqslant v_B\}$, where $\Lambda(\alpha)$ is the deviation function of the vector ξ. Further, let $\alpha_{(B)}$ be such that $\Lambda(B) = \Lambda(\alpha_{(B)})$, $\lambda_{(B)} := \lambda(\alpha_{(B)})$. Since $\Lambda(B) < \infty$, we have $\Lambda(B) = \Lambda([B])$, $\alpha_{(B)} \in \partial B$ (see Corollary 1.2.3).

Theorem 2.8.7. *Suppose that the condition* $[\mathbf{C}_0]$ *holds and* $B \subset \mathbb{R}^d$ *is an open convex set which does not contain the point* $\mathbf{E}\xi$, $\Lambda(B) < \infty$. *Then*

$$\mathbf{P}\left(\frac{S_{0,n}}{n} \in B\right) \leqslant \psi_0(\lambda_{(B)})\, e^{-n\Lambda(B)}. \tag{2.8.11}$$

The condition $\Lambda(B) < \infty$ can be relaxed and replaced with $\Lambda(B) = \Lambda([B])$ (see Corollary 1.2.3), but then the proof will become more complicated. If $\Lambda(B) = \infty$ then the point $\alpha_{(B)}$ is not defined and generally the theorem ceases to be true, as the following example demonstrates.

Example 2.8.8. Suppose that $\mathbf{P}(\xi = 0) = \mathbf{P}(\xi = -1) = 1/2$, $\mathbf{P}(\xi_0 = 1) = 1$, $B = (0, \infty)$. Then $\mathcal{A} = \mathcal{A}_0 = \mathbb{R}$, $\mathbf{E}\xi = -1/2 \notin B$,

$$\Lambda(B) = -\ln \mathbf{P}(S_n > 0) = \infty, \qquad -\frac{1}{n} \ln \mathbf{P}(S_n \geqslant 0) = \ln 2 = \Lambda([B]).$$

Since in this case

$$\mathbf{P}(S_{0,n} > 0) = \mathbf{P}(S_n = 0) = 2^{-n},$$

one has

$$\lim_{n\to\infty} \frac{1}{n} \ln \mathbf{P}(S_{0,n} > 0) = -\ln 2 \neq -\Lambda(B) = -\infty.$$

Proof of Theorem 2.8.7. As in section 1.3, the proof is based on the 'one-dimensional' inequality for the probability that $S_{0,n}$ hits the subspace

$$\Pi_> := \{\alpha : \langle e, \alpha \rangle > b\}$$

which contains the set B. We have

$$\mathbf{P}(S_{0,n} \in nB) \leqslant \mathbf{P}(S_{0,n} \in \Pi_>) = \mathbf{P}\left(\left\langle e, \frac{S_{0,n}}{n} \right\rangle > b\right)$$
$$= \mathbf{P}\big(\langle e, \xi_0 \rangle + \langle e, S_n \rangle > nb\big). \tag{2.8.12}$$

As before, let ψ and ψ_0 be the Laplace transforms of the distributions of ξ and ξ_0 respectively. Denote by Λ_e, λ_e the deviation function and its derivative for the random variable $\langle e, \xi \rangle$ and let

$$\psi_{\mathrm{e}}(\mu) := \mathbf{E}e^{\mu\langle \mathrm{e},\xi\rangle}, \qquad \psi_{0,\mathrm{e}}(\mu) := \mathbf{E}e^{\mu\langle \mathrm{e},\xi_0\rangle}.$$

Then by (2.8.7), (2.8.8) and (2.8.12),

$$\mathbf{P}(S_{0,n} \in nB) \leqslant \psi_{0,\mathrm{e}}\big(\lambda_{\mathrm{e}}(b)\big)\, e^{-n\Lambda_{\mathrm{e}}(b)}. \tag{2.8.13}$$

In section 1.3, it was shown that

$$\Lambda_{\mathrm{e}}(b) = \Lambda(B), \qquad \lambda_{(B)} = \mathrm{e}\lambda_{\mathrm{e}}(b).$$

Clearly,

$$\psi_{\mathrm{e}}(\mu) = \mathbf{E}\exp\langle \mu\mathrm{e}, \xi\rangle = \psi(\mu\mathrm{e})$$

and

$$\psi_{\mathrm{e}}\big(\lambda_{\mathrm{e}}(b)\big) = \psi\big(\lambda_{\mathrm{e}}(b)\mathrm{e}\big) = \psi(\lambda_{(B)}),$$
$$\psi_{0,\mathrm{e}}\big(\lambda_{\mathrm{e}}(b)\big) = \psi_0(\lambda_{(B)}).$$

This implies (2.8.11). The theorem is proved. □

2.8.4 Large deviation principles

We shall begin with a 'special' l.d.p. for open convex sets.

Theorem 2.8.9. *For any open convex set B such that $\Lambda(B) < \infty$ and $\lambda_{(B)} \in [\mathcal{A}_0]$, where*

$$\mathcal{A}_0 := \big\{\lambda : \psi_0(\lambda) < \infty\big\},$$

we have

$$\lim_{n\to\infty} \frac{1}{n} \ln \mathbf{P}(S_{0,n} \in nB) = -\Lambda(B). \tag{2.8.14}$$

Proof. The upper bound. If $\lambda_{(B)} \in \mathcal{A}_0$ then the necessary bound follows from Theorem 2.8.7 and from $\psi_0(\lambda_{(B)}) < \infty$.

Now suppose that $\lambda_{(B)} \in \partial\mathcal{A}_0$, $\psi_0(\lambda_{(B)}) = \infty$. We will use the arguments from the proof of Theorem 2.8.7. Without loss of generality, one can assume that $\Lambda_{\mathrm{e}}(b) = \Lambda(B) > 0$, so

$$b > \mathbf{E}\langle \mathrm{e}, \xi\rangle, \qquad \lambda_{\mathrm{e}}(b) > 0.$$

Therefore, for any $\lambda_{\,\mathrm{e}} \in \big(0, \lambda_{\mathrm{e}}(b)\big)$ one has

$$\mathrm{e}\lambda_{\,\mathrm{e}} \in \mathcal{A}_0, \qquad \psi_{0,\mathrm{e}}(\lambda_{\,\mathrm{e}}) = \psi_0(\mathrm{e}\lambda_{\,\mathrm{e}}) < \infty.$$

Instead of inequality (2.8.13), which loses its meaning in our case, consider the Chebyshev inequality for the right-hand side of (2.8.12):

$$\mathbf{P}\big(\langle \mathrm{e}, \xi_0\rangle + \langle \mathrm{e}, S_n\rangle > nb\big) \leqslant \psi_{0,\mathrm{e}}(\lambda_{\,\mathrm{e}})\psi_{\mathrm{e}}^n(\lambda_{\,\mathrm{e}})e^{-nb\lambda_{\mathrm{e}}}.$$

This inequality and (2.8.12) imply that

$$\overline{\lim_{n\to\infty}} \frac{1}{n} \ln \mathbf{P}(S_{0,n} \in nB) \leqslant -\big(b\lambda_e - \ln \psi_e(\lambda_e)\big). \tag{2.8.15}$$

Since the deviation function Λ_e is continuous in the interior of its set of finiteness, we can let λ_e tend to $\lambda_e(b)$ and make the right-hand side of (2.8.15) arbitrarily close to $-\Lambda_e(b) = -\Lambda(B)$. Then we get

$$\overline{\lim_{n\to\infty}} \frac{1}{n} \ln \mathbf{P}(S_{0,n} \in nB) \leqslant -\Lambda(B). \tag{2.8.16}$$

The lower bound. For any point $\beta \in B$ and sufficiently small ε we have, $(\beta)_\varepsilon \in B$,

$$\mathbf{P}(S_{0,n} \in nB) \geqslant \mathbf{P}\big(S_{0,n} \in n(\beta)_\varepsilon\big).$$

Choose $N > 0$ such that $\mathbf{P}\big(|\xi_0| < N\big) > q > 0$. Then, for all $n > 2N/\varepsilon$,

$$\mathbf{P}\left(\frac{S_{0,n}}{n} \in (\beta)_\varepsilon\right) \geqslant q\mathbf{P}\left(\frac{S_n}{n} \in (\beta)_{\varepsilon/2}\right).$$

Hence, according to the l.d.p. in the one-dimensional case we obtain

$$\begin{aligned}\underline{\lim_{n\to\infty}} \frac{1}{n} &\ln \mathbf{P}(S_{0,n} \in nB) \\ &\geqslant \underline{\lim_{n\to\infty}} \frac{1}{n} \ln \mathbf{P}\big(S_n \in (\beta)_{\varepsilon/2}\big) \geqslant -\Lambda\big((\beta)_{\varepsilon/2}\big) \geqslant -\Lambda(\beta).\end{aligned} \tag{2.8.17}$$

Since the left-hand side of this inequality does not depend on β, one has

$$\underline{\lim_{n\to\infty}} \frac{1}{n} \ln \mathbf{P}(S_{0,n} \in nB) \geqslant -\inf_{\beta\in B} \Lambda(\beta) = -\Lambda(B). \tag{2.8.18}$$

Now the required statement (2.8.14) follows from inequalities (2.8.16), (2.8.18). The theorem is proved. □

Before we formulate a local l.d.p. for $S_{0,n}$, let us emphasise the following fact. Denote by Π_α the subspace

$$\Pi_\alpha := \big\{\beta : \langle \beta - \alpha, \lambda(\alpha)\rangle \geqslant 0\big\}.$$

Its boundary $\partial\Pi_\alpha$ is a hyperplane which touches the surface $\big\{\beta : \Lambda(\beta) = \Lambda(\alpha)\big\}$ (the level surface of the deviation function Λ of the level $\Lambda(\alpha)$). By the exponential Chebyshev inequality, we have

$$\begin{aligned}\mathbf{P}\left(\frac{S_n}{n} \in \Pi_\alpha\right) &= \mathbf{P}\left(\left\langle\frac{S_n}{n} - \alpha, \lambda(\alpha)\right\rangle \geqslant 0\right) = \mathbf{P}\big(\langle S_n, \lambda(\alpha)\rangle \geqslant n\langle\alpha, \lambda(\alpha)\rangle\big) \\ &\leqslant e^{-n\langle\alpha,\lambda(\alpha)\rangle}\mathbf{E}e^{\langle S_n,\lambda(\alpha)\rangle} = \exp\Big\{-n\langle\alpha, \lambda(\alpha)\rangle + A(\lambda)\Big\} \\ &= \exp\big\{-n\Lambda(\alpha)\big\}.\end{aligned} \tag{2.8.19}$$

The local l.d.p. has the following form.

Theorem 2.8.10 (An analogue of Theorem 1.2.1). (i) *For all* $\alpha \in \mathbb{R}^d$ *and* $\varepsilon > 0$,

$$\lim_{n\to\infty} \frac{1}{n} \ln \mathbf{P}\left(\frac{S_{0,n}}{n} \in (\alpha)_\varepsilon\right) \geqslant -\Lambda\big((\alpha)_\varepsilon\big). \tag{2.8.20}$$

(ii) *Suppose that* ξ_0, ξ *satisfy the condition* $[\mathbf{C}_0]$. *Then, for all* $\alpha \in \mathbb{R}^d$ *such that* $\lambda(\alpha) \in [\mathcal{A}_0]$, *one has*

$$\lim_{\varepsilon\to 0} \overline{\lim_{n\to\infty}} \frac{1}{n} \ln \mathbf{P}\left(\frac{S_{0,n}}{n} \in (\alpha)_\varepsilon\right) \leqslant -\Lambda(\alpha). \tag{2.8.21}$$

(iii) *Conversely, if for given* α *the distribution of* $S_{0,n}$ *satisfies the upper bound*

$$\overline{\lim_{n\to\infty}} \frac{1}{n} \ln \mathbf{P}\left(\frac{S_{0,n}}{n} \in \Pi_\alpha\right) \leqslant -\Lambda(\alpha) \tag{2.8.22}$$

(see (2.8.19)), *then* $\lambda(\alpha) \in [\mathcal{A}_0]$.

It is not difficult to see that (2.8.20), (2.8.21) imply the existence of a sequence $\varepsilon_n \to 0$ sufficiently slowly as $n \to \infty$ such that

$$\lim_{n\to\infty} \frac{1}{n} \ln \mathbf{P}\left(\frac{S_{0,n}}{n} \in (\alpha)_{\varepsilon_n}\right) = -\Lambda(\alpha)$$

(more details are provided in section 4.1).

For the validity of the condition (2.8.22) it is enough to require that the bound (2.8.21) holds in the local l.d.p. for the projection $\langle S_{0,n}, \lambda(\alpha)\rangle$: i.e. for any $\beta > 0$,

$$\lim_{\varepsilon\to 0} \overline{\lim_{n\to\infty}} \frac{1}{n} \ln \mathbf{P}\left(\frac{\langle S_{0,n}, \lambda(\alpha)\rangle}{n} \in (\beta)_\varepsilon\right) \leqslant -\Lambda(\beta).$$

Proof of Theorem 2.8.10. (i) *The lower bound* (2.8.20) follows immediately from (2.8.18) for $B = (\alpha)_\varepsilon$.

(ii) *The upper bound.* First observe that by condition $[\mathbf{C}_0]$ the set $\mathcal{A}_0$ contains the ball $(0)_\delta$ for some $\delta > 0$. Since $\lambda(\alpha) \in [\mathcal{A}_0]$, the set $[\mathcal{A}_0]$ includes the minimal 'half-cone' which contains the point $\lambda(\alpha)$ and the ball $(0)_\delta$, so the half-interval $\big[0, \lambda(\alpha)\big)$ belongs to $\mathcal{A}_0$.

Consider the half-space

$$\Pi_{\alpha,q} := \big\{\beta : \langle \lambda(\alpha), \beta\rangle > q\langle\lambda(\alpha), \alpha\rangle\big\}.$$

Clearly, for $q \in (0,1)$ and sufficiently small ε, it contains the neighbourhood $(\alpha)_\varepsilon$ and, by the above argument, $q\lambda(\alpha) \in \mathcal{A}_0$. Thus, for $q < 1$ and sufficiently small $\varepsilon > 0$,

$$P_n := \mathbf{P}\left(\frac{S_{0,n}}{n} \in (\alpha)_\varepsilon\right) \leqslant \mathbf{P}\left(\frac{S_{0,n}}{n} \in \Pi_{\alpha,q}\right)$$
$$= \mathbf{P}\big(q\langle\lambda(\alpha), \xi_0 + S_n\rangle > q^2 n\langle\lambda(\alpha), \alpha\rangle\big).$$

By the exponential Chebyshev inequality we have

$$P_n \leqslant e^{-q^2 n\langle\lambda(\alpha),\alpha\rangle}\mathbf{E}e^{q\langle\lambda(\alpha),\xi_0+S_n\rangle},$$

where $\mathbf{E}e^{q\langle\lambda(\alpha),\xi_0\rangle} = A_0\big(q\lambda(\alpha)\big) < \infty$. Therefore

$$\overline{\lim_{n\to\infty}} \frac{1}{n} \ln P_n \leqslant -q^2\langle\lambda(\alpha),\alpha\rangle + qA\big(\lambda(\alpha)\big).$$

Obviously, the same inequality is also satisfied by

$$\lim_{\varepsilon\to 0} \overline{\lim_{n\to\infty}} \frac{1}{n} \ln P_n.$$

However, the right-hand sides of these inequalities converge to $-\Lambda(\alpha)$ as $q \uparrow 1$. This proves (2.8.21).

(iii) *Necessity.* First consider the one-dimensional case $d = 1$. Without loss of generality, one can assume that $\mathbf{E}\xi = 0, \alpha > 0$. Suppose the contrary, that (2.8.22) holds but $\lambda(\alpha) \notin [\mathcal{A}_0]$, i.e.

$$\lambda(\alpha) > \lambda_{0,+} := \sup\big\{\lambda : \psi_0(\lambda) < \infty\big\}.$$

Since necessarily $\lambda_{0,+} < \infty$, for any $\delta > 0$ we have

$$\overline{\lim_{x\to\infty}} \frac{1}{x} \ln \mathbf{P}(\xi_0 \geqslant x) \geqslant -\lambda_{0,+} - \delta. \tag{2.8.23}$$

Indeed, if one supposes that (2.8.23) is not true then, for some $\delta > 0$, the left-hand side of (2.8.23) will be less than $-\lambda_{0,+} - \delta$. But this implies the finiteness of $\psi_0(\lambda_{0,+} + \delta/2)$, which is impossible.

Inequality (2.8.23) implies that for $\delta < (\lambda(\alpha) - \lambda_{0,+})/2$ there exists a sequence $x_k \to \infty$ as $k \to \infty$ such that, for all sufficiently large k,

$$\ln \mathbf{P}(\xi_0 \geqslant x_k) \geqslant -x_k\big(\lambda(\alpha) - \delta\big). \tag{2.8.24}$$

If condition (2.8.22) is met then, for $\alpha > 0$ and $n \to \infty$, one has

$$\begin{aligned} Q_n &:= \frac{1}{n} \ln \mathbf{P}(S_{0,n} \in \Pi_\alpha) = \frac{1}{n} \ln \mathbf{P}(S_{0,n} \geqslant \alpha n) \\ &= \frac{1}{n} \ln \int \mathbf{P}(\xi_0 \in dy)\mathbf{P}(S_n \geqslant \alpha n - y) = -\Lambda(\alpha) + o(1). \end{aligned} \tag{2.8.25}$$

To prove the necessity, we will show that the inequality $\lambda(\alpha) > \lambda_{0,+}$ contradicts (2.8.25). Indeed,

$$Q_n \geqslant \frac{1}{n} \ln \mathbf{P}(\xi_0 \geqslant x_k)\mathbf{P}(S_n \geqslant \alpha n - x_k).$$

Let $n = [x_k/\gamma] + 1$ for small γ, which will be chosen later. Then, by (2.8.24) and the l.d.p. for S_n

$$\begin{aligned} Q_n &\geqslant \frac{1}{n}\Big[-n\Lambda(\alpha-\gamma) - \gamma n\big(\lambda(\alpha) - \delta\big) + o(n)\Big], \\ \lim_{n\to\infty} Q_n &\geqslant -\Big[\Lambda(\alpha-\gamma) + \gamma\big(\lambda(\alpha) - \delta\big)\Big], \end{aligned}$$

where, for small γ,

$$\Lambda(\alpha-\gamma) + \gamma\big(\lambda(\alpha) - \delta\big) = \Lambda(\alpha) - \gamma\delta + o(\gamma^2).$$

Choosing $\gamma = \delta^2$, we obtain that for sufficiently small δ,

$$\underline{\lim}_{n\to\infty} Q_n \geqslant -\Lambda(\alpha) + \frac{\delta^3}{2}.$$

This contradicts (2.8.25) and proves that the inequality $\lambda(\alpha) > \lambda_{0,+}$ is impossible if (2.8.22) holds.

Now suppose that $d > 1$. Consider the random variables $\langle\xi, \lambda(\alpha)\rangle$, $\langle\xi_0, \lambda(\alpha)\rangle$. For them one has

$$\psi_\alpha(v) := \mathbf{E}e^{v\langle\xi,\lambda(\alpha)\rangle} = \psi\big(v\lambda(\alpha)\big),$$
$$\psi_{0,\alpha}(v) := \mathbf{E}e^{v\langle\xi_0,\lambda(\alpha)\rangle} = \psi_0\big(v\lambda(\alpha)\big).$$

The relation $\lambda(\alpha) \notin [\mathcal{A}_0]$ means that

$$1 > v_{0,+} := \sup\big\{v : \psi_{0,\alpha}(v) < \infty\big\}. \tag{2.8.26}$$

However, for the deviation function $\Lambda_\alpha(\beta)$ of the random variable $\langle\xi, \lambda(\alpha)\rangle$, for $\beta = \langle\alpha, \lambda(\alpha)\rangle$, $A_\alpha(v) = \ln\psi_\alpha(v)$, one has

$$\Lambda_\alpha(\beta) = \sup_v \big(v\beta - A_\alpha(v)\big) = \sup_v \Big(v\langle\alpha, \lambda(\alpha)\rangle - A\big(v\lambda(\alpha)\big)\Big).$$

The l.d.ps. for S_n and $\langle S_n, \lambda(\alpha)\rangle$ imply that $\sup_v$ is attained here at $v = 1$, so

$$\lambda_\alpha(\beta) = \Lambda'_\alpha(\beta) = 1, \qquad \Lambda_\alpha(\beta) = \Lambda(\alpha).$$

Therefore, relation (2.8.22) can be written as

$$\overline{\lim_{n\to\infty}} \frac{1}{n} \ln \mathbf{P}\bigg(\frac{\langle S_{0,n}, \lambda(\alpha)\rangle}{n} > \beta\bigg) \leqslant -\Lambda_\alpha(\beta). \tag{2.8.27}$$

Since $\lambda_\alpha(\beta) = 1$, relation (2.8.26) means that $\lambda_\alpha(\beta) > v_{0,+}$. According to statement (iii), which has already been proved in the one-dimensional case, this contradicts (2.8.27). The theorem is proved. □

The values of $\lambda(\alpha)$ fill the whole set $\mathcal{A}$ when α runs through the set $\mathbb{R}^d$, so (2.8.10) implies the following.

Corollary 2.8.11. *Suppose that ξ_0 and ξ satisfy condition* **[C$_0$]**. *Then the condition $\mathcal{A} \subset [\mathcal{A}_0]$ is necessary and sufficient for the validity of the l.d.ps.* (2.8.20), (2.8.21) *for all $\alpha \in \mathbb{R}^d$.*

It is not difficult to see that the condition $\mathcal{A} \subset [\mathcal{A}_0]$ is equivalent to the condition $(\mathcal{A}) \subset \mathcal{A}_0$.

Theorem 2.8.12 (The 'usual' l.d.p.). *Suppose that the condition* $[\mathbf{C}_0]$ *holds for some* ξ *and* ξ_0. *If* $\mathcal{A} \subset [\mathcal{A}_0]$ *then, for any Borel set* $B \subset \mathbb{R}^d$, *one has the inequalities*

$$\varlimsup_{n\to\infty} \frac{1}{n} \ln \mathbf{P}\left(\frac{S_{0,n}}{n} \in B\right) \leqslant -\Lambda\big([B]\big), \tag{2.8.28}$$

$$\varliminf_{n\to\infty} \frac{1}{n} \ln \mathbf{P}\left(\frac{S_{0,n}}{n} \in B\right) \geqslant -\Lambda\big((B)\big) \tag{2.8.29}$$

Conversely, if inequalities (2.8.28), (2.8.29) *are valid then* $\mathcal{A} \subset [\mathcal{A}_0]$.

The proof of the sufficiency is based on Theorem 2.8.10 and is similar to the proofs of Theorems 2.7.3 and 2.7.4. The necessity follows from Theorem 2.8.10, since the usual l.d.p. implies the local l.d.p.

The statement about the sufficiency in Theorem 2.8.12, apparently, can be somewhat extended, if one does not require the condition $\mathcal{A} \subset [\mathcal{A}_0]$ but requires that inequalities (2.8.28), (2.8.29) hold for sets B such that $\lambda_{(B)} \in [\mathcal{A}_0]$.

Along with the l.l.d.p. in Theorem 2.8.10, it is possible to provide a stronger statement about the asymptotics of the probability that $S_{0,n}$ enters a narrower zone, a half-open rectangular parallelepiped $\Delta[n\alpha)$ with fixed side lengths $\Delta_{(1)}, \ldots, \Delta_{(d)}$, for vectors ξ with coordinates having various distributions, either non-lattice or arithmetical.

Theorem 2.8.13. *Suppose that* $\mathcal{A} \subset [\mathcal{A}_0]$ *and the coordinates of the vector* ξ *can have either a non-lattice or arithmetic distribution (not necessarily the same), and if* $\xi_{(k)}$, $1 \leqslant k \leqslant d$, *is an arithmetic coordinate then the coordinate* $\xi_{0(k)}$ *must be integer-valued. Then, for any fixed* $\Delta_{(k)}$ *for the non-lattice coordinates* $\xi_{(k)}$ *and for integer* $\Delta_{(k)} \geqslant 1$ *for the arithmetic coordinates* $\xi_{(k)}$, *there exists*

$$\lim_{n\to\infty} \frac{1}{n} \ln \mathbf{P}\big(S_{0,n} \in \Delta[n\alpha)\big) = -\Lambda(\alpha).$$

Proof. The upper bound follows from Theorem 2.8.10. The lower bound can be proved as in Theorem 1.2.1 (see Section 2.7.3 and Remark 2.8.4).

2.9 Asymptotics of the renewal function and related problems. The second deviation function

2.9.1 Introduction

As before, let

$$S_n = \sum_{k=1}^{n} \xi_k$$

be a sum of independent identically distributed vectors ξ_k. We will assume that the distribution of ξ is non-degenerate (i.e. it is not concentrated on some hyperplane). Consider a Borel set $B \subset \mathbb{R}^d$ and define the renewal function (measure) H by the equality

$$H(B) = \sum_{n=1}^{\infty} \mathbf{P}(S_n \in B). \tag{2.9.1}$$

If B does not contain a neighbourhood of the point 0 then, for a sequence of sets TB, moving away as $T \to \infty$, we will be interested in the analytical properties of the function $H(TB)$.

Often in our considerations the set B will be a half-open cube $\Delta[\alpha)$ with side Δ. In the arithmetic case, we will also consider single point sequences $TB = T\alpha = x$, $|x| \to \infty$, with integer coordinates.

The renewal function plays an important role in the study of many problems (see e.g. section 4.10). Let

$$\eta(TB) = \min\{n \geqslant 1 : S_n \in TB\}.$$

Along with the asymptotics of the function $H(TB)$, it is also of interest to study the asymptotics of the function

$$H^{(r)}(TB) := \mathbf{E}\big(\eta^r(TB);\ \eta(TB) < \infty\big)$$

as $T \to \infty$, $r = 0, 1, \ldots$ For $r = 0$ the function

$$P(TB) := H^{(0)}(TB) = \mathbf{P}\big(\eta(TB) < \infty\big) = \mathbf{P}\Big(\bigcup_{n=1}^{\infty}\{S_n \in TB\}\Big)$$

is the probability that a trajectory $\{S_n\}$ will enter the set TB at some moment of time.

A large number of results are devoted to the study of the asymptotic behaviour of the one-dimensional renewal function (see e.g. [92], [171], [173], [22], [91]). A multidimensional variant of a local renewal theorem was obtained for the first time, apparently, by R.A. Doney [86]: *if Cramér's condition on a characteristic function holds and* $\mathbf{E}|\xi|^r < \infty$ *for sufficiently large r then, for the unit vector* e *and a bounded measurable set B such that $\mu(\partial B) = 0$ (μ is the Lebesgue measure in $\mathbb{R}^d$), the following equalities hold as $T \to \infty$:*

$$H(T\mathrm{e} + B) = \begin{cases} cT^{(d-1)/2}\mu(B)\big(1 + o(1)\big), & \textit{if} \quad \mathrm{e} = \mathbf{E}\xi/|\mathbf{E}\xi|, \\ o(T^{(d-1)/2}, & \textit{if} \quad \mathrm{e} \neq \mathbf{E}\xi/|\mathbf{E}\xi|. \end{cases} \tag{2.9.2}$$

A.J. Stam [172] obtained the minimal sufficient conditions for the validity of (2.9.2): $\mathbf{E}|\xi|^r < \infty$ for $r = \min\{2, (d-1)/2\}$; later the same result was obtained by A.V. Nagaev [143]. One should also mention the paper [10], where the asymptotics of $H\big(\mathcal{K}(T, T + b)\big)$ for the ring $\mathcal{K}(T, T + b)$ in the plane R^2 of constant width b and growing radius $T \to \infty$ was obtained.

In [50], a rather complete description of the asymptotics of $H\big(T\Delta[\alpha)\big)$ and $H^{(r)}\big(T\Delta[\alpha)\big)$, $r = 0, 1, \ldots$ and $T \to \infty$, was obtained, which in many respects improved the above-mentioned results. Great attention was devoted in [50] to the case when the condition **[C]** holds. Below we provide a number of results from [50]. A more complete list of references can be found in that paper and also in [101].

2.9.2 The second deviation function and its properties

Let

$$B_{T,\alpha} = T\Delta[\alpha),$$

where $\Delta = \Delta_T \to 0$ sufficiently slowly as $T \to \infty$, and consider the asymptotics of the function $H(B_{T,\alpha})$ for $T \to \infty$ and fixed α. According to (2.9.1) and the l.d.p. for S_n, the value of $H(B_{T,\alpha})$ is the sum over n of terms with the exponential part

$$e^{-n\Lambda(\alpha T/n)} = e^{-T\theta\Lambda(\alpha/\theta)}, \qquad \theta = \frac{n}{T}.$$

Therefore, one can naturally expect that the exponential part of $H(B_{T,\alpha})$ as $T \to \infty$ will be of the form

$$\exp\{-TD_\Lambda(\alpha)\},$$

where

$$D_\Lambda(\alpha) = \inf_{\theta>0} \theta\Lambda\left(\frac{\alpha}{\theta}\right). \tag{2.9.3}$$

This is the so-called *second deviation function*. Along with the function $\theta\Lambda(\alpha/\theta)$, it plays a crucial role in the description of the asymptotics of the functions $H(TB)$, $H^{(r)}(TB)$ and also in a number of other problems; in Chapter 3, for example, in the construction of the level curves of a random walk for finding the asymptotics $\mathbf{P}(\sup S_n > T) \sim ce^{-TD_\Lambda(1)}$ as $T \to \infty$ in the case $d = 1$, $\mathbf{E}\xi < 0$ (see below), and in many other problems.

It follows directly from the definition (2.9.3) that *the function $D_\Lambda(\alpha)$ is linear along each ray, i.e. for $t > 0$ and any $\alpha \in \mathbb{R}^d$,*

$$D_\Lambda(t\alpha) = tD_\Lambda(\alpha). \tag{2.9.4}$$

Since $\Lambda(\alpha) = 0$ for $\alpha = a = \mathbf{E}\xi$, then, for $t \geqslant 0$,

$$D_\Lambda(ta) = tD_\Lambda(a) = 0. \tag{2.9.5}$$

From the definition (2.9.3), it also follows that

$$D_\Lambda(\alpha) \leqslant \Lambda(\alpha). \tag{2.9.6}$$

Suppose that $a \neq 0$, the condition $[\mathbf{C}_0]$ holds and the coordinates $\xi_{(i)}$ may assume positive as well as negative values. In this case $\Lambda(0) = -\inf A(\lambda) \in (0,\infty)$, the function $\Lambda(\alpha)$ is finite, positive and continuous in a neighbourhood of the point 0 and $\lim_{\theta\to\infty} \theta\Lambda(\alpha/\theta) = \infty$. Then, by the property $(\overrightarrow{\Lambda}9)$,

$$\varliminf_{\theta\to 0} \theta\Lambda\left(\frac{\alpha}{\theta}\right) \geqslant c > 0.$$

Hence, if α is not collinear to a and $\Lambda(\alpha\theta) < \infty$ for some $\theta > 0$ then $D_\Lambda(\alpha) \in (0,\infty)$. This means that $D_\Lambda(t\alpha)$ grows linearly in t. This fact makes the second deviation function considerably different from the deviation function $\Lambda(\alpha)$. Let us prove the following result.

Lemma 2.9.1. *The function $D_\Lambda(\alpha) \geqslant 0$ is convex.*

Proof. We need to show that, for any $p \geqslant 0, q \geqslant 0, p + q = 1, \alpha, \beta \in \mathbb{R}^d$,

$$D_\Lambda(p\alpha + q\beta) \leqslant pD_\Lambda(\alpha) + qD_\Lambda(\beta). \tag{2.9.7}$$

Since the function $\Lambda(\alpha)$ is convex for

$$p' := \frac{t}{t+u}, \qquad q' := \frac{u}{t+u}$$

as $t, u > 0$, we have

$$\begin{aligned} t\Lambda\left(\frac{p\alpha}{t}\right) + u\Lambda\left(\frac{q\beta}{u}\right) &= (t+u)\left[p'\Lambda\left(\frac{p\alpha}{t}\right) + q'\Lambda\left(\frac{q\beta}{u}\right)\right] \\ &\geqslant (t+u)\Lambda\left(\frac{p'p\alpha}{t} + \frac{q'q\beta}{u}\right) \\ &= (t+u)\Lambda\left(\frac{p\alpha + q\beta}{t+u}\right) \geqslant D_\Lambda(p\alpha + q\beta). \end{aligned}$$

Since the right-hand side of the latter inequality does not depend on $t > 0, u > 0$, this inequality will remain true if we replace its left-hand side with the supremum over $t > 0$ and $u > 0$. This will give inequality (2.9.7). The lemma is proved. □

The above arguments imply that the surface $y = D_\Lambda(\alpha)$ in the space $\mathbb{R}^{d+1}$ of variables (y, α) is the boundary of the convex cone with vertex at the origin, which touches the hyperplane $y = 0$ only along the ray $\{y = 0;\ t\alpha,\ t \geqslant 0\}$.

Denote by $\mathcal{D} = \{\alpha \in \mathbb{R}^d : D_\Lambda(\alpha) < \infty\}$ the zone, where $D_\Lambda(\alpha) < \infty$. Inside $\mathcal{D}$, the function D_Λ is continuous since it is convex. It is clear that if $B \cap \mathcal{D} = \varnothing$ then the random walk $\{S_n\}$ will never enter the zone $\bigcup_{t>0}(tB)$, since in that case $\Lambda(tB) = \infty$ and $\mathbf{P}(S_n \in tB) = 0$ for all $n \geqslant 1, t > 0$. This implies that the set $\mathcal{D}$ is the interior of a convex cone with vertex at the origin. It is clear that if, for example, $\mathbf{P}(\xi_{(1)} \geqslant 0) = 1$ then the half-space $\alpha_{(1)} < 0$ does not intersect $\mathcal{D}$.

As noted above, one can naturally expect that the function $D_\Lambda(\alpha)$ will describe the asymptotics of of the sequence $-\frac{1}{T}\ln H(B_{T,\alpha})$ as $T \to \infty$. However, this is not precisely true. It is not difficult to see that the function of the variable α equal to

$$-\varliminf_{T\to\infty} \frac{1}{T} \ln H(B_{T,\alpha})$$

is lower semicontinuous (see also section 4.1). At the same time, the function $D_\Lambda(\alpha)$, defined in (2.9.3), in general is not lower semicontinuous. If, for instance, $\mathbf{P}(\xi_{(1)} > 0) = 1$ then $a \neq 0$, $\Lambda(0) = \infty$ and $D_\Lambda(0) = \infty$, but according to (2.9.5)

$$\lim_{t\to 0} D_\Lambda(ta) = 0 < \infty = D_\Lambda(0).$$

If at the boundary points $\partial\mathcal{D}$ of the set $\mathcal{D}$ (i.e. at the points of discontinuity of the function D_Λ if such points exist) we 'correct' the function $D_\Lambda(\alpha)$ by changing $D_\Lambda(\alpha)$, using continuity, to the value

$$D(\alpha) := \lim_{\varepsilon\to 0} D_\Lambda\big((\alpha)_\varepsilon\big), \tag{2.9.8}$$

where $D_\Lambda(B) = \inf_{\alpha\in B} D_\Lambda(\alpha)$, and at other points α ($\alpha \notin \partial\mathcal{D}$) put

$$D(\alpha) = D_\Lambda(\alpha)$$

then we will obtain the function $D(\alpha)$, which has all the required properties.

Henceforth the 'corrected' function D will be called the *second deviation function*. It is precisely this function that will be used in the formulation of a local l.d.p. for the renewal function H (see Theorem 2.9.7).

Denote

$$\mathcal{A}^{\leqslant 0} = \big\{\lambda : A(\lambda) \leqslant 0\big\} = \big\{\lambda : \psi(\lambda) \leqslant 1\big\}.$$

This is a non-empty convex set since the function $\psi(\lambda)$ is convex and continuous from below and $0 \in \mathcal{A}^{\leqslant 0}$.

Theorem 2.9.2. *The function $D(\alpha)$ is convex and lower semicontinuous,*

$$D(\alpha t) = tD(\alpha), \qquad D(\alpha + \beta) \leqslant D(\alpha) + D(\beta). \tag{2.9.9}$$

For all $\alpha \in \mathbb{R}^d$, it holds that

$$D(\alpha) = \sup_{\lambda\in\mathcal{A}^{\leqslant 0}} \langle\lambda,\alpha\rangle = \sup_{\lambda\in\partial\mathcal{A}^{\leqslant 0}} \langle\lambda,\alpha\rangle, \tag{2.9.10}$$

where $\partial\mathcal{A}^{\leqslant 0}$ is the boundary of $\mathcal{A}^{\leqslant 0}$.

Proof.[2] Let us prove (2.9.10). For a non-negative convex function $G(\lambda)$, $\lambda \in \mathbb{R}^d$, denote by $G^*(\alpha)$, $\alpha \in \mathbb{R}^d$, the Legendre transform of $G(\lambda)$:

$$G^*(\alpha) = \sup_{\lambda} \big(\langle\lambda,\alpha\rangle - G(\lambda)\big),$$

which is necessarily a convex lower semicontinuous function (see e.g. [161]). Introduce the function

$$A^{(\leqslant 0)}(\lambda) = \begin{cases} 0, & \text{if } A(\lambda) \leqslant 0 \quad (\lambda \in \mathcal{A}^{\leqslant 0}), \\ \infty, & \text{if } A(\lambda) > 0 \quad (\lambda \notin \mathcal{A}^{\leqslant 0}). \end{cases} \tag{2.9.11}$$

Then

$$\sup_{\lambda\in\mathcal{A}^{\leqslant 0}} \langle\lambda,\alpha\rangle = \sup_{\lambda\in\mathbb{R}^d} \big(\langle\lambda,\alpha\rangle - A^{(\leqslant 0)}(\lambda)\big) = A^{(\leqslant 0)*}(\alpha).$$

We need to prove that

$$D(\alpha) = A^{(\leqslant 0)*}(\alpha). \tag{2.9.12}$$

[2] There is an error in the proof of (2.9.10) in Theorem 1 in [50].

Denote

$$\Lambda_\theta(\alpha) = \theta\Lambda\left(\frac{\alpha}{\theta}\right).$$

Then

$$\begin{aligned} D^*_\Lambda(\lambda) &= \sup_\alpha \left\{\langle\lambda,\alpha\rangle - \inf_{\theta>0}\Lambda_\theta(\alpha)\right\} = \sup_\alpha \sup_{\theta>0}\left\{\langle\lambda,\alpha\rangle - \Lambda_\theta(\alpha)\right\} \\ &= \sup_{\theta>0}\sup_\alpha \left\{\langle\lambda,\alpha\rangle - \Lambda_\theta(\alpha)\right\} = \sup_{\theta>0}\theta \sup_{\alpha/\theta}\left\{\langle\lambda,\alpha/\theta\rangle - \Lambda(\alpha/\theta)\right\} \\ &= \sup_{\theta>0}\theta\Lambda^*(\lambda) = \sup_{\theta>0}\theta A(\lambda) = A^{(\leqslant 0)}(\lambda), \end{aligned} \tag{2.9.13}$$

where the function $A^{(\leqslant 0)}(\lambda)$ is defined in (2.9.11). In the last equality we used the well-known identity

$$\Lambda^*(\lambda) = A^{**}(\lambda) = A(\lambda),$$

which follows since the double Legendre transform converts a convex lower semicontinuous function into itself. Applying the Legendre transform to both sides of the relation (2.9.13), by (2.9.12) we obtain

$$D^{**}_\Lambda(\alpha) = A^{(\leqslant 0)*}(\alpha). \tag{2.9.14}$$

As has been observed, the double Legendre transform 'corrects' a convex function at points of discontinuity in a way such that it becomes lower semicontinuous while at the same time preserving the property of convexity. Therefore $D^{**}_\Lambda(\alpha) = D(\alpha)$ and hence, in view of (2.9.14),

$$D(\alpha) = A^{(\leqslant 0)*}(\alpha).$$

This implies (2.9.10) and also the convexity and lower semicontinuity of the function D. However, the last property does not need to be proved since it has been already established (see the definition of the function D). The properties (2.9.9) follow from (2.9.4) and the convexity of D. Theorem 2.9.2 is proved. □

From the properties (2.9.9), it follows that surfaces at level v,

$$\Gamma_v = \left\{\alpha : D(\alpha) = v\right\},$$

form a family of concentric unbounded surfaces which contain the convex sets $D_v := \left\{\alpha : D(\alpha) \leqslant v\right\}$.

Then, from Theorem 2.9.2, it follows that

$$D(\alpha) \equiv 0 \quad \text{if and only if} \quad \Lambda(0) = 0. \tag{2.9.15}$$

Indeed, the condition $\Lambda(0) = 0$ is equivalent to $\inf\psi(\lambda) = 1$, which in turn is equivalent to the condition $\mathcal{A}^{\leqslant 0} = \{0\}$, and the latter one is equivalent to the condition $D(\alpha) \equiv 0$ by Theorem 2.9.2.

Example 2.9.3. Let a non-degenerate vector ξ have a normal distribution with mean $a = \mathbf{E}\xi \neq 0$ and matrix of second central moments M. Then

$$\Lambda(\alpha) = \frac{1}{2}(\alpha - a)M^{-1}(\alpha - a)^T \tag{2.9.16}$$

(see section 1.2.2) and

$$\theta\Lambda\left(\frac{\alpha}{\theta}\right) = \frac{1}{2}\left[\frac{1}{\theta}\alpha M^{-1}\alpha^T - 2\alpha M^{-1}a^T + \theta a M^{-1}a^T\right].$$

Clearly, the function $\theta\Lambda(\alpha/\theta)$ increases unboundedly as $\theta \to 0$ and $\theta \to \infty$ and attains its minimal value at the point θ_α which is the unique solution of the equation

$$\frac{\partial}{\partial t}\left(\theta\Lambda\left(\frac{\alpha}{\theta}\right)\right) = 0 = -\frac{1}{\theta^2}\alpha M^{-1}\alpha^T + aM^{-1}a^T. \tag{2.9.17}$$

If we set

$$Q^2(\alpha) := \alpha M^{-1}\alpha^T, \qquad Q(\alpha, a) := \alpha M^{-1}a^T, \tag{2.9.18}$$

then (2.9.16) implies that

$$\theta_\alpha = \frac{Q(\alpha)}{Q(a)}.$$

Substituting this value into (2.9.16), we find

$$D(\alpha) = Q(\alpha)Q(a) - Q(\alpha, a). \tag{2.9.19}$$

If one denotes $M^{-1} =: Q$ then $Q(\alpha, a)$ can be written as $\langle\alpha Q, aQ\rangle$, so that

$$D(\alpha) = |\alpha Q|\,|aQ| - \langle\alpha Q, aQ\rangle. \tag{2.9.20}$$

This implies that $D(\alpha) = 0$ if and only if αQ and aQ are collinear, i.e. when α and a are collinear.

Let us return to the case of an arbitrary distribution and study the behaviour of the function $D(\alpha)$ in a neighbourhood of the point $a = \mathbf{E}\xi$.

Theorem 2.9.4. *Suppose that the condition* $[\mathbf{C_0}]$ *holds,* $a = \mathbf{E}\xi \neq 0$ *and* M *is the matrix of second central moments of* ξ. *Then, with the notation of* (2.9.18), *for* $\alpha = a + \delta$, $\delta \to 0$ *the following representation is true:*

$$D(\alpha) = \frac{1}{2}\left[\frac{Q^2(\delta)Q^2(a) - Q^2(\delta, a)}{Q^2(a)}\right] + o(|\delta|^3). \tag{2.9.21}$$

The quadratic form in δ, *which appears on the right-hand side of* (2.9.21), *can be represented as* $c(\delta/|\delta|, a, M)|\delta|^2$, *where the coefficient* $c(\cdot) \geqslant 0$ *vanishes only if* δ *and* a *are collinear.*

Proof. Since $D(\alpha) \leqslant \Lambda(\alpha)$, $D(\alpha)$ is close to 0 for $\alpha = a + \delta$ and small $|\delta|$. At the same time, the value of θ_α for which

$$D(\alpha) = \theta_\alpha\Lambda\left(\frac{\alpha}{\theta_\alpha}\right)$$

is necessarily close to 1. Indeed, the cases $|\theta_\alpha| \to 0$ and $|\theta_\alpha| \to \infty$ are impossible in view of the property ($\vec{\Lambda}$ 9) and the relation $\Lambda(0) > 0$. 'Proper' values θ_α which are not close to 1 can be also excluded since for them $\Lambda(\alpha/\theta_\alpha)$ is separated from 0.

Consequently, $D(\alpha) \sim \Lambda(\alpha/\theta_\alpha)$ as $\alpha \to a$ and the asymptotics of the functions $D(\alpha)$ and θ_α near the point a are completely determined by the asymptotic behaviour of the function $\Lambda(\alpha)$ in a neighbourhood of the point $\alpha = a$. However, in such a neighbourhood, the function $\Lambda(\alpha)$ behaves like the similar deviation function for normally distributed ξ with parameters (a, M) (see the property ($\vec{\Lambda}$ 2)), i.e.

$$\Lambda(a+\delta) = \frac{1}{2}\,\delta M^{-1}\delta + o\big(|\delta|^3\big) = \frac{Q^2(\delta)}{2} + o\big(|\delta|^3\big).$$

This implies also that the functions $D(\alpha)$ and θ_α, asymptotically as $\delta \to 0$, behave like the corresponding functions for the normal law with parameters (a, M). Therefore we can use Example 2.9.3 and conclude that

$$D(\alpha) = Q(\alpha)Q(a) - Q(\alpha, a) + o\big(|\delta|^3\big). \tag{2.9.22}$$

Hence, for $\alpha = a + \delta$, we find

$$\begin{aligned} Q(\alpha) &= [\alpha M^{-1}\alpha^T]^{1/2} = \big[Q^2(a) + 2Q(\delta,a) + Q^2(\delta)\big]^{1/2} \\ &= Q(a)\left[1 + \frac{Q(\delta,a)}{Q^2(a)} + \frac{Q^2(\delta)}{2Q^2(a)} - \frac{1}{2}\frac{Q^2(\delta,a)}{Q^4(a)} + o\big(|\delta|^3\big)\right], \\ Q(\alpha,a) &= \alpha M^{-1}a^T = Q^2(a) + Q(\delta,a). \end{aligned}$$

Substituting these into (2.9.22), we obtain (2.9.21). The last claim of the theorem follows from (2.9.20). The theorem is proved. □

In the one-dimensional case let

$$\lambda_{1+} = \sup\{\lambda : \psi(\lambda) \leqslant 1\}, \qquad \lambda_{1-} = \inf\{\lambda : \psi(\lambda) \leqslant 1\}.$$

From Theorem 2.9.2 follows

Corollary 2.9.5. *In the one-dimensional case the second deviation function is of the form*

$$D(\alpha) = \begin{cases} \alpha\lambda_{1+}, & \text{if } \alpha \geqslant 0, \\ \alpha\lambda_{1-}, & \text{if } \alpha \leqslant 0. \end{cases}$$

In [50] a number of other properties of the function $D(\alpha)$ were also established.

2.9.3 Limit theorems

We will need the following condition.

[E] *Either*

(i) *the expectation* $\mathbf{E}\xi \neq 0$ *exists* ($\mathbf{E}\xi_i$ *may be equal to* $\pm\infty$)

or

(ii) $d \geqslant 3$, $\mathbf{E}\xi = 0$, *and there exists the matrix of second moments* $M = \mathbf{E}\xi^T\xi$.

The following result is an analogue of the large deviation principle.

Theorem 2.9.6. *Suppose that the condition* **[E]** *holds and* $\Delta = \Delta_T \to 0$ *sufficiently slowly as* $T \to \infty$*. Then, for any* $\alpha \in \mathbb{R}^d$,

$$\lim_{T\to\infty} \frac{1}{T} \ln H\big(T\Delta[\alpha)\big) = -D(\alpha). \tag{2.9.23}$$

The same statement is true for the probability $P\big(T\Delta[\alpha)\big) = H^{(0)}\big(T\Delta[\alpha)\big)$*. If additionally the condition*

$$\sum n^r \mathbf{P}\big(|S_n| \leqslant N\big) < \infty$$

holds for any $N < \infty$*, then the relation* (2.9.23) *holds also for the function* $H^{(r)}\big(T\Delta[\alpha)\big)$.

Theorem 2.9.6 follows from Theorem 4 in [50]. As has already been noted, the essence of the proof of Theorem 2.9.6 is rather simple. The value $H\big(T\Delta[\alpha)\big)$ is the sum $\sum \mathbf{P}\big(S_n \in T\Delta[\alpha)\big)$, in which the exponential part of the terms, according to the local l.d.p., has the form for large n and T

$$e^{-n\Lambda(\alpha T/n)} = e^{-T\theta\Lambda(\alpha/\theta)}$$

where $\theta = n/T$. Therefore, the exponential part of the sum $H\big(T\Delta[\alpha)\big)$ will be determined by the value

$$\inf_{\theta>0} \theta\Lambda\left(\frac{\alpha}{\theta}\right) = D(\alpha).$$

For the function $P\big(T\Delta[\alpha)\big) < H\big(T\Delta[\alpha)\big)$, a lower bound can be obtained from the inequality $P\big(T\Delta[\alpha)\big) \geqslant \mathbf{P}\big(S_{n_T} \in T\Delta[\alpha)\big)$, where one should take the value $[T\theta_\alpha]$ as n_T and the point where the infimum is attained in the function $D(\alpha)$ as θ_α.

Let us now consider theorems on the exact asymptotics of the function $H\big(T\Delta[\alpha)\big)$.

Theorem 2.9.7. *Let* ξ *be non-arithmetic,* $\mathbf{E}\xi \neq 0$, $T = |x| \to \infty$, $d \leqslant 2$ *and* $\Delta = \Delta_T \to 0$ *sufficiently slowly as* $T \to \infty$*. Then*

$$H\big(\Delta[x)\big) \sim \frac{\Delta^d c(\mathrm{e})}{T^{(d-1)/2}}\, e^{-TD(\mathrm{e})}, \tag{2.9.24}$$

where $\mathrm{e} = \mathrm{e}(x) = x/|x|$ *and the function* $c(\mathrm{e})$ *is found in an explicit form. For integer* $\Delta \geqslant 1$ *the claim* (2.9.24) *remains true in the arithmetic case as well.*

This statement and a similar one for $d > 2$ follow from Theorem 5 in [50], if instead of using Cramér's condition on a characteristic function one uses the Stone–Shepp integro-local theorem. When Cramér's condition holds, along with (2.9.24), an asymptotic expansion for $H\big(\Delta[x)\big)$ in powers of $T^{-1/2}$ was also obtained in Theorem 5 in [50].

In Theorem 2.9.7, along with the cube $\Delta[x)$ one can also consider a parallelepiped $\prod_{k=1}^{d} \Delta_{(k)}[x_{(k)})$ with vertex at a point $x = (x_{(1)}, \dots, x_{(k)})$ and sides of length $\Delta_{(1)}, \dots, \Delta_{(k)}$ (see Remark 2.8.4).

In the one-dimensional case, from Theorem 2.9.7 one can easily obtain the following result.

Corollary 2.9.8. *Suppose the* $\mathbf{E}\xi < 0$ *and there exists a solution* $\lambda_1 = \lambda_{1+} > 0$ *of the equation* $\psi(\lambda) = 1$, $\psi'(\lambda_1) < \infty$. *Then, as* $T \to \infty$,

$$H\big([T,\infty)\big) = ce^{-\lambda_1 T}(1+\varepsilon_T), \qquad \varepsilon_T = o(1), \tag{2.9.25}$$

where in the non-arithmetic case

$$c = \frac{1}{\lambda_1 \psi'(\lambda_1)}$$

and in the arithmetic case

$$c = \frac{1}{(1-e^{-\lambda_1})\psi'(\lambda)}$$

for integer T.

If certain additional conditions are satisfied then it becomes possible to estimate the order of the remainder ε_T in (2.9.25) and also to find the asymptotics of $H\big([T,\infty)\big)$ when the condition $\psi(\lambda_1) = 1$, $\lambda_1 > 0$ is not satisfied (see Theorem 8 in [50]).

Also true is the following.

Theorem 2.9.9 (Theorem 9 in [50]). *Suppose that the condition* **[E]** *holds. Then*

$$H\big(\Delta[x)\big) = P\big(\Delta[x)\big)\big(1 + H(\{0\}) + \varepsilon(\Delta,x)\big),$$

where $\lim_{\Delta\to 0} \sup_{x\in\mathbb{R}^d} \big|\varepsilon(\Delta,x)\big| = 0$.

Thus, for small Δ and $H\big(\{0\}\big) = 0$, the functions $P\big(\Delta[x)\big)$ and $H\big(\Delta[x)\big)$ behave asymptotically in the same way. Unfortunately, this fact is not useful for the study of the asymptotics of $P(TB)$ for 'large' sets TB since the function P is not additive.

2.9.4 The non-homogeneous case

We will now consider sums $S_{0,n} = \xi_0 + S_n$, $n \geqslant 0$, of the *non-homogeneous* terms studied in section 2.8. It was established that they satisfy the l.l.d.p. under the condition $\lambda(\alpha) \in \mathcal{A}_0$ (see Theorem 2.8.10). For the renewal function

$$H_0(B) = \sum_{n=0}^{\infty} \mathbf{P}(S_{0,n} \in B)$$

the analogue of Theorem 2.9.6 given below is true. Assume, as before, that

$$\psi_0(\lambda) = \mathbf{E}e^{\langle\lambda,\xi_0\rangle}, \quad A_0(\lambda) = \ln\psi_0(\lambda), \quad \Lambda_0(\alpha) = \sup\big\{\langle\lambda,\alpha\rangle - A_0(\lambda)\big\},$$
$$\mathcal{A}_0 = \big\{\lambda : A_0(\lambda) < \infty\big\}, \qquad \mathcal{A}^{\leqslant 0} := \big\{\lambda : A(\lambda) \leqslant 0\big\}.$$

It is not difficult to see that the conditions

$$(\mathcal{A}^{\leqslant 0}) \subset \mathcal{A}_0 \quad \text{and} \quad \mathcal{A}^{\leqslant 0} \subset [\mathcal{A}_0]$$

are equivalent.

Theorem 2.9.10. *Suppose that* $\mathbf{E}\xi \neq 0$ *and the random variables* ξ_0 *and* ξ *satisfy the condition* $[\mathbf{C}_0]$*. Then the condition*

$$\mathcal{A}^{\leqslant 0} \subset [\mathcal{A}_0] \tag{2.9.26}$$

is necessary and sufficient for the validity of the equality

$$\lim_{T\to\infty} \frac{1}{T} \ln H_0\big(T\Delta[\alpha)\big) = -D(\alpha) \tag{2.9.27}$$

for any $\alpha \in \mathbb{R}^d$ *and for* $\Delta = \Delta_T \to 0$ *sufficiently slowly as* $T \to \infty$.

In order to prove this theorem, we will need the following auxiliary result. Along with the set $\mathcal{A}^{\leqslant 0}$, for given $R > 0$ introduce the sets

$$\mathcal{A}_R^{\leqslant 0} := \{\lambda \in \mathcal{A}^{\leqslant 0} : A_0(\lambda) < R\},$$

and denote

$$D^{(R)}(\alpha) = \sup_{\lambda\in\mathcal{A}_R^{\leqslant 0}} \langle \lambda, \alpha\rangle.$$

It is clear that $\mathcal{A}_R^{\leqslant 0}$ is a sequence of imbedded sets,

$$\bigcup_R (\mathcal{A}_R^{\leqslant 0}) = (\mathcal{A}^{\leqslant 0}), \qquad D^{(R)}(\alpha) = \sup_{\lambda\in(\mathcal{A}_R^{\leqslant 0})} \langle \lambda, \alpha\rangle. \tag{2.9.28}$$

The following analogue of Lemma 2.9.1 is true.

Lemma 2.9.11. *The function* $D^{(R)}(\alpha)$ *has all the properties of the function* $D(\alpha)$*: for any R it is convex, semiadditive, so that*

$$D^{(R)}(\alpha+\beta) \leqslant D^{(R)}(\alpha) + D^{(R)}(\beta), \tag{2.9.29}$$

linear, so that $D^{(R)}(c\alpha) = cD^{(R)}(\alpha)$, $c \geqslant 0$, *and lower semicontinuous. Moreover,*

$$D^{(R)}(\alpha) \uparrow D(\alpha) \quad \text{as} \quad R \uparrow \infty, \tag{2.9.30}$$

$$\Lambda_0(\alpha) \geqslant D^{(R)}(\alpha) - R. \tag{2.9.31}$$

Proof. The convexity of the function $D^{(R)}$ follows from the fact that it is the Legendre transform of the function $A_R^{\leqslant 0}(\lambda)$ with argument λ. The function is equal to 0 for $\lambda \in \mathcal{A}_R^{\leqslant 0}$ and equal to ∞ for all other λ. This means that, for any $\alpha, \beta \in \mathbb{R}^d$, $p \geqslant 0, q \geqslant 0, p+q = 1$,

$$pD^{(R)}(\alpha) + qD^{(R)}(\beta) \geqslant D^{(R)}(p\alpha + q\beta), \qquad \alpha, \beta \in \mathbb{R}^d. \tag{2.9.32}$$

The linearity of the function $D^{(R)}$ is obvious. The semiadditivity (2.9.29) follows from (2.9.32) and also the linearity. Furthermore, from the condition $(\mathcal{A}^{\leqslant 0}) \subset \mathcal{A}_0$ (see (2.9.26)), the properties (2.9.28) and the equality $D(\alpha) = \sup_{\alpha \in (\mathcal{A}^{\leqslant 0})} \langle \lambda, \alpha \rangle$, we obtain (2.9.30). Finally,

$$\begin{aligned}\Lambda_0(\alpha) &= \sup_{\lambda}\{\langle \lambda, \alpha\rangle - A_0(\lambda)\} \\ &\geqslant \sup_{\lambda \in \mathcal{A}_R^{\leqslant 0}} \{\langle \lambda, \alpha\rangle - A_0(\lambda)\} \geqslant \sup_{\lambda \in \mathcal{A}_R^{\leqslant 0}} \{\langle \lambda, \alpha\rangle\} - R = D^{(R)}(\alpha) - R.\end{aligned}$$

The lemma is proved. □

Proof of Theorem 2.9.10. Sufficiency. The upper bound. For brevity, let $T\Delta[\alpha) = B = B_{T,\alpha}$. Let us prove that, for $T \to \infty$ and any fixed sequence $\Delta = o(1)$,

$$\frac{1}{T} \ln H_0(B) \leqslant -D\big(\Delta[\alpha)\big) + o(1), \tag{2.9.33}$$

where $D(B) = \inf_{\alpha \in B} D(\alpha)$. For arbitrary $M > 0$, use the inequality

$$H_0(B) \leqslant (MT + 1) \max_{0 \leqslant n \leqslant MT} \mathbf{P}(S_{0,n} \in B) + \sum_{n > MT} \mathbf{P}(S_{0,n} \in B). \tag{2.9.34}$$

Let us bound from above the probabilities $\mathbf{P}(S_{0,n} \in B)$ for $n > MT$. In view of the condition $[\mathbf{C}_0]$ for ξ_0 and the properties of the function Λ_0, for sufficiently large M we have

$$\mathbf{P}\big(|\xi_0| > \sqrt{MT}\big) \leqslant \exp\{-c_1\sqrt{MT} + c_2\} \leqslant e^{-2D(\alpha)T}.$$

Therefore,

$$\begin{aligned}\mathbf{P}(S_{0,n} \in B) &\leqslant e^{-2D(\alpha)T} + \mathbf{P}\big(S_n \in B - \xi_0;\ |\xi_0| \leqslant \sqrt{MT}\big) \\ &\leqslant e^{-2D(\alpha)T} + \mathbf{P}(S_n \in B_M),\end{aligned} \tag{2.9.35}$$

where

$$B_M = \bigcup_{|y| \leqslant \sqrt{MT}} \{B - y\}$$

is a convex set. By the exponential Chebyshev inequality,

$$\mathbf{P}(S_n \in B_M) \leqslant e^{-n\Lambda(B_M/n)}. \tag{2.9.36}$$

For $n \geqslant MT$, $|y| \leqslant \sqrt{MT}$, points of the set $B_M/n = T\Delta[\alpha)/n - y/n$ are located no further than $1/\sqrt{M} + O(1/M)$ from the origin as $M \to \infty$. Since $\mathbf{E}\xi \neq 0$, we have $\Lambda(0) > 0$, and, for any $n > MT$ and sufficiently large M and T,

$$n\Lambda\left(\frac{B_M}{n}\right) \geqslant n\frac{\Lambda(0)}{2}.$$

Hence, when as $n > MT$ grows, the probabilities (2.9.36) decrease faster than a geometric series, and the second term on the right-hand side of (2.9.34) (the sum over $n > MT$) is not greater than

$$e^{-2D(\alpha)T} + ce^{-MT\Lambda(0)/2} \leqslant 2e^{-2D(\alpha)T} \tag{2.9.37}$$

for sufficiently large M.

Let us now consider the first term in the right-hand side of (2.9.34), and bound the probability $\mathbf{P}(S_{0,n} \in B)$ uniformly in $n \leqslant MT$. Here, as before, the first inequality in (2.9.35) holds, and it is sufficient to bound the probability

$$P_T := \int_{|y|\leqslant NT} \mathbf{P}(\xi_0 \in dy)\mathbf{P}(S_n \in B - y) \tag{2.9.38}$$

for $N = \sqrt{M}$. To simplify the computations, first consider the case when the random vectors ξ_0 and ξ are arithmetic. Since

$$\mathbf{P}(S_n \in B - y) \leqslant (\Delta T)^d \max_{z\in B} \mathbf{P}(S_n = z - y),$$

it is enough to estimate, uniformly in $n \leqslant MT$,

$$p_T := \sum_{|y|\leqslant NT} \mathbf{P}(\xi_0 = y) \max_{z\in B} \mathbf{P}(S_n = z - y),$$

where $z \in B$ has the property $z/T \to \alpha$ as $T \to \infty$. By (2.9.31), we have

$$\mathbf{P}(\xi_0 = y) \leqslant e^{-\Lambda_0(y)} \leqslant e^{R-D^{(R)}(y)}, \tag{2.9.39}$$

and, by (2.9.30), the inequality $\Lambda(\alpha) \geqslant D^{(R)}(\alpha)$ and the linearity of the function $D^{(R)}(\alpha)$, we obtain

$$\mathbf{P}(S_n = z - y) \leqslant e^{-n\Lambda((z-y)/n)} \leqslant e^{-D^{(R)}(z-y)}. \tag{2.9.40}$$

Using the semiadditivity (2.9.29) of the function $D^{(R)}$, we obtain

$$\begin{aligned} p_T &\leqslant (2NT+1)^d \max_{x\in B} \exp\{R - D^{(R)}(y) - D^{(R)}(z-y)\} \\ &\leqslant (2NT+1)^d \max_{x\in B} \exp\{R - D^{(R)}(z)\}, \end{aligned}$$

$$P_T \leqslant (\Delta T)^d p_T.$$

Owing to the lower semicontinuity of the function $D^{(R)}$, uniformly in $z \in B = T\Delta[\alpha)$ it holds that

$$\frac{1}{T}D^{(R)}(z) = D^{(R)}\left(\frac{z}{T}\right) \geqslant D^{(R)}(\alpha) + o(1)$$

for $\Delta = \Delta_T \to 0$, $T \to \infty$. Therefore

$$\overline{\lim_{T\to\infty}} \frac{1}{T} \ln P_T \leqslant -D^{(R)}(\alpha).$$

Since the left-hand side of the last inequality does not depend on R, the right-hand side can be replaced with the limit as $R \to \infty$. Therefore

$$\overline{\lim_{T\to\infty}} \frac{1}{T} \ln P_T \leqslant -D(\alpha).$$

In the general case (not necessarily the arithmetic case), estimation of the probability P_T in (2.9.34) can be done in a similar way:

$$P_T \leqslant \sum_{|y|\leqslant NT,\ y\in\mathbb{Z}^d} \mathbf{P}(\xi_0 \in \mathrm{e}[y))\mathbf{P}(S_n \in (B)_1 - y),$$

where $\mathrm{e}[y)$ is the cube $\Delta[y)$ for $\Delta = 1$, $(B)_1$ is the δ-neighbourhood of the set B for $\delta = 1$ and

$$\mathbf{P}(S_n \in (B)_1 - y) = \sum_{z\in[(B)_1],\ z\in\mathbb{Z}^d} \mathbf{P}(S_n \in \mathrm{e}[z) - y).$$

Here, to estimate $\mathbf{P}(\xi_0 \in \mathrm{e}[y))$ and $\mathbf{P}(S_n \in \mathrm{e}[z) - y)$, one should use analogues of the inequalities (2.9.39) and (2.9.40) with the same right-hand sides. The remaining details of the required upper bound in (2.9.27) are the same up to obvious changes.

The lower bound. For any n we have

$$H_0(B) \geqslant \mathbf{P}(S_{0,n} \in B).$$

Let θ_α be the point where

$$\inf_{\theta>0} \theta\Lambda\left(\frac{\alpha}{\theta}\right) = D(\alpha)$$

is attained. By Theorem 2.9.2, the function $D(\alpha)$ is determined by the set $\mathcal{A}^{\leqslant 0}$ of the values λ for which $\psi(\lambda) \leqslant 1$. This means that the infimum in (2.9.3) is necessarily attained for θ such that $\psi\big(\lambda(\alpha/\theta)\big) \leqslant 1$ and, consequently,

$$\psi\left(\lambda\left(\frac{\alpha}{\theta_\alpha}\right)\right) \leqslant 1, \qquad \lambda\left(\frac{\alpha}{\theta_\alpha}\right) \in \mathcal{A}^{\leqslant 0} \subset [\mathcal{A}_0].$$

Let $n = [T\theta_\alpha] + 1$. Then

$$\widetilde{\theta}_\alpha := \frac{n}{T} \geqslant \theta_\alpha, \qquad \widetilde{\theta}_\alpha = \theta_\alpha + O\left(\frac{1}{T}\right)$$

as $T \to \infty$ and $\widetilde{\theta}_\alpha$ has the property $\lambda(\alpha/\widetilde{\theta}_\alpha) \in \mathcal{A}^{\leqslant 0} \subset [\mathcal{A}_0]$. Therefore, one can apply Theorem 2.8.10 to the probability

$$\mathbf{P}\big(S_{0,n} \in T\Delta[\alpha)\big) = \mathbf{P}\left(\frac{S_{0,n}}{n} \in \Delta\left[\frac{\alpha}{\widetilde{\theta}_\alpha}\right)\right);$$

which yields for $T \to \infty$

$$\begin{aligned}\frac{1}{T}\ln H_0(B) &\geqslant \frac{1}{T}\ln \mathbf{P}\left(\frac{S_{0,n}}{n} \in \Delta\left[\frac{\alpha}{\widetilde{\theta}_\alpha}\right)\right)\\ &\sim -\Lambda\left(\frac{\alpha}{\theta_\alpha}\right) = -\theta_\alpha D\left(\frac{\alpha}{\theta_\alpha}\right) = -D(\alpha).\end{aligned}$$

The required lower bound is proved.

Necessity. First consider the one-dimensional case $d = 1$. Suppose, for specificity, $\mathbf{E}\xi < 0$. Then the statement of Theorem 2.9.10 makes sense only if $\alpha > 0$. For such α we have

$$D(\alpha) = \lambda_1 \alpha, \quad \text{where} \quad \lambda_1 := \sup\{\lambda : \psi(\lambda) \leqslant 1\},$$

and, according to (2.9.27),

$$\frac{1}{T} \ln H_0(T\Delta[\alpha)) \sim -\lambda_1 \alpha. \tag{2.9.41}$$

The condition $\mathcal{A}^{\leqslant 0} \subset [\mathcal{A}_0]$ is equivalent to the inequality $\lambda_1 \leqslant \lambda_{0,+}$, where $\lambda_{0,+} := \sup\{\lambda : A_0(\lambda) \leqslant \infty\}$. Suppose the contrary: that (2.9.41) holds but

$$\lambda_1 > \lambda_{0,+}. \tag{2.9.42}$$

From (2.9.41), it follows that

$$\ln H_0([T, \infty)) \sim -\lambda_1 T. \tag{2.9.43}$$

However, there exists a sequence $T_k \to \infty$ such that

$$\varliminf_{k \to \infty} \frac{1}{T_k} \ln \mathbf{P}(\xi_0 \geqslant T_k) \geqslant -\lambda_{0,+}. \tag{2.9.44}$$

Therefore, for $T = T_k$, by (2.9.43) and (2.9.44),

$$-\lambda_1 T \sim \ln H_0([T, \infty)) \geqslant \ln \mathbf{P}(\xi_0 \geqslant T) \geqslant -\lambda_{0,+} T + o(T).$$

We have obtained the inequality $\lambda_1 \leqslant \lambda_{0,+}$, which contradicts (2.9.42).

Now suppose that $d > 1$. If $\mathcal{A}^{\leqslant 0} \not\subset [\mathcal{A}_0]$ then there exists a unit vector e such that the point of intersection $\mathrm{e}t_1$ of the vector $\mathrm{e}t$ and the boundary $\partial \mathcal{A}^{\leqslant 0}$ and the point of intersection $\mathrm{e}t_{0,+}$ of $\mathrm{e}t$ and the boundary $\partial \mathcal{A}_0$ are such that $t_1 > t_{0,+}$ (cf. (2.9.42)). Then the random variables

$$\xi_{\mathrm{e}} := \langle \mathrm{e}, \xi \rangle \quad \text{and} \quad \xi_{0,\mathrm{e}} := \langle \mathrm{e}, \xi_0 \rangle$$

satisfy the conditions of the case $d = 1$, so the problem reduces to the one-dimensional case. Theorem 2.9.10 is proved. □

It is also possible to consider an alternative version of the sufficient conditions in Theorem 2.9.10 where it is assumed that the condition **[C]** holds (which is weaker than the condition **[C$_0$]**) and the condition $\mathcal{A}^{\leqslant 0} \subset [\mathcal{A}_0]$ is replaced with the stronger condition $\mathcal{A} \subset [\mathcal{A}_0]$ or a 'pointwise' (condition (i.e. one that depends on α), according to which the curve $\{\lambda(t\alpha), t > 0\}$ lies in $[\mathcal{A}_0]$. Then the proof of the l.l.d.p. for H_0 can be considerably simplified, since the bounds of the terms $\mathbf{P}(S_{0,n} \in T\Delta[\alpha))$ will differ from the corresponding bounds for the terms $\mathbf{P}(S_n \in T\Delta[\alpha))$ only by a finite factor $\psi_0(\lambda(\alpha/\theta))$ for corresponding θ. Therefore, the asymptotics of $\frac{1}{T} H_0(T\Delta[\alpha))$ for $T \to \infty$ will remain the same as that for $\frac{1}{T} H(T\Delta[\alpha))$.

2.10 Sums of non-identically distributed random variables in the triangular array scheme

2.10.1 Introduction

Let $\xi_{1,n}, \dots, \xi_{n,n}$ be independent random variables in the triangular array scheme

$$\mathbf{E}\xi_{j,n} = 0, \quad 0 < \sigma_{j,n}^2 := \mathbf{D}\xi_{j,n} < \infty, \quad B_n^2 := \sum_{j=1}^{n} \sigma_{j,n}^2,$$

$$\overline{\sigma}_n^2 := \max_{j \leqslant n} \sigma_{j,n}^2 \quad S_n := \sum_{j=1}^{n} \xi_{j,n}.$$

We will study the asymptotic behaviour of

$$\mathbf{P}\left(\frac{S_n}{\overline{\sigma}_n} \in \Delta[x)\right) \tag{2.10.1}$$

as $n \to \infty$ for values of x from both the normal and large deviation zones, under the assumption that

$$0 < \overline{\sigma}_n = o(B_n). \tag{2.10.2}$$

Moreover, when studying the large deviation probabilities it will be assumed that

$$\psi_{j,n}(\lambda_0) := \mathbf{E}e^{\lambda_0 \xi_{j,n}} < \infty \tag{2.10.3}$$

for some $\lambda_0 > 0$ and all $j \leqslant n$.

In this section a number of results obtained in the case when the $\xi_{j,n} \underset{d}{=} \xi$ are uniformly distributed and do not depend on n, will be generalised to the case of a series of random variables with different distributions, including the zone of large deviations. Some modifications of the theorems of Gnedenko and Stone and Shepp and their extensions to identically distributed random variables were considered in [54]. Integral theorems for probabilities of large deviations of sums S_n of differently distributed random variables beyond a scheme of series (mainly, for moderately large deviations) were obtained in [151], [165] and other papers. We will not consider the results of those works in details, since their statements and methodology are only weakly related to the results presented below.

2.10.2 Extension of the Stone–Shepp and Gnedenko theorems to sums of non-identically-distributed random variables in the normal deviation zone

To make the exposition more convenient and simplify our statements, instead of the probability (2.10.1) we will study the probabilities

$$\mathbf{P}\big(S_n \in \Delta[x)\big) \tag{2.10.4}$$

for sums of random variables $\xi_{j,n}$ such that

$$\mathbf{E}\xi_{j,n} = 0 \tag{2.10.5}$$

and there exist constants $0 < c_1 < c_2 < \infty$ independent of n such that

$$c_1 \leqslant \overline{\sigma}_n \leqslant c_2. \tag{2.10.6}$$

Under this convention, we will need the condition

$$B_n \to \infty \quad \text{as} \quad n \to \infty. \tag{2.10.7}$$

Conditions (2.10.6) and (2.10.7) are equivalent to (2.10.2). In fact, (2.10.2) follows immediately from (2.10.6) and (2.10.7). To establish the converse implication, it is sufficient to consider the scaled random variables $\xi_{j,n}/\overline{\sigma}_n$ instead of $\xi_{j,n}$.

Note that in fact the first inequality in (2.10.6) is not used hereafter in our proofs, but it is nevertheless essential in the sense that it specifies the 'scale' of the variables $\xi_{j,n}$ (which is not present in (2.10.2)). This is important in integro-local theorems. When the conditions (2.10.6) are met, Δ-values 'comparable with 1' (such values are considered in Theorem 2.10.1) prove to be, generally speaking, the smallest possible values for which Theorem 2.10.1 holds true (if $\overline{\sigma}_n \to 0$ as $n \to \infty$ then one can take Δ to be of the same order of magnitude as $\overline{\sigma}_n$ (see (2.10.1))).

We will also need the following uniform integrability condition, which is close to Lindeberg's condition.

[UI] *The sequence* $\{\xi_{j,n}^2/\sigma_{j,n}^2\}$ *is uniformly integrable; i.e., there exists a function* $h(N) \downarrow 0$ *as* $N \uparrow \infty$ *such that*

$$\max_{j \leqslant n} \mathbf{E}\left(\frac{\xi_{j,n}^2}{\sigma_{j,n}^2};\ \frac{|\xi_{j.n}|}{\sigma_{j,n}} > N\right) \leqslant h(N). \tag{2.10.8}$$

Since under conditions (2.10.6) and (2.10.7), for any $\tau > 0$ one has $N = N_n := \tau B_n/\sigma_{1,n} \to \infty$ as $n \to \infty$, from condition **[UI]** we obtain

$$\begin{aligned}\sum_{j=1}^{n} \mathbf{E}\big(\xi_{j,n}^2;\ |\xi_{j,n}| > \tau B_n\big) &\leqslant \sum_{j=1}^{n} \sigma_{j,n}^2\, \mathbf{E}\left(\frac{|\xi_{j,n}^2|}{\sigma_{j,n}^2};\ \frac{|\xi_{j,n}|}{\sigma_{j,n}} > \frac{\tau B_n}{\sigma_{1,n}}\right)\\ &\leqslant \sum_{j=1}^{n} \sigma_{j,n}^2 h(N) = o(B_n^2).\end{aligned}$$

This means that conditions (2.10.6), (2.10.7) and **[UI]** imply Lindeberg's condition. It is not hard to see that the converse assertion, that Lindeberg's condition implies **[UI]** under the assumption that (2.10.6) and (2.10.7) are met, is wrong. Therefore, provided that (2.10.6) and (2.10.7) are satisfied, condition **[UI]** is stronger.

Now we turn to non-lattice conditions. For a random variable ξ the condition that it is non-lattice can be written as $\big|\varphi(t)\big| < 1$ for any $t \neq 0$, where

$$\varphi(t) := \mathbf{E}e^{it\xi}.$$

For the sums S_n we will need condition **[R]** below to ensure that they are *asymptotically non-lattice*. For given $\varepsilon > 0$ and $N > 0$, we denote

$$q_{j,n} = q_{j,n}(\varepsilon, N) := \sup_{\varepsilon \leqslant |t| \leqslant N} \left|\varphi_{j,n}(t)\right|, \qquad \varphi_{j,n}(t) := \mathbf{E}e^{it\xi_{j,n}}. \tag{2.10.9}$$

Condition **[R]** has the following form:

[R] *For any fixed* $\varepsilon > 0$ *and* $N > 0$,

$$B_n \prod_{j=1}^{n} q_{j,n} \to 0 \tag{2.10.10}$$

as $n \to \infty$.

Condition **[R]** ensures the that the sum S_n does not 'degenerate' into an integer-valued random variable, i.e. it ensures the absence of the quick convergence (in distribution) of the random variables $\xi_{j,n}$ to integer-valued random variables.

Suppose there exist $m = m(n) \leqslant n$ random variables $\xi_{j,n}$ (we can assume that they are $\xi_{1,n}, \ldots, \xi_{m,n}$), for which the following 'relaxed' non-lattice -type condition holds uniformly in j and n: *for any fixed* $\varepsilon > 0$ *and* $N < \infty$,

$$\max_{j \leqslant m} q_{j,n} \leqslant q < 1, \tag{2.10.11}$$

where q *does not depend on* n. In this case, the product in (2.10.10) does not exceed $q^m B_n \to 0$ as $n \to \infty$, provided that, say, $m \geqslant -2 \ln B_n / \ln q$. Thus, condition **[R]** always holds for $m \gg \ln B_n$ whenever (2.10.11) holds true. The remaining summands $\xi_{j,n}$ for $j > m$ can be arbitrary (provided that (2.10.7) holds).

Now we can state the following integro-local theorem.

Theorem 2.10.1. *Let* $\xi_{1,n}, \ldots, \xi_{n,n}$ *be independent random variables,* $\mathbf{E}\xi_{j,n} = 0$, $S_n = \sum_{j=1}^{n} \xi_{j,n}$. *Assume that conditions* (2.10.6), (2.10.7), **[UI]** *and* **[R]** *are satisfied. Then, for any fixed* $\Delta > 0$,

$$\mathbf{P}\big(S_n \in \Delta[x)\big) = \frac{\Delta}{B_n} \phi\left(\frac{x}{B_n}\right) + o\left(\frac{1}{B_n}\right), \tag{2.10.12}$$

where

$$\phi(t) = \frac{1}{\sqrt{2\pi}} e^{-t^2/2}$$

is the standard normal density and the remainder term $o(1/B_n)$ *is uniform in* x.

Example 2.10.2. Let $\xi_{j,n} = \xi_j = \zeta_j g(j)$, where $\zeta_j \stackrel{d}{=} \zeta$ are independent non-lattice identically distributed random variables, $\mathbf{E}\zeta = 0$, $\mathbf{E}\zeta^2 = 1$ and $g(j)$ is a regularly varying function of index β, i.e. a function admitting a representation of the form $g(j) = j^\beta l(j)$, where $l(j)$ is a slowly varying function.

If $\beta < 0$ then $g(j)$ is 'asymptotically' decreasing, $\overline{\sigma}_n \in \big[g(1), \max_{j\geqslant 1} g(j)\big]$ and

$$B_n^2 = \sum_{j=1}^{n} g^2(j) \sim \frac{1}{2\beta+1}\, n^{2\beta+1} l^2(n) \to \infty \tag{2.10.13}$$

for $\beta > -1/2$, so that conditions (2.10.6), (2.10.7) are met for such βs. Since the random variables $\xi_j^2/\mathbf{D}\xi_j = \zeta_j^2$ are uniformly integrable, condition **[UI]** holds as well. To verify condition **[R]**, observe that, for some $q < 1$ and $\varphi_{(\zeta)}(t) := \mathbf{E}e^{it\zeta}$, one has

$$\begin{aligned} q_{j,n} &= \sup_{\varepsilon\leqslant t\leqslant N} \Big|\varphi_{(\zeta)}\big(tg(j)\big)\Big| \leqslant \sup_{\varepsilon g(j))\leqslant |u|\leqslant Ng(j)} \Big|\varphi_{(\zeta)}(u)\Big| \\ &= \max\Big(q, \sup_{\varepsilon g(j)\leqslant |u|\leqslant \varepsilon} \big|\varphi_{(\zeta)}(u)\big|\Big), \end{aligned}$$

where, for small enough $\varepsilon > 0$ and $|u| \leqslant \varepsilon$,

$$\big|\varphi_{(\zeta)}(u)\big| \leqslant 1 - \frac{u^2}{3}, \qquad \frac{1-\varepsilon^2}{3} \geqslant q,$$

so that

$$\prod_{j=1}^{n} q_{j,n} \leqslant \prod_{j=1}^{n}\left(1 - \frac{\varepsilon^2 g^2(j)}{3}\right) \approx \exp\left\{-\frac{\varepsilon^2 l^2(n) n^{2\beta+1}}{3(2\beta+1)}\right\} = o(B_n^{-1}).$$

This means that condition **[R]** is met and $\mathbf{P}\big(S_n \in \Delta[x)\big)$ is described by relation (2.10.12), with B_n specified by (2.10.13).

If $\beta > 0$ then the $\sigma_{j,n} = g(j)$ grow together with j, so that condition (2.10.6) is not met and Theorem 2.10.1 is not directly applicable. However, we can apply it to new random variables given by

$$\xi_{j,n}^* = \frac{\xi_{n-j+1}}{\sigma_{n,n}} = \frac{\xi_{n-j+1}}{g(n)}, \quad j = 1, \ldots, n.$$

For these variables, condition (2.10.6) will be satisfied. Endowing with the superscript $*$ the symbols for the quantities introduced above but defined for the sequence $\{\xi_{j,n}^*\}$, we obtain (see (2.10.13))

$$(B_n^*)^2 := \frac{B_n^2}{g^2(n)} \sim \frac{n^{2\beta+1}}{2\beta+1}\,\frac{l^2(n)}{n^{2\beta} l^2(n)} = \frac{n}{2\beta+1} \to \infty$$

as $n \to \infty$, which means that (2.10.7) holds. The validity of **[UI]** follows from the uniform integrability of $(\xi_{j,n}^*)^2$. To verify condition **[R]**, one can use the criterion (2.10.11), as the first $m = n/2$ random variables $\xi_{j,n}^*$ are uniformly non-lattice and (2.10.11) is met for them. Therefore, the asymptotic behaviour of

$$\mathbf{P}\big(S_n^* \in \Delta[x)\big) = \mathbf{P}\left(\frac{S_n}{g(n)} \in \Delta[x)\right)$$

is now described by Theorem 2.10.1 but with the quantity B_n replaced by $B_n^* \sim \sqrt{n/(2\beta+1)}$.

Thus, if $\sigma_{j,n}$ and $\sigma_{1,n}$ are increasing and comparable with 1, then the behaviour of the integro-local probabilities $\mathbf{P}(S_n \in \Delta[x))$ can be described only for large values of Δ that are comparable with $\sigma_{n,n}$ (which is quite natural). Such values of Δ are still small relative to B_n (they are $B_n/\overline{\sigma}_n$ times smaller).

Now, we will state a *local* theorem on the asymptotic behaviour of the density f_n of the sum S_n when the density exists. For given $M \geqslant 1$ we will need the following modification of condition **[R]**.

[R$_\mathrm{M}$] *For any fixed* $\varepsilon > 0$

$$B_n \sum_{j=M+1}^{n} q_{j,n}(\varepsilon, \infty) \to 0$$

as $n \to \infty$, *where* $q_{j,n}(\varepsilon, N)$ *are defined as in* (2.10.9).

Remark 2.10.3. If among the random variables $\xi_{j,n}$ with $j \geqslant M+1$ there are $m = m(n)$ (assume that they are the $\xi_{j,n}$ with $M < j \leqslant M+m$) for which

$$q_{j,n}(\varepsilon, \infty) \leqslant q(\varepsilon) < 1;$$

then, as in the case of (2.10.11), it is not hard to verify that the condition **[R$_\mathrm{M}$]** will be met provided that $m \gg \ln B_n$.

Theorem 2.10.4. *Let conditions* (2.10.6) *and* (2.10.7) *and* **[UI]** *be satisfied, and let there exist* $M \geqslant 1$ *such that condition* **[R$_\mathrm{M}$]** *and at least one of the following two conditions are satisfied:*

(i) *The density of the distribution of the sum* $S_M = \sum_{j=1}^{M} \xi_{j,n}$ *is uniformly bounded in n (i.e. bounded by a constant independent of n).*
(ii) *The distribution of* S_M *has a density which is square integrable uniformly in n, and the characteristic function* $\varphi_{(S_M)}(t) := \prod_{j=1}^{M} \varphi_{j,n}(t)$ *of the sum* S_M *is uniformly integrable in n.*

Then, for $n \geqslant M$, *the distribution of the sum* S_n *has a density* f_n *for which the following relation holds uniformly in x as* $n \to \infty$,

$$f_n(x) = \frac{1}{B_n}\phi\left(\frac{x}{B_n}\right) + o\left(\frac{1}{B_n}\right).$$

Remark 2.10.5. (1) Conditions (i) and (ii) are equivalent (with possibly different values of M). For instance, the relation $f_M \in L_2$ in condition (ii) implies the boundedness of the density f_{2M}; for details, see e.g. [39], § 8.7.

(2) Each of the following two conditions is sufficient for the integrability of $\varphi_{(S_M)}$:

(ii1) There exists $M_1 < M$ such that the sums S_{M_1} and $S_M - S_{M_1}$ have bounded densities.

(ii2) There exists $M_1 < M$ such that the sums S_{M_1} and $S_M - S_{M_1}$ have square integrable densities.

Indeed, it is not hard to see that (ii1)$\Rightarrow$(ii2). Therefore we need only verify that condition (ii2) is sufficient for the integrability of $\varphi_{(S_M)}$. For simplicity, let $M_1 = M - M_1 = 1$. Since $f_{\xi_{1,n}}$ and $f_{\xi_{2,n}}$ are square integrable, the functions $\varphi_{(\xi_{1,n})}$ and $\varphi_{(\xi_{2,n})}$ are also square integrable and

$$\left(\int \left|\varphi_{(\xi_{1,n})}\varphi_{(\xi_{2,n})}\right|\right)^2 \leqslant \int \varphi^2_{(\xi_{1,n})} \int \varphi^2_{(\xi_{2,n})} < \infty.$$

This means that the function

$$\left|\varphi_{(\xi_{1,n})}\varphi_{(\xi_{2,n})}\right| = \left|\varphi_{(S_2)}\right|$$

is integrable.

Now consider the arithmetic case, when all the summands $\xi_{j,n}$ are *integer-valued*. The condition that the distribution of a given random variable ξ is arithmetic (i.e. the greatest common divisor of its possible values is equal to 1) can be written as

$$\left|\varphi(t)\right| < 1 \quad \text{for any} \quad t \in (0, 2\pi).$$

We will need the condition **[Z]** below, asserting that the distributions of the sums S_n are *asymptotically arithmetic*. For a given $\varepsilon > 0$, we set, by analogy with the aforesaid,

$$q_{j,n} := q_{j,n}(\varepsilon, 2\pi - \varepsilon) = \sup_{t \in [\varepsilon, 2\pi - \varepsilon]} \left|\varphi_{j,n}(t)\right|.$$

Condition **[Z]** has the following form.

[Z] *For any fixed* $\varepsilon > 0$,

$$B_n \prod_{j=1}^{n} q_{j,n} \to 0 \tag{2.10.14}$$

as $n \to \infty$.

Condition **[Z]** ensures that the sums S_n 'do not degenerate' into lattice random variables with lattice size greater than 1, i.e. it ensures the absence of fast convergence (in distribution) of the random variables $\xi_{j,n}$ to random variables which are multiples of some number $k > 1$.

As before, it is easy to verify that if m summands $\xi_{j,n}$ with $j \leqslant m \leqslant n$ are arithmetic uniformly in n, i.e. $\max_{j \leqslant m} q_{j,n} \leqslant q < 1$, where q does not depend on n, then condition **[Z]** will be met for $m \geqslant -2\ln B_n/\ln q$. The remaining $\xi_{j,n}$ with $j > m$ can be arbitrary integer-valued random variables (provided that (2.10.6) is satisfied).

Note that for arithmetic random variables $\xi_{j,n}$ we cannot assume without loss of generality that $a_{j,n} := \mathbf{E}\xi_{j,n} = 0$. This means that the condition **[UI]** will have the following form in the arithmetic case:

[UI] *The sequence*

$$\left\{\frac{(\xi_{j,n} - a_{j,n})^2}{\sigma_{j,n}^2}\right\}$$

is uniformly integrable.

Set $A_n = \sum_{j=1}^n a_{j,n}$.

Theorem 2.10.6. *Let the conditions of Theorem* 2.10.1 *be met, except for the condition* $a_{j,n} = 0$, *and let condition* **[R]** *be replaced with* **[Z]**. *Then, for integer-valued x,*

$$\mathbf{P}(S_n = x) = \frac{1}{B_n}\phi\left(\frac{x - A_n}{B_n}\right) + o\left(\frac{1}{B_n}\right), \tag{2.10.15}$$

where the remainder term is uniform in x.

2.10.3 The proof of Theorem 2.10.1

Let us use the method of smoothing to consider the sums

$$Z_n = S_n + \theta\eta,$$

where $\theta = \text{const.}$, and η is independent of S_n and has an integrable characteristic function

$$\varphi_{(\eta)}(t) = \max\big(0, 1 - |t|\big),$$

so that, by the inversion formula,

$$\begin{aligned}\mathbf{P}\big(Z_n \in \Delta[x)\big) &= \frac{1}{2\pi}\int e^{-itx}\varphi_n(t)\,\frac{1 - e^{-it\Delta}}{it}\,\varphi_{(\eta)}(\theta t)dt\\ &= \frac{\Delta}{2\pi}\int_{|t|\leqslant 1/\theta} e^{-itx}\varphi_n(t)\varphi(t)dt,\end{aligned} \tag{2.10.16}$$

where

$$\varphi_n(t) := \varphi_{(S_n)}(t) = \prod_{j=1}^n \varphi_{j,n}(t), \quad \varphi(t) := \varphi^{(\Delta)}(t)\varphi_{(\eta)}(\theta t),$$

$$\varphi^{(\Delta)} := \frac{1 - e^{-it\Delta}}{it\Delta}.$$

Here $\varphi(t)$ is the characteristic function of the sum of the two independent random variables $\theta\eta$ and $\eta_{(\Delta)}$, where $\eta_{(\Delta)}$ is uniformly distributed over $[-\Delta, 0]$.

Split the integral on the right-hand side of (2.10.16) into two integrals, I_1 and I_2, over the interval $|t| < \gamma$, for some $\gamma < 1/\theta$, and over its complement, respectively.

First consider I_1. Represent the function $\varphi_n(t)$ in the form

$$\varphi_n(t) = \exp\left\{\sum_{j=1}^{n} \ln\Big[1 - \big(\varphi_{j,n}(t) - 1\big)\Big]\right\}$$

and make use of the identity

$$\varphi_{j,n}(t) - 1 = \left(\varphi_{j,n}(t) - 1 + \frac{t^2\sigma_{j,n}^2}{2}\right) - \frac{t^2\sigma_{j,n}^2}{2},$$

where, for a fixed $N > 0$,

$$\varphi_{j,n}(t) - 1 + \frac{t^2\sigma_{j,n}^2}{2} = \int \left(e^{itz} - 1 - itz + \frac{t^2z^2}{2}\right) \mathbf{P}(\xi_{j,n} \in dz)$$
$$= \int_{|z|<N\sigma_{j,n}} + \int_{|z|\geqslant N\sigma_{j,n}}.$$

By the inequality

$$\left|e^{itz} - 1 - itz + \frac{t^2z^2}{2}\right| \leqslant \frac{|t^3z^3|}{6}$$

and condition (2.10.6), we see that for $|t| \leqslant \gamma$ one has (see (2.10.6))

$$\int_{|z|<N\sigma_{j,n}} \leqslant \frac{N\gamma\, t^2\sigma_{j,n}}{6} \int_{|z|<N\sigma_{j,n}} z^2\mathbf{P}(\xi_{j,n} \in dz) \leqslant \frac{c_2N\gamma t^2\sigma_{j,n}^2}{6}.$$

Further, from the inequality

$$|e^{itz} - 1 - itz| \leqslant \frac{t^2z^2}{2}$$

and condition **[UI]** we have (see (2.10.8)

$$\int_{|z|>N\sigma_{j,n}} \leqslant t^2 \int_{|z|>N\sigma_{j,n}} z^2\mathbf{P}(\xi_{j,n} \in dz) \leqslant t^2\sigma_{j,n}^2 h(N).$$

Put $\gamma = h(N)/N$ and, for a given $\varepsilon > 0$, choose N large enough that

$$h(N)\left(1 + \frac{c_2}{6}\right) \leqslant \frac{\varepsilon}{2}.$$

Then, for the chosen values of γ and N, we obtain

$$\varphi_{j,n}(t) - 1 = -\frac{t^2\sigma_{j,n}^2}{2}\,(1 + \varepsilon_{j,n}), \tag{2.10.17}$$

where $|\varepsilon_{j,n}| \leqslant \varepsilon$. This implies that

$$\ln\Big[1 - \big(\varphi_{j,n}(t) - 1\big)\Big] = -\frac{t^2\sigma_{j,n}^2}{2}\,(1 + r_{j,n}),$$

where $|r_{j,n}| \leqslant 2\varepsilon$, provided that γ and ε are small enough. Assuming without loss of generality that $\varepsilon \leqslant 1/4$, we obtain

$$\ln \varphi_n(t) = \sum_{j=1}^{n} \ln\Big[1 - \big(\varphi_{j,n}(t) - 1\big)\Big] = -\frac{t^2 B_n^2}{2}(1 + r_n), \quad |r_n| \leqslant 2\varepsilon, \tag{2.10.18}$$

$$\operatorname{Re} \ln \varphi_n(t) \leqslant -\frac{t^2 B_n^2}{4}. \tag{2.10.19}$$

For fixed u, letting $t = u/B_n$ in (2.10.18) we obtain that, as $n \to \infty$ (and so $B_n \to \infty$; see (2.10.7))

$$\ln \varphi_n\left(\frac{u}{B_n}\right) \to -\frac{u^2}{2}.$$

(This also follows from the classical proofs of the central limit theorem since the Lindeberg condition is implied by condition **[UI]**.) Thus,

$$\varphi_n\left(\frac{u}{B_n}\right) \to e^{-u^2/2}$$

and, for $|u| < \gamma B_n$, by virtue of (2.10.19) one has

$$\left|\varphi_n\left(\frac{u}{B_n}\right)\right| \leqslant e^{-u^2/4}.$$

Moreover, since for any fixed u one has $|\varphi(u/B_n)| \leqslant 1$ and $\varphi(u/B_n) \to 1$ as $n \to \infty$, we obtain from the dominated convergence theorem that, for $x = vB_n, t = u/B_n$,

$$\lim_{n\to\infty} B_n I_1 = \lim_{n\to\infty} \int_{|u|<\gamma B_n} e^{-iuv} \varphi_n\left(\frac{u}{B_n}\right) \varphi\left(\frac{u}{B_n}\right) du = \int e^{-iuv-u^2/2} du \tag{2.10.20}$$

holds uniformly in v since the integral on the right-hand side is equicontinuous in v. However, that integral (up to a factor $1/2\pi$) is simply the inversion formula applied to the normal characteristic function, so that

$$\lim_{n\to\infty} B_n I_1 = \sqrt{2\pi} e^{-v^2/2}. \tag{2.10.21}$$

Now consider the integral I_2:

$$|I_2| = \left|\int_{\gamma\leqslant|t|\leqslant 1/\theta} e^{-itx} \varphi_n(t)\varphi(t)dt\right| \leqslant \int_{\gamma\leqslant t\leqslant 1/\theta} \big|\varphi_n(t)\big| dt.$$

By virtue of condition **[R]** for any fixed γ and θ one has

$$\sup_{\gamma\leqslant t\leqslant 1/\theta} \big|\varphi_n(t)\big| = o(B_n^{-1})$$

as $n \to \infty$, and, therefore, uniformly in x,

$$\lim_{n\to\infty} B_n |I_2| = 0.$$

From this and (2.10.21) it follows that, letting $x = vB_n$, $n \to \infty$, the following relations hold uniformly in v (and thus in x):

$$I_1 + I_2 = \frac{\sqrt{2\pi}}{B_n} e^{-v^2/2} + o\left(\frac{1}{B_n}\right),$$

$$\mathbf{P}(Z_n \in \Delta[x)) = \frac{\Delta}{\sqrt{2\pi}B_n} e^{-x^2/(2B_n^2)} + o\left(\frac{1}{B_n}\right).$$

The proof of that this implies the desired assertion (2.10.15) is contained in [174] (see also [39], § 8.7). The theorem is proved.

2.10.4 The proofs of Theorems 2.10.4 and 2.10.6

The proofs of Theorems 2.10.4, 2.10.6 differ little from the above argument, and where they do, they are simpler than the latter.

Proof of Theorem 2.10.4. By virtue of one of the conditions (i), (ii), the characteristic function $\varphi_n(t)$ is integrable, and therefore one can use the inversion formula

$$f_n(x) = \frac{1}{2\pi} \int e^{-itx} \varphi_n(t) dt.$$

Here the integral on the right-hand side does not 'qualitatively' differ from the integral on the right-hand side of (2.10.16) for $\theta = 0$; one should just put $\varphi(t) \equiv 1$ in part I_1 of this integral (over the set $|t| < \gamma$), while in part I_2 of the integral (over the set $|t| \geqslant \gamma$) the integrable function $\varphi(t)$ should be replaced with the integrable characteristic function $\varphi_{(S_M)}(t)$ and the function $\varphi_n(t)$ should be replaced with $\prod_{j=M+1}^n \varphi_{j,n}(t)$. After this substitution, all that remains is to follow the argument in the proof of Theorem 2.10.1. The theorem is proved. □

Proof of Theorem 2.10.6. Under the conditions of Theorem 2.10.6, the inversion formula takes the form

$$\mathbf{P}(S_n = x) = \frac{1}{2\pi} \int_{-\pi}^{\pi} e^{-itx} \varphi_n(t) dt. \tag{2.10.22}$$

Here the integral on the right-hand side should be split, as before, into two integrals I_1 and I_2 over the regions $|t| \leqslant \gamma$ and $|t| \in [\gamma, \pi]$, respectively; arguments from the proofs of Theorems 2.10.1 and 2.10.4 are then applied, replacing condition **[R]** with condition **[Z]**. The integrand in (2.10.22) should be rewritten as $e^{-it(x-A_n)}\varphi_n^*(t)$, where $\varphi_n^*(t)$ is the characteristic function of the random variable $S_n - A_n$ with zero mean. The theorem is thus proved. □

Remark 2.10.7. Note that integro-local and local theorems differ from the integral theorems not only by the presence of 'structural' conditions **[R]**, **[Z]**, **[R$_\mathbf{M}$]** but also because the classical Lindeberg condition from the central limit theorem for sums of non-identically-distributed random variables in the triangular array scheme should now be replaced by the stronger condition **[UI]**. Without condition, proving the above assertions is not feasible.

2.10.5 Limit theorems for large deviation probabilities. The problem statement and the main integro-local theorem

In this subsection, we will use the notation from the previous section. We will assume that the $\xi_{j,n}$ satisfy the conditions of Theorem 2.10.1 and, moreover, that the following Cramér moment condition is met:

$\mathbf{[C]}_{\lambda_0}$ *For some* $\lambda_0 > 0$ *and all* $j \leqslant n$,

$$\psi_{j,n}(\lambda) := \mathbf{E}e^{\lambda\xi_{j,n}} < \infty \quad \text{for} \quad \lambda \in [0, \lambda_0], \qquad \psi''_{j,n}(\lambda_0) < \infty.$$

Put

$$\psi_n(\lambda) := \prod_{j=1}^{n} \psi_{j,n}(\lambda) = \varphi_n(-i\lambda), \qquad A_{j,n}(\lambda) := \ln \psi_{j,n}(\lambda).$$

Since the function

$$A(\lambda) := B_n^{-2} \sum_{j=1}^{n} A_{j,n}(\lambda) = B_n^{-2} \ln \psi_n(\lambda)$$

is strictly convex, the function

$$A'(\lambda) = B_n^{-2} \frac{\psi_n'(\lambda)}{\psi_n(\lambda)} = B_n^{-2} \sum_{j=1}^{n} A'_{j,n}(\lambda) = B_n^{-2} \sum_{j=1}^{n} \frac{\psi'_{j,n}(\lambda)}{\psi_{j,n}(\lambda)}$$

is strictly increasing on $[0, \lambda_0]$ from $A'(0) = A_n = 0$ (provided that $\mathbf{E}S_n = 0$) to $A'(\lambda_0)$. Hence if

$$0 \leqslant \alpha := \frac{x}{B_n^2} \leqslant A'(\lambda_0)$$

then the equation

$$A'(\lambda) = \alpha \tag{2.10.23}$$

has a unique solution $\lambda(\alpha)$ (this is the function inverse to $A'(\lambda)$). Put $\alpha_0 = A'(\lambda_0)$. Then $\lambda(\alpha)$ will be a function analytic on $(0, \alpha_0)$ and differentiable on $[0, \alpha_0]$ from 'inside' this segment.

Now we consider the Cramér transforms of the distributions of $\xi_{j,n}$ and S_n and introduce the random variables $\xi_{j,n}^{(\lambda)}$ and $S_n^{(\lambda)}$, $\lambda \in [0, \lambda_0]$, distributed according to the relations

$$\begin{aligned} \mathbf{P}(\xi_{j,n}^{(\lambda)} \in dz) &= \frac{e^{\lambda z}\mathbf{P}(\xi_{j,n} \in dz)}{\psi_{j,n}(\lambda)}, \\ \mathbf{P}(S_n^{(\lambda)} \in dz) &= \frac{e^{\lambda z}\mathbf{P}(S_n \in dz)}{\psi_n(\lambda)}. \end{aligned} \tag{2.10.24}$$

Clearly,

$$
\begin{aligned}
\mathbf{E}e^{\mu\xi_{j,n}^{(\lambda)}} &= \frac{\psi_{j,n}(\mu+\lambda)}{\psi_{j,n}(\lambda)}, \qquad \mathbf{E}\xi_{j,n}^{(\lambda)} = \frac{\psi'_{j,n}(\lambda)}{\psi_{j,n}(\lambda)} = A'_{j,n}(\lambda),\\
\mathbf{E}(\xi_{j,n}^{(\lambda)})^2 &= \frac{\psi''_{j,n}(\lambda)}{\psi_{j,n}(\lambda)}, \qquad (\sigma_{j,n}^{(\lambda)})^2 := \mathbf{D}\xi_{j,n}^{(\lambda)} = \frac{\psi''_{j,n}(\lambda)}{\psi_{j,n}(\lambda)} - \big(A'_{j,n}(\lambda)\big)^2.
\end{aligned}
\tag{2.10.25}
$$

One of the main assumptions in this section is that the Cramér transform does not strongly affect the distributions of $\xi_{j,n}$ in the following sense:

(i) *Relations* (2.10.6) (2.10.7) *remain true; i.e. there exist constants* $0 < c_1 < c_2 < \infty$, *independent of n, such that*

$$
c_1 \leqslant \overline{\sigma}_n^{(\lambda)} := \max_{1\leqslant j\leqslant n} \sigma_{j,n}^{(\lambda)} \leqslant c_2, \qquad (B_n^{(\lambda)})^2 := \sum_{j=1}^{n} (\sigma_{j,n}^{(\lambda)})^2 \to \infty \tag{2.10.26}
$$

as $n \to \infty$, *for all* $\lambda \in [0, \lambda_0]$.

It is not difficult to see that for (2.10.26) to hold it suffices to assume that these relations hold at the extreme points $\lambda = 0$ and $\lambda = \lambda_0$.

(ii) *The uniform integrability condition remains satisfied. Let* $a_{j,n}^{(\lambda)} := \mathbf{E}\xi_{j,n}^{(\lambda)} = A'_{j,n}(\lambda)$, *so that* $a_{j,n}^{(0)} = a_{j,n} = \mathbf{E}\xi_{j,n}$.

$[\mathbf{UI}]_{\lambda_0}$ *The sequence*

$$
\left(\frac{\xi_{j,n}^{(\lambda)} - a_{j,n}^{(\lambda)}}{\sigma_{j,n}^{(\lambda)}}\right)^2
$$

is uniformly integrable for all $\lambda \in [0, \lambda_0]$ (i.e. (2.10.8) holds with $\xi_{j,n}$ replaced by $\xi_{j,n}^{(\lambda)} - a_{j,n}^{(\lambda)}$ and $\sigma_{j,n}$ replaced by $\sigma_{j,n}^{(\lambda)}$, respectively).

(iii) *The structural conditions* **[R]** *and* **[Z]** are satisfied. Put

$$
\begin{aligned}
\varphi_{j,n}^{(\lambda)}(t) &:= \frac{\psi_{j,n}(it+\lambda)}{\psi_{j,n}(\lambda)},\\
q_{j,n}^{(\lambda)} = q_{j,n}^{(\lambda)}(\varepsilon, N) &:= \sup_{\varepsilon\leqslant t\leqslant N} \big|\varphi_{j,n}^{(\lambda)}(t)\big|.
\end{aligned}
\tag{2.10.27}
$$

The modified condition **[R]** will have the following form:

$[\mathbf{R}]_{\lambda_0}$ *For any fixed* $\varepsilon > 0$ *and* $N < \infty$, *one has*

$$
\sup_{\lambda\in[0,\lambda_0]} B_n^{(\lambda)} \prod_{j=1}^{n} q_{j,n}^{(\lambda)} \to 0
$$

as $n \to \infty$.

It appears that for $[\mathbf{UI}]_{\lambda_0}$ and $[\mathbf{R}]_{\lambda_0}$ to hold, it also suffices that the respective relation holds only at the extreme points $\lambda = 0$ and $\lambda = \lambda_0$ (or even at just one of these points).

Now we can formulate the main integro-local assertion. Introduce the deviation function

$$\Lambda(\alpha) := \sup_{\lambda} \big(\alpha\lambda - A(\lambda)\big) = \alpha\lambda(\alpha) - A\big(\lambda(\alpha)\big).$$

Theorem 2.10.8. *Let $\xi_{1,n}, \dots, \xi_{n,n}$ be independent random variables, $\mathbf{E}\xi_{j,n} = 0$, $S_n = \sum_{j=1}^{n} \xi_{j,n}$. Assume that Cramér's condition $[\mathbf{C}]_{\lambda_0}$ and conditions (2.10.26), $[\mathbf{UI}]_{\lambda_0}$ and $[\mathbf{R}]_{\lambda_0}$ are all met. If $\Delta = \Delta_n \to 0$ slowly enough as $n \to \infty$ and $x = \alpha B_n^2$ then*

$$\mathbf{P}\big(S_n \in \Delta[x)\big) = \frac{\Delta e^{-B_n^2 \Lambda(\alpha)}}{\sqrt{2\pi} B_n^{(\lambda(\alpha))}} \big(1 + o(1)\big), \tag{2.10.28}$$

where $\big(B_n^{(\lambda)}\big)^2$ is defined in (2.10.26), and the remainder term $o(1)$ is uniform in $\alpha \in [0, \alpha_0]$.

Moreover, the following representation holds true:

$$\Lambda(\alpha) = \int_0^{\alpha} \lambda(v) dv. \tag{2.10.29}$$

Remark 2.10.9. Note that in the above conditions there is a certain arbitrariness in the choice of the scaling of the large deviations. In Theorem 2.10.8 we chose the 'traditional' scale and measured the large deviations in terms of the variance B_n^2 of the sum S_n, setting the scaled large deviations equal to $\alpha = x/B_n^2$. However, one should keep in mind that the boundary of the deviations x for the large deviation probabilities that can be studied using the Cramér transform is specified (see the proof of Theorem 2.10.8 in what) by the value

$$\frac{\psi_n'(\lambda)}{\psi_n(\lambda)} = \sum_{j=1}^{n} \frac{\psi_{j,n}'(\lambda)}{\psi_{j,n}(\lambda)} =: H(\lambda) \tag{2.10.30}$$

for $\lambda = \lambda_0$, and this quantity does not need to have the same growth rate as B_n^2. Hence in the cases where x is comparable with $H(\lambda_0)$, it may be more natural to consider the scaled deviations

$$\beta := \frac{x}{H(\lambda_0)} \leqslant 1.$$

In this case, to state the result we will need a solution $\mu(v)$ of the equation

$$\frac{H(\lambda)}{H(\lambda_0)} = v, \quad v \leqslant 1, \tag{2.10.31}$$

and the deviation function

$$M(\beta) = \sup_{\lambda} \left(\beta\lambda - \frac{H(\lambda)}{H(\lambda_0)}\right) = \beta\mu(\beta) - \frac{H(\mu(\beta))}{H(\lambda_0)}.$$

An analogue of Theorem 2.10.8 under the new scaling will have the following form.

Theorem 2.10.4A. *Let the conditions of Theorem* 2.10.8 *be met and* $\beta = x/H(\lambda_0) \leqslant 1$. *Then*

$$\mathbf{P}\big(S_n \in \Delta[x)\big) = \frac{\Delta e^{-H(\lambda_0)M(\beta)}}{\sqrt{2\pi}\, B_n^{(\mu(\beta))}}\big(1+o(1)\big),$$

the functions $H(\lambda)$, $M(\beta)$ *are defined by* (2.10.30) *and* (2.10.31) *and the remainder term* $o(1)$ *is uniform in* $\beta \in [0,1]$.

Moreover,

$$M(\beta) = \int_0^\beta \mu(v)dv.$$

It is not hard to see that $\mu(\beta) = \lambda(\alpha)$ and $H(\lambda_0)M(\beta) = B_n^2\Lambda(\alpha)$. The latter value can also be considered as the deviation function of the sum S_n at a point $x \leqslant H(\lambda_0)$, i.e. as the value $\sup_\lambda\{\lambda x - \ln\psi_n(\lambda)\}$. The point $\lambda(\alpha)$ is scale-invariant.

Proof of Theorem 2.10.4A. This is essentially the same as the proof of Theorem 2.10.8 presented below. □

Proof of Theorem 2.10.8. In what follows, for brevity we will sometimes omit the subscript n referring to the row number in the triangular array scheme, and we will use the notation $\psi(\lambda) := \prod_{j=1}^n \psi_j(\lambda)$. Put

$$S_n(\lambda) := \sum_{j=1}^n \xi_j^{(\lambda)}.$$

We will need the following lemma.

Lemma 2.10.10. *Under the conditions of Theorem* 2.10.8,

$$\mathbf{P}\big(S_n \in \Delta[x)\big) = e^{-B_n^2\Lambda(\alpha)}\int_0^\Delta e^{-\lambda(\alpha)z}\mathbf{P}\Big(S_n\big(\lambda(\alpha)\big) - x \in dz\Big) \tag{2.10.32}$$

Proof. The Laplace transform of the distribution of the sum $S_n(\lambda)$ is equal to

$$\mathbf{E}e^{\mu S_n(\lambda)} = \prod_{j=1}^n \frac{\psi_j(\lambda+\mu)}{\psi_j(\lambda)}.$$

By virtue of (2.10.24) this coincides with $\mathbf{E}e^{\mu S_n^{(\lambda)}}$. Hence the distributions of $S_n(\lambda)$ and $S_n^{(\lambda)}$ also coincide with each other. Therefore, owing to (2.10.24) one has

$$\mathbf{P}\big(S_n(\lambda) \in dz\big) = \frac{e^{\lambda z}\mathbf{P}(S_n \in dz)}{\psi(\lambda)}.$$

Hence for $\lambda = \lambda(\alpha)$ we have

$$\begin{aligned}\mathbf{P}(S_n \in dz) &= e^{-\lambda(\alpha)z + B_n^2 A(\lambda(\alpha))}\mathbf{P}\Big(S_n\big(\lambda(\alpha)\big) \in dz\Big)\\ &= e^{-B_n^2\Lambda(\alpha)+\lambda(\alpha)(x-z)}\mathbf{P}\Big(S_n\big(\lambda(\alpha)\big) \in dz\Big).\end{aligned}$$

Integrating this from x to $x+\Delta$ and making the change of variable $z-x=v$, we get

$$\mathbf{P}\big(S_n \in \Delta[x)\big) = e^{-B_n^2\Lambda(\alpha)} \int_x^{x+\Delta} e^{\lambda(\alpha)(x-z)} \mathbf{P}\Big(S_n\big(\lambda(\alpha)\big) \in dz\Big)$$
$$= e^{-B_n^2\Lambda(\alpha)} \int_0^{\Delta} e^{-\lambda(\alpha)v} \mathbf{P}\Big(S_n\big(\lambda(\alpha)\big) - x \in dv\Big).$$

The lemma is proved. □

Let us return to the proof of Theorem 2.10.8. We will use Lemma 2.10.10, in which we take $\Delta = \Delta_n$ to be a quantity depending on n and converging to 0 slowly enough as $n \to \infty$. Since $\lambda(\alpha) \leqslant \lambda_0$, we have $\Delta_n\lambda(\alpha) \to 0$ as $n \to \infty$, and in the integral in (2.10.32) one has $e^{-\lambda(\alpha)z} \to 0$ uniformly in $z \in [0, \Delta]$. Therefore, as $n \to \infty$,

$$\mathbf{P}\Big(S_n \in \Delta_n[x)\Big) = e^{-B_n^2\Lambda(\alpha)} \mathbf{P}\Big(S_n\big(\lambda(\alpha)\big) - x \in \Delta[0)\Big)\big(1+o(1)\big), \qquad (2.10.33)$$

where the remainder term is uniform in $\alpha \in [0, \alpha_0]$. Now we will show that one can apply Theorem 2.10.1 to $S_n\big(\lambda(\alpha)\big) - x$. Indeed, since

$$x = \alpha B_n^2 \equiv B_n^2 A'\big(\lambda(\alpha)\big) = \sum_{j=1}^{n} A_j'\big(\lambda(\alpha)\big) = \sum \mathbf{E}\xi_j^{(\lambda(\alpha))},$$

we see that

$$S_n\big(\lambda(\alpha)\big) - x = \sum \big(\xi_j^{(\lambda(\alpha))} - \mathbf{E}\xi_j^{(\lambda(\alpha))}\big)$$

is a sum of random variables with zero means. As conditions (2.10.26), $[\mathbf{UI}]_{\lambda_0}$ and $[\mathbf{R}]_{\lambda_0}$ imply that the conditions of Theorem 2.10.1 applied to the random variables $\xi_j^{(\lambda(\alpha))} - \mathbf{E}\xi_j^{(\lambda(\alpha))}$ are satisfied, we conclude that for any fixed $\Delta > 0$ one has in (2.10.33) the representation

$$\mathbf{P}\Big(S_n\big(\lambda(\alpha)\big) - x \in \Delta[x)\Big) = \frac{\Delta}{B_n^{(\lambda(\alpha))}}\phi(0) + o\left(\frac{1}{B_n^{(\lambda(\alpha))}}\right). \qquad (2.10.34)$$

It is clear that this relation will also hold true for $\Delta = \Delta_n$, where $\Delta_n \to 0$ slowly enough as $n \to \infty$. Substituting (2.10.34) with such a $\Delta = \Delta_n$ into (2.10.33), we obtain the first assertion of the theorem.

Equality (2.10.29) follows from the relations

$$\Lambda(v) = v\lambda(v) - A\big(\lambda(v)\big),$$
$$\Lambda'(v) = \lambda(v) + v\lambda'(v) - A'\big(\lambda(v)\big)\lambda'(v) \equiv \lambda(v).$$

Since $\Lambda(0) = \lambda(0) = 0$, the last relation integrated from 0 to α yields (2.10.29).

The theorem is proved. □

2.10.6 Moderately large deviations

If $x = o(B_n^2)$ ($\alpha = o(1)$) then for the right-hand side of (2.10.28) it is possible to find an explicit expression (in terms of the cumulants of ξ_j; again we will write for brevity ξ_j instead of $\xi_{j,n}$), using a segment of the so-called Cramér series. For this we will need an additional condition. To formulate it, consider the kth-order cumulants γ_{kj} of the random variables ξ_j (i.e. the expansion coefficients of the function

$$A_j(\lambda) = \sum_{k=1}^{s} \frac{\gamma_{kj}\lambda^k}{k!} + o(\lambda^s),$$

as $\lambda \to 0$, in powers of λ, provided that $\mathbf{E}|\xi_j^s| < \infty$). It is not hard to verify that (see e.g. (1.1.1), (1.1.2) or [151])

$$\gamma_{1j} = \mathbf{E}\xi_j = 0, \qquad \gamma_{2j} = \mathbf{D}\xi_j = \sigma_j^2, \qquad \gamma_{3j} = \mathbf{E}\xi_j^3, \quad \text{etc.}$$

If ξ_j has $s \geqslant 2$ finite moments then, for all $r \leqslant s$ one has the following representation as $\lambda \to 0$:

$$A_j'(\lambda) = \sum_{k=2}^{r} \frac{\gamma_{kj}\lambda^{k-1}}{(k-1)!} + o(\lambda^{r-1}). \tag{2.10.35}$$

Put

$$\Gamma_k := B_n^{-2} \sum_{j=1}^{n} \gamma_{kj};$$

hence

$$\Gamma_1 = 0, \quad \Gamma_2 = 1, \quad \Gamma_3 = B_n^{-2} \sum_{j=1}^{n} \mathbf{E}\xi_j^3, \quad \text{etc.}$$

Then, for all n, along with (2.10.35) one has, as $\lambda \to 0$,

$$A'(\lambda) = \sum_{k=2}^{r} \frac{\Gamma_k\lambda^{k-1}}{(k-1)!} + o(\lambda^{r-1}), \quad r \leqslant s. \tag{2.10.36}$$

The aforementioned additional condition has the following form:

[U] *The remainders $o(\lambda^{r-1})$ in* (2.10.36) *are uniform in n for all $r \leqslant s$.*

It is not hard to see that condition **[U]** is equivalent to the assumption that the sth derivative

$$A^{(s)}(\lambda) = \big(\ln \psi_n(\lambda)\big)^{(s)} B_n^{-2}$$

is equicontinuous at $\lambda = 0$.

The following condition is sufficient for **[U]**: *all the summands ξ_j have finite moments of order $s+1$ and, moreover, for all $\lambda \in [0, \delta]$ and sufficiently small $\delta > 0$ the derivatives*

$$A^{(s+1)}(\lambda) = \big(\ln \psi_n(\lambda)\big)^{(s+1)} B_n^{-2}$$

are uniformly bounded in n. The sufficiency of this condition follows from the fact that, for all $\lambda \in [0, \delta]$, $r \leqslant s$,

$$A'(\lambda) = \sum_{k=2}^{r} \frac{\Gamma_k \lambda^{k-1}}{(k-1)!} + \frac{A^{(r+1)}(\widetilde{\lambda})\lambda^r}{r!},$$

where $\widetilde{\lambda} \in [0, \lambda] \subset [0, \delta]$ and $A^{(r+1)}(\widetilde{\lambda})\lambda \to 0$ as $\lambda \to 0$ uniformly in n.

It is also not hard to see that if we assume in Example 2.10.1 that $g(j) \leqslant 1$ and that $\psi(\lambda) := \mathbf{E}e^{\lambda\zeta}$ satisfies the condition $[\mathbf{C}]_{\lambda_0}$, $\mathbf{E}|\zeta|^{s+1} < \infty$, then condition **[U]** will be met. If the functions $g(j)$ are increasing then condition **[U]** will also hold, when applied to the random variables $\xi^*_{j,n} = \xi_{n-j+1}/g(n)$.

We will need the following auxiliary assertion.

Lemma 2.10.11. *Let all ξ_j have $s \geqslant 2$ finite moments. If condition* **[U]** *is met then the function $\Lambda(\alpha)$ admits the representation*

$$\Lambda(\alpha) = \sum_{k=2}^{s} \Lambda_k \alpha^k + o(\alpha^s), \tag{2.10.37}$$

as $\alpha \to 0$, where the remainder term is uniform in n; an algorithm for calculating the coefficients Λ_k is contained in the proof. In particular,

$$\Lambda_2 = \frac{1}{2}, \quad \Lambda_3 = -\frac{\Gamma_3}{6} \quad (\text{for } s \geqslant 3), \quad \Lambda_4 = \frac{1}{8}\left(\Gamma_3^2 - \frac{\Gamma_4}{3}\right) \quad (\text{for } s \geqslant 4). \tag{2.10.38}$$

The algorithm for calculating Λ_k is presented (in a somewhat different form) in [151], where one can also find a closed-form representation for Λ_5.

Integral theorems on the asymptotic behaviour of $\mathbf{P}(S_n \geqslant x)$ as $x = o(B_n^2)$ can be found in [151] and [165].

Proof of Lemma 2.10.11. First we will obtain an asymptotic representation for the function $\lambda(\alpha)$ inverse to $A'(\lambda)$. That function is a solution to the equation $A'(\lambda) = \alpha$, which can be written, under the assumption that $\lambda \to 0$, as

$$\sum_{k=2}^{s} \frac{\Gamma_k \lambda^{k-1}}{(k-1)!} + o(\lambda^{s-1}) = \alpha. \tag{2.10.39}$$

Since the function $A'(\lambda)$ is $s-1$ times continuously differentiable and $\Gamma_1 = 1$, the well-known implicit function theorems imply that the function $\lambda(\alpha)$ (which satisfies the identity $A'(\lambda(\alpha)) \equiv \alpha$) has the same smoothness properties, i.e. it is $s-1$ times continuously differentiable and admits a representation of the form

$$\lambda(\alpha) = \sum_{k=1}^{s-1} \lambda_k \alpha^k + o(\alpha^{s-1}). \tag{2.10.40}$$

Substituting $\lambda = \lambda(\alpha)$ in the form (2.10.40) into (2.10.39) and equating the coefficients of powers of α on the right- and left-hand sides of (2.10.39), we obtain $s-1$ recursive equations for the coefficients λ_k, from which, in particular, it will follow that

$$\lambda_1 = 1, \quad \lambda_2 = -\frac{\Gamma_3}{2} \quad (\text{for } s \geqslant 3), \quad \lambda_3 = \frac{\Gamma_3^2}{2} - \frac{\Gamma_4}{6} \quad (\text{for } s \geqslant 4), \quad \text{etc.} \tag{2.10.41}$$

Integrating representation (2.10.40) with respect to α we obtain by virtue of (2.10.29) the required representation (2.10.37), in which

$$\Lambda_k = \frac{\lambda_{k-1}}{k}, \quad k \geqslant 2.$$

The assertion on the uniformity of the remainder term in (2.10.40) follows from the uniformity, for all $r \leqslant s$, of the remainders in the relations (2.10.39) with s replaced by r (see (2.10.36) and of condition **[U]**). The latter uniformity means that, as $\lambda \to 0$, for all $r \leqslant s$ one has

$$\sum_{k=r+1}^{s} |\Gamma_k|\lambda^{k-1} = o\big(|\Gamma_r|\lambda^r\big)$$

uniformly in n. These relations enable one to bound the remaining terms (those that we did not use when forming the equations for λ_k) in the expansion (2.10.39), in which we substituted (2.10.40), and also the coefficient ε of α^{s-1} in the representation for the remainder $o(\alpha^{s-1})$ in (2.10.40), of form $\varepsilon\alpha^{s-1}$. The bound for ε proves to be uniformly small in n provided that the coefficient δ in the representation of form $\delta\lambda^{s-1}$ for the remainder term in (2.10.39) is also uniformly small in n. When verifying these statements, one should use the fact that $\lambda(\alpha) \sim \alpha$ uniformly in n as $\alpha \to 0$ (see condition **[U]** and equation (2.10.39) for $s = 2$). The lemma is proved. □

Let us return to the problem of describing the asymptotic behaviour of moderately large deviations when $x = o(B_n^2)$. Since $B_n^{(\lambda(\alpha))} \sim B_n$ as $\alpha \to 0$ (and so $\lambda(\alpha) \to 0$), we have that $n \to \infty$, Theorem 2.10.8 and Lemma 2.10.11 imply the following result.

Theorem 2.10.12. *Assume that* $\max_j \mathbf{E}|\xi_{j,n}|^s < \infty$ *and condition* **[U]** *and the conditions of Theorem* 2.10.8 *are satisfied for some* $\lambda_0 > 0$. *Then, for* $\alpha = x/B_n^2 = o(1)$, *we have the representation* (2.10.28), *in which* $B_n^{(\lambda(\alpha))}$ *is to be replaced by* B_n *and* $\Lambda(\alpha)$ *by the right-hand side of* (2.10.37) *with the coefficients* Λ_k *specified in* (2.10.38). *In particular, for* $s = 3$, $\alpha^3 \leqslant cB_n^{-2}$, $c = const. > 0$ *one has*

$$\mathbf{P}\big(S_n \in \Delta[x)\big) \sim \frac{1}{\sqrt{2\pi}B_n} \exp\left\{-\frac{x^2}{2B_n^2} + \frac{\Gamma_3 x^3}{6B_n^6}\right\}.$$

For $s = 4$, $\alpha^4 \leqslant cB_n^{-2}$ *one has*

$$\mathbf{P}\big(S_n \in \Delta[x)\big) \sim \frac{1}{\sqrt{2\pi}B_n} \exp\left\{-\frac{x^2}{2B_n^2} + \frac{\Gamma_3 x^3}{6B_n^4} + \frac{\Lambda_4 x^4}{B_n^6}\right\}$$

(Λ_4 *was defined in* (2.10.38)*).*

2.10.7 Local theorems

One can see from subsection 2.8.5 that a similar approach can be used to obtain local large deviation theorems. To formulate these local theorems, we will need modifications of conditions $[\mathbf{R_M}]$, $[\mathbf{Z}]$ for the Cramèr transforms of the variables $\xi_j = \xi_{j,n}$ (see conditions $[\mathbf{R}]$ and $[\mathbf{R}]_{\lambda_0}$). These modifications have the following form.

$[\mathbf{R_M}]_{\lambda_0}$ *There exists a number $M \geqslant 1$ such that, for any fixed $\varepsilon > 0$,*

$$\sup_{\lambda \in [0,\lambda_0]} B_n^{(\lambda)} \sum_{j=M+1}^{n} q_{j,n}^{(\lambda)}(\varepsilon, \infty) \to 0$$

as $n \to \infty$, where $q_{j,n}^{(\lambda)}(\varepsilon, N)$ are specified in (2.10.27).

The modification of condition $[\mathbf{Z}]$ has the following form.

$[\mathbf{Z}]_{\lambda_0}$ *For any fixed $\varepsilon > 0$,*

$$\sup_{\lambda \in [0,\lambda_0]} B_n^{(\lambda)} \prod_{j=1}^{n} q_{j,n}^{(\lambda)}(\varepsilon, 2\pi - \varepsilon) \to 0$$

as $n \to \infty$.

Now we can state the local theorems.

Theorem 2.10.13. *Let the conditions* (2.10.26), $[\mathbf{C}]_{\lambda_0}$, $[\mathbf{UI}]_{\lambda_0}$, $[\mathbf{R}]_{\lambda_0}$ *be satisfied, and at least one of the conditions* (i) *or* (ii) *from Theorem* 2.10.4 *be met for the random variables $\xi_j^{(\lambda(\alpha))}$. Then for $n \geqslant M$ the distribution of the sum S_n has density f_n that admits the following representation as $n \to \infty$, which is uniform in x:*

$$f_n(x) = \frac{e^{-B_n^2 \Lambda(\alpha)}}{\sqrt{2\pi} B_n^{(\lambda(\alpha))}} \big(1 + o(1)\big), \qquad \alpha = \frac{x}{B_n^2}.$$

When considering *the arithmetic* case, by the scaled deviation α we will mean the ratio $\alpha = (x - A_n)/B_n^2$, $A_n = \mathbf{E}S_n$, and the function $\Lambda(\alpha)$ will be constructed for the centered random variables $\xi_{j,n} - a_{j,n}$, $a_{j,n} = \mathbf{E}\xi_{j,n}$.

Theorem 2.10.14. *Let the conditions of Theorem* 2.10.8 *be satisfied, except for the condition $a_{j,n} = 0$ and with the condition $[\mathbf{R}]_{\lambda_0}$ replaced by $[\mathbf{Z}]_{\lambda_0}$. Then, for integer-valued x, one has the following representation as $n \to \infty$:*

$$\mathbf{P}(S_n = x) = \frac{e^{-B_n^2 \Lambda(\alpha)}}{\sqrt{2\pi} B_n^{(\lambda(\alpha))}} \big(1 + o(1)\big), \qquad \alpha = \frac{x - A_n}{B_n^2}.$$

The proofs of Theorems 2.10.13 and 2.10.14 differ little from that of Theorem 2.10.8 and are actually simpler. The reader can obtain them with the help of

Theorems 2.10.4, 2.10.6, respectively, by using the Cramèr transform in exactly the same way as we did when proving Theorem 2.10.8.

Remark 2.10.9 is applicable to Theorems 2.10.13 and 2.10.14 as well.

Making use of Lemma 2.10.11 and arguments similar to those above, one can obtain under condition **[U]**, corollaries of Theorems 2.10.12 and 2.10.13 for deviations $x = o(B_n^2)$ (cf. Theorem 2.10.12).

The results of this section can be found in [37].

3

Boundary crossing problems for random walks

By the term boundary crossing problems we mean problems related to the crossings of boundaries of certain regions by trajectories $\{S_0, S_1, S_2, \ldots\}$ of sequential sums of random variables $\xi_1, \xi_2, \ldots$ Basic examples of such problems concern the distribution of the maximum of $\overline{S}_n = \max(S_0, S_1, \ldots, S_n)$ (in the one-dimensional case), the first entry time to a given set (the first passage of a given boundary), etc.

Significant progress in the study of the asymptotics of distributions of the functionals occurring in boundary crossing problems (known as boundary functionals) for *one-dimensional* random walks was achieved in the 1960–1970s. The advance was achieved thanks to the analytic approach to problems with *linear boundaries*. For such problems, the, distributions of boundary functionals (such as those for the time and place of first passage of a boundary) satisfy integro-differential equations which induce appropriate integral equations on the semi-axis for double Laplace transforms (with respect to time and space) over the original distributions. Using a generalisation (a modification) of the Wiener–Hopf method it is possible to solve those equations in terms of factorisation components of a known function of a complex variable. Further, it has become possible to 'invert' asymptotically those double transforms and obtain a rather complete description of the asymptotic behaviour of the distributions considered, including asymptotic expansions.

This analytic approach to boundary crossing problems is quite complicated technically. It will be summarized in section 3.9.

Recently (see [25], [26], [53]) it was discovered that the conditional distributions of the jumps ξ_k of a walk $\{S_1, \ldots, S_n\}$ under the condition that the sum S_n of the random variables $\xi_1, \ldots, \xi_n$ is in a given remote set have a quite simple nature and are expressed in terms of the Cramér transform. Using the simple probabilistic approaches presented in this chapter allows to solve a number of boundary crossing problems without resorting to analytic methods.

In the *multidimensional* case, analytic methods, which play an important role in the one-dimensional case, do not lead to the desired results. This is primarily due to the fact that there are no analogues of the Wiener–Hopf method for solving 'multidimensional' integral equations, and to the absence of relevant results on the factorisation of functions of several complex variables. Attempts to use analytic methods in the multidimensional case involved only simple walks (see [125], [79]), but even in those cases these methods did not lead to the desired results. An exception is the results of the paper [70], where limit theorems for the distribution of the maximum of some two-dimensional random walks (compound renewal processes) were derived using analytic methods.

In contrast with analytic approaches, direct probabilistic approaches can be generalised to the multidimensional case for a wide class of problems (see [27], [53]). This is illustrated in section 3.8.

Thus, in this chapter, in sections 3.1–3.8 we consider a 'direct' probabilistic approach to boundary crossing problems that is based on asymptotic properties of the jumps $\xi_1, \dots, \xi_n$ of a trajectory $\{S_1, \dots, S_n\}$ with a 'fixed' end S_n. In section 3.9 the analytic approach will be briefly introduced.

3.1 Limit theorems for the distribution of jumps when the end of a trajectory is fixed. A probabilistic interpretation of the Cramér transform

As before, let

$$\psi(\lambda) = \mathbf{E}e^{\langle \lambda, \xi \rangle}, \quad A(\lambda) = \ln \psi(\lambda),$$
$$\mathcal{A} = \{\lambda : A(\lambda) < \infty\}, \quad \mathcal{A}' = \{A'(\lambda) : \lambda \in \mathcal{A}\},$$

and let $\Lambda(\alpha)$ be the deviation function corresponding to ξ, $\lambda(\alpha) = \Lambda'(\alpha)$. Recall that ${}^{\alpha}\xi = \xi^{(\lambda(\alpha))}$ denotes the Cramér transform of ξ, i.e. the random vector with distribution

$$\mathbf{P}({}^{\alpha}\xi \in dv) = \frac{e^{\langle \lambda(\alpha), v \rangle} \mathbf{P}(\xi \in dv)}{\psi(\lambda(\alpha))},$$

so that there exists the following density of the distribution of ${}^{\alpha}\xi$ with respect to the distribution of ξ:

$$p^{(\lambda(\alpha))}(v) := \frac{\mathbf{P}({}^{\alpha}\xi \in dv)}{\mathbf{P}(\xi \in dv)} = \frac{e^{\langle \lambda(\alpha), v \rangle}}{\psi(\lambda(\alpha))}. \tag{3.1.1}$$

For an arbitrary fixed integer $m \geqslant 1$ let $\mathcal{A}^*$ and K denote arbitrary compacts from $(\mathcal{A}')$ and $\mathbb{R}^{dm}$, respectively; let $\Delta[v]$ with $v = \big(v_{(1)}, \dots, v_{(d)}\big)$, as before, denote the half-open cube

$$\Delta[v) = \big\{u \in \mathbb{R}^d : v_{(1)} \leqslant u_{(1)} < v_{(1)} + \Delta, \dots, v_{(d)} \leqslant u_{(d)} < v_{(d)} + \Delta\big\}$$

with side length Δ. Further, let $\vec{v} = (v_1, \dots, v_m)$, $v_i \in \mathbb{R}^d$.

Theorem 3.1.1. *Suppose that ξ is a non-lattice random variable and condition* **[C]** *is satisfied, $\alpha = x/n$. Then, for any $k_1 < \dots < k_m$, $k_1 \geqslant 1$, $k_m \leqslant n$, the conditional distribution of $\overline{\xi} = (\xi_{k_1}, \dots, \xi_{k_m})$ under the condition $\{S_n \in \Delta[x)\}$*

is absolutely continuous with respect to the distribution of $^{\alpha}\vec{\xi} = ({}^{\alpha}\xi_1, \dots, {}^{\alpha}\xi_m)$*, where the* ${}^{\alpha}\xi_j$ *are independent copies of* ${}^{\alpha}\xi$*.*

The corresponding density is

$$p_{n,x}(\vec{v}) = \frac{\mathbf{P}\big(\vec{\xi} \in d\vec{v} \,|\, S_n \in \Delta[x)\big)}{\mathbf{P}({}^{\alpha}\vec{\xi} \in d\vec{v})}$$

where $\vec{v} = (v_1, \dots, v_m)$ *converges to* 1 *as* $n \to \infty$ *uniformly in* $\alpha \in \mathcal{A}^*$, $\vec{v} \in K \subset \mathbb{R}^{dm}$, $\Delta \in [\Delta_n, \Delta_0]$*, and where* $\Delta_n \to 0$ *slowly enough as* $n \to \infty$ *and* Δ_0 *is a fixed number.*

In the arithmetic case, instead of $\{S_n \in \Delta[x)\}$ *one should consider the event* $\{S_n = x\}$ *for integer* x $(\Delta = 1)$*.*

Theorem 3.1.1 means that convergence in variation, of the conditional distributions of $\vec{\xi}$ under the condition $\{S_n \in \Delta[x)\}$ to the distribution of ${}^{\alpha}\vec{\xi}$, takes place. In particular, it implies the following.

Corollary 3.1.2. *Suppose that the conditions of Theorem* 3.1.1 *are satisfied. Then, for any Borel sets* $B_1, \dots, B_m$ *from* $\mathbb{R}^d$ *and any* $k_1, \dots, k_m$*,*

$$\mathbf{P}\big(\xi_{k_1} \in B_1, \dots, \xi_{k_m} \in B_m \,|\, S_n \in \Delta[x)\big) = \prod_{i=1}^{m} \mathbf{P}({}^{\alpha}\xi_i \in B_i) + \varepsilon_n, \tag{3.1.2}$$

where $\varepsilon_n \to 0$ *as* $n \to \infty$ *uniformly in* $\alpha \in \mathcal{A}^*$, $\Delta \in [\Delta_n, \Delta_0]$*.*

In the arithmetic case one should take the event $\{S_n = x\}$, for x from the integer lattice, as the condition.

Thus if $\alpha = x/n \to \alpha_0 \in \mathcal{A}^*$ when $n \to \infty$, we obtain that *the Cramér transform of* ${}^{\alpha_0}\mathbf{F}$ *of* $\mathbf{F}$ *is the limiting conditional distribution of the variable* ξ_1 *under the condition* $S_n \in \Delta[x)$:

$$\mathbf{P}({}^{\alpha_0}\xi \in B) = \lim_{n \to \infty} \mathbf{P}\big(\xi_1 \in B \,|\, S_n \in \Delta[x)\big).$$

This limiting distribution does not depend on Δ.

In [25], in the one-dimensional case the same relation was obtained for $\alpha_0 = \alpha_+ \in \partial\mathcal{A}'$ if $\alpha_+ < \infty$ and some additional conditions hold.

Proof of Theorem 3.1.1. According to Theorem 2.3.2,

$$\mathbf{P}\big(S_n \in \Delta[x)\big) = \frac{e^{-n\Lambda(\alpha)}}{\sigma(\alpha)(2\pi n)^{d/2}} \int_{\Delta[0)} e^{-\langle\lambda(\alpha),u\rangle} du (1 + \varepsilon_n), \tag{3.1.3}$$

where $\varepsilon_n \to 0$ as $n \to \infty$ uniformly in $\alpha = x/n \in \mathcal{A}^*$, $\Delta \in [\Delta_n, \Delta_0]$. A similar relation holds for $\mathbf{P}\big(S_{n-m} \in \Delta[x - y)\big)$ for fixed m and y, and with α replaced by $(x - y)/(n - m) = \alpha + O(1/n)$. Therefore, assuming without loss of generality that $k_i = i$, $1 \leqslant i \leqslant m$, and letting $y = v_1 + \dots + v_m$, we obtain

$$p^0_{n,x}(\vec{v}) := \frac{\mathbf{P}(\vec{\xi} \in d\vec{v} \mid S_n \in \Delta[x))}{\mathbf{P}(\vec{\xi} \in d\vec{v})} = \frac{\mathbf{P}(\vec{\xi} \in d\vec{v},\ S_n \in \Delta[x))}{\mathbf{P}\big(S_n \in \Delta[x)\big)\mathbf{P}(\vec{\xi} \in d\vec{v})}$$
$$= \frac{\mathbf{P}\big(S_{n-m} \in \Delta[x-y)\big)}{\mathbf{P}\big(S_n \in \Delta[x)\big)}.$$

According to (3.1.3), the last ratio is equal to

$$\exp\left\{-(n-m)\Lambda\left(\frac{x-y}{n-m}\right) + n\Lambda\left(\frac{x}{n}\right)\right\}(1+\varepsilon'_n), \tag{3.1.4}$$

where $\varepsilon'_n \to 0$ to $n \to \infty$ uniformly in $\alpha \in \mathcal{A}^*$, $\vec{v} \in K \subset \mathbb{R}^{dm}$, $\Delta \in [\Delta_n, \Delta_0]$. Since

$$\frac{x-y}{n-m} = \alpha - \frac{y}{n} + \frac{\alpha m}{n} + o\left(\frac{1}{n}\right),$$

the main part of the argument of the exponent in (3.1.4) is equal to

$$-(n-m)\Lambda\left(\frac{x-y}{n-m}\right) + n\Lambda\left(\frac{x}{n}\right)$$
$$= m\Lambda(\alpha) + \langle\lambda(\alpha), y\rangle - m\langle\lambda(\alpha), \alpha\rangle + o(1)$$
$$= \sum_{j=1}^{m}\Big[\langle\lambda(\alpha), v_j\rangle - A\big(\lambda(\alpha)\big)\Big] + o(1),$$

and therefore (see (3.1.1))

$$p^0_{n,x}(\vec{v}) = \prod_{j=1}^{m} p^{(\lambda(\alpha))}(v_j)\big(1+o(1)\big),$$

where the remainder term $o(1)$ is uniform in $\alpha \in \mathcal{A}^*$, $\vec{v} \in K$, $\Delta \in [\Delta_n, \Delta_0]$. However, owing to (3.1.1) we have

$$\frac{\mathbf{P}(^{\alpha}\vec{\xi} \in d\vec{v})}{\mathbf{P}(\vec{\xi} \in d\vec{v})} = \prod_{j=1}^{m} p^{(\lambda(\alpha))}(v_j). \tag{3.1.5}$$

The sought-for density $p_{n,x}(\vec{v})$ is obviously equal to the ratio of the densities $p^0_{n,x}(\vec{v})$ and (3.1.5). □

3.2 The conditional invariance principle and the law of the iterated logarithm

3.2.1 The conditional invariance principle

In this and the following sections, we will consider *one-dimensional* random walks $\{S_1, \ldots, S_n\}$.

Let $\mathbb{C}(0,1)$, $\mathbb{D}(0,1)$ denote the spaces of continuous functions and of functions without discontinuities of the second kind, respectively. We will say that *the distributions of a sequence of processes* $\zeta_n(t)$, $t \in [0,1]$, $\mathbb{C}$*-converge to the distribution of a process* $\zeta(t) \in \mathbb{C}(0,1)$ if, for any measurable functional f which is continuous at the points of the space $\mathbb{C}(0,1)$ with respect to the uniform metric, for $n \to \infty$ we have

$$\mathbf{P}\big(f(\zeta_n) < t\big) \Longrightarrow \mathbf{P}\big(f(\zeta) < t\big),$$

where the sign $\Longrightarrow$ denotes the weak convergence of a distribution.

It is clear that the $\mathbb{C}$-convergence generalises the weak convergence of distributions in $\mathbb{C}(0,1)$. See [23], [24], [72] for more details about the $\mathbb{C}$-convergence.

In the case $d = 1$, it is known that if

$$\mathbf{E}\xi = 0, \qquad \sigma^2 = \mathbf{E}\xi^2 < \infty$$

and the distribution of ξ either has a density or is arithmetic then the so-called conditional invariance principle holds: *the conditional distribution of the process* $S_{[nt]}/\sigma\sqrt{n}$, $0 \leqslant t \leqslant 1$, *under the condition* $S_n \in \Delta[0)$, $\Delta = O(\sqrt{n})$, $\mathbb{C}$*-converges to the distribution of a Brownian bridge* $w^0(t) = w(t) - tw(1)$, *where* $w(t)$ *is a standard Wiener process* (see e.g. [14], [115]).

This statement can be generalised in the following way. In order to use Theorems 2.2.1, 3.1.1 and Corollaries 2.2.4, 3.1.2, we will again assume that $\Delta_n \leqslant \Delta \leqslant \Delta_0 = \text{const.}$ (see Theorem 2.2.1), with $\Delta_n \to 0$ slowly enough as $n \to \infty$.

Theorem 3.2.1. *Let* ξ *be a non-lattice random variable and assume that condition* **[C]** *holds. Then the conditional distribution of the process*

$$\zeta_n(t) = \frac{S_{[nt]} - xt}{\sigma(\alpha)\sqrt{n}}, \qquad 0 \leqslant t \leqslant 1,$$

under the condition $S_n \in \Delta[x)$ $\mathbb{C}$*-converges uniformly in* $\alpha = x/n \in \mathcal{A}^*$, $\Delta \in [\Delta_n, \Delta_0]$, *to the distribution of a Brownian bridge* $w^0(t) = w(t) - tw(1)$, *where* $w(t)$ *is a standard Wiener process.*

In the arithmetic case one should consider integer x.

Theorem 3.2.1 means that for any Borel set B in the space $\mathbb{C}(0,1)$ of continuous functions such that $\mathbf{P}\big(w^0(\cdot) \in \partial B\big) = 0$, where ∂B is the boundary of B in the uniform topology, we have

$$\mathbf{P}\big(\zeta_n(\cdot) \in B \,|\, S_n \in \Delta[x)\big) \to \mathbf{P}\big(w^0(\cdot) \in B\big).$$

From Theorem 3.2.1, one can obtain a statement about the weak convergence of the distributions of continuous analogues $\widetilde{\zeta}_n(t)$ of the processes $\zeta_n(t)$ which are defined as continuous polygons passing through the points $\big(k/n, \zeta_n(k/n)\big)$, $k = 0, 1, \ldots, n$. If $\rho_{\mathbb{C}}$ is the uniform metric then, for any $\varepsilon > 0$, we have

$$\begin{aligned}\mathbf{P}\big(\rho_{\mathbb{C}}(\widetilde{\zeta}_n, \zeta_n) > \varepsilon \,|\, S_n \in \Delta[x)\big) &= \mathbf{P}\Big(\max_{k\leqslant n} |\xi_k| > \varepsilon\sqrt{n} \,|\, S_n \in \Delta[x)\Big)\\ &\leqslant n\mathbf{P}\big(|\xi_1| > \varepsilon\sqrt{n} \,|\, S_n \in \Delta[x)\big).\end{aligned}$$

Using the properties of the function Λ and the exponential Chebyshev inequality, it is possible to verify that the right-hand side of this equation converges to 0 as $n \to \infty$. This and Theorem 3.2.1 easily imply

Corollary 3.2.2. *The conditional distributions of the processes $\widetilde{\zeta}_n(t)$ under the condition $S_n \in \Delta[x)$ and as $n \to \infty$ weakly converge in the space $\mathbb{C}(0,1)$ with uniform metric to the distribution of the process $w^0(t)$, uniformly in $\alpha = x/n \in \mathcal{A}^*$, $\Delta \in [\Delta_n, \Delta_0]$.*

Proof of Theorem 3.2.1. According to well-known results on the convergence of random processes (see e.g. [11], [23], [24], [72], [94]) it is necessary to prove the convergence of the finite-dimensional distributions of $\zeta_n(t)$ and the tightness (compactness) of the family of distributions of those processes.

Let us prove the convergence of the one-dimensional distributions. For simplicity, let nt be integer. Then, for the half-interval $\delta_n[v) := \{u : v \leqslant u < v + \delta_n\}$ of length δ_n, where $\delta_n\sqrt{n} \to 0$ sufficiently slowly as $n \to \infty$, we have

$$\begin{aligned} &\mathbf{P}\left(\frac{S_{nt} - xt}{\sqrt{n}\sigma(\alpha)} \in \delta_n[v) \mid S_n \in \Delta[x)\right) \\ &\quad = \Big[\mathbf{P}\big(S_n \in \Delta[x)\big)\Big]^{-1} \int_{\delta_n[v)} \mathbf{P}\big(S_{nt} - xt \in \sigma(\alpha)\sqrt{n}du\big) \\ &\quad \times \mathbf{P}\big(S_{n(1-t)} + xt + u\sigma(\alpha)\sqrt{n} \in \Delta[x)\big). \end{aligned} \tag{3.2.1}$$

Here, owing to Theorem 2.2.1 and the mean value theorem, we have the following representation for the value of the integral in (3.2.1) for some $u \in \delta_n[v)$:

$$\begin{aligned} &\frac{\Delta \exp\left\{-n(1-t)\Lambda\left(\frac{x(1-t)-u\sigma(\alpha)\sqrt{n}}{n(1-t)}\right)\right\}}{\sqrt{2\pi n(1-t)}\,\sigma(\alpha)} (1+\varepsilon_n)\big(1+r(\Delta)\big) \\ &\quad \times \frac{\delta_n}{\sqrt{2\pi t}} \exp\left\{-nt\Lambda\left(\frac{xt + v\sigma(\alpha)\sqrt{n}}{nt}\right)\right\} (1+\varepsilon_n'), \end{aligned} \tag{3.2.2}$$

where ε_n , ε_n' have all the properties of uniform convergence to 0 as $n \to \infty$ specified in Theorem 2.2.1 and

$$1 + r(\Delta) := \frac{(1 - e^{-\lambda(\alpha)\Delta})}{\lambda(\alpha)\Delta}.$$

Let us again use Theorem 2.2.1, to compute $\mathbf{P}\big(S_n \in \Delta[x)\big)$ in (3.2.1). According to (3.2.2), for (3.2.1) we obtain the value

$$\begin{aligned} &\frac{\delta_n}{\sqrt{2\pi t(1-t)}} \exp\left\{-nt\Lambda\left(\alpha + \frac{v\sigma(\alpha)}{t\sqrt{n}}\right)\right. \\ &\qquad \left. -n(1-t)\Lambda\left(\alpha - \frac{u\sigma(\alpha)}{(1-t)\sqrt{n}}\right) + n\Lambda(\alpha)\right\} (1+\varepsilon_n''), \end{aligned} \tag{3.2.3}$$

where $\varepsilon''_n \to 0$ as $n \to \infty$ uniformly in $\alpha \in \mathcal{A}^*$, $\Delta \in [\Delta_n, \Delta_0]$ (see Theorem 3.2.1). For the functions in the exponent in (3.2.3) we have

$$\Lambda\left(\alpha + \frac{v\sigma(\alpha)}{t\sqrt{n}}\right) = \Lambda(\alpha) + \frac{\lambda(\alpha)v\sigma(\alpha)}{t\sqrt{n}} + \frac{\Lambda''(\alpha)\big(v\sigma(\alpha)\big)^2}{2t^2 n} + o(1/n),$$

$$\Lambda\left(\alpha - \frac{u\sigma(\alpha)}{(1-t)\sqrt{n}}\right) = \Lambda(\alpha) - \frac{\lambda(\alpha)u\sigma(\alpha)}{(1-t)\sqrt{n}} + \frac{\Lambda''(\alpha)\big(u\sigma(\alpha)\big)^2}{2(1-t)^2 n} + o(1/n),$$

where in the last equality u can be replaced by v with addition of the remainder term $o(1/n)$. Inserting these decompositions into (3.2.3), we obtain the value

$$\frac{\delta_n}{\sqrt{2\pi t(1-t)}} \exp\left\{-\frac{v^2}{2t(1-t)}\right\}(1+\varepsilon_n), \tag{3.2.4}$$

with $\varepsilon_n \to 0$ uniformly in $\alpha \in \mathcal{A}^*$, $\Delta \in [\Delta_n, \Delta_0]$. But the probability $\mathbf{P}\big(w^0(t) \in \delta_n[v)\big)$ has the same form as (3.2.4). This also implies, owing to the uniformity in (3.2.4), the convergence

$$\mathbf{P}\big(\zeta_n(t) \in \delta[v) \,|\, S_n \in \Delta[x)\big) \to \mathbf{P}\big(w^0(t) \in \delta[v)\big)$$

as $\delta[v) := [v, v+\delta)$ and for any fixed δ.

If the number nt is not integer then S_{nt} in (3.2.1) should be replaced by $S_{[nt]}$ (according to the definition of the process $\zeta_n(t)$), and $S_{n(1-t)}$ by $S_{n-[nt]}$. In (3.2.2), nt and $n(1-t)$ should be replaced by $[nt]$ and $n-[nt]$, respectively. Similar obvious changes should be made in the following computations, the essence of which remains intact.

The convergence of finite-dimensional distributions can be considered in a completely similar way.

From the above arguments, we obtain

Corollary 3.2.3. *We have the convergence*

$$\mathbf{P}\Big(\big|\zeta_n(t)\big| > N\sqrt{t(1-t)} \,|\, S_n \in \Delta[x)\Big) \to 0 \tag{3.2.5}$$

as $n \to \infty$, $N \to \infty$ uniformly in $\alpha = x/n \in \mathcal{A}^*$, $\Delta \in [\Delta_n, \Delta_0]$ and $t \in (0,1)$.

The convergence (3.2.5) for t such that $tn \to \infty$, $(1-t)n \to \infty$, follows from the arguments above. For fixed $k = tn$ (or $(1-t)n$) this convergence follows from Theorem 3.1.1. The uniformity in α and Δ follows from the uniformity of the remainder terms in α and Δ in the theorems used.

Let us return to the proof of Theorem 3.2.1. It remains to verify that the family of conditional distributions of $\zeta_n(\cdot)$ is tight. For that, it is sufficient to show (see e.g. [11], [23], [24], [72], [94]) that, for any $\varepsilon > 0$,

$$\lim_{\gamma\to 0} \overline{\lim_{n\to\infty}} \frac{1}{\gamma}\mathbf{P}\Big(\sup_{t\leqslant s\leqslant t+\gamma} \big|\zeta_n(s) - \zeta(t)\big| \geqslant \varepsilon \,|\, S_n \in \Delta[x)\Big) = 0$$

for all t or, equivalently,

$$\lim_{\gamma\to 0}\overline{\lim_{n\to\infty}}\frac{1}{\gamma}\mathbf{P}\Big(\max_{l\leqslant n\gamma}\frac{1}{\sqrt{n}}\Big|\big(S_{k+l}-S_k-\alpha l\big)\sigma^{-1}(\alpha)\Big|\geqslant\varepsilon\,\Big|\,S_n\in\Delta[x)\Big)=0 \tag{3.2.6}$$

for all k, $0\leqslant k<n(1-\gamma)$.

By symmetry, it is sufficient to prove (3.2.6) for $k = 0$. Without loss of generality, we may assume that $\sigma(\alpha)=1$. We will need

Lemma 3.2.4 (an analogue of Kolmogorov's inequality). *Let $\alpha = x/n \in \mathcal{A}^*$, $\Delta\in[\Delta_n,\Delta_0]$, $q\in(0,1)$, $m\leqslant qn$. Then there exists $b>0$, which does not depend on n, such that, for all $y>0$ and n sufficiently large,*

$$\mathbf{P}\big(\max_{k\leqslant m}|S_k-\alpha k|\geqslant y\,|\,S_n\in\Delta[x)\big)$$
$$\leqslant 2\mathbf{P}\Big(|S_m-\alpha m|\geqslant y(1-q)-b\sqrt{m}\,\Big|\,S_n\in\Delta[x)\Big)+ce^{-\beta n},$$

where $c<\infty$, $\beta>0$ do not depend on n and y.

Proof. It is enough to prove this result for one-sided deviations only. Let

$$S_k^0=S_k-\alpha k,\qquad \eta(y)=\min\{k:\ S_k^0\geqslant y\},\qquad \overline{\xi}_k=\max_{j\leqslant k}\xi_j.$$

Then, for any $\varepsilon>0$,

$$\mathbf{P}\Big(\max_{k\leqslant m}S_k^0\geqslant y\,\Big|\,S_n\in\Delta[x)\Big)$$
$$\leqslant\mathbf{P}\Big(\max_{k\leqslant m}S_k^0\geqslant y,\ \overline{\xi}_m<\varepsilon n\,\Big|\,S_n\in\Delta[x)\Big)+\mathbf{P}\big(\overline{\xi}_m\geqslant\varepsilon n\,|\,S_n\in\Delta[x)\big).$$

First, let us estimate the first term on the right-hand side. We have

$$\mathbf{P}\big(S_m^0\geqslant y(1-q)-z\,|\,S_n\in\Delta[x)\big)$$
$$\geqslant\sum_{k=1}^{m}\int_{u=y}^{\varepsilon n}\mathbf{P}\big(\eta(y)=k,\ S_k^0\in du,\ S_m^0>y(1-q)-z,\ S_n\in\Delta[x)\big)$$
$$=\sum_{k=1}^{m}\int_{y}^{\varepsilon n}\mathbf{P}\big(\eta(y)=k,\ S_k^0\in du\big)$$
$$\times\mathbf{P}\big(u+S_{m-k}^0\geqslant y(1-q)-z,\ S_{n-k}\in\Delta[x-u-\alpha k)\big). \tag{3.2.7}$$

Let $n'=n-k$, $x'=x-u-\alpha k$ and $\alpha'=x'/n'=\alpha-u/(n-k)$. Then

$$u+S_{m-k}^0=S_{m-k}-\alpha'(m-k)+u\Big(1-\frac{m-k}{n-k}\Big),$$

and the second factor on the right-hand side of (3.2.7) can be written as

$$\mathbf{P}\big(S_{n'} \in \Delta[x')\big) - \mathbf{P}\bigg(S_{n'} \in \Delta[x'),\ S_{m-k} - \alpha'(m-k) < -u\Big(1 - \frac{m-k}{n-k}\Big) + y(1-q) - z\bigg). \tag{3.2.8}$$

Since $(m-k)/(n-k) < q$, we have $u\big(1-(m-k)/(n-k)\big) > y(1-q)$, and the last probability does not exceed

$$\mathbf{P}\big(S_{n'} \in \Delta[x'),\ S_{m-k} - \alpha'(m-k) < -z\big). \tag{3.2.9}$$

But $\alpha - \alpha' = u/(n-k)$ and $\varepsilon n/n(1-q) = \varepsilon/(1-q)$, and we can assume that for sufficiently small ε, along with the relation $\alpha \in \mathcal{A}^*$, it also holds that $\alpha' \in \mathcal{A}^*$ (where $\mathcal{A}^*$ is a bounded closed set inside the open set $(\mathcal{A}') = (\alpha_-, \alpha_+)$ such that the ε-neighbourhood of the original set $\mathcal{A}^*$ again belongs to $(\mathcal{A}')$ for sufficiently small $\varepsilon > 0$). Therefore, from the arguments used in the first part of the proof of Theorem 3.2.1 (see also Corollary 3.2.3), we obtain that

$$\mathbf{P}\big(S_{m-k} - \alpha'(m-k) < -z \,|\, S_n \in \Delta[x')\big) \to 0$$

for $z = b\sqrt{m}$, $b \to \infty$, so that for b large enough the probability (3.2.9) does not exceed $\frac{1}{2}\mathbf{P}\big(S_{n'} \in \Delta[x')\big)$ for all $\alpha \in \mathcal{A}^*$ and m, k within the bounds specified above. This means that for the selected b the second multiplier on the right-hand side of (3.2.7) will be greater than or equal to (see (3.2.8))

$$\frac{1}{2}\mathbf{P}\big(S_{n'} \in \Delta[x')\big).$$

Therefore

$$\begin{aligned}\mathbf{P}\big(S_m^0 \geqslant y(1-q) - b\sqrt{m},\ S_n \in \Delta[x)\big) &\geqslant \frac{1}{2}\sum_{k=1}^{m}\int_y^{\varepsilon n} \mathbf{P}\big(\eta(y) = k,\ S_k^0 \in du\big)\mathbf{P}\big(S_{n'} \in \Delta[x')\big)\\ &\geqslant \frac{1}{2}\mathbf{P}\Big(\max_{k\leqslant m} S_k^0 \geqslant y,\ \overline{\xi}_m < \varepsilon n,\ S_n \in \Delta[x)\Big).\end{aligned}$$

This gives the required upper bound on the right-hand side of the last inequality.

To finish the proof, it remains to estimate

$$\mathbf{P}\big(\overline{\xi}_m \geqslant \varepsilon n \,|\, S_n \in \Delta[x)\big).$$

Let $\alpha > a = \mathbf{E}\xi$. Then

$$\mathbf{P}\big(\xi_1 > \varepsilon n,\ S_n \in \Delta[x)\big) \leqslant \int_{\varepsilon n}^{(\alpha-a)n} \mathbf{P}\big(\xi_1 \in du,\ S_n \in \Delta[x)\big) + \mathbf{P}\big(\xi_1 > (\alpha - a)n\big). \tag{3.2.10}$$

Here, the last probability on the right-hand side does not exceed

$$\exp\Big\{-\Lambda\big((\alpha-a)n\big)\Big\},$$

where, due owing the strict convexity of the function Λ at the point α,

$$\Lambda\big((\alpha-a)n\big) > n\big(\Lambda(\alpha)-\Lambda(a)\big)+\beta n = n\Lambda(\alpha)+\beta n, \qquad \beta>0.$$

The first term on the right-hand side of (3.2.10)) is equal to

$$I := \int_{\varepsilon n}^{(\alpha-a)n} \mathbf{P}(\xi_1 \in du)\mathbf{P}\big(S_{n-1} \in \Delta[x-u)\big).$$

The value a always belongs to the zone $(\mathcal{A}')$ or its boundary. Suppose that the former is true. Then $\alpha - u/n \in (\mathcal{A}')$ for $u \in \big(\varepsilon n, (\alpha-a)n\big)$ and

$$I \leqslant \frac{c\Delta}{\sqrt{n}} \int_{\varepsilon n}^{(\alpha-a)n} \mathbf{P}(\xi \in du)\ \exp e^{-(n-1)\Lambda((x-u)/(n-1))},$$

where, owing to the convexity of the function Λ,

$$(n-1)\Lambda\left(\frac{x-u}{n-1}\right) > n\Lambda(\alpha)+\lambda(\alpha)(\alpha-u),$$

$$I \leqslant \frac{c\Delta}{\sqrt{n}} e^{-n\Lambda(\alpha)-\lambda(\alpha)\alpha} \int_{\varepsilon n}^{\infty} \mathbf{P}(\xi \in du)e^{u\lambda(\alpha)}.$$

The last integral is not greater than

$$e^{-\beta\varepsilon n}\int_{\varepsilon n}^{\infty} \mathbf{P}(\xi \in du)e^{u(\lambda(\alpha)+\beta)} \leqslant e^{-\beta\varepsilon n}\psi\big(\lambda(\alpha)+\beta\big),$$

where $\psi\big(\lambda(\alpha)+\beta\big) < \infty$ for sufficiently small $\beta>0$ and $\alpha \in \mathcal{A}^*$.

As a result, we obtain

$$\mathbf{P}\big(\xi > \varepsilon n,\ S_n \in \Delta[x)\big) \leqslant ce^{-n\Lambda(\alpha)-\beta n}.$$

Since the asymptotic behaviour of $\mathbf{P}\big(S_n \in \Delta[x)\big)$ is known (it has the form $(c\Delta/\sqrt{n})e^{-n\Lambda(\alpha)}$), we obtain the existence of $\beta'>0$, $\beta''>0$, $c'<\infty$, $c''<\infty$ such that

$$\mathbf{P}\big(\xi > \varepsilon n \,|\, S_n \in \Delta[x)\big) \leqslant c'e^{-\beta' n},$$
$$\mathbf{P}\big(\overline{\xi}_m > \varepsilon n \,|\, S_n \in \Delta[x)\big) \leqslant c'me^{-\beta' n} \leqslant c''e^{-\beta'' n}.$$

This gives the required bound. If a lies on the boundary of $(\mathcal{A}')$, then instead of a we should consider a value a' which is sufficiently close to a. The lemma is proved. □

Let us return to the proof of Theorem 3.2.1. It remains to prove (3.2.6) for $k = 0$. According to Lemma 3.2.4 and the first part of the proof of the theorem, for $m=[n\gamma]$ we have

$$\overline{\lim_{n\to\infty}}\, \mathbf{P}\left(\max_{k\leqslant m} \frac{1}{\sqrt{n}}\,|S_k-\alpha k| \geqslant \varepsilon \,\Big|\, S_n \in \Delta[x)\right)$$
$$\leqslant \overline{\lim_{n\to\infty}}\, 2\mathbf{P}\Big(|S_m-\alpha m| \geqslant \varepsilon(1-q)\sqrt{n}-b\sqrt{m} \,|\, S_n \in \Delta[x)\Big)$$

$$= \overline{\lim_{n\to\infty}}\, 2\mathbf{P}\Big(\big|\zeta_n(\gamma) - \alpha\gamma\big| \geqslant \varepsilon(1-q) - b\sqrt{\gamma} + O\Big(\frac{1}{\sqrt{n}}\Big) \,\Big|\, S_n \in \Delta[x)\Big)$$

$$= 2\mathbf{P}\Big(\big|w^0(\gamma)\big| \geqslant \varepsilon(1-q) - b\sqrt{\gamma}\Big) \leqslant 2\mathbf{P}\Big(\big|w^0(\gamma)\big| \geqslant \frac{\varepsilon(1-q)}{2}\Big),$$

if γ is small enough that $b\sqrt{\gamma} \leqslant \varepsilon(1-q)/2$. To prove (3.2.6), it remains to verify that

$$\lim_{\gamma\to 0} \frac{1}{\gamma}\mathbf{P}\Big(\big|w^0(\gamma)\big| \geqslant \frac{\varepsilon(1-q)}{2}\Big) = 0.$$

This follows from the fact that $w^0(\gamma)$ has the normal distribution with parameters $\big(0, \gamma(1-\gamma)\big)$, so, for any fixed $v > 0$,

$$\frac{1}{\gamma}\mathbf{P}\Big(\big|w^0(\gamma)\big| \geqslant v\Big) = \frac{1}{\gamma}\mathbf{P}\Big(\big|w(1)\big| > \frac{v}{\sqrt{\gamma(1-\gamma)}}\Big) \to 0$$

for $\gamma \to 0$. Relation (3.2.6), and therefore Theorem 3.2.1, is proved. □

A multidimensional version of Theorem 3.2.1 was proved in [25].

3.2.2 The conditional law of the iterated logarithm

For $\delta > 0$, $N \in \mathbb{Z}$, let

$$\begin{aligned} \theta(k) &:= (1+\delta)\sqrt{2k\ln\ln k}, \\ B_{N,n}(\alpha) &:= \big\{|S_k - \alpha k|\sigma^{-1}(\alpha) < \theta(k) \text{ for all } k = N,\dots,n\big\}. \end{aligned} \tag{3.2.11}$$

By $\overline{B}$ we will denote the complement of B.

In what follows, we will use the symbol c with or without indices to denote constants, which may be different in different formulas.

Theorem 3.2.5 (Conditional law of the iterated logarithm). *Suppose that the condition* **[C]** *is satisfied and* $\alpha = x/n \in \mathcal{A}^*$. *Then for any given* $\varepsilon > 0$, $\delta > 0$ *there exist* N *and* n_0 *such that, for all* $n \geqslant n_0$,

$$\mathbf{P}\big(\overline{B}_{N,n}(\alpha) \mid S_n \in \Delta[x)\big) < \varepsilon. \tag{3.2.12}$$

Proof. Without loss of generality, we may assume that $\sigma(\alpha) = 1$. Let

$$r > 1, \qquad q < 1, \qquad M = \frac{\ln N}{\ln r}, \qquad L = \frac{\ln n + \ln q}{\ln r} - 1,$$

so that $r^M = N$, $r^{L+1} = qn$.

We have

$$\overline{B}_{N,n} \subset \bigcup_{M\leqslant l\leqslant L} D_l + D^{(L)}, \tag{3.2.13}$$

where

$$D_l = \Big\{\max_{1\leqslant j<r^{l+1}} |S_j - \alpha j| \geqslant \theta(r^l)\Big\},$$

$$D^{(L)} = \Big\{\max_{1\leqslant j\leqslant n} |S_j - \alpha j| \geqslant \theta(r^{L+1})\Big\}.$$

The conditional invariance principle (see Theorem 3.2.1) implies that

$$\mathbf{P}\big(D^{(L)} \mid S_n \in \Delta[x)\big) \to 0$$

as $n \to \infty$ ($q < 1$ is fixed). Furthermore,

$$\mathbf{P}\Big(\bigcup_{M \leqslant l \leqslant L} D_l \,\Big|\, S_n \in \Delta[x)\Big) \leqslant \sum_{M \leqslant l \leqslant L} \mathbf{P}\big(D_l \mid S_n \in \Delta[x)\big). \tag{3.2.14}$$

By virtue of Lemma 3.2.4, when

$$m = [r^{l+1}], \qquad m_1 = [r^l],$$

we have

$$\mathbf{P}\big(D_l \mid S_n \in \Delta[x)\big)$$
$$\leqslant 2\mathbf{P}\big(|S_m - \alpha m| > \theta(m_1)(1-q) - b\sqrt{m} \mid S_n \in \Delta[x)\big) + c^{-\beta n},$$

where $\theta(m_1)(1-q) = h\sqrt{2m \ln\ln m}$ and we can put $h = 1 + \delta/2$ for sufficiently small q, $r - 1$ and sufficiently large M (and, hence, m). Denoting for brevity $h\sqrt{2\ln\ln m} = H$ and, for simplicity, passing to the consideration of one-sided deviations for S_m, we arrive at the requirement to estimate the probabilities

$$\mathbf{P}\big(S_m - \alpha m > H\sqrt{m} \mid S_n \in \Delta[x)\big) = P_1 + P_2 + P_3, \tag{3.2.15}$$

where

$$P_1 = \mathbf{P}\big(S_m - \alpha m > \varepsilon n \mid S_n \in \Delta[x)\big),$$
$$P_2 = \mathbf{P}\big(\varepsilon n \geqslant S_m - \alpha m > \varepsilon m \mid S_n \in \Delta[x)\big)$$
$$P_3 = \mathbf{P}\big(\varepsilon m \geqslant S_m - \alpha m > H\sqrt{m} \mid S_n \in \Delta[x)\big)$$

for some $\varepsilon > 0$. Let us start by estimating the probability P_1. Since the magnitude of the deviations of $S_m \in \Delta[v)$, $S_{n-m} \in \Delta[x - v)$ for $v > \varepsilon n$ may not correspond to the zone $\mathcal{A}^*$, we cannot apply here 'regular' large deviation theorems such as Theorem 2.2.1. Let us use the inequalities

$$\mathbf{P}\big(S_k \in \Delta[y)\big) \leqslant \exp\Big\{-k\Big[\Lambda\Big(\frac{y}{k}\Big) \bigvee \Lambda\Big(\frac{y+\Delta}{k}\Big)\Big]\Big\}.$$

Assuming for simplicity that $\Delta = 1$, for

$$\mathbf{P}\big(S_m - \alpha m > \varepsilon n,\ S_n \in \Delta[x)\big)$$
$$= \int_{\varepsilon n}^{\infty} \mathbf{P}\big(S_m - \alpha m \in dv\big)\mathbf{P}\big(S_{n-m} \in \Delta[x - \alpha m - v)\big) \tag{3.2.16}$$

we get the upper bound

$$\sum_{j \geqslant \varepsilon n} \exp\Big\{-m\Big[\Lambda\Big(\alpha + \frac{j}{m}\Big) \bigvee \Lambda\Big(\alpha + \frac{j+1}{m}\Big)\Big]$$
$$-(n-m)\Big[\Lambda\Big(\alpha - \frac{j-1}{n-m}\Big) \bigvee \Lambda\Big(\alpha - \frac{j+1}{n-m}\Big)\Big]\Big\}. \tag{3.2.17}$$

However, for $j > \varepsilon m$, owing to the strict convexity of the function Λ there exists $c\,(j/m,\alpha) > 0$ such that

$$\frac{m}{n}\Lambda\left(\alpha+\frac{j}{m}\right)+\left(\frac{n-m}{n}\Lambda\left(\alpha-\frac{j}{n-m}\right)\right)$$
$$\geqslant \Lambda(\alpha)+c\left(\frac{j}{m},\alpha\right)\frac{j}{n} \geqslant \Lambda(\alpha)+c(\varepsilon,\alpha)\frac{j}{n}, \qquad (3.2.18)$$

where the last inequality is valid because the function $c(y,\alpha)$ can be assumed to be increasing in y. Since m will be assumed to be unboundedly increasing, $n-m \to \infty$, the same inequality as (3.2.18) will obviously be valid for the argument of the exponent in (3.2.17). Using Theorem 2.2.1 to estimate $\mathbf{P}\big(S_n \in \Delta[x)\big)$, we finally obtain that

$$P_1 \leqslant c_1\sqrt{n}\sum_{j\geqslant\varepsilon n} e^{-jc(\varepsilon,\alpha)} \leqslant c_2\sqrt{n}\,e^{-\varepsilon c(\varepsilon,\alpha)n}. \qquad (3.2.19)$$

In a similar way, one can consider the probability

$$P_2 = \mathbf{P}\big(\varepsilon n \geqslant S_m - \alpha m > \varepsilon m \,|\, S_n \in \Delta[x)\big).$$

Here, in order to bound the second multiplier in the integrand in (3.2.16) we can use Theorem 2.2.1 for sufficiently small ε. This will allow us to get rid of the multiplier $\sqrt{n}$ in the relation (3.2.18) for the bound of P_2 (inequalities like (3.2.17) are preserved):

$$P_2 \leqslant c_1\sum_{j\geqslant\varepsilon m} e^{-jc(\varepsilon,\alpha)} \leqslant c_2 e^{-\varepsilon c(\varepsilon,\alpha)n}.$$

Now we estimate the probability

$$P_3 = \mathbf{P}\big(\varepsilon m \geqslant S_m - \alpha m > H\sqrt{m} \,|\, S_n \in \Delta[x)\big).$$

To simplify the argument, assume first that the distribution of ξ has a density and that condition [**D**] of Section 2.3.1 is satisfied. Then, using Theorem 2.3.1, we find that, for all sufficiently large M (and, hence, m),

$$P_3 = \frac{1}{\mathbf{P}(S_n \in \Delta[x))}\int_{H\sqrt{m}}^{\varepsilon m} \mathbf{P}\big(S_m - \alpha m \in dv\big)\mathbf{P}\big(S_{n-m} \in \Delta(x-\alpha m - v)\big)$$
$$\leqslant \frac{c}{\sqrt{m}}\int_{H\sqrt{m}}^{\varepsilon m} dv\,\exp\left\{-m\Lambda\left(\alpha+\frac{v}{m}\right)-(n-m)\Lambda\left(\alpha-\frac{v}{n-m}\right)+n\Lambda(\alpha)\right\}, \qquad (3.2.20)$$

where the argument of the exponent is equal to

$$-\frac{v^2}{2}\left(\frac{1}{m}+\frac{1}{n-m}\right)+O\left(\frac{v^3}{m^2}\right) \quad \text{as} \quad m\to\infty, \quad n-m\to\infty$$

(we have used the power series expansion of the function Λ at the point α and the equalities $\lambda(\alpha) = \Lambda'(\alpha)$, $\lambda'(\alpha) = \Lambda''(\alpha) = \sigma^{-1}(\alpha) = 1$). The substitution

$x/\sqrt{m} = u$ gives us the following upper bound for the right-hand side of (3.2.20):

$$c_1 \int_H^{\varepsilon\sqrt{m}} e^{-u^2/2(1-\varepsilon_1)} du \leqslant c_2\Big(1 - \Phi\big(h\sqrt{2(1-\varepsilon_1)\ln\ln m}\big)\Big), \tag{3.2.21}$$

where Φ is the standard normal distribution function and ε_1 can be made arbitrarily small by appropriate choices of $q < 1$ and ε. Using the bound

$$1 - \Phi(y) = \frac{1}{y\sqrt{2\pi}} e^{-y^2/2}\big(1 + o(1)\big)$$

for $y \to \infty$, it is relatively easy to obtain (see the standard arguments to be found in the proof of the classic law of the iterated logarithm for the Wiener process, in e.g. [39]) that (3.2.21) is not greater than $c_3 l^{-1-\delta_1}$, where $\delta_1 > 0$ for appropriate r, q and ε (recall that $l \sim \ln m/\ln r$ as $m \to \infty$).

In the case when the distribution of ξ does not have a density, we obtain the same bound,

$$P_3 \leqslant c_3 l^{-1-\delta_1}, \qquad \delta_1 > 0,$$

if we bound the integral on the left-hand side of (3.2.20) by a sum, using

$$\mathbf{P}\big(S_{n-m} \in \Delta[x - \alpha m - v)\big) < \frac{c}{\sqrt{n}} \exp e^{-(n-m)\Lambda(\alpha - j/(n-m))}$$

for $v \in [j, j+1]$, so that the integral does not exceed

$$c\frac{1}{\sqrt{n}} \sum_{H\sqrt{m}<j<\varepsilon m} \mathbf{P}\big(S_m - \alpha m \in [j, j+1)\big)\, e^{-(n-m)\Lambda(\alpha - j/(n-m))}$$
$$\leqslant \frac{c_1}{\sqrt{nm}} \sum_{H\sqrt{m}<j<\varepsilon m} \exp\left\{-m\Lambda\left(\alpha + \frac{j}{m}\right) - (n-m)\Lambda\left(\alpha - \frac{j}{n-m}\right)\right\},$$

which is essentially an inequality of the type (3.2.20).

Thus, for the one-sided deviations of the probabilities considered in (3.2.14) we have obtained the bound

$$c\big(\sqrt{n}\, e^{-\varepsilon c(\varepsilon,\alpha)n} + e^{-\varepsilon c(\varepsilon,\alpha)r^l} + l^{-1-\delta_1}\big). \tag{3.2.22}$$

The same bound is true for the event $\{S_m - \alpha m < -H\sqrt{m}\}$ (see (3.2.15)). Therefore, the quantity (3.2.22), if we replace c by $2c$ and add $ce^{-\beta n}$ to the right-hand side, will also bound from above the probability $\mathbf{P}\big(D_l \,|\, S_n \in \Delta[x)\big)$, and hence

$$\mathbf{P}\left(\bigcup_{M\leqslant l\leqslant L} D_l \,\middle|\, S_n \in \Delta[x)\right) \leqslant c_1 \sum_{l\geqslant M} l^{1+\delta_1} < c_2 M^{-\delta_1} \to 0$$

for $M \to \infty$. This and (3.2.13), (3.2.14) imply (3.2.12) for sufficiently large N. The theorem is proved. □

One can also prove the second part of the conditional law of the iterated logarithm, which states that for any fixed $N > 2$ and $n \to \infty$ we have $\mathbf{P}\big(B_{N,n}(\alpha)\,|\,S_n \in \Delta[x)\big) \to 0$, if $\theta(k)$ in (3.2.11) is replaced by the sequence $(1-\delta)\sqrt{2k\ln\ln k}$, $\delta > 0$. However we will not need this second part.

3.2.3 The conditional law of the iterated logarithm in the backward direction

Along with $B_{N,n}(\alpha)$ in (3.2.11), consider the event

$$C_{N,n}(\alpha) := \Big\{ \big|S_{n-k} - (n-k)\alpha\big|\,\sigma^{-1}(\alpha) < \theta(k) \quad \text{for all } k = N,\ldots,n \Big\},$$

which is an analogue of the event $B_{N,n}(\alpha)$ with respect to a trajectory in the backward direction, starting from a point (n,x), directed towards the origin $(0,0)$ and obtained by summation of the increments

$$\overleftarrow{\xi}_1 = -\xi_n, \qquad \overleftarrow{\xi}_2 = -\xi_{n-1}, \ \ldots, \ \overleftarrow{\xi}_n = -\xi_1,$$

so that

$$\big\{S_n \in \Delta[x)\big\} = \big\{\overleftarrow{S}_n \in \Delta[-x)\big\} = \big\{x + \overleftarrow{S}_n \in \Delta[0)\big\},$$

where $\overleftarrow{S}_k = \sum_{j=1}^{k} \overleftarrow{\xi}_j$.

Theorem 3.2.6. *Suppose that condition* **[C]** *holds and* $\alpha = x/n \in \mathcal{A}^*$. *Then for any given* $\varepsilon > 0$, $\delta > 0$ *there exist* N *and* n_0 *such that*

$$\mathbf{P}\big(\overline{C}_{N,n}(\alpha)\,|\,S_n \in \Delta[x)\big) < \varepsilon$$

for all $n \geqslant n_0$.

Proof. The statement of the theorem is a result of applying Theorem 3.2.5 to the random walk $\{\overleftarrow{S}_k\}$. One should note that the terms $\overleftarrow{\xi}_k$ are independent and distributed as $-\xi$, and the condition $\alpha \in \mathcal{A}^*$ means that the normalised deviation $\overleftarrow{\alpha} = -x/n = -\alpha$ for $\overleftarrow{S}_k$ belongs to the compact $-\mathcal{A}^* \in (\overleftarrow{\mathcal{A}}')$, where the set $\overleftarrow{\mathcal{A}}'$ corresponds to the random variable $\overleftarrow{\xi}_1$. The theorem is proved. □

Theorems 3.2.5 and 3.2.6 imply that when $n \to \infty$, for a trajectory $\{S_k\}_{k=1}^{n}$ the event $B_{N,n}(\alpha) \bigcap C_{N,n}(\alpha)$ holds with high probability.

Many results in Sections 3.1 and 3.2 can be found in [25] and [26].

3.3 The boundary crossing problem

3.3.1 Asymptotic closeness of distributions

We will now consider asymptotically close distributions. For such distributions, it turns out to be inconvenient to use the weak convergence of distributions and to fix the limit distribution (in the same way as we did not to fix the limiting value of the parameter $\alpha = x/n$ in the theorems of section 2.2). In view of this, we will use the following.

Definition 3.3.1. Let X_n, Y_n be sequences of random vectors with distribution functions $G_n = G_n(t)$ and $H_n = H_n(t)$, $n = 1, 2, \ldots$, respectively. We say that *the distributions G_n and H_n are asymptotically close* if, for any $u > 0$,

$$\sup_{t \in K_u} \left| G_n(t) - H_n(t) \right| \to 0 \tag{3.3.1}$$

as $n \to \infty$, where K_u is the ball $K_u = \{t : |t| \leqslant u\}$.

To simplify the exposition, we omit the tightness condition for the distributions $\mathbf{G}_n$ and $\mathbf{H}_n$ corresponding to the distribution functions G_n and H_n, i.e. the condition $\inf_n \mathbf{G}_n(K_u) \to 1$ as $u \to \infty$. This condition is not essential here, but it will be satisfied in an obvious way in further applications of Definition 3.3.1.

Relation (3.3.1) will be denoted by the symbol

$$G_n \approx H_n \qquad (X_n \approx Y_n).$$

This relation follows from the weak convergence of G_n and H_n to a distribution function G, if the latter is continuous. This property of G will be satisfied in many further applications of Definition 3.3.1 in an implicit way, so the requirement (3.3.1) will not be excessive with respect to the notion of weak convergence.

The convergence (3.3.1) will follow from the convergence in variation when the distributions G_n and H_n are mutually continuous. For example, ${}^{\alpha}\xi \approx {}^{\beta}\xi$ if $|\alpha - \beta| \to 0$.

Relation (3.3.1) is always satisfied if G_n is the conditional distribution function of $(\xi_{k_1}, \ldots, \xi_{k_N})$, $1 \leqslant k_i \leqslant n$, given the condition $\{S_n \in \Delta[x)\}$, and H_n is the distribution function of the vector $({}^{\alpha}\xi_1, \ldots, {}^{\alpha}\xi_N)$ with independent components ${}^{\alpha}\xi_j$ distributed as ${}^{\alpha}\xi$, $\alpha = x/n \in \mathcal{A}^*$. According to Theorem 3.1.1, the convergence will be uniform in $\alpha \in \mathcal{A}^*$ and $\Delta \in [\Delta_n, \Delta_0]$, where $\Delta_n \to 0$ sufficiently slowly as $n \to \infty$, $\Delta_0 < \infty$.

Let us provide one more example of the asymptotic closeness of distributions. Theorem 2.2.1 implies the following.

Corollary 3.3.2. *If the distribution of ξ is non-lattice then for $y \leqslant \Delta$ and as $n \to \infty$ one has*

$$\mathbf{P}\big(S_n - x < y \mid S_n \in \Delta[x)\big) \approx \mathbf{P}(\tau < y) \tag{3.3.2}$$

uniformly in $\alpha = x/n \in \mathcal{A}^*$, $\Delta \in [\Delta_1, \Delta_2]$, $0 < \Delta_1 < \Delta_2 < \infty$, where τ has the truncated exponential distribution

$$\mathbf{P}(\tau < y) = E_{\alpha,\Delta}(y) := \frac{1 - e^{-\lambda(\alpha)y}}{1 - e^{-\lambda(\alpha)\Delta}}. \tag{3.3.3}$$

It is not hard to see that $E_{\alpha,\Delta}$ is the Cramér transform with parameter $-\lambda(\alpha)$ of the uniform distribution on $[0, \Delta]$.

Proof of Corollary 3.3.2. Put $y_k = k\Delta_n$, $k = 0,1,\dots,N$, $\Delta_n = \Delta/N$, where $N = N(n) \to \infty$ sufficiently slowly as $n \to \infty$. The event $\{S_n - x \in [0, y_k)\}$ is the union of the disjoint events $\{S_n \in \Delta_n[x + y_j)\}$, $j = 0,\dots,k$, for which, by virtue of Theorem 2.2.1,

$$\mathbf{P}\big(S_n \in \Delta_n[x + y_j)\big) \sim \frac{\Delta_n e^{-n\Lambda((x+y_j)/n)}}{\sqrt{2\pi n}\,\sigma(\alpha)}$$

uniformly in $\alpha = x/n \in \mathcal{A}^*$. Here, as $n \to \infty$,

$$n\Lambda\left(\frac{x+y_j}{n}\right) = n\Lambda(\alpha) + y_j\lambda(\alpha) + o(1).$$

Furthermore,

$$\begin{aligned}
&\mathbf{P}\big(S_n - x < y_k \,|\, S_n \in \Delta[x)\big)\\
&= \left[\sum_{j=0}^{k} \mathbf{P}\big(S_n \in \Delta_n[x + y_j)\big)\right] \times \left[\sum_{j=0}^{N} \mathbf{P}\big(S_n \in \Delta_n[x + y_j)\big)\right]^{-1}\\
&\sim \left[\sum_{j=0}^{k} \Delta_n e^{-\lambda(\alpha)y_j}\right]\left[\sum_{j=0}^{N} \Delta_n e^{-\lambda(\alpha)y_j}\right]^{-1} \sim \left[\int_0^{y_k} e^{-\lambda(\alpha)v}dv\right]\left[\int_0^{\Delta} e^{-\lambda(\alpha)v}dv\right]^{-1}\\
&= \frac{1 - e^{-\lambda(\alpha)y_k}}{1 - e^{-\lambda(\alpha)\Delta}}.
\end{aligned}$$

This implies (3.3.2). The corollary is proved. □

3.3.2 A boundary crossing problem

Let $g = \{g_k = g_{k,n}\}_{k\geqslant 1}$ be an arbitrary sequence of numbers g_k, in general depending on n, which satisfies the following condition:

$[\mathbf{g}]_\alpha$ *There exist numbers $\varepsilon > 0$, M, N (not depending on n) such that*

$$g_k > -\alpha k + \varepsilon k \quad \textit{for } k \geqslant M, \tag{3.3.4}$$

$$g_k > -k\min(s_+ - \varepsilon, N) \quad \textit{for all } k \geqslant 1, \tag{3.3.5}$$

where s_+ is the upper bound of the support of the distribution $\mathbf{F}$ (the number N is introduced for the case $s_+ = \infty$).

We will need the following statement about the conditional probability (under the condition $\{S_n \in \Delta[x)\}$) that the random walk does not intersect the boundary $x + g_k$ in the backward direction. For comparison, consider also the probability

$$P(g,\alpha) := \mathbf{P}({}^{\alpha}S_k > \tau - g_k \quad \text{for all } k \geqslant 1) \tag{3.3.6}$$

that the sequence $\{{}^{\alpha}S_k\}$ does not intersect the boundary $\tau - g_k$ in the forward direction, where τ is defined in (3.3.3) and does not depend on $\{{}^{\alpha}S_k\}$. In Lemma 3.3.4 below, we show that the probability $P(g,\alpha)$ is separated from zero uniformly in n and $\alpha \in \mathcal{A}^*$.

Theorem 3.3.3. *Suppose that condition* $[\mathbf{g}]_\alpha$ *is satisfied for* $\alpha \in (\mathcal{A}')$. *Then, for* $n \to \infty$,

$$\mathbf{P}\big(S_{n-k} < x + g_k \quad \text{for all } k = 1, \dots, n-1 \mid S_n \in \Delta[x)\big) \sim P(g, \alpha), \tag{3.3.7}$$

$$\mathbf{P}\big(S_{n-k} < x + g_k \quad \text{for all } k = 1, \dots, n-1 \mid S_n \in \Delta[x - \Delta)\big) \sim P(g + \Delta, \alpha) \tag{3.3.8}$$

uniformly in $\alpha = x/n \in \mathcal{A}^*$, *where* $g + \Delta$ *is the sequence* $\{g_k + \Delta\}$.

First, we will prove an auxiliary result.

Lemma 3.3.4. *If condition* $[\mathbf{g}]_\alpha$ *is satisfied then*

$$\inf_{\alpha \in \mathcal{A}^*} P(g, \alpha) > \delta > 0, \tag{3.3.9}$$

where δ *does not depend on* n.

Proof. Let $h = \min(s_+ - \varepsilon, N)$. By the condition $[\mathbf{g}]_\alpha$,

$$\begin{aligned} P(g, \alpha) &\geqslant \mathbf{P}\big({}^\alpha S_k > kh \quad \text{for} \quad k \in [1, M]; \ {}^\alpha S_k > k(\alpha - \varepsilon) \quad \text{for} \quad k > M\big) \\ &= \mathbf{P}(B_M)\mathbf{P}\big({}^\alpha S_k > (\alpha - \varepsilon)k \quad \text{for} \quad k > M \mid B_M\big), \end{aligned}$$

where

$$B_M = \big\{{}^\alpha S_k > kh \quad \text{for} \quad k \in [1, M]\big\}.$$

Clearly, the multiplier $\mathbf{P}(B_M)$ on the right-hand side for all $\alpha \in \mathcal{A}^*$ is not less than some positive number $\delta_{\varepsilon,M} > 0$. The second multiplier is not less than the unconditional probability

$$\begin{aligned} \mathbf{P}\big({}^\alpha S_k > (\alpha - \varepsilon)k \quad \text{for } k > M\big) &> \mathbf{P}\big({}^\alpha S_k > (\alpha - \varepsilon)k \quad \text{for all } k \geqslant 1\big) \\ &= \mathbf{P}\Big(\inf_{k \geqslant 1} ({}^\alpha S_k - \alpha k + \varepsilon k) > 0\Big) =: P_\alpha. \end{aligned}$$

It remains to show that $\inf_{\alpha \in \mathcal{A}^*} P_\alpha > 0$. Let $\mathcal{A}^* = [\alpha_*, \alpha^*]$. Divide $[\alpha_*, \alpha^*]$ into $K > 2/\varepsilon$ segments $[\alpha_{j-1}, \alpha_j]$, $j = 1, \dots, K$, with lengths not less than $\varepsilon/2$. Then, on a segment $[\alpha_j, \alpha_{j+1}]$, owing to the stochastic monotonicity of ${}^\alpha S_k$ in α (${}^\alpha S_k \underset{st}{<} {}^\beta S_k$ for $\alpha < \beta$; see section 2.1), we have $\alpha - \varepsilon < \alpha_j - \varepsilon/2$,

$$\begin{aligned} P_\alpha &> \mathbf{P}\Big(\inf_{k \geqslant 1} ({}^{\alpha_j} S_k - \alpha k + \varepsilon k) > 0\Big) \\ &> \mathbf{P}\left(\inf_{k \geqslant 1}\left({}^{\alpha_j} S_k - \alpha_j k + \frac{\varepsilon k}{2} > 0\right)\right) =: p_{j,\varepsilon} > 0 \end{aligned}$$

(see e.g. [39], Chapter 12). Since $\inf_{\alpha \in \mathcal{A}^*} P_\alpha \geqslant \min_{j \leqslant k} p_{j,\varepsilon} > \delta > 0$, the lemma is proved. □

Proof of Theorem 3.3.3. For brevity, let $\big\{S_n \in \Delta[x)\big\} = A$. For any fixed $N < n$ we have

$$\mathbf{P}\big(S_{n-k} < x + g_k,\ k = 1, \ldots, n-1 \,|\, A\big) = \mathbf{P}\big(B_1^N B_{N+1}^{n-1} \,|\, A\big)$$
$$= \mathbf{P}\big(B_1^N \,|\, A\big) - \mathbf{P}\Big(B_1^N \overline{B}_{N+1}^{n-1} \,|\, A\Big), \tag{3.3.10}$$

where

$$B_i^j = \bigcap_{k=i}^{j} \{S_{n-k} < x + g_k\}.$$

The distribution of $S_n - x$ under the condition A is asymptotically close to the distribution of τ (see (3.3.2), (3.3.3)). Therefore, according to Corollaries 3.3.2 and 3.1.2 and Lemma 3.3.4, we have (as $n \to \infty$)

$$\mathbf{P}\big(B_1^N \,|\, A\big) \sim \mathbf{P}\big(S_{n-k} - S_n + \tau < g_k,\ k = 1, \ldots, N \,|\, A\big)$$
$$\sim \mathbf{P}\big(-{}^{\alpha}S_k + \tau < g_k,\ k = 1, \ldots, N\big)$$
$$= \mathbf{P}\big({}^{\alpha}S_k > \tau - g_k,\ k = 1, \ldots, N\big) \geqslant \delta > 0 \tag{3.3.11}$$

uniformly in n and $\alpha \in \mathcal{A}^*$. Let

$$B_i^{j,\alpha} = \bigcap_{k=i}^{j} \{{}^{\alpha}S_k > \tau - g_k\}.$$

Then the right-hand side of (3.3.11) is equal to

$$\mathbf{P}(B_1^{N,\alpha}) = \mathbf{P}(B_1^{\infty,\alpha}) + \mathbf{P}(B_1^{N,\alpha}\, \overline{B}_{N+1}^{\infty,\alpha}),$$

where for N sufficiently large we have

$$\mathbf{P}(B_1^{N,\alpha}\, \overline{B}_{N+1}^{\infty,\alpha}) \leqslant \mathbf{P}(\overline{B}_{N+1}^{\infty,\alpha}) = \mathbf{P}\Big(\inf_{k>N}({}^{\alpha}S_k + g_k) \leqslant \tau\Big)$$
$$< \mathbf{P}\Big(\inf_{k>N}({}^{\alpha}S_k - \alpha k + \varepsilon k) \leqslant \Delta\Big). \tag{3.3.12}$$

Since $\mathbf{E}({}^{\alpha}\xi - \alpha) = 0$ and ${}^{\alpha}\xi$ satisfy condition $[\mathbf{C}_0]$ for all $\alpha \in \mathcal{A}^*$, using the exponential Chebyshev inequality we obtain, for $k > 2\Delta/\varepsilon$,

$$\mathbf{P}({}^{\alpha}S_k - \alpha k < -\varepsilon k + \Delta) \leqslant \mathbf{P}\big({}^{\alpha}S_k - \alpha k < -\varepsilon k/2\big) \leqslant q_{\alpha}^k,$$

where $q = \sup_{\alpha \in \mathcal{A}^*} q_{\alpha} < 1$. This gives for (3.3.12) the upper bound

$$\frac{q^N}{1-q}.$$

Now we return to relation (3.3.10). By virtue of condition $[\mathbf{g}]_{\alpha}$ and the conditional law of the iterated logarithm (see Theorem 3.2.6), we have

$$\mathbf{P}\big(B_1^N \overline{B}_{N+1}^{n-1} \,|\, A\big) \leqslant \mathbf{P}\big(\overline{B}_{N+1}^{n-1} \,|\, A\big) < \delta(N),$$

where $\delta(N) \to 0$ for $N \to \infty$. Thus, taking into account that $\mathbf{P}(B_1^{\infty,\alpha}) = P(g, \alpha)$, we finally obtain for $n \to \infty$

$$\mathbf{P}(S_{n-k} < x + g_k,\ k = 1, \ldots, n-1 \,|\, A) = P(g, \alpha)\big(1 + o(1)\big) + r_1 + r_2, \tag{3.3.13}$$

where

$$|r_1| < \delta(N), \qquad |r_2| < \frac{q^N}{1-q}.$$

Since the left-hand side of (3.3.13) does not depend on N, this relation can be true only if (3.3.7) is satisfied. Relation (3.3.8) can be proved in a similar way. The theorem is proved. □

Remark 3.3.5. It is not hard to see that condition $[\mathbf{g}]_\alpha$ in Theorem 3.3.3 can be relaxed to the following.

$[\mathbf{g}_1]_\alpha$ *There exist numbers δ, M, N such that*

$$g_k > -\alpha k + \sigma(\alpha)(1+\delta)\sqrt{2k \ln \ln k} \quad \textit{for } k \geqslant M$$

and (3.3.5) *is satisfied for all* $k \geqslant 1$.

3.4 The first passage time of a trajectory over a high level and the magnitude of overshoot

3.4.1 Local theorems

In this section we will assume that

$$\alpha \geqslant \alpha_0 > 0.$$

Let

$$\overline{S}_n = \max_{k \leqslant n} S_k, \qquad \eta(x) = \min\{k : S_k \geqslant x\}, \qquad \chi(x) = S_{\eta(x)} - x,$$

so that

$$\begin{aligned} &\{\eta(x) = n,\ \chi(x) < \Delta\} = \{\overline{S}_{n-1} < x,\ S_n \in \Delta[x)\},\\ &\mathbf{P}(\eta(x) = n,\ \chi(x) < \Delta) = \mathbf{P}(S_n \in \Delta[x))\mathbf{P}(\overline{S}_{n-1} < x \mid S_n \in \Delta[x)). \end{aligned} \tag{3.4.1}$$

We know the asymptotics of $\mathbf{P}(S_n \in \Delta[x))$. Recall that if ξ is non-lattice, $\alpha \in \mathcal{A}^*$, then (see (3.3.5) and Theorem 2.2.3)

$$\mathbf{P}(S_n \in \Delta[x)) \sim f_{(n)}(x)\, \frac{1 - e^{-\lambda(\alpha)\Delta}}{\lambda(\alpha)}, \tag{3.4.2}$$

where

$$f_{(n)}(x) = \frac{e^{-n\Lambda(\alpha)}}{\sigma(\alpha)\sqrt{2\pi n}} \tag{3.4.3}$$

is the asymptotic density of S_n (see section 2.2). Thus, the study of the asymptotics of the probability (3.4.1) reduces to studying the asymptotics of $\mathbf{P}(\overline{S}_{n-1} < x \mid S_n \in \Delta[x))$, i.e. to the problem considered in the previous section, with boundary g_k, $g_k \equiv 0$, $k = 1, 2, \ldots$ (see Theorem 3.3.3). According to Theorem 3.3.3, for $n \to \infty$,

$$\mathbf{P}(\overline{S}_{n-1} < x \mid S_n \in \Delta[x)) \sim P(0, \alpha) = \mathbf{P}({}^{\alpha}\underline{S} > \tau),$$

where

$$^{\alpha}\underline{S} = \inf_{k\geqslant 1} {}^{\alpha}S_k.$$

Since

$$P(0,\alpha) = \frac{\lambda(\alpha)}{1-e^{-\lambda(\alpha)\Delta}} \int_0^{\Delta} e^{-\lambda(\alpha)v}\mathbf{P}({}^{\alpha}\underline{S} > v)dv,$$

according to Theorem 3.3.3 and (3.4.1), we obtain the following statement.

Corollary 3.4.1. *For non-lattice ξ, when $n \to \infty$,*

$$\mathbf{P}\big(\eta(x) = n,\ \chi(x) < \Delta\big) = f_{(n)}(x)I(\alpha,\Delta)\big(1+o(1)\big) \tag{3.4.4}$$

uniformly in $\alpha = x/n \in \mathcal{A}^* \bigcap [\alpha_0,\infty)$, where $\alpha_0 > 0$ and

$$I(\alpha,\Delta) = \int_0^{\Delta} e^{-\lambda(\alpha)v}\mathbf{P}({}^{\alpha}\underline{S} > v)dv. \tag{3.4.5}$$

It is not difficult to see that

$$I(\alpha,\infty) < \infty. \tag{3.4.6}$$

Indeed, for $\lambda(\alpha) \geqslant \lambda_0 > 0$ this is obvious. If $\lambda(\alpha) \to 0$ then $\alpha \to \mathbf{E}\xi \in \mathcal{A}^*$ and hence, for sufficiently small $\delta > 0$, we have $\delta \in (\mathcal{A}^*)$ $(\psi(\delta) < \infty)$ and $\delta - \lambda(\alpha) > 0$ starting from some value. Therefore, according to the exponential Chebyshev inequality,

$$\mathbf{P}({}^{\alpha}\xi > v) \leqslant e^{-(\delta-\lambda(\alpha))v}\frac{\psi(\delta)}{\psi(\lambda(\alpha))}.$$

Since

$$\mathbf{P}({}^{\alpha}\underline{S} > v) < \mathbf{P}({}^{\alpha}\xi > v),$$

we have

$$I(\alpha,\infty) \leqslant \int_0^{\infty} e^{-\delta v}\frac{\psi(\delta)dv}{\psi(\lambda(\alpha))} < \infty.$$

If $\lambda(\alpha) < 0$, integrating

$$\int_0^{\infty} e^{-\lambda(\alpha)v}\mathbf{P}({}^{\alpha}\xi > v)dv$$

by parts, we can see that this integral is finite if and only if

$$\int_0^{\infty} e^{-\lambda(\alpha)v}\mathbf{P}({}^{\alpha}\xi \in dv) \leqslant \int_{-\infty}^{\infty} e^{-\lambda(\alpha)v}\mathbf{P}({}^{\alpha}\xi \in dv) = \frac{\psi(0)}{\psi(\lambda(\alpha))} < \infty.$$

Inequality (3.4.6) is proved.

The above arguments and Corollary 3.4.1 imply

Corollary 3.4.2. *For non-lattice ξ, when $n \to \infty$,*

$$\mathbf{P}\big(\eta(x) = n\big) = f_{(n)}(x) I(\alpha, \infty)\big(1 + o(1)\big), \tag{3.4.7}$$

$$\mathbf{P}\big(\chi(x) < \Delta \mid \eta(x) = n\big) = \frac{I(\alpha, \Delta)}{I(\alpha, \infty)}\big(1 + o(1)\big)$$

uniformly in $\alpha \in \mathcal{A}^* \bigcap [\alpha_0, \infty)$.

Thus, the asymptotics of $\mathbf{P}\big(\eta(x) = n\big)$ for $\alpha \in \mathcal{A}^* \bigcap [\alpha_0, \infty)$ is the same (up to a positive factor $I(\alpha, \infty)$) as that of the asymptotic density $f_{(n)}(x)$. In Theorem 3.4.6 below, it will be shown that $I(\mathbf{E}\xi, \infty) = \mathbf{E}\xi$ when $\mathbf{E}\xi > 0$.

The statements obtained can be easily carried over to the case of arithmetic ξ.

3.4.2 Integral limit theorems for the first passage time over a high level and for the maximum of sequential sums

Now we will obtain several integral limit theorems for $\eta(x)$ and for

$$\overline{S}_n = \max_{k \leqslant n} S_k,$$

for $\alpha \geqslant \alpha_0 > 0$. An important role will be played by the mutual location of the parameters α and

$$\alpha_1 = \psi'(\lambda_1), \quad \text{where } \lambda_1 = \sup\big\{\lambda : \psi(\lambda) \leqslant 1\big\}.$$

Note that α_1 is the value of the function $\psi'(\lambda)/\psi(\lambda)$ inverse to $\lambda(\alpha)$ at the point λ_1. If $\mathbf{E}\xi \geqslant 0$ then $\lambda_1 = 0$ and $\alpha_1 = \mathbf{E}\xi$. If $\mathbf{E}\xi < 0$ and $\lambda_+ \geqslant \lambda_1$ then

$$\alpha_1 = \mathbf{E}\xi^{(\lambda_1)} > 0.$$

The limit theorems mentioned in the subsection heading will be obtained by summation of the local probabilities (3.4.4), (3.4.7), so that the asymptotics

$$\mathbf{P}\big(\eta(x) = n + k\big) \sim f_{(n+k)}(x) I(\infty, \alpha)$$

is defined by the function

$$(n+k)\Lambda\left(\frac{x}{n+k}\right) = n\left(1 + \frac{k}{n}\right)\Lambda\left(\frac{\alpha}{1 + k/n}\right), \qquad \alpha = \frac{x}{n}.$$

Hence we will need the properties of the function

$$G(\alpha, t) := (1+t)\Lambda\left(\frac{\alpha}{1+t}\right) = \alpha\,\frac{\Lambda(v)}{v}, \qquad \text{where} \qquad v = \frac{\alpha}{1+t}. \tag{3.4.8}$$

We will distinguish the following three possibilities:

(i) $\alpha > \alpha_1 + \delta$ for some $\delta > 0$;
(ii) $0 < \alpha_0 \leqslant \alpha < \alpha_1 - \delta$;
(iii) α is located in the $1/\sqrt{n}$-neighbourhood of the point α_1.

One can see that the third possibility is equivalent (from the point of view of computations and the content of the results) to n being located in the $\sqrt{m}$-neighbourhood of the point $m = x/\alpha_1 \sim n$, i.e. $n = m + k$ where $k = z\sqrt{m}$, $z = O(1)$. Indeed, in the latter case,

$$\begin{aligned}\alpha = \frac{x}{n} &= \frac{x}{m + z\sqrt{m}} = \frac{x}{m(1 + z/\sqrt{m})} \\ &= \alpha_1\left(1 - \frac{z}{\sqrt{m}} + O\left(\frac{1}{m}\right)\right) = \alpha_1 - \frac{\alpha_1 z}{\sqrt{n}} + O\left(\frac{1}{n}\right).\end{aligned}$$

In a similar way, the converse statement can be obtained.

Lemma 3.4.3. *Suppose that* $\alpha \in \mathcal{A}^* \subset (\mathcal{A}')$. *Then*

$$G'_t(\alpha, 0) = -\ln \psi(\lambda(\alpha)). \tag{3.4.9}$$

(i) *If* $\alpha > \alpha_1$ *then*

$$\lambda_+ > \lambda(\alpha) > \lambda(\alpha_1) = \lambda_1 \geqslant 0, \qquad G'_t(\alpha, 0) < 0. \tag{3.4.10}$$

(ii) *If* $0 < \alpha < \alpha_1$ *then* $\mathbf{E}\xi \neq 0$,

$$\lambda_- < \lambda(\alpha) < \lambda(\alpha_1) = \lambda_1, \qquad G'_t(\alpha, 0) > 0. \tag{3.4.11}$$

(iii) *If* $\alpha_1 > 0$ *then* $\mathbf{E}\xi \neq 0$,

$$\begin{aligned}G(\alpha_1, 0) &= \alpha_1\lambda_1, \\ G'_t(\alpha_1, 0) &= 0, \\ G''_{tt}(\alpha_1, 0) &= \alpha_1^2 \Lambda''(\alpha_1) > 0.\end{aligned}$$

Proof. By virtue of (3.4.8),

$$G'_t(\alpha, t) = \Lambda(v) - v\lambda(v) = -\ln \psi(\lambda(v)), \quad \text{where } v = \frac{\alpha}{1+t}, \tag{3.4.12}$$

$$G'_t(\alpha, 0) = -\ln \psi(\lambda(\alpha)). \tag{3.4.13}$$

(i) Suppose that $\alpha > \alpha_1$. Since $\alpha \in \mathcal{A}^*$, we have $\alpha_+ > \alpha$,

$$\lambda_+ > \lambda(\alpha) > \lambda(\alpha_1) = \lambda_1,$$

and also $\psi(\lambda(\alpha)) > \psi(\lambda_1) = 1$. This and (3.4.13) imply (3.4.10).

(ii) Suppose that $0 < \alpha < \alpha_1$. If $\mathbf{E}\xi = 0$ then $\alpha_1 = 0$ and we get a contradiction. Hence $\mathbf{E}\xi \neq 0$. Since $\alpha \in \mathcal{A}^*$, we have $\alpha_- < \alpha$,

$$\lambda_- < \lambda(\alpha) < \lambda(\alpha_1) = \lambda_1, \qquad \psi(\lambda_1) \leqslant 1;$$

hence on the one hand $\ln \psi(\lambda_1 - \varepsilon) < 0$ for a sufficiently small $\varepsilon > 0$. On the other hand,

$$\ln \psi(\lambda(0)) = -\Lambda(0) < 0.$$

Since the function $\ln\psi(\lambda)$ is convex, we have $\ln\psi(\lambda) < 0$ on the whole segment $[0, \lambda_1 - \varepsilon]$ for any sufficiently small $\varepsilon > 0$. This and (3.4.13) imply (3.4.11).

(iii) If $\alpha_1 > 0$ then, as before, we see that $\mathbf{E}\xi \neq 0$. Furthermore,

$$G(\alpha_1, 0) = \Lambda(\alpha_1) = \alpha_1\lambda_1.$$

By virtue of (3.4.12), for $v = \alpha_1/(1+t)$, $t = 0$, we find

$$G'_t(\alpha_1, 0) = -\ln\psi(\lambda_1) = 0,$$
$$G''_{tt}(\alpha_1, 0) = v\Lambda''(v)\alpha_1 = \alpha_1^2\Lambda''(\alpha_1) > 0.$$

The lemma is proved. □

Now we will establish integral theorems for $\eta(x)$.

Theorem 3.4.4. *Suppose that $\alpha \geqslant \alpha_1 + \delta$ for some $\delta > 0$. Then, for $n \to \infty$,*

$$\mathbf{P}\big(\overline{S}_n \geqslant x,\ \chi(x) < \Delta\big) = \mathbf{P}\big(\eta(x) \leqslant n,\ \chi(x) < \Delta\big) \sim \frac{f_{(n)}(x)I(\alpha, \Delta)}{1 - q_\alpha}, \qquad (3.4.14)$$

where

$$q_\alpha = \frac{1}{\psi(\lambda(\alpha))} < \frac{1}{\psi(\lambda(\alpha_1 + \delta))} < 1$$

uniformly in $\alpha \in \mathcal{A}^* \bigcap [\alpha_1 + \delta, \infty)$.

Proof. We have

$$\mathbf{P}\big(\overline{S}_n \geqslant x,\ \chi(x) < \Delta\big) = \sum_{k=0}^{n-1} \mathbf{P}\big(\eta(x) = n - k,\ \chi(x) < \Delta\big)$$
$$= \Sigma_1 + \mathbf{P}\big(\eta(x) < n(1-\varepsilon),\ \chi(x) < \Delta\big), \qquad (3.4.15)$$

where

$$\Sigma_1 = \sum_{k=0}^{[n\varepsilon]} \mathbf{P}\big(\eta(x) = n - k,\ \chi(x) < \Delta\big).$$

Since $\alpha \in \mathcal{A}^* \bigcap [\alpha_1 + \delta, \infty)$, for a sufficiently small ε and all sufficiently large n we have

$$\frac{\alpha}{1-\varepsilon} < \alpha_+.$$

Hence, for $k \in [0, \varepsilon n]$, in order to find the probabilities $\mathbf{P}\big(\eta(x) = n - k,$ $\chi(x) < \Delta\big)$, we can use Corollary 3.4.1. According to (3.4.9), for $k = o(n)$,

$$(n-k)\Lambda\left(\frac{x}{n-k}\right) = nG\left(\alpha, -\frac{k}{n}\right) = nG(\alpha, 0) + k\ln\psi\big(\lambda(\alpha)\big) + o(k). \qquad (3.4.16)$$

This means that for $k < n\varepsilon$ and sufficiently small ε, the function $e^{-(n-k)\Lambda(x/(n-k))}$ (and also $f_{(n-k)}(x)$) decreases in k asymptotically as a geometric sequence with ratio

$$q_\alpha = e^{-\ln\psi(\lambda(\alpha))} = \frac{1}{\psi\big(\lambda(\alpha)\big)} < \frac{1}{\psi\big(\lambda(\alpha_1+\delta)\big)} < 1.$$

Since $G(\alpha, 0) = \Lambda(\alpha)$, together with (3.4.4) this proves that for sufficiently small ε the value $\sum_1$ is asymptotically equivalent to the right-hand side of (3.4.14).

The second term on the right-hand side of (3.4.15) can be bounded from above by the probability

$$\begin{aligned}\mathbf{P}\big(\eta(x) < n\varepsilon\big) &= \mathbf{P}\big(\overline{S}_{[n(1-\varepsilon)]+1} \geqslant x\big) \\ &\leqslant e^{-n(1-\varepsilon)\Lambda(x/(n(1-\varepsilon)))} + O(1) \end{aligned} \tag{3.4.17}$$

(see inequality (1.1.23)). Owing to (3.4.10), for sufficiently small $\varepsilon > 0$ there exists $\gamma > 0$ such that

$$n(1-\varepsilon)\ln\frac{x}{n(1-\varepsilon)} > nG(\alpha, 0) + \gamma n = n\Lambda(\alpha) + \gamma n.$$

This means that the probability (3.4.17) is $o(\Sigma_1)$ for $n \to \infty$. The theorem is proved. □

Theorem 3.4.5. *Suppose that ξ is non-lattice, $0 < \alpha_0 \leqslant \alpha < \alpha_1 - \delta$ for some $\delta > 0$. Then $\mathbf{E}\xi \neq 0$ and, for $n \to \infty$*

$$\mathbf{P}\big(\infty > \eta(x) \geqslant n,\ \chi(x) < \Delta\big) \sim \frac{f_{(n)}(x)I(\alpha, \Delta)}{1-q_\alpha}, \tag{3.4.18}$$

$$\mathbf{P}(\overline{S}_n < x) = \mathbf{P}\big(\eta(x) \geqslant n+1\big) \sim \frac{f_{(n+1)}(x)I(\alpha, \infty)}{1-q_\alpha} \tag{3.4.19}$$

uniformly in $\alpha \in \mathcal{A}^* \bigcap [\alpha_0, \alpha_1 - \delta]$, where

$$q_\alpha = \psi\big(\lambda(\alpha)\big) < \psi\big(\lambda(\alpha_1 - \delta)\big) < 1.$$

Proof. The inequality $\mathbf{E}\xi \neq 0$ follows from Lemma 3.4.3. Further, we have a similar expression to (3.4.15):

$$\mathbf{P}\big(\infty > \eta(x) \geqslant n,\ \chi(x) < \Delta\big) = \sum_{k=0}^{\infty}\mathbf{P}\big(\eta(x) = n+k,\ \chi(x) < \Delta\big) = \Sigma_1 + \Sigma_2,$$

where, for $\varepsilon > 0$,

$$\Sigma_1 := \sum_{k=0}^{[n\varepsilon]}\mathbf{P}\big(\eta(x) = n+k,\ \chi(x) < \Delta\big).$$

For $k = o(n)$ we obtain a similar expression to (3.4.16):

$$(n+k)\Lambda\left(\frac{x}{n+k}\right) = nG\left(\alpha, \frac{k}{n}\right) = nG(\alpha, 0) - k\ln\psi\big(\lambda(\alpha)\big) + o(k).$$

According to Lemma 3.4.3(ii) this means that for $k \leqslant \varepsilon n$ and sufficiently small ε the function $e^{-(n+k)\Lambda(x/(n+k))}$ (and also the function $f_{(n+k)}(x)$) decreases in k as a geometric sequence (asymptotically) with ratio

$$q_\alpha = e^{\ln\psi(\lambda(\alpha))} = \psi\big(\lambda(\alpha)\big) < \psi\big(\lambda(\alpha_1-\delta)\big) < 1.$$

This means that Σ_1 is asymptotically equivalent to the right-hand side of (3.4.18).

Let us now estimate Σ_2. Note that

$$\mathbf{P}\big(\eta(x)=n+k\big) \leqslant \mathbf{P}(S_{n+k}\geqslant x) \leqslant e^{-(n+k)\Lambda(x/(n+k))} \quad \text{for } \mathbf{E}\xi < 0, \tag{3.4.20}$$

$$\mathbf{P}\big(\eta(x)=n+k+1\big) \leqslant \mathbf{P}(S_{n+k} < x) \leqslant e^{-(n+k)\Lambda(x/(n+k))} \quad \text{for } \mathbf{E}\xi > 0. \tag{3.4.21}$$

In order to estimate the exponents on the right-hand sides of these inequalities, we use the fact, that, by virtue of (3.4.12), the derivative $G'_t(\alpha,t)$ at the point $t=\varepsilon$ is equal to

$$G'_t(\alpha,\varepsilon) = -\ln\psi\left(\lambda\left(\frac{\alpha}{1+\varepsilon}\right)\right) > 0.$$

For $t\to\infty$ (which corresponds to $k\gg n$), we have $G'(\alpha,t)\to -\ln\psi\big(\lambda(0)\big) = \Lambda(0) > 0$. But the function $-\ln\psi(\lambda)$ is concave and hence $\ln\psi\big(\lambda(\alpha/(1+t))\big) < 0$ for all $t\in(0,\infty)$, so the derivative $G'(\alpha,t)$ remains strictly positive for all $t\in(0,\infty)$. This means that the terms of Σ_2 decrease faster than some geometric sequence,

$$\Sigma_2 = O\big(e^{-nG(\alpha,\varepsilon)}\big) = o(\Sigma_1).$$

Relation (3.4.17) is proved. Relation (3.4.19) follows from (3.4.18). The theorem is proved. □

Now we consider the case when α is located in a neighbourhood of the point α_1, i.e. when

$$n \sim m = \frac{x}{\alpha_1} \quad \text{as} \quad n\to\infty$$

(see also the remark before Lemma 3.3.4).

Theorem 3.4.6. *Let $0<\alpha_1\in\mathcal{A}^*$. Then $\mathbf{E}\xi\neq 0$ and, for $x\to\infty$, $z=o(x^{1/6})$, we have*

$$\mathbf{P}\left(\infty > \eta(x) \geqslant \frac{x}{\alpha_1} + \frac{z\sigma(\alpha_1)\sqrt{x}}{\alpha_1^{3/2}},\ \chi(x) < \Delta\right) \sim \frac{e^{-x\lambda_1}I(\alpha_1,\Delta)}{\alpha_1}\big[1-\Phi(z)\big], \tag{3.4.22}$$

$$\mathbf{P}\left(\eta(x) < \frac{x}{\alpha_1} + \frac{z\sigma(\alpha_1)\sqrt{x}}{\alpha_1^{3/2}}\right) \sim \frac{e^{-x\lambda_1}I(\alpha_1,\infty)}{\alpha_1}\Phi(z),$$

where

$$\Phi(z) = \frac{1}{\sqrt{2\pi}}\int_{-\infty}^{z} e^{-u^2/2}du.$$

The variable $S = \overline{S}_\infty = \sup_{k\geqslant 0} S_k$ *satisfies the relation*

$$\mathbf{P}\big(S \geqslant x,\ \chi(x) < \Delta\big) \sim \frac{e^{-x\lambda_1} I(\alpha_1, \Delta)}{\alpha_1}. \tag{3.4.23}$$

If $\mathbf{E}\xi > 0$ *then* $\lambda_1 = 0$, $\alpha_1 = \mathbf{E}\xi = I(\alpha_1, \infty)$,

$$\frac{I(\alpha_1, \Delta)}{\alpha_1} = \lim_{x\to\infty} \mathbf{P}\big(\chi(x) < \Delta\big)$$

is the limiting distribution of the first overshoot over an infinitely distant barrier (see e.g. [39]).

Clearly, it follows from (3.4.22) and (3.4.23) that, as $x \to \infty$,

$$\mathbf{P}(S \geqslant x) \sim \frac{e^{-x\lambda_1} I(\alpha_1, \infty)}{\alpha_1}. \tag{3.4.24}$$

This relation is also obtained in [16], [22] (Theorem 21.11) and [39] (Theorem 12.7.4), but in those publications the multipliers of $e^{-x\lambda_1}$ on the right-hand side are expressed in different terms.

Indeed, the following statement holds true. Let $\chi^{(\lambda_1)}$ be the magnitude of the first overshoot of $\{S_k^{(\lambda_1)}\}$ over an infinitely distant boundary barrier.

Theorem 3.4.7 (Theorem 12.7.4 in [39]). *If* $\lambda_1 < \lambda_+$ *then*

$$\mathbf{P}(S \geqslant x) \sim pe^{-x\lambda_1},$$

where $p = \mathbf{E}e^{-\lambda_1 \chi^{(\lambda_1)}}$.

The asymptotics of $\mathbf{P}(S \geqslant x)$ in the case $\lambda_1 = \lambda_+$, $\psi(\lambda_1) < 1$, was found in [22] (Theorem 21.12).

Here we also provide (in addition to Theorem 1.1.1) the following exact inequalities for $\mathbf{P}(S \geqslant x)$ in the case $a < 0$.

Theorem 3.4.8 (Theorem 15.3.5 in [39]). *If* $a = \mathbf{E}\xi < 0$, $\lambda_1 < \lambda_+$, *then*

$$\psi_+^{-1} e^{-\lambda_1 x} \leqslant \mathbf{P}(S \leqslant x) \leqslant \psi_-^{-1} e^{-\lambda_1 x}, \tag{3.4.25}$$

where

$$\psi_+ = \sup_{t>0} \mathbf{E}\big(e^{\lambda_1(\xi - t)} \,|\, \xi > t\big), \qquad \psi_- = \inf_{t>0} \mathbf{E}\big(e^{\lambda_1(\xi - t)} \,|\, \xi > t\big).$$

If, for instance, $\mathbf{P}(\xi > t) = ce^{-\lambda_+ t}$, $c = \text{const.}$, $t > 0$, then

$$\mathbf{P}(\xi - t > v \,|\, \xi > t) = \frac{\mathbf{P}(\xi > t + v)}{\mathbf{P}(\xi > t)} = e^{-\lambda_+ v},$$

$$\psi_+ = \psi_- = \frac{\lambda_+}{\lambda_+ - \lambda_1},$$

and, hence, in this case we have the exact inequality

$$\mathbf{P}(S \geqslant x) = \frac{\lambda_+ - \lambda_1}{\lambda_+} e^{-\lambda_1 x}. \tag{3.4.26}$$

This mean that the inequalities (3.4.25) cannot be improved in any known sense. Clearly, the right-hand side of (3.4.26) can also be used as an approximation in cases when $\mathbf{P}(\xi > t)$ is close to $ce^{-\lambda_+ t}$, for $t \in [0, T]$, sufficiently large T, and the inequality

$$c_1 e^{-\lambda_+ t} < \mathbf{P}(\xi \geqslant t) < c_2 e^{-\lambda_+ t}, \qquad c_i = \text{const.}, \quad i = 1, 2,$$

is satisfied for all $t > 0$.

Proof of Theorem 3.4.6. The inequality $\mathbf{E}\xi \neq 0$ follows from Lemma 3.4.3. Let

$$m = \frac{x}{\alpha_1}, \qquad z_1 = \frac{z\sigma(\alpha_1)}{\alpha_1}.$$

In order to avoid too cumbersome notation including the integer and fractional parts of the number m, we will assume for simplicity that m is integer. The case of arbitrary m will merely increase the length of the formulas without changing their essence. Moreover, for simplicity assume that $\Delta = \infty$. Passage to the case $\Delta < \infty$ does not present any difficulty. For integer $m = x/\alpha_1$, we have

$$\mathbf{P}\Big(\infty > \eta(x) \geqslant \frac{x}{\alpha_1} + \frac{z\sigma(\alpha_1)\sqrt{x}}{\alpha_1^{3/2}}\Big) = \mathbf{P}\big(\infty > \eta(x) \geqslant m + z_1\sqrt{m}\big) = \Sigma_1 + \Sigma_2,$$

where, for $\varepsilon > 0$,

$$\Sigma_1 := \sum_{k \geqslant z_1\sqrt{m}}^{[\varepsilon m]} \mathbf{P}\big(\eta(x) = m + k\big), \quad \Sigma_2 = \sum_{k > \varepsilon m} \mathbf{P}\big(\eta(x) = m + k\big).$$

For $k = o(n)$, by virtue of Corollary 3.4.2 we have

$$\mathbf{P}\big(\eta(x) = m + k\big) \sim \frac{I(\alpha_1, \infty)}{\sqrt{2\pi m}\,\sigma(\alpha_1)} e^{-(m+k)\Lambda(x/(m+k))}.$$

According to Lemma 3.4.3(iii), $mG(\alpha_1, 0) = x\lambda_1$,

$$\begin{aligned}(m+k)\Lambda\Big(\frac{x}{m+k}\Big) &= mG\Big(\alpha_1, \frac{k}{m}\Big) \\ &= mG(\alpha_1, 0) + \frac{1}{2}\alpha_1^2\Lambda''(\alpha_1)\frac{k^2}{m} + O\Big(\frac{k^3}{m^2}\Big),\end{aligned}$$

where the remainder term is $o(1)$ as $k = o(m^{2/3})$. So, for sufficiently small ε and $|z_1| = o(m^{1/6})$, it follows that

$$\Sigma_1 \sim \frac{I(\alpha_1, \infty)e^{-x\lambda_1}}{\sqrt{2\pi m}\,\sigma(\alpha_1)} \sum_{k \geqslant z_1\sqrt{m}}^{[\varepsilon m]} \exp\Big\{-\frac{\alpha_1^2 k^2}{2\sigma^2(\alpha_1)m}\Big\},$$

where the sum on the right-hand side is asymptotically equivalent to the integral

$$\int_{z_1\sqrt{m}}^{\infty} \exp\Big\{-\frac{\alpha_1^2 t^2}{2\sigma^2(\alpha_1)m}\Big\}\,dt = \frac{\sqrt{m}\sigma(\alpha_1)}{\alpha_1}\int_z^{\infty} e^{-u^2/2}du,$$

so that

$$\Sigma_1 \sim \alpha_1^{-1} I(\alpha_1, \infty) e^{-x\lambda_1} \big[1 - \Phi(z)\big].$$

The quantity Σ_2 is equal to the probability $\mathbf{P}\big(\infty > \eta(x) > m(1+\varepsilon)\big)$, where

$$\frac{x}{m(1+\varepsilon)} = \frac{\alpha_1}{1+\varepsilon} < \alpha_1,$$

which is estimated in Theorem 3.4.5. According to that theorem, for $n = m(1+\varepsilon)$, $\alpha = x/n < \alpha_1$, we have

$$\Sigma_2 = O\left(\frac{1}{\sqrt{m}} e^{-n\Lambda(x/n)}\right) = O\left(\frac{1}{\sqrt{m}} e^{-x\Lambda(\alpha)/\alpha}\right).$$

Here $R(\alpha) := \Lambda(\alpha)/\alpha > \alpha_1\lambda_1$ for sufficiently small $\varepsilon > 0$. The latter inequality can be extracted either from the preceding arguments or by observing that

$$R(\alpha_1) = \alpha_1\lambda_1, \qquad R'(\alpha_1) = \frac{\alpha_1\lambda_1 - \Lambda(\alpha_1)}{\alpha_1^2} = 0, \qquad R''(\alpha_1) = \alpha_1\Lambda''(\alpha_1) > 0.$$

Thus, $\Sigma_2 = o(\Sigma_1)$ as $x \to \infty$ and relation (3.4.22) is proved.

In a completely similar way, using Theorem 3.4.4 one can establish that, as $x \to \infty$,

$$\mathbf{P}\left(\eta(x) < \frac{x}{\alpha_1} + \frac{z\sigma(\alpha_1)\sqrt{x}}{\alpha_1^{3/2}}\right) \sim \frac{\alpha_1^{-1} I(\alpha_1, \infty)\, e^{-x\lambda_1}}{\alpha_1} \Phi(z).$$

From this and (3.4.22), we obtain (3.4.23).

If $\mathbf{E}\xi > 0$ then $\lambda_1 = 0$, $\alpha_1 = \mathbf{E}\xi$, $\mathbf{P}(S \geqslant x) = 1$ and in (3.4.23) necessarily $\alpha_1 = I(\alpha_1, \infty)$ and

$$\frac{I(\alpha_1, y)}{\alpha_1} = \lim_{x\to\infty} \mathbf{P}\big(\chi(x) \leqslant y\big).$$

The theorem is proved. □

Corollary 3.4.9. *If* $0 < \alpha_1 \in \mathcal{A}^*$ *then*

$$\mathbf{E}\big(\eta(x) \mid \eta(x) < \infty\big) \sim \frac{x}{\alpha_1} = \frac{x}{\mathbf{E}\xi^{(\lambda_1)}},$$

as $x \to \infty$.

Proof. The statement of the corollary follows from Theorem 3.4.6, the properties of the function $\Lambda(\alpha)$ and the inequalities

$$\mathbf{P}\big(\eta(x) = n\big) < \mathbf{P}(S_n \geqslant x) < e^{-n\Lambda(x/n)},$$

where

$$\min_n n\Lambda\left(\frac{x}{n}\right) = x \min_\alpha \frac{\Lambda(\alpha)}{\alpha}$$

is attained at the point $\alpha = \alpha_1$. □

If in the case $\mathbf{E}\xi > 0$ we assume that

$$\frac{x}{\alpha_1} + \frac{z\sigma(\alpha_1)\sqrt{x}}{\alpha_1^{3/2}} = n$$

(here, $\alpha_1 = \mathbf{E}\xi$ and $\sigma(\alpha_1) = \sigma$) and solve this equation for x, we obtain

$$x = \alpha_1 n - z\sigma\sqrt{n} + o(\sqrt{n}),$$

so that, by (3.4.22),

$$\mathbf{P}\big(\infty > \eta(x) \geqslant n\big) = \mathbf{P}\big(\overline{S}_{n-1} < \alpha_1 n - z\sigma\sqrt{n} + o(\sqrt{n})\big) \sim 1 - \Phi(z) = \Phi(-z).$$

This means that the limiting distribution $(\overline{S}_n - \alpha_1 n)/\sigma\sqrt{n}$ is normal and hence coincides with the limiting distribution of $(S_n - \alpha_1 n)/\sigma\sqrt{n}$. This can also be established in a different way, assuming only the existence of $\mathbf{E}\xi^2$ and using the fact that the distribution of $\overline{S}_n - S_n$ converges as $n \to \infty$ to the distribution of a proper random variable $\inf_{k\geqslant 0} S_k \geqslant 0$ (see [39], Lemma 10.5.3).

A considerably more detailed and complete asymptotic analysis of the joint distribution of the variables $\overline{S}_n$, $\eta(x)$, $\chi(x)$ (including the zone $\alpha = o(1)$) under the existence of an absolutely continuous component of the distribution of ξ can be found in [16]. For arithmetically bounded ξ, see [15]. For more details, see also section 3.9.

3.4.3 The arithmetic case

All the results of the previous sections related to the non-lattice case can be carried over without any difficulty to the case when ξ has an arithmetic distribution. The proofs of the main statements have a similar structure and differ only by simplifications. So we do not provide them in full detail.

An analogue of Corollary 3.3.2 is the following statement.

Corollary 3.4.10. *For $n \to \infty$ and integer x,*

$$\frac{\mathbf{P}(S_n = x+1)}{\mathbf{P}(S_n = x)} \sim \frac{f_{(n)}(x+1)}{f_{(n)}(x)} \sim e^{-\lambda(\alpha)} \tag{3.4.27}$$

uniformly in $\alpha = x/n \in \mathcal{A}^*$, where, as above,

$$f_{(n)}(x) = \frac{e^{-n\Lambda(\alpha)}}{\sigma(\alpha)\sqrt{2\pi n}}.$$

The proof follows in an obvious way from Theorem 2.2.3. From (3.4.27) it is easy to obtain a counterpart of relation (3.3.2) in which τ has a truncated geometric distribution.

When considering problems with boundaries (see subsection 3.3.2) one should assume that the values $g_k = g_{k,n}$ are integer. The condition $[\mathbf{g}]_\alpha$ can be left unchanged. Denote

$$p(g,\alpha,y) = \mathbf{P}\big({}^{\alpha}S_k > -g_k - y \quad \text{for all } k \geqslant 1\big).$$

As in Lemma 3.3.4, it is not difficult to verify that if the condition $[\mathbf{g}]_\alpha$ is satisfied then

$$\inf_{\alpha\in\mathcal{A}^*} p(g,\alpha,y) > 0$$

for each integer $y \geqslant 0$.

An analogue of Theorem 3.3.3 has the following form.

Theorem 3.4.11. *Suppose that the condition $[\mathbf{g}]_\alpha$ is satisfied for $\alpha \in (\mathcal{A}')$. Then, for $n \to \infty$ and each fixed y,*

$$\mathbf{P}\big(S_{n-k} < x + g_k \quad \text{for all } k = 1,\ldots,n-1 \,|\, S_n = x+y\big) \sim p(g,\alpha,y)$$

uniformly in $\alpha = x/n \in \mathcal{A}^*$.

The proof of this theorem is similar to the proof of Theorem 3.3.3.

When considering the first passage time of a high level x, as in subsection 3.4.1, we will assume that

$$\alpha = \frac{x}{n} \geqslant \alpha_0 > 0.$$

Clearly, for integer x and y,

$$\begin{aligned}
\big\{\eta(x) = n,\ \chi(x) = y\big\} &= \{\overline{S}_{n-1} < x,\ S_n = x+y\},\\
\mathbf{P}\big(\eta(x) = n,\ \chi(x) = y\big) &\sim f_{(n)}(x+y)\mathbf{P}\big(\overline{S}_{n-1} < x \,|\, S_n = x+y\big)\big).
\end{aligned}$$

From Theorem 3.4.11 for $g_k \equiv 0$ and Corollaries 3.4.1 and 3.4.10 follows

Corollary 3.4.12. *As $n \to \infty$,*

$$\begin{aligned}
\mathbf{P}\big(\eta(x) = n, \chi(x) = y\big) &= f_{(n)}(x+y)p(0,\alpha,y)\big(1+o(1)\big)\\
&= f_{(n)}(x)e^{-y\lambda(\alpha)}p(0,\alpha,y)\big(1+o(1)\big),\\
\mathbf{P}\big(\eta(x) = n\big) &= f_{(n)}(x)I(\alpha)\big(1+o(1)\big)
\end{aligned}$$

uniformly in $\alpha \in \mathcal{A}^*$, where

$$I(\alpha) = \sum_{y=0}^{\infty} e^{-y\lambda(\alpha)}p(0,\alpha,y) < \infty.$$

For $\lambda(\alpha) \geqslant 0$, the last inequality is obvious. For $\lambda(\alpha) < 0$, it is proved in the same way as inequality (3.4.8).

It is known that for random walks that are continuous from above (i.e. for random walks with jumps $\xi \leqslant 1$, which corresponds to $s_+ = 1$), for any $x \geqslant 1$ it always holds (without condition **[C]**) that

$$\mathbf{P}\big(\eta(x) = n\big) = \frac{x}{n}\mathbf{P}(S_n = x) \sim \alpha f_{(n)}(x)$$

(see e.g. [39], § 12.8.2). This means that for $s_+ = 1$,

$$p(0,\alpha,0) = I(\alpha) = \alpha.$$

Integral limit theorems, as well as their proofs, for arithmetic $\eta(x)$ (without taking into account $\chi(x)$) do not differ from the corresponding theorems in the non-lattice case. Theorems about the conditional distribution of $\eta(x)$ and $\chi(x)$ (or $\overline{S}_n$ and $\chi(x)$) follow from Theorems 3.4.4–3.4.6 on making obvious changes. For example, the counterpart of statement (3.4.14) in the case $\alpha > \alpha_1 + \delta$ is

$$\mathbf{P}\big(\overline{S}_n \geqslant x,\ \ \chi(x) = y\big) \sim \frac{f_{(n)}(x)p(0,\alpha,y)}{1 - q_\alpha},$$

where $q_\alpha < 1$ has its previous meaning.

Similar changes should be made in Theorems 3.4.5 and 3.4.6. The value $I(\alpha,\infty)$ should be replaced by $I(\alpha)$ and $I(\alpha,\Delta)$ by $p(0,\alpha,y)$ if the left-hand sides of the probabilities contain the event $\{\chi(x) = y\}$ instead of $\{\chi(x) < \Delta\}$.

For $\mathbf{E}\xi > 0$, $\alpha = \alpha_1 = \mathbf{E}\xi$, it holds that $I(\alpha_1) = \alpha_1$.

For bounded arithmetic ξ, a more detailed asymptotic analysis of the joint distribution of $\overline{S}_n$, $\eta(x)$, $\chi(x)$, S_n can be found in [15].

For the sake of completeness, let us present results which do not require Cramér's condition.

3.4.4 Supplement. Probabilities of large deviations of $\overline{S}_n$ outside the Cramér zone (Cramér's condition may not be satisfied)

Everywhere so far in this section we have assumed that $\alpha \in \mathcal{A}^*$, which means that Cramér's condition is satisfied. If this condition is not satisfied then, unfortunately, it is not possible to study the asymptotics of $\overline{S}_n$ to the same extent as for the sums S_n. However, in two special cases it is possible; one can find the asymptotics of the probabilities

(a) $\mathbf{P}(\overline{S}_n \geqslant x)$ in the case $\alpha > 0$, $\alpha > \alpha_+ = a = \mathbf{E}\xi$ $(\lambda_+ = 0)$,
(b) $\mathbf{P}(\overline{S}_n < x)$ in the case $0 < \alpha < \alpha_- = a = \mathbf{E}\xi$ $(\lambda_- = 0)$.

Of course, one has to impose additional conditions on the regularity of the tails of $\mathbf{P}(\xi \geqslant t)$ and $\mathbf{P}(\xi < -t)$, respectively, as $t \to \infty$.

First consider case (a).

We will need the assumption that $\mathbf{P}(\xi \geqslant t)$ is a regularly varying function (r.v.f.):

$$\mathbf{P}(\xi \geqslant t) =: F_+(t) = t^{-\beta_+} L_+(t), \tag{3.4.28}$$

where $\beta_+ > 1$ and $L_+(t)$ is a slowly varying function (s.v.f.) as $t \to \infty$.

Theorem 3.4.13. *Assume that* (3.4.28) *is satisfied and either* $\beta_+ \in (1,2)$, $\mathbf{P}(\xi < -t) < cF_+(t)$ *for some* $c < \infty$, *or* $\beta_+ > 2$, $\mathbf{E}\xi^2 < \infty$. *Then*

(i) *If* $a > 0$, $x - (a+\delta)n \to \infty$, $\delta > 0$, *we have*

$$\mathbf{P}(\overline{S}_n \geqslant x) \sim nF_+(x - an)$$

as $x \to \infty$.

(ii) *If* $a < 0$ *and* $x \to \infty$, *we have*

$$\mathbf{P}(\overline{S}_n \geqslant x) \sim \begin{cases} nF_+(x) & \text{if } n \ll x, \\ \dfrac{1}{|a|(\beta_+ - 1)}\big[xF_+(x) - (x-an)F_+(x-an)\big] & \text{if } n \sim x, \\ \dfrac{xF_+(x)}{|a|(\beta_+ - 1)} & \text{if } n \gg x. \end{cases}$$

(iii) *Suppose that* $a = 0$. *Assume that* $x \to \infty$, $x \ll x^2/\ln x$, *in the case* $\mathbf{E}\xi^2 < \infty$ *and that* $nF_+(x) \to 0$ *in the case* $\beta_+ \in (1,2)$. *Then*

$$\mathbf{P}(\overline{S}_n \geqslant x) \sim nF_+(x).$$

This theorem follows from Theorems 3.4.1, 3.6.1, 3.6.2, 4.4.1, 4.6.1 and 4.6.2 in [42]. It can be extended to the asymptotics of the probability that a trajectory $\{S_k\}_{k=1}^n$ crosses an arbitrary boundary (see [42], §§ 3, 4).

Now consider case (b), when

$$0 < \alpha < a = \alpha_-.$$

Here we will assume that the right tail is an r.v.f., i.e.

$$F_-(t) := \mathbf{P}(\xi < -t) = t^{-\beta_-}L_-(t), \tag{3.4.29}$$

where $\beta_- > 1$, $L_-(t)$ is a s.v.f. as $t \to \infty$. Moreover, in the case $\beta_- \in (1,2)$ we may need the condition

$$\mathbf{P}(\xi \geqslant t) \leqslant t^{-\beta_+}L(t), \tag{3.4.30}$$

where $\beta_+ > 1$, L is a s.v.f. This condition is, obviously, always satisfied if condition $[\mathbf{C}_+]$ $(\lambda_+ > 0)$ is satisfied. Denote

$$z = an - x.$$

Theorem 3.4.14. *Let* $a > 0$, $x \to \infty$, $x < an$ *and suppose that* (3.4.29) *is satisfied. Then*

(i) *If* $\mathbf{E}\xi^2 < \infty$, $\beta_- > 2$ *and* $z \gg \sqrt{n \ln n}$ *then*

$$\mathbf{P}(\overline{S}_n < x) \sim \frac{x}{a}F_-(z). \tag{3.4.31}$$

(ii) *Relation* (3.4.31) *remains valid if* $\beta_- \in (1,2)$, (3.4.30) *is satisfied and* n, z *are such that*

$$nF_-(z) \to 0, \qquad nF_+\left(\frac{z}{\ln z}\right) \to 0. \tag{3.4.32}$$

The meaning of (3.4.31) is quite simple: the main contribution to the probability $\mathbf{P}(\bar{S}_n < x)$ is made by paths such that one of the first x/a jumps (before the boundary x is crossed by the 'drift line' $\mathbf{E}S_k = ak$) has an overshoot of magnitude less than $x - an$.

It follows from the theorem that

$$\mathbf{P}(\bar{S}_n < x) \sim \frac{x}{an} \mathbf{P}(S_n < x).$$

The proof of Theorem 3.4.14 can be found in [35] and in [42] (Theorem 8.3.4).

3.5 Asymptotics of the distribution of the first passage time through a fixed horizontal boundary

In this section we present some results about the asymptotics of the distribution of the first passage time through the zero level and a fixed positive level, which are somewhat different from the main exposition in this chapter. A review of these results can be found in [34] and [35].

In addition to the notation introduced above we will also use the following:

$$\eta_+(x) := \min\{k : S_k > x\}, \quad \eta_-(x) := \min\{k \geqslant 1 : S_k \leqslant -x\}, \quad x \geqslant 0,$$

where we assume $\eta_+(x) = \infty$ ($\eta_-(x) = \infty$), if all $S_k \leqslant x$ ($S_k > -x$), $k = 1, 2, \ldots$

We will study the asymptotics of the probabilities $\mathbf{P}(\eta_-(x) = n)$ and $\mathbf{P}(\eta_+(x) = n)$ or their integral counterparts $\mathbf{P}(n < \eta_-(x) < \infty)$ and $\mathbf{P}(n < \eta_+(x) < \infty)$ as $n \to \infty$. Special attention will be paid to the random variables $\eta_\pm = \eta_\pm(0)$.

For certain reasons, in this section it will be more convenient for us to study these random variables but not the following closely related variables:

$$\eta_+^0 := \eta(0) = \min\{k \geqslant 1 : S_k \geqslant 0\},$$
$$\eta_-^0 := \min\{k \geqslant 1 : S_k < 0\}.$$

If the distribution $\mathbf{F}$ is continuous on the half-axis $[0, \infty)$ (or $(-\infty, 0]$) then $\eta_\pm$ and $\eta_\pm^0$ coincide with probability 1. A small inconvenience in using $\eta_\pm^0$ is that some statements and notation for the pair $\eta_\pm^0$ (see e.g. the first statement of Theorem 3.5.1 given below) will be slightly more cumbersome. However, for all the statements for the variables $\eta_\pm$ presented below, there exist dual counterparts in terms of the variables $\eta_\pm^0$ (see e.g. Remark 3.5.2).

The definitive role in studying the above-mentioned asymptotics is played by the 'drift' of the random walk, which we will characterise using the values

$$D = D_+ := \sum_{k=1}^{\infty} \frac{\mathbf{P}(S_k > 0)}{k} \quad \text{and} \quad D_- := \sum_{k=1}^{\infty} \frac{\mathbf{P}(S_k \leqslant 0)}{k},$$

where, obviously $D_+ + D_- = \infty$.

These characteristics of the drift, which are more general than $\mathbf{E}\xi$, are always well defined. We use them because the validity of the condition **[C]** does not guarantee the existence of $\mathbf{E}\xi$ in the case when $\lambda = 0$ is a boundary point of the interval (λ_-, λ_+).

It is well known (see e.g. [39], [92]) that

$$\begin{aligned}\{D_+ < \infty,\, D_- = \infty\} &\Leftrightarrow \{\eta_- < \infty \text{ a.s.},\, p := \mathbf{P}(\eta_+ = \infty) > 0\}\\ &\Leftrightarrow \{\underline{S} = -\infty,\, \overline{S} < \infty \text{ a.s.}\}, \qquad (3.5.1)\\ \{D_+ = \infty,\, D_- = \infty\} &\Leftrightarrow \{\eta_- < \infty,\, \eta_+ < \infty \text{ a.s.}\} \Leftrightarrow \{\underline{S} = -\infty,\, \overline{S} = \infty \text{ a.s.}\},\end{aligned}$$

where

$$\overline{S} = \sup_{k\geqslant 0} S_k, \qquad \underline{S} = \inf_{k\geqslant 0} S_k.$$

Obviously, in the case $\{D_- < \infty, D_+ = \infty\}$, a relation 'dual' to (3.5.1) is also valid. Taking into account this symmetry, it is enough to consider the case

$$A_0 = \{D_- = \infty,\, D_+ = \infty\}$$

and only one of the possibilities

$$A_- = \{D_- = \infty,\, D_+ < \infty\} \quad \text{and} \quad A_+ = \{D_- < \infty,\, D_+ = \infty\}.$$

If the expectation $\mathbf{E}\xi = a$ exists then

$$A_0 = \{a = 0\}, \qquad A_- = \{a < 0\}, \qquad A_+ = \{a > 0\}.$$

Further results will be classified by the following three main characteristics:

(1) the value of x, distinguishing the cases $x = 0$ and fixed $x > 0$;
(2) the direction of the drift (one of the possibilities $A_0, A_\pm$);
(3) the character of the distribution of ξ.

To make the picture complete, we will not exclude cases when the positive tail of the distribution $\mathbf{P}(\xi \geqslant t)$ is not rapidly decreasing (a regularly varying function is such a case). This assumption does not contradict the assumption that condition **[C]** is satisfied (in the case when the point $\lambda = 0$ is a boundary of the interval (λ_-, λ_+)). A comprehensive asymptotic analysis of random walks with slowly varying tails can be found in the monograph [42].

3.5.1 General properties of the variables $\eta_\pm$ in the case $D_- = \infty, D < \infty$

Here we will mainly study the asymptotics of

$$\mathbf{P}(\eta_- > n) \qquad \text{and} \qquad \mathbf{P}(\eta_+ = n), \qquad (3.5.2)$$

and the closely related asymptotics of $\mathbf{P}(\theta = n)$ for the variable $\theta = \min\{n : S_n = \overline{S}\}$ representing the first time of reaching the maximum of $\overline{S}$. This connection has the simple form

$$\mathbf{P}(\theta = n) = \mathbf{P}(\overline{S}_{n-1} < S_n = \overline{S}_n)\, \mathbf{P}\big(\max_{k \geqslant n}(S_k - S_n) = 0\big),$$

(the first multiplier on the right-hand side in equal to 1 when $n = 0$), where, as before, $\overline{S}_n = \max_{k \leqslant n} S_k$,

$$\mathbf{P}\big(\overline{S}_{n-1} < S_n = \overline{S}_n\big) = \mathbf{P}\big(\underline{S}_n > \underline{S}^0_n = 0\big),$$

where $\underline{S}_n = \min_{1 \leqslant k \leqslant n} S_k$, $\underline{S}^0_n \equiv \min_{0 \leqslant k \leqslant n} S_k$, so that

$$\mathbf{P}(\theta = n) = p\, \mathbf{P}\big(\underline{S}_n > \underline{S}^0_n = 0\big) = p\, \mathbf{P}(\eta_- > n), \tag{3.5.3}$$

for $p = \mathbf{P}\left(\overline{S} = 0\right) = \mathbf{P}\big(\eta_+ = \infty\big) = 1/\mathbf{E}\eta_-$ (for the latter equalities see below or [22], [39], [92]).

We will consider the following classes of distributions and their extensions: $\mathcal{R}$, *the class of regularly varying distributions*, for which

$$\mathbf{P}(\xi \geqslant x) =: F_+(x) = x^{-\beta_+} L(x), \qquad \beta_+ > 1, \tag{3.5.4}$$

$L(x)$ being a regularly varying function; $\mathcal{S}e$, *the class of semi-exponential distributions*, for which

$$\mathbf{P}(\xi \geqslant x) = F_+(x) = e^{-l(x)}, \qquad l(x) = x^{\beta_+} L(x), \qquad \beta_+ \in (0,1), \tag{3.5.5}$$

where L is an s.v.f. such that, for $\Delta = o(x)$, $x \to \infty$ and any fixed $\varepsilon > 0$,

$$\begin{aligned} l(x+\Delta) - l(x) &\sim \frac{\beta_+ \Delta l(x)}{x} \quad \text{if } \frac{\Delta l(x)}{x} > \varepsilon, \\ l(x+\Delta) - l(x) &\to 0 \quad \text{if } \frac{\Delta l(x)}{x} \to 0. \end{aligned} \tag{3.5.6}$$

A sufficient condition for the validity of (3.5.6) is that L is differentiable 'at infinity', $L'(x) = o\left(L(x)/x\right)$. As can be seen from the definition, one can include in the class $\mathcal{S}e$ all the functions $F_+(x) = e^{-\widetilde{l}(x)}$, where $\widetilde{l}(x) = l(x) + o(1)$ as $x \to \infty$ and l satisfies (3.5.6), so that a function L for $F_+ \in \mathcal{S}e$ does not necessarily need to be differentiable.

Distributions from the classes $\mathcal{R}$ and $\mathcal{S}e$ are *subexponential*, i.e., they satisfy the properties

$$F_+^{*(2)}(x) \sim 2F_+(x), \qquad \frac{F_+(x+v)}{F_+(x)} \to 1 \tag{3.5.7}$$

for $x \to \infty$ and any fixed v, where $F_+^{*(2)}$ is the convolution of the tail of F_+ with itself:

$$F^{*(2)}(x)_+ := \int_{-\infty}^{\infty} \mathbf{P}(\xi \in dt) F_+(x-t) = -\int_{-\infty}^{\infty} F_+(x-t)\, dF_+(t).$$

The properties of the tails of $F_+ \in \mathcal{R}$ and $F_+ \in \mathcal{S}e$, which will be needed below, are preserved if the classes $\mathcal{R}$ and $\mathcal{S}e$ are extended by allowing a function F_+ to 'oscillate slowly' (see [42], § 4.9). These extensions can also be considered.

An alternative to $\mathcal{R}$ and $\mathcal{S}e$ is the class of *distributions decreasing exponentially fast on the positive semi-axis*, i.e. distributions for which Cramér's condition **[C$_+$]** is satisfied:

$$\lambda_+ =: \sup\{\lambda : \psi(\lambda) < \infty\} > 0.$$

Here we will distinguish two possibilities:

$$\begin{array}{lll} (1) & \lambda(0) \leqslant \lambda_+, & \psi'(\lambda(0)) = 0, \\ (2) & \lambda(0) = \lambda_+, & \psi'(\lambda_+) < 0, \end{array} \tag{3.5.8}$$

where the function $\lambda(\alpha)$ is defined in section 1.2 and $\lambda(0)$ is the point where $\min \psi(\lambda) = \psi_0$ is attained. It is clear that we always have $\psi'(\lambda(0)) = 0$ if $\lambda(0) \in (\lambda_-, \lambda_+)$ $(\alpha \in (\alpha_-, \alpha_+))$.

Let us now consider the known results. By the symbol c, with or without indices, as before we will denote constants, which may be different if they appear in different formulas.

As has been already noted, a complete asymptotic analysis of the distributions $\mathbf{P}(\eta_\pm(x) = n)$ for all x and $n \to \infty$ for bounded lattice variables ξ was carried out in [15].

In [16], a similar analysis, but for $x \to \infty$, was carried out under the condition **[C$_0$]** (or **[C]** in some cases) and under the existence of an absolutely continuous component of the distribution of ξ (see section 3.9 for details).

For an extension of the class $\mathcal{R}$ of regularly varying distributions, the asymptotics $\mathbf{P}(\theta = n) \sim cF_+(-an)$, and hence the asymptotics of $\mathbf{P}(\eta_- > n)$, was found in [22]. In that monograph it was also established that when **[C$_+$]** is satisfied, $\lambda_0 < \lambda_+$,

$$\mathbf{P}(\theta = n) \sim \frac{c\psi_0^n}{n^{3/2}} \;\big(\sim p\mathbf{P}(\eta_- > n)\big). \tag{3.5.9}$$

Rather complete results on the asymptotics of $\mathbf{P}(\eta_-(x) > n)$ and $\mathbf{P}(n < \eta_+(x) < \infty)$ for fixed $x \geqslant 0$ for the class $\mathcal{R}$ or when **[C$_+$]** is satisfied were obtained in [138], [8], [12], [87] and [90]. Here, we present some of those results for the sake of completeness of exposition.

Necessary and sufficient conditions for the finiteness of $\mathbf{E}(\eta_-)^\gamma$, $\gamma \geqslant 1$, and closely related problems were studied in [99], [97], [104] and [107]. It was found, in particular, that $\mathbf{E}\eta_-^\gamma < \infty$, $\gamma > 0$, if and only if $\mathbf{E}(\xi^+)^\gamma < \infty$, where $\xi^+ = \max(0, \xi)$ (see e.g. [99]). Unimprovable bounds for $\mathbf{P}(\eta_- > n)$ can be found in [29], § 43.

In view of the established connection (3.5.3) between the distributions of θ and η_-, in what follows we limit our attention to the distribution of $\eta_\pm$.

We begin by clarifying the connection of the distributions of η_- and η_+, which is of interest in its own right. Introduce a random variable ζ with generating function

$$H(z) := \mathbf{E}z^\zeta = \frac{1 - \mathbf{E}z^{\eta_-}}{(1 - z)\mathbf{E}\eta_-},$$

so that

$$\mathbf{P}(\zeta = k) = \frac{\mathbf{P}(\eta_- > k)}{\mathbf{E}\eta_-}, \qquad k = 0, 1, \ldots,$$

and denote $p(z) = \mathbf{E}\big(z^{\eta_+} \big| \, \eta_+ < \infty\big)$, $p = \mathbf{P}(\eta_+ = \infty) = \mathbf{P}\big(\overline{S} = 0\big)$ and $q = 1 - p = \mathbf{P}(\eta_+ < \infty)$.

Theorem 3.5.1. *Let $D_- = \infty$, $D < \infty$. Then the following relations are valid.*

(i)

$$H(z) = \frac{1-q}{1-qp(z)} \tag{3.5.10}$$

and, hence, the distribution of η_+ completely determines the distribution of η_- and vice versa.

(ii) *For all $n = 1, 2, \ldots$,*

$$\mathbf{P}(\eta_+ = n) < \mathbf{P}(\eta_- > n). \tag{3.5.11}$$

(iii) *If the sequence $\mathbf{P}(\eta_+ = k)/q$ is subexponential then, for $n \to \infty$,*

$$\mathbf{P}(\eta_- > n) \sim \frac{\mathbf{P}(\eta_+ = n)}{p^2}, \qquad p = \frac{1}{\mathbf{E}\eta_-}. \tag{3.5.12}$$

If the sequence

$$b_n := \frac{\mathbf{P}(S_n > 0)}{nD} \tag{3.5.13}$$

is subexponential, then

$$\mathbf{P}(\eta_- > n) \sim e^D \frac{\mathbf{P}(S_n > 0)}{n}, \qquad \mathbf{P}(\eta_+ = n) \sim e^{-D} \frac{\mathbf{P}(S_n > 0)}{n}. \tag{3.5.14}$$

(iv) *The random variable ζ has an infinitely divisible distribution and can be represented as*

$$\zeta = \omega_1 + \cdots + \omega_\nu, \tag{3.5.15}$$

where $\omega_1, \omega_2, \ldots$ are independent copies of a random variable ω such that

$$\mathbf{P}(\omega = k) = \frac{\mathbf{P}(S_k > 0)}{kD},$$

$$D = \sum \frac{\mathbf{P}(S_k > 0)}{k} = -\ln p, \qquad p = \frac{1}{\mathbf{E}\eta_-}, \tag{3.5.16}$$

ν does not depend on $\{\omega_i\}$ and has the Poisson distribution with parameter D.

From (3.5.15), it follows that

$$\mathbf{P}(\eta_- > n) = \mathbf{E}\eta_- \mathbf{P}(\omega_1 + \cdots + \omega_\nu = n). \tag{3.5.17}$$

Remark 3.5.2. (1) For all the statements of Theorem 3.5.1 and the subsequent theorems, it is not difficult to obtain dual analogues in terms of the variables $\eta^0_\pm$. For example, along with (3.5.12), the following relation is valid:

$$\mathbf{P}(\eta^0_- > n) \sim \frac{\mathbf{P}(\eta^0_+ = n)}{p_0^2}, \qquad p_0 = \frac{1}{\mathbf{E}\eta^0_-} \quad \text{as} \quad n \to \infty.$$

The proofs of these analogues are very similar to the proofs provided below.

(2) The statement of the theorem that the distributions $\eta_\pm$ determine themselves looks somewhat unexpected, since a similar statement about the first positive and the first non-positive sums $\chi_+ = S_{\eta_+}$ and $\chi_- = S_{\eta_-}$ is false. Indeed, if $\mathbf{P}(\xi \leqslant -x) = ce^{-hx}$ for $x \geqslant 0$, $c < 1$, $h > 0$, then for any distribution of ξ on the positive half-axis $(0, \infty)$ we have $\mathbf{P}(\chi_- < -x) = e^{-hx}$, but at the same time

$$1 - \mathbf{E}(e^{i\lambda\chi_+};\, \eta_+ < \infty) = \frac{(1 - \psi(i\lambda))\,(i\lambda + h)}{i\lambda}.$$

Therefore, the distributions $\mathbf{P}(\chi_+ > x;\, \eta_+ < \infty)$ will be different for different distributions of $\mathbf{P}(\xi > x)$, for $x > 0$.

Proof of Theorem 3.5.1. The proof of Theorem 3.5.1 will follow the lines of analysis of the asymptotics of $\mathbf{P}(\theta = n)$ in [22], § 21. It is based on the factorisation identities

$$1 - \mathbf{E}z^{\eta_-} = \exp\left\{-\sum_{k=1}^{\infty} \frac{z^k}{k}\mathbf{P}(S_k \leqslant 0)\right\}, \tag{3.5.18}$$

$$1 - \mathbf{E}\big(z^{\eta_+};\, \eta_+ < \infty\big) = \exp\left\{-\sum_{k=1}^{\infty} \frac{z^k}{k}\mathbf{P}(S_k > 0)\right\}, \quad |z| \leqslant 1 \tag{3.5.19}$$

(see e.g. [22], [39], [92]). Since

$$(1 - z) = e^{\ln(1-z)} = e^{-\sum_{k=1}^{\infty} z^k/k},$$

the identities (3.5.18) and (3.5.19) can be represented in the forms

$$\frac{1 - \mathbf{E}z^{\eta_-}}{1 - z} = \sum_{k=0}^{\infty} z^k\mathbf{P}(\eta_- > n) = \exp\left\{\sum_{k=1}^{\infty} \frac{z^k}{k}\mathbf{P}(S_k > 0)\right\}, \tag{3.5.20}$$

$$\sum_{k=0}^{\infty} z^k\mathbf{P}(\eta_+ = n) = 1 - \exp\left\{-\sum_{k=0}^{\infty} \frac{z^k}{k}\mathbf{P}(S_k > 0)\right\}. \tag{3.5.21}$$

Assuming that here $z = 1$, we obtain $D = -\ln p$, $p = 1/\mathbf{E}\eta_-$.

Relation (3.5.10) follows from a comparison of (3.5.20) and (3.5.21). Inequality (3.5.11) follows immediately from (3.5.10).

Let us now prove the third statement of the theorem. Introduce the entire functions

$$Q_-(v) = e^{vD} \qquad \text{and} \qquad Q_+(v) = 1 - e^{-vD}.$$

Then (3.5.20) and (3.5.21) can be written as

$$\sum_{n=0}^{\infty} z^n \mathbf{P}(\eta_- > n) = Q_-\big(b(z)\big), \tag{3.5.22}$$

$$\sum_{n=1}^{\infty} z^n \mathbf{P}(\eta_+ = n) = Q_+\big(b(z)\big), \tag{3.5.23}$$

where

$$b(z) = \frac{1}{D}\sum_{n=1}^{\infty} \frac{z^n \mathbf{P}(S_n > 0)}{n} =: \sum_{n=1}^{\infty} z^n b_n, \qquad b_n := \frac{\mathbf{P}(S_n > 0)}{nD}.$$

Since the sequence b_n is subexponential, it remains to use well-known theorems (see e.g. [42], Chapter 7; [78]) about functions of distributions (here defined by the functions $Q_\pm$ in (3.5.22), (3.5.23)).

Since the functions $Q_\pm$ are entire,

$$Q_-'(1) = De^D, \qquad Q_+'(1) = De^{-D},$$

by virtue of (3.5.22), (3.5.23) and Theorem 1.4.4 in [42] (see also [33], [78]), and

$$\mathbf{P}(\eta_- > n) \sim b_n Q_-'(1) = De^D b_n, \qquad \mathbf{P}(\eta_+ = n) \sim b_n Q_+'(1) = De^{-D} b_n.$$

This proves (3.5.14). Relation (3.5.12) can be proved in the same way, taking into account that the function $Q(v) = (1-q)/(1-qv)$ is analytic in the zone $|v| < 1/q$ and that $Q'(1) = q/(1-q)$ by virtue of (3.5.10), so that $H(z) = Q\big(p(z)\big)$. Therefore

$$\frac{\mathbf{P}(\eta_- > n)}{\mathbf{E}\eta_-} \sim \frac{q}{1-q}\,\frac{\mathbf{P}(\eta_+ = n)}{q},$$

which is equivalent to (3.5.12).

Statement (3.5.15) can be obtained if one notices that

$$H(z) = \exp\left\{\sum_{k=1}^{\infty} \frac{z^k \mathbf{P}(S_k > 0)}{k} - D\right\} = \exp\big\{D\big(M(z) - 1\big)\big\}, \tag{3.5.24}$$

where

$$M(z) := \sum_{k=1}^{\infty} \frac{z^k \mathbf{P}(S_k > 0)}{kD}.$$

It is clear that the first part of (3.5.24) is the generating function of $\omega_1 + \cdots + \omega_\nu$. The theorem is proved. □

Now we proceed with explicit asymptotic representations for the distributions of $\eta_\pm$ in terms of the distribution of ξ.

3.5.2 Asymptotic properties of the variables $\eta_\pm$ in the case $a = \mathbf{E}\xi < 0$, when the right tail of the distribution of the variable ξ belongs to the classes $\mathcal{R}$ or $\mathcal{S}e$

Theorem 3.5.3. *Let the distribution of ξ belong to one of the classes $\mathcal{R}$ or $\mathcal{S}e$, while in the case $F_+ \in \mathcal{S}e$ it is assumed that $\beta_+ < 1/2$ in* (3.5.5). *Then, as $n \to \infty$,*

$$\mathbf{P}(\eta_- > n) \sim F_+(|a|n)e^D, \tag{3.5.25}$$

$$\mathbf{P}(\eta_+ = n) \sim F_+(|a|n)e^{-D}, \tag{3.5.26}$$

where D is defined in (3.5.16),

$$e^D = \mathbf{E}\eta_-, \qquad e^{-D} = \mathbf{P}(\overline{S} = 0) = \mathbf{P}(\eta_+ = \infty). \tag{3.5.27}$$

The same statement is true for the dual pair of random variables

$$\eta_+^0 = \min\{k \geqslant 1 : S_k \geqslant 0\}, \qquad \eta_-^0 = \min\{k : S_k < 0\}.$$

Theorem 3.5.3A. *Suppose that the conditions of Theorem* 3.5.3 *are satisfied. Then, as $n \to \infty$,*

$$\mathbf{P}(\eta_-^0 > n) \sim F_+(|a|n)e^{D^0}, \qquad \mathbf{P}(\eta_+^0 = n) \sim F_+(|a|n)e^{-D^0},$$

where

$$D^0 := \sum_{k=1}^{\infty} \frac{\mathbf{P}(S_k \geqslant 0)}{k}, \qquad e^{D^0} = \mathbf{E}\eta_-^0 = \frac{1}{\mathbf{P}(\eta_+^0 = \infty)}.$$

Remark 3.5.4. Both statements of Theorem 3.5.3 admit a simple 'physical' interpretation. Note, beforehand, that under the conditions of the theorem, the probability that during a time m a large jump of magnitude $x \sim cn$ or more occurs is equal to $\approx mF_+(x)$. Furthermore, the rare event $\{\eta_- > n\}$ (n is large) occurs, roughly speaking, when during a time η_- (with mean $\mathbf{E}\eta_-$) a large jump of magnitude $|a|n$ occurs (so that, afterwards, on average, a time n is needed to reach the negative half-axis). Therefore, it is natural to expect that $\mathbf{P}(\eta_- > n) \sim \mathbf{E}\eta_- F_+(|a|n)$.

A similar interpretation can be provided for the result of [4], that $\mathbf{P}(\overline{S}_{\eta_-} > x) \sim F_+(x)\mathbf{E}\eta_-$.

The rare event $\{\eta_+ = n\}$ occurs when first (before a time n) a trajectory does not leave the negative half-axis (the probability of that is close to $\mathbf{P}(\overline{S} = 0)$; at time $n-1$ the path will be in a neighbourhood of the point an), and then at time n a large jump of magnitude $\geqslant -an$ occurs. This idea explains (3.5.26) to some extent.

Proof of Theorem 3.5.3. Theorem 3.5.3 is a simple corollary from the third statement of Theorem 3.5.1 (see (3.5.13), (3.5.14)). We need to find the asymptotics of b_n and verify that this sequence is subexponential.

The probability $\mathbf{P}(S_n > 0)$ can be written in the form $\mathbf{P}(S_n^0 > |a|n)$, where $S_k^0 = S_k - ak$, $\mathbf{E}S_k^0 = 0$. It is known that if $F_+ \in \mathcal{R}$ then

$$\mathbf{P}(S_n^0 > |a|n) \sim nF_+(|a|n) \tag{3.5.28}$$

(see e.g. [42], Chapters 3 and 4; [145]). The same relation is valid if $F_+ \in \mathcal{S}e$ (i.e. (3.5.5) and (3.5.6) hold true) and $\beta_+ < 1/2$ (see e.g. [42], [32]; for $\beta_+ \geqslant 1/2$ the deviation $x = |a|n$ does not belong to the zone, where where (3.5.28)) is true). Thus, if the conditions of Theorem 3.5.3 are satisfied,

$$b_n \sim \frac{F_+(|a|n)}{D}$$

and the sequence b_n is subexponential, i.e. $\sum_{k=0}^{n} b_k b_{n-k} \sim 2b_n$ as $n \to \infty$. The theorem is proved. □

If follows from Theorem 3.5.3 that, as $n \to \infty$, (3.5.25) and (3.5.26) are satisfied and, for any increasing function g and $G(t) = \int_0^t g(u)du$, it holds that

$$\{\mathbf{E}g(\eta_-) < \infty\} \iff \left\{\mathbf{E}\left(g\left(\frac{\xi}{|a|}\right); \xi > 0\right) < \infty\right\},$$

$$\{\mathbf{E}\big(g(\eta_+); \eta_+ < \infty\big) < \infty\} \iff \left\{\mathbf{E}\left(G\left(\frac{\xi}{|a|}\right); \xi > 0\right) < \infty\right\}.$$

Theorem 3.5.3A can be proved in the same way using the dual identities

$$1 - \mathbf{E}z^{\eta_-^0} = \exp\left\{-\sum_{k=1}^{\infty} \frac{z^k}{k} \mathbf{P}(S_k < 0)\right\},$$

$$1 - \mathbf{E}\big(z^{\eta_+^0}; \eta_+^0 < \infty\big) = \exp\left\{-\sum \frac{z^k}{k} \mathbf{P}(S_k \geqslant 0)\right\}.$$

The duality presented in Theorems 3.5.3 and 3.5.3A is preserved in the subsequent exposition. However, we will often skip its description because it is evidently true.

The conditions guaranteeing the validity of (3.5.25) and (3.5.26) can be extended (see [34], Theorems 3, 3A).

3.5.3 Asymptotic properties of the variables $\eta_\pm$ in the case $a = \mathbf{E}\xi < 0$, when condition $[\mathbf{C}_+]$ is satisfied

First, consider the first of the two possibilities in (3.5.8) in the case $\lambda_+ > 0$. Recall that $\psi(\lambda) = \mathbf{E}e^{\lambda\xi}$, $\lambda(0) > 0$ is the point where $\psi_0 = \min \psi(\lambda)$ is attained, $\lambda_+ = \sup\{\lambda : \psi(\lambda) < \infty\}$.

The case $\lambda(0) \leqslant \lambda_+$, $\psi'(\lambda(0)) = 0$ under consideration includes two considerably different subcases:

(a) $\lambda(0) < \lambda_+$, $\psi'(\lambda(0)) = 0$;
(b) $\lambda(0) = \lambda_+$, $\psi'(\lambda_+) = 0$.

In case (b), the finiteness of $\psi(\lambda_+)$ and $\psi'(\lambda_+)$ means that

$$\mathbf{P}(\xi \geqslant t) = e^{-\lambda_+ t} F_+(t), \qquad \int_0^\infty t F_+(t)dt < \infty. \tag{3.5.29}$$

In this case we will require that one of the following conditions is satisfied: either
(b_2) $\psi''(\lambda_+) < \infty$
or
($b_{(1,2)}$) $\psi''(\lambda_+) = \infty$ and

$$F_+^I(t) := \int_t^\infty F_+(u)du \in \mathcal{R}; \quad \text{more precisely,} \tag{3.5.30}$$

$$F_+^I(t) = t^{-\beta} L(t), \quad \text{where} \quad \beta \in (1,2), \quad L \text{ is an s.v.f.} \tag{3.5.31}$$

Now we return to the main case, $\lambda(0) \leqslant \lambda_+$. Suppose, as before, $\mathbf{F}(B) = \mathbf{P}(\xi \in B)$ and

$$^0\mathbf{F}(dt) = \mathbf{F}^{(\lambda(0))}(dt) = \frac{e^{\lambda(0)t}\mathbf{F}(dt)}{\psi_0} \tag{3.5.32}$$

is the Cramér transform of $\mathbf{F}$.

Let $\xi_i^{(\lambda(0))} = {}^0\xi_i$ be independent random variables with distribution $^0\mathbf{F}$. Then, obviously, $\mathbf{E}\,{}^0\xi_i = 0$. If $\lambda(0) < \lambda_+$ or $\lambda(0) = \lambda_+$, $\psi''(\lambda_+) < \infty$, then $\sigma^2(0) = \mathbf{E}({}^0\xi_i)^2 < \infty$ and ${}^0S_n/(\sigma(0)\sqrt{n})$ converges in distribution to the normal law. If $\lambda(0) = \lambda_+$ and condition ($b_{(1,2)}$) is met, then

$$^0V(t) := \mathbf{P}({}^0\xi_1 > t) = \int_t^\infty \frac{e^{\lambda_+ u}\mathbf{F}(du)}{\psi_0} = \frac{1}{\psi_0}\int_t^\infty \big(\lambda_+ F_+(u)du - dF_+(u)\big)$$
$$\sim \frac{\lambda_+}{\psi_0} F_+^I(t). \tag{3.5.33}$$

This means that if the conditions $\lambda(0) = \lambda_+$, ($b_{(1,2)}$) are met and $\beta \in (1,2)$ then ${}^0\xi_i$ belong to the domain of attraction of the stable law $F_{\beta,1}$ with exponents $(\beta, 1)$, $\beta \in (1,2)$ (the left tail of the distribution $\mathbf{F}$ decreases exponentially). The distribution $F_{\beta,1}$ has continuous density ϕ. We will use the same symbol ϕ to denote the density of the normal law in the case $\sigma(0) < \infty$.

Let

$$D_* = \sum \frac{\mathbf{P}(S_k > 0)\psi_0^{-k}}{k},$$

$$\sigma_n = \begin{cases} \sigma(0)\sqrt{n} & \text{if } \sigma^2(0) < \infty, \\ {}^0V^{(-1)}\left(\dfrac{1}{n}\right) & \text{if } \sigma^2(0) = \infty, \end{cases}$$

where ${}^0V^{(-1)}(u) = \inf\{v : {}^0V(v) \leqslant u\}$ is the inverse function of ${}^0V$.

Theorem 3.5.5. *If $0 < \lambda(0) \leqslant \lambda_+$, $\psi'(\lambda(0)) = 0$ and in the case $\lambda(0) = \lambda_+$ one of the conditions* (b_2) *or* ($b_{(1,2)}$) *is met, then in the non-lattice case*

$$\mathbf{P}(\eta_- > n) \sim e^{D_*} \frac{\phi(0)\psi_0^n}{\lambda(0) n\sigma_n}, \tag{3.5.34}$$

$$\mathbf{P}(\eta_+ = n) \sim e^{-D_*} \frac{\phi(0)\psi_0^n}{\lambda(0) n\sigma_n}, \tag{3.5.35}$$

where $\phi(0)$ is the density of the limit law $F_{\beta,1}$ in the case $\sigma(0) = \infty$ and is the density of the normal law at the point 0 in the case $\sigma(0) < \infty$; $\psi_0 = e^{-\Lambda(0)}$.

If ξ has an arithmetic distribution then the factor $1/\lambda(0)$ on the right-hand sides of (3.5.34), (3.5.35) *should be replaced by $e^{-\lambda(0)}/(1 - e^{-\lambda(0)})$.*

Proof. Between the distributions of S_n and ${}^0S_n$ there exists a connection similar to (3.5.32):

$$\mathbf{P}(S_n \in dt) = \psi_0^n e^{-\lambda(0)t} \mathbf{P}({}^0S_n \in dt),$$

so that

$$\begin{aligned} \mathbf{P}(S_n > 0) &= \psi_0^n \int_0^\infty e^{-\lambda(0)t} \mathbf{P}({}^0S_n \in dt) \\ &= \psi_0^n \lambda(0) \int_0^\infty e^{-\lambda(0)\Delta} \mathbf{P}({}^0S_n \in \Delta[0)) d\Delta. \end{aligned} \tag{3.5.36}$$

If the ${}^0\xi_i$ are non-lattice then, under the conditions $\lambda(0) < \lambda_+$ or $\lambda(0) = \lambda_+$, (b_2), by virtue of the integro-local theorem 1.5.1,

$$\mathbf{P}({}^0S_n \in \Delta[0)) = \mathbf{P}\left(\frac{{}^0S_n}{\sigma_n} \in \frac{\Delta[0)}{\sigma_n}\right) \sim \frac{\Delta}{\sigma_n}\phi(0)$$

uniformly in $\Delta \in (\Delta_1, \Delta_2)$ for any fixed $0 < \Delta_1 < \Delta_2$; moreover $\mathbf{P}({}^0S_n \in \Delta[0)) < c\Delta/\sigma_n$ for all Δ. Therefore, by virtue of (3.5.36) and the dominated convergence theorem,

$$\mathbf{P}(S_n > 0) \sim \frac{\lambda(0)\psi_0^n\phi(0)}{\sigma_n} \int_0^\infty \Delta e^{-\lambda(0)\Delta} d\Delta = \frac{\psi_0^n \phi(0)}{\lambda(0)\sigma_n}.$$

Next, we will use again the factorisation identities (3.5.20), (3.5.21), making the change of variables $s = z\psi_0$ and following the proof of Theorem 3.5.3, but instead of using the functions $b(z)$ in (3.5.22), (3.5.23) we will use the functions

$$\widetilde{b}(s) = b\left(\frac{s}{\psi_0}\right) = \sum_{k=1}^\infty \widetilde{b}_k s^k.$$

The sequence

$$\widetilde{b}_k = \frac{\mathbf{P}(S_k > 0)\psi_0^{-k}}{D_* k} \sim \frac{\phi(0)}{\lambda(0) k\sigma_k D_*}, \qquad D_* = \sum \frac{\mathbf{P}(S_k > 0)\psi_0^{-k}}{k}, \tag{3.5.37}$$

is subexponential. Hence, applying again the results of [42], Chapter 7, and [78], we obtain

$$\mathbf{P}(\eta_- > n)\psi_0^{-n} \sim e^{D_*} \frac{\phi(0)}{\lambda(0)k\sigma_k D_*}.$$

In the arithmetic case one should use the local limit theorem 1.5.1 (see also [96], Theorem 4.2.2), from which, similarly to the above, we obtain

$$\mathbf{P}(S_n > 0) \sim \frac{\psi_0^n e^{-\lambda(0)}\phi(0)}{(1 - e^{-\lambda(0)})\sigma_n}.$$

The proof of the second statement (3.5.35) of the theorem can be carried over in a similar way using Theorem 3.5.1. □

Note that in the cases $\lambda(0) < \lambda_+$ or $\lambda(0) = \lambda_+$, $\psi''(\lambda_+) < \infty$ in (3.5.34) and (3.5.35) we have

$$\phi(0) = \frac{1}{\sqrt{2\pi}}, \qquad \sigma_n = \sigma(0)\sqrt{n}, \qquad \sigma^2(0) = \frac{\psi''(\lambda(0))}{\psi_0}.$$

Now consider in (3.5.8) the second possibility, $\lambda(0) = \lambda_+$, $\psi'(\lambda_+) < 0$. Here, (3.5.29) is still valid. Assume that $F_+ \in \mathcal{R}$ in (3.5.29), i.e.

$$F_+(t) = t^{-\beta-1}L(t), \qquad \beta > 1, \tag{3.5.38}$$

where L is an s.v.f.

Theorem 3.5.6. *If* $0 < \lambda(0) = \lambda_+$, $\psi'(\lambda_+) < 0$ *and relations* (3.5.29), (3.5.38) *are met then*

$$\mathbf{P}(\eta_- > n) \sim e^{D_*}F_+(a_+n)\psi_0^{n-1}, \qquad \mathbf{P}(\eta_+ = n) \sim e^{-D_*}F_+(a_+n)\psi_0^{n-1},$$

where

$$a_+ = -\mathbf{E}\xi^{(\lambda_+)} = -\frac{\psi'(\lambda_+)}{\psi_0} > 0.$$

Proof. Under the conditions of Theorem 3.5.6, equality (3.5.36) is still valid for $\mathbf{P}(S_n > 0)$, where $\lambda(0)$ can be replaced by λ_+, so that our problem again reduces to finding the asymptotics of $\mathbf{P}\big(S_n^{(\lambda_+)} \in \Delta[0)\big) = \mathbf{P}\big(S_n^{(\lambda_+)} + a_+n \in [a_+n, a_+n+\Delta)\big)$. We have

$$\mathbf{P}\big(\xi^{(\lambda_+)} \in \Delta[t)\big) = \int_t^{t+\Delta} \frac{e^{\lambda_+ v}\mathbf{F}(dv)}{\psi_0} = \int_t^{t+\Delta} \frac{\lambda_+ F_+(u)}{\psi_0}\,dv - \int_t^{t+\Delta} \frac{dF_+(v)}{\psi_0}$$
$$\sim \frac{\lambda_+\Delta F_+(t)}{\psi_0}.$$

This means that for every fixed Δ the conditions of Theorem 3.7.1 in [42] are satisfied, and for deviations of order n (for the centered sums $S_n^{(\lambda_+)} + a_+n$) we have

$$\mathbf{P}\big(S_n^{(\lambda_+)} \in \Delta[0)\big) \sim \frac{\lambda_+\Delta n F_+(a_+n)}{\psi_0\beta}. \tag{3.5.39}$$

By virtue of (3.5.36), from this, as before, we find

$$\mathbf{P}(S_n > 0)\psi_0^{-n} \sim \lambda_+ \int_0^\infty e^{-\lambda_+ u} u du \, \frac{\lambda_+ n F_+(a_+ n)}{\psi_0 \beta} = \frac{n F_+(a_+ n)}{\psi_0 \beta}.$$

Thus, with the notation of (3.5.37),

$$\widetilde{b}_k = \frac{\mathbf{P}(S_k > 0)\psi_0^{-k}}{D_* k} \sim \frac{F_+(a_+ n)}{\psi_0 \beta D_*}.$$

The sequence $\widetilde{b}_k$, as in Theorem 3.5.5, will be subexponential and, hence, all the remaining arguments of the previous proof remain the same.

The theorem is proved. □

3.5.4 Properties of the variables $\eta_-(x)$ with fixed $x \geqslant 0$ in the case $D_+ = D_- = \infty$

First, note that here we have the following analogue of Theorem 3.5.1.

Theorem 3.5.7. *If $D = D_- = \infty$ then*

$$\frac{1 - \mathbf{E}z^{\eta_-}}{1 - z} = \exp\left\{\sum_{k=1}^{\infty} \frac{z^k}{k} \mathbf{P}(S_k > 0)\right\} = \frac{1}{1 - \mathbf{E}z^{\eta_+}},$$

$\mathbf{E}\eta_\pm = \infty$ and the distribution of η_- is uniquely determined by the distribution of η_+ and vice versa.

The proof of the theorem obviously follows from (3.5.18) and (3.5.19) and the fact that $\mathbf{P}(\eta_+ < \infty) = 1$.

Moreover, the following results are valid. Denote

$$r_n = \mathbf{P}(S_n \leqslant 0) - \gamma, \qquad R(z) = \exp\left\{-\sum_{k=1}^{\infty} \frac{z^n r_n}{n}\right\}.$$

Theorem 3.5.8. *In the case $D = D_- = \infty$,*

(i) *It holds that*

$$\mathbf{P}(\eta_- > n) \sim \frac{n^{-\gamma} L(n)}{\Gamma(1-\gamma)}, \qquad \gamma \in (0, 1), \tag{3.5.40}$$

as $n \to \infty$, where L is an s.v.f., if and only if

$$n^{-1} \sum_{k=1}^{n} \mathbf{P}(S_k \leqslant 0) \to \gamma; \tag{3.5.41}$$

(ii) *if (3.5.40) or (3.5.41) are satisfied then $L(n) \sim R(1 - 1/n)$,*

$$\frac{\mathbf{P}(\eta_-(x) > n)}{\mathbf{P}(\eta_- > n)} \to r_-(x)$$

for $n \to \infty$ and any fixed $x \geqslant 0$, where the function $r_-(x)$ is found in an explicit form;

(iii) *if* $\mathbf{E}\xi = 0$, $\mathbf{E}\xi^2 < \infty$, *then* $\gamma = 1/2$ *and*

$$\sum \frac{|r_n|}{n} < \infty. \tag{3.5.42}$$

This means that $L(n)$ in (3.5.40) *can be replaced with $R(1)$, $0 < R(1) < \infty$.*

The proof of Theorem 3.5.8 can be found in [12], pp. 381, 382. (See also the full reference list provided there.) Obviously, a statement symmetric to (3.5.40) is valid for η_+, $\eta_+(x)$ (replacing γ and r_n by, respectively, $1-\gamma$ and $-r_n$):

$$\mathbf{P}(\eta_+ > n) \sim \frac{n^{\gamma-1}L^{-1}(n)}{\Gamma(\gamma)}.$$

If $\mathbf{E}\xi^2 = \infty$ (if $\mathbf{E}\xi$ exists, then $\mathbf{E}\xi = 0$), then, under known conditions on the regularity of the tails of the distribution of ξ, we have

$$\mathbf{P}(S_n \leqslant 0) \to F_{\beta,\theta}(0) \equiv \gamma,$$

where $F_{\beta,\theta}$ is the distribution function of the stable law with parameters $\beta \in (0,2]$, $\theta \in [-1,1]$.

For symmetric ξ we have $\gamma = 1/2$,

$$\mathbf{P}(S_n \leqslant 0) - \frac{1}{2} = \frac{1}{2}\mathbf{P}(S_n = 0) < \frac{c}{\sqrt{n}},$$

so that the series (3.5.42) converges,

$$R(z) = \exp\left\{-\frac{1}{2}\sum \frac{z^n \mathbf{P}(S_n = 0)}{n}\right\}.$$

From the above discussion one can see that, in contrast with the case $D_- = \infty$, $D < \infty$, here the influence of the tails of the distribution of ξ on the asymptotics (3.5.2) is less significant.

There exists a voluminous literature on the speed of convergence of $F^{(n)}(t) = \mathbf{P}(S_n/\sigma_n < t)$ to $F_{\beta,\theta}(t)$ under an appropriate normalization σ_n. There one can find, in particular, necessary conditions for the convergence of the series (3.5.42) in the case $\mathbf{E}\xi^2 = \infty$. Here, as an example, we provide just one of the known results:

Suppose that $v_r = \int \big|x^r(\mathbf{F}(dx) - dF_{\beta,\theta}(x))\big| < \infty$ *for* $r > \beta$ *and* $\int x\big(\mathbf{F}(dx) - dF_{\beta,\theta}(x)\big) = 0$ *in the case* $\beta \geqslant 1$. *Then*

$$\sup_t \big|F^{(n)}(t) - F_{\alpha,\theta}(t)\big| \leqslant c v_r n^{1-r/\beta}$$

(see [147], [148]).

Obviously, in this case, the convergence (3.5.42) takes place.

Now we will obtain a refinement of Theorem 3.5.8 concerning *local* theorems for η_-. In what follows, the relation $g_n \sim cf_n$ for $c = 0$ will be understood as $g_n = o(f_n)$ for $n \to \infty$.

Theorem 3.5.9. (i) *Let $D = D_- = \infty$ and*

$$r_n \sim cn^{-\gamma}, \qquad 0 \leqslant c < \infty, \qquad \gamma > 0, \tag{3.5.43}$$

as $n \to \infty$. Then

$$\mathbf{P}(\eta_- = n) \sim \frac{\gamma n^{-\gamma-1}}{\Gamma(1-\gamma)} R(1), \qquad 0 < R(1) < \infty. \tag{3.5.44}$$

(ii) *If $\mathbf{E}\xi = 0$, $\mathbf{E}|\xi|^3 < \infty$ and either the distribution of ξ is arithmetic or $\limsup_{|t|\to\infty} |\psi(it)| < 1$, then $\gamma = 1/2$ and* (3.5.43), (3.5.44) *are satisfied.*

Analogous relations are valid for $\mathbf{P}(\eta_+ = n)$.

Proof. By virtue of (3.5.18),

$$\mathbf{E}z^{\eta_-} = 1 - \exp\left\{-\gamma \sum_{k=1}^{\infty} \frac{z^k}{k} - \sum \frac{z^k r_k}{k}\right\} = 1 - (1-z)^{\gamma} R(z). \tag{3.5.45}$$

The asymptotics of the coefficients a_n in the expansion of the function

$$a(z) := -(1-z)^{\gamma} = \sum_{n=0}^{\infty} a_n z^n$$

is well known:

$$a_n \sim \frac{\gamma n^{-1-\gamma}}{\Gamma(1-\gamma)}. \tag{3.5.46}$$

To find the asymptotics of the coefficients r_k in the expansion

$$R(z) = \sum_{k=0}^{\infty} r_k z^k,$$

we will need the following auxiliary result.

Let $d_n \sim cn^{-\beta}$, $\beta > 1$, $0 \leqslant c < \infty$,

$$d(z) = \sum_{k=0}^{\infty} d_k z^k, \qquad \widehat{d}_n = |d_n|, \qquad \widehat{d}(z) = \sum_{k=0} \widehat{d}_k z^k.$$

Below, let $Q(\lambda)$ be an analytic function in the zone $|\lambda| \leqslant \widehat{d}(1)$, so that

$$Q(\lambda) = \sum_{k=0}^{\infty} Q_k \lambda^k, \qquad |Q_k| \leqslant c_1 \big(\widehat{d}(1)(1+\varepsilon)\big)^{-k} \tag{3.5.47}$$

for some $\varepsilon > 0$, $c_1 < \infty$. Since the series $d(z)$ converges absolutely, according to the Levý–Wiener theorem, the function $Q\big(d(z)\big)$ can be represented as an absolutely convergent series:

$$Q\big(d(z)\big) = \sum_{k=0}^{\infty} q_k z^k.$$

Then, by virtue of Theorem 1.4.4 in [42] (see also [33], [78]),

$$q_n \sim Q'\big(d(1)\big)d_n.$$

Let us use this claim with $d_n = -r_n/n$, $Q(\lambda) = e^{\lambda}$. Then $R(z) = Q\big(d(z)\big)$, $r_k = q_k$ and (3.5.43) imply that the assumed conditions are satisfied for $\beta = 1 + \gamma$. From (3.5.45) and (3.5.46), we get

$$\mathbf{P}(\eta_- = n) = \sum_{k=0}^{n} a_k r_{n-k} = \sum_{k\leqslant n/2} + \sum_{k>n/2},$$

where

$$\sum_{k\leqslant n/2} := r_n a(1) + o(r_n) = o(r_n),$$

$$\sum_{k>n/2} := a_n R(1) + o(a_n).$$

This proves (3.5.44).

The second statement of the theorem follows because, under the conditions of the second part of the theorem, we have

$$r_n = \mathbf{P}(S_n \leqslant 0) - \frac{1}{2} = \frac{c\mathbf{E}\xi^3}{\sqrt{n}} + o\left(\frac{1}{\sqrt{n}}\right)$$

(see e.g. [151]) and, hence, (3.5.43) is satisfied for $\gamma = 1/2$.

The theorem is proved. □

3.5.5 Properties of the variables $\eta_\pm(x)$ with a fixed level $x \geqslant 0$

The asymptotics of the distributions of $\eta_\pm(x)$ with fixed $x > 0$ differ from the corresponding asymptotics of $\eta_\pm$ by a factor depending only on x. Namely, in [12], [90], [138], [87], [8], it was established that if the conditions of Theorems 3.5.3 (for the class $\mathcal{R}$), 3.5.6, 3.5.7 and 3.5.8 are met, then the following theorem is valid.

Theorem 3.5.10. *In the cases $D_- = \infty$, $D \leqslant \infty$ for $n \to \infty$ and any fixed $x \geqslant 0$,*

$$\frac{\mathbf{P}(\eta_-(x) > n)}{\mathbf{P}(\eta_- > n)} \to r_-(x),$$

$$\frac{\mathbf{P}(n < \eta_+(x) < \infty)}{\mathbf{P}(n < \eta_+ < \infty)} \to r_+(x),$$

where the functions $r_\pm(x)$ can be found explicitly.

The exact form of the functions $r_\pm(x)$ turns out to be quite difficult.

More 'advanced' results can be obtained in the case

$$\mathbf{E}\xi = 0, \qquad \mathbf{E}\xi^2 < \infty.$$

Denote

$$R_+ := \frac{1}{2\Gamma(1/2)} \exp\left\{-\sum_{n=1}^{\infty} \frac{\mathbf{P}(S_n > 0) - 1/2}{n}\right\},$$
$$R := \frac{1}{2\Gamma(1/2)} \exp\left\{-\sum_{n=1}^{\infty} \frac{\mathbf{P}(S_n \geqslant 0) - 1/2}{n}\right\}. \tag{3.5.48}$$

It is well-known (see [92], Theorem 1, p. 687) that in the case under consideration (i.e. for $\mathbf{E}\xi = 0$, $\sigma^2 < \infty$) the series appearing in the definitions of R and R_+ converge and

$$\mathbf{E}\chi = R\sqrt{2\pi}\,\sigma, \qquad \mathbf{E}\chi_+ = R_+\sqrt{2\pi}\,\sigma,$$

where $\chi = \chi(0) = S_{\eta(0)} = S_{\eta_+^0}$, $\chi_+ = S_{\eta_+}$.

First consider the case $x = 0$.

Theorem 3.5.11. *Let*

$$\mathbf{E}\xi = 0, \qquad \sigma^2 < \infty. \tag{3.5.49}$$

Then

$$0 < \varliminf_{n\to\infty} n^{3/2}\mathbf{P}(\eta_+ = n) \leqslant \varlimsup_{n\to\infty} n^{3/2}\mathbf{P}(\eta_+ = n) < \infty.$$

The limit

$$\lim_{n\to\infty} n^{3/2}\mathbf{P}(\eta_+ = n) \tag{3.5.50}$$

exists if and only if the distribution of ξ is either non-lattice or arithmetic (up to the size of the lattice step, i.e. ξ assumes values kb for any $b > 0$ and $k = \dots - 1, 0, 1, \dots$). If the limit exists, it is equal to the value of R_+ defined in (3.5.48).

This result remains valid if the random variable η_+ is replaced by η_+^0, and R_+ by R.

Consider now local theorems for $\eta_+(x)$, $\eta(x)$ when $x > 0$.

Let $H_{\chi_+}(B)$, $B \subset \mathbb{R}$, be the renewal measure corresponding to the random variable χ_+, so that

$$\int_0^\infty e^{-\lambda t} H_{\chi_+}(dt) = \left(1 - \mathbf{E}e^{-\lambda\chi_+}\right).$$

Let

$$D_0 := \sum_{k=1}^{\infty} \frac{1}{k}\mathbf{P}(S_k = 0),$$
$$H(x) = R\,e^{D_0} H_{\chi_+}\big([0,x)\big),$$
$$H_+(x) = R_+ H_{\chi_+}\big([0,x]\big).$$

It is known that we have $D_0 < \infty$ always (see e.g. [39], Corollary 12.2.5). If the distribution of the variable ξ is continuous on the half-axis $[0,\infty)$ (or $(-\infty,0]$), then obviously $D_0 = 0$.

Theorem 3.5.12. *Suppose that the distribution of ξ is non-lattice or arithmetic (up to the size of the lattice step) and condition* (3.5.49) *is satisfied. Then, for any $x \geqslant 0$ and $n \to \infty$,*

$$\mathbf{P}\big(\eta(x) = n\big) \sim H(x)n^{-3/2},$$
$$\mathbf{P}\big(\eta_+(x) = n\big) \sim H_+(x)n^{-3/2}.$$

Theorems 3.5.11 and 3.5.12 are proved in [134]. Detailed comments on the bibliography can be also found there.

Clearly, from Theorem 3.5.12, one can easily obtain integral theorems for

$$\mathbf{P}\big(\eta(x) \geqslant n\big), \qquad \mathbf{P}\big(\eta_+(x) \geqslant n\big).$$

3.6 Asymptotically linear boundaries

Let us return to arbitrary boundaries from section 3.3 and extend the class of boundaries considered in sections 3.4 and 3.5.

We will call a boundary $g = \{g_{k,n}\}_{k\geqslant 1}$ *asymptotically linear* if, for any fixed $N \geqslant 0$ and some $\varepsilon > 0$, it satisfies condition $\mathbf{[g]}_\alpha$ from subsection 3.3.2 and

$$g_{k,n} \sim \beta k \quad \text{for} \quad n \to \infty \quad \text{and all} \quad k = 1,\dots,N, \qquad \beta \geqslant -\alpha + \varepsilon, \tag{3.6.1}$$

where the parameter β can depend on n but $\beta \leqslant \beta_0$, where $\beta_0 < \infty$ does not depend on n. Clearly, for the asymptotics of a linear boundary, inequalities (3.3.4) in condition $\mathbf{[g]}_\alpha$ can be assumed to be satisfied for all $k \geqslant 1$.

Denote by $g^{(\beta)}$ the linear sequence

$$g^{(\beta)} := \{\beta k\}_{k\geqslant 0},$$

so that (see (3.3.6))

$$\begin{aligned} P\big(g^{(\beta)},\alpha\big) &= \mathbf{P}({}^\alpha S_k > \tau - \beta k \quad \text{for all} \quad k \geqslant 1) \\ &= \mathbf{P}\Big(\inf_{k\geqslant 1}\big({}^\alpha S_k + \beta k\big) > \tau\Big), \\ P\big(g^{(\beta)} + \Delta,\alpha\big) &= \mathbf{P}({}^\alpha S_k > \tau - \Delta - \beta k \quad \text{for all} \quad k \geqslant 1) \\ &= \mathbf{P}\Big(\inf_{k\geqslant 1}\big({}^\alpha S_k + \beta k\big) > \tau - \Delta\Big). \end{aligned} \tag{3.6.2}$$

Theorem 3.6.1. *For an asymptotically linear boundary g we have*

$$\mathbf{P}\big(S_{n-k} < x + g_k \textit{ for all} \quad k = 1,\dots,n-1 \mid S_n \in \Delta[x)\big) \sim P(g^{(\beta)},\alpha), \tag{3.6.3}$$

$$\mathbf{P}\big(S_{n-k} < x + g_k \quad \textit{for all} \quad k = 1,\dots,n-1;\ S_n \in \Delta[x)\big) \sim f_{(n)}(x)I(\alpha,\beta,\Delta) \tag{3.6.4}$$

as $n \to \infty$ uniformly in $\alpha \in \mathcal{A}^*$, $\beta \in [-\alpha+\varepsilon, \beta_0]$ for any fixed $\beta_0 < \infty$, where

$$I(\alpha,\beta,\Delta) := \int_0^{\Delta} e^{-\lambda(\alpha)v}\mathbf{P}\big(\inf_{k\geqslant 1}({}^{\alpha}S_k+\beta k) > v\big)dv, \tag{3.6.5}$$

and the asymptotic density $f_{(n)}(x)$ was defined in section 2.2.

If in the left-hand sides of relations (3.6.3) and (3.6.4) $\Delta[x)$ is replced by $\Delta[x-\Delta)$ then on the right-hand side of (3.6.3) one should put $P(g^{(\beta)}+\Delta,\alpha)$, and in the right hand-side of (3.6.4) one should put

$$I(\alpha,\beta,-\Delta) = \int_{-\Delta}^{0} e^{-\lambda(\alpha)v}\,\mathbf{P}\Big(\inf_{k\geqslant 1}({}^{\alpha}S_k+\beta k) > v\Big)dv. \tag{3.6.6}$$

Proof. The proof of the theorem follows from Theorem 3.3.3 and some estimates in its proof. Since g satisfies condition $[\mathbf{g}]_\alpha$, (3.3.7) is valid. As can be seen from the estimates in the proof of Theorem 3.3.3, the main contribution to the probability $P(g,\alpha)$ comes from the probability of the first N events $B_k := \{{}^{\alpha}S_k > \tau - g_k\}$ (with large N). The influence of the 'tail' $\{g_k\}_{k>N}$ of the sequence g on this probability for large N is infinitisimaly small. The same is true for the probability $P\big(g^{(\beta)},\alpha\big)$. Moreover, $\mathbf{P}\Big(\bigcap_{k=1}^{N} B_k\Big)$ is a continuous function of $g_1,\dots,g_N$, since the variable τ has a density. Therefore, for $n\to\infty$,

$$P(g,\alpha) \sim P(g^{(\beta)},\alpha).$$

By virtue of Theorem 3.3.3, relation (3.6.3) is proved. Relation (3.6.4) can be obtained if one uses the formula (3.4.2) for $\mathbf{P}\big(S_n \in \Delta[x)\big)$ and the identity

$$P(g^{(\beta)},\alpha) = \frac{\lambda(\alpha)}{1-e^{-\lambda(\alpha)}}\,I(\alpha,\beta,\Delta),$$

which follows from (3.6.2) and the explicit form of the distribution of τ. The theorem is proved. □

Note that

$$P(g^{(\beta)},\alpha) = \int_{-\beta}^{\infty}\mathbf{P}({}^{\alpha}\xi-\tau\in du)\mathbf{P}(\underline{Z}_0 > -u),$$

where

$$\underline{Z}_0 := \inf_{k\geqslant 0}({}^{\alpha}S_k+\beta k),$$

and that

$$P(g^{(\beta')},\alpha) \sim P(g^{(\beta)},\alpha) \tag{3.6.7}$$

for $\beta'-\beta\to 0$.

3.7 Crossing of a curvilinear boundary by a normalised trajectory of a random walk

Consider a normalised trajectory

$$s_n(t) = \frac{S_{[nt]}}{x}, \qquad t \in [0, 1].$$

Let $g(t)$ be a measurable function on $[0, 1]$ such that $g(t) \geqslant g_0 > 0$ in a neighbourhood of the point $t = 0$. We will be interested in the distribution of the time

$$\nu_g = \min\left\{\frac{k}{n} : s_n\left(\frac{k}{n}\right) \geqslant g\left(\frac{k}{n}\right)\right\}$$

of first passage of the trajectory $s_n(t)$ over the curvilinear boundary $g(t)$, and also the asymptotics of the probability

$$\mathbf{P}\left(\max_{1\leqslant k\leqslant n}\left(s_n\left(\frac{k}{n}\right) - g\left(\frac{k}{n}\right)\right) \geqslant 0\right) \tag{3.7.1}$$

that the trajectory $s_n(t)$ crosses the boundary $g(t)$ at least once. As in section 3.4, we will assume that $\alpha = x/n \geqslant \alpha_0 > 0$ is bounded away from zero as $n \to \infty$.

3.7.1 A local limit theorem for the first passage time

When we study the probability $\mathbf{P}(\nu_g = m/n)$, $1 \leqslant m \leqslant n$, we will assume that the function $g(\cdot)$ and a point $z > 0$, in a neighbourhood of which the point m/n is located, satisfy the condition

[g, z] *There exists $\delta > 0$ such that the function $g(\cdot)$ is continuously differentiable in the δ-neighbourhood of a point z and for $t = m/n$, $|t - z| < \delta$, the following inequalities are valid:*

$$g'(t) \leqslant \frac{g(t)}{t} - \varepsilon, \tag{3.7.2}$$

$$g(u) \geqslant \frac{ug(t)}{t} + (t - u)\varepsilon \quad \textit{for all} \quad u \leqslant t = m/n, \tag{3.7.3}$$

for some $\varepsilon = \varepsilon(\delta) > 0$.

Conditions (3.7.2) and (3.7.3) mean that the curve $g(u)$ and the ray $ug(t)/t$, containing the points $(0, 0)$ and $\big(t, g(t)\big)$, cannot approach each other on the interval $(0, t)$ except in a neighbourhood of the point t where the approach is not faster than linear. Denote

$$\chi_g = S_{\nu_g n} - xg(\nu_g), \qquad \beta = -g'(m/n)\alpha, \qquad x' = xg(m/n), \qquad \alpha' = \frac{x'}{m}. \tag{3.7.4}$$

Theorem 3.7.1. *Let the function g and the point z satisfy the condition* **[g, z]**. *Then, for $n \to \infty$ and $|m/n - z| < \varepsilon$,*

$$\mathbf{P}\left(\nu_g = \frac{m}{n},\ \chi_g < \Delta\right) = f_{(m)}(x')I(\alpha', \beta, \Delta)\big(1 + o(1)\big)$$

uniformly in $\alpha' \in \mathcal{A}^*$, $\beta \in [-\alpha' + \varepsilon', \beta']$ *for some* $\varepsilon' > 0$ *and any fixed* $\beta' < \infty$. *The functions* $f_{(n)}(x)$ *and* $I(\alpha, \beta, \Delta)$ *were defined in, respectively,* (3.4.3) *and* (3.6.5).

Proof. It is not hard to see that if condition **[g, z]** is satisfied then we are in a situation when Theorem 3.6.1 can be applied to the sequence $\xi_1, \dots, \xi_m$. Let us show that the sequence

$$g_k = g_{k,m} = \left(g\left(\frac{m-k}{n}\right) - g\left(\frac{m}{n}\right)\right)x, \qquad k = 0, 1, \dots, m,$$

is asymptotically linear (see (3.6.1)). Indeed, for $k \leqslant N$, $n \to \infty$,

$$g_k = -kg'\left(\frac{m}{n}\right)\alpha + o(1) = -k\beta + o(1).$$

The inequality $\beta \geqslant -\alpha' + \varepsilon$ follows from (3.7.2), since, by virtue of (3.7.2),

$$\begin{aligned}\beta = -g'\left(\frac{m}{n}\right)\alpha &\geqslant -\frac{g(m/n)n\alpha}{m} + \varepsilon = -\frac{g(m/n)x}{m} + \varepsilon \\ &= -\frac{x'}{m} + \varepsilon = -\alpha' + \varepsilon.\end{aligned}$$

In a similar way we can verify that (3.7.3) implies that condition $[\mathbf{g}]_\alpha$ is satisfied for $\alpha = \alpha'$ (see (3.3.4), (3.3.5)): for $u = (m-k)/n$ and $t - u = k/n$, by virtue of (3.7.3) we have

$$\begin{aligned}g_k &= \left(g\left(\frac{m-k}{n}\right) - g\left(\frac{m}{n}\right)\right)x \geqslant -\frac{k}{m}g\left(\frac{m}{n}\right)x + \frac{\varepsilon kx}{n} \\ &= -k\alpha' + \varepsilon k\alpha \geqslant -k\alpha' + \varepsilon k\alpha_0.\end{aligned}$$

This proves (3.3.4) with regard to our conditions and also under the condition $[\mathbf{g}]_{\alpha'}$. Applying Theorem 3.6.1, we obtain

$$\begin{aligned}&\mathbf{P}\left(\nu_g = \frac{m}{n},\ \chi_g < \Delta\right) \\ &\quad = \mathbf{P}\big(S_{m-k} < x' + g_{k,m},\ k = 1, \dots, m-1;\ S_m \in \Delta[x')\big) \\ &\quad \sim f_{(m)}(x')I(\alpha', \beta, \Delta)\end{aligned}$$

uniformly in $\alpha' \in \mathcal{A}^*$, $\beta \in [-\alpha' + \varepsilon', \beta']$. The theorem is proved. □

In order to obtain the asymptotics of the boundary crossing probability (i.e. integral limit theorems for the first passage time) using the local theorem 3.7.1, we will need the notion of a *level line*.

3.7.2 Level lines

According to Theorem 3.7.1, the main part of the asymptotics of $\mathbf{P}(\nu_g = m/n)$ or, equivalently, of the asymptotic density

$$f_{(m)}(x') = f_{(m)}\left(xg\left(\frac{m}{n}\right)\right)$$

(see (3.4.3)) is determined for $t = m/n$ by the function

$$e^{-nt\Lambda((\alpha/t)\,g(t))} = e^{-m\Lambda((x/m)\,g(m/n))}. \tag{3.7.5}$$

One can ask a natural question: for what function $\alpha g(t) = \mathfrak{l}(t)$ does the asymptotics (3.7.5) remain the same for all $t \in (0,1]$? Consider the parametric family of functions $\mathfrak{l}_\omega(t)$ depending on the parameter ω such that

$$\mathfrak{l}_\omega(1) = \omega, \qquad t\Lambda\left(\frac{\mathfrak{l}_\omega(t)}{t}\right) = \Lambda\big(\mathfrak{l}_\omega(1)\big) \equiv \Lambda(\omega). \tag{3.7.6}$$

By applying a linear transformation to $s_n(t)$ and $g(t)$ (i.e. adding the function ct), without loss of generality, we may assume that

$$\mathbf{E}\xi = 0$$

(we will assume that $\mathbf{E}\xi$ exists). In this case, $\Lambda(\alpha)$ is a strictly increasing continuous function of α on $(0,\infty)$, except $[s_+,\infty)$ with $s_+ < \infty$, $\Lambda(s_+) < \infty$, where $\Lambda(\alpha) = \infty$ for $\alpha > s_+$. Therefore, for $v \in [0, s_+]$, the continuous increasing inverse function $\Lambda^{(-1)}(v)$ exists, so that (see (3.7.6))

$$\mathfrak{l}_\omega(t) = t\Lambda^{(-1)}\left(\frac{\Lambda(\omega)}{t}\right). \tag{3.7.7}$$

The curve $\mathfrak{l}_\omega(t)$ is called *a level line* of level ω. At the point $t = 0$, we define $\mathfrak{l}_\omega(t)$ by continuity: $\mathfrak{l}_\omega(0) = \mathfrak{l}_\omega(+0)$. The properties of level lines are summarised in the following theorem.

Theorem 3.7.2. *Let* $\mathbf{E}\xi = 0$, $\omega < \alpha_+$. *Then*

(i) *The solution* $\mathfrak{l}_\omega(t)$ *of equation* (3.7.6) *always exists except in the case*

$$s_+ < \infty, \quad \Lambda(s_+) \equiv -\ln \mathbf{P}(\xi = s_+) < \infty, \quad t < t_\omega := \frac{\Lambda(\omega)}{\Lambda(s_+)}. \tag{3.7.8}$$

If the relation (3.7.8) *is met then* $\alpha_+ = s_+$ *and, for* $t < t_\omega$, *there exists a 'generalised' solution* $\mathfrak{l}_\omega(t) = ts_+$, *which has the properties*

$$t\Lambda\left(\frac{\mathfrak{l}_\omega(t)}{t}\right) < \Lambda(\omega), \qquad \Lambda\left(\frac{\mathfrak{l}_\omega(t)+0}{t}\right) = \infty. \tag{3.7.9}$$

In the subsequent assertions of the theorem the case (3.7.8), (3.7.9) *is excluded and it is assumed that* $\lambda_+ > 0$.

(ii)

$$\mathfrak{l}_\omega(0) = \frac{\Lambda(\omega)}{\lambda_+}.$$

(iii) *The functions* $\mathfrak{l}_\omega(t)$ *increase on* $[0,1]$ *and are concave* ($\mathfrak{l}''_\omega(t) \leqslant 0$). *For each* t *the values of* $\mathfrak{l}_\omega(t)$ *increase in* ω.

(iv) *If* $\Lambda(\alpha_+) = \infty$ *then the functions* $\mathfrak{l}_\omega(t)$ *are analytic on* $(0,1)$ *and strictly convex* ($\mathfrak{l}''_\omega(t) < 0$).

(v) *If* $\Lambda(\alpha_+) < \infty$ *then at the point* $t_\omega = \Lambda(\omega)/\Lambda(\alpha_+)$ *the curve* $\mathfrak{l}_\omega(t)$ *is coupled from two parts: a linear part*

$$\mathfrak{l}_\omega(t) = \frac{\Lambda(\omega)}{\lambda_+} + t\left(\alpha_+ - \frac{\Lambda(\alpha_+)}{\lambda_+}\right) \quad \textit{for} \quad t \leqslant t_\omega,$$

and a strictly convex analytic part on the interval $(t_\omega, 1)$. *At the point* t_ω, *the function* $\mathfrak{l}_\omega(t)$ *is continuously differentiable.*

Proof. In the proof of the theorem we will use many properties of deviation functions presented in section 1.1. If $s_+ = \infty$ or $\Lambda(s_+) = \infty$ then on $[0,\infty)$ the function $\Lambda(\alpha)$ strictly and continuously increases from 0 to ∞ (recall that $\mathbf{E}\xi = 0$). Hence the positive branch of the inverse function $\Lambda^{(-1)}(v)$ exists and is unique on the whole half-axis $(0,\infty)$. The function $\Lambda(\alpha)$ is coupled, in general, from two parts: a strictly convex analytic part on $(0,\alpha_+)$, and a linear part

$$\Lambda(\alpha) = \Lambda(\alpha_+) + (\alpha - \alpha_+)\lambda_+ \quad \text{for} \quad \alpha \geqslant \alpha_+,$$

(if $\alpha_+ < \infty$ then $\lambda_+ < \infty$ under our conditions). Moreover, the derivative $\Lambda'(\alpha) = \lambda(\alpha)$ at the point α_+ is continuous. Therefore, the inverse function $\Lambda^{(-1)}(v)$ is also coupled from the two parts in a continuously differentiable way: there is a strictly convex analytic part on $\big(0, \Lambda(\alpha_+)\big)$ and a linear part

$$\Lambda^{(-1)}(v) = \alpha_+ + \big(v - \Lambda(\alpha_+)\big)\lambda_+^{-1} \quad \text{for} \quad v \geqslant \Lambda(\alpha_+).$$

According to this and relation (3.7.7), the coupling point $\theta = \theta_\omega$ is determined by the identity

$$\theta_\omega = \frac{\Lambda(\omega)}{\Lambda(\alpha_+)}$$

and we have

$$\mathfrak{l}_\omega(t) = \frac{\Lambda(\omega)}{\lambda_+} + t\left(\alpha_+ - \frac{\Lambda(\alpha_+)}{\lambda_+}\right) \quad \text{for} \quad t \leqslant \theta_\omega.$$

It is clear that $\theta_\omega = 0$ in the case $\Lambda(\alpha_+) = \infty$, and the function $\mathfrak{l}_\omega(t)$ is analytic on $(0,1)$.

The property that $\mathfrak{l}_\omega(t)$ is increasing and strictly convex on $(\theta_\omega, 1)$ follows from the following relations. Differentiating identity (3.7.6) in t and writing for brevity $\mathfrak{l}_\omega(t) = \mathfrak{l}$, we get

$$\begin{aligned} \Lambda\left(\frac{\mathfrak{l}}{t}\right) &= -t\lambda\left(\frac{\mathfrak{l}}{t}\right)\left(\frac{\mathfrak{l}'}{t} - \frac{\mathfrak{l}}{t^2}\right) \\ &= \frac{\mathfrak{l}}{t}\lambda\left(\frac{\mathfrak{l}}{t}\right) - A\left(\lambda\left(\frac{\mathfrak{l}}{t}\right)\right), \quad \text{where} \quad A(\lambda) = \ln\psi(\lambda). \end{aligned} \tag{3.7.10}$$

From this we find

$$\mathfrak{l}' = \frac{A(\lambda(\mathfrak{l}/t))}{\lambda(\mathfrak{l}/t)} > 0, \tag{3.7.11}$$

since $\psi(\lambda) > 1$ for $\lambda > 0$.

From (3.7.10), it follows that

$$\frac{\mathfrak{l}'}{t} - \frac{\mathfrak{l}}{t^2} = \left(\frac{\mathfrak{l}}{t}\right)' < 0$$

and, hence, the function $\mathfrak{l}/t$ decreases as t increases. Since $A(\lambda)$ is a convex function, $A(\lambda)/\lambda$ increases and, consequently, by virtue of (3.7.11), the derivative $\mathfrak{l}'(t)$ decreases in t, $\mathfrak{l}''(t) < 0$.

Now we can find the value $\mathfrak{l}_\omega(0) = \mathfrak{l}_\omega(+0) = \lim_{t \downarrow 0} \mathfrak{l}_\omega(t)/t$. Since $\Lambda(\alpha) \sim \lambda_+ \alpha$ for $\lambda_+ < \infty$, $\alpha \to \infty$ $(\Lambda(\alpha) \gg \alpha$ for $\lambda_+ = \infty$ and $\alpha \to \infty)$, letting $t \to 0$ in (3.7.6) we obtain

$$\Lambda(\omega) = \mathfrak{l}_\omega(0)\lambda_+, \qquad \mathfrak{l}_\omega(0) = \frac{\Lambda(\omega)}{\lambda_+}.$$

The statement (3.7.9) of the theorem concerning the previously excluded case (3.7.8) is verified in an obvious way. The theorem is proved. □

If ξ has the normal distribution with parameters $(0, \sigma^2)$ then

$$\Lambda(u) = \frac{u^2}{2\sigma^2}, \qquad \Lambda^{(-1)}(v) = \sigma\sqrt{2v}$$

and, by virtue of (3.7.7),

$$\mathfrak{l}_\omega(t) = \omega\sqrt{t}.$$

Since for any distribution satisfying condition **[C$_0$]**, for small u and v it holds that

$$\Lambda(u) \sim \frac{u^2}{2\sigma^2}, \qquad \Lambda^{(-1)}(v) \sim \sigma\sqrt{2v};$$

by virtue of (3.7.7), for small ω the level lines $\mathfrak{l}_\omega(t)$ will be close to $\omega\sqrt{t}$ (as can be seen from Theorem 3.7.2, this claim in a neighbourhood of the point $t = 0$ needs to be refined).

3.7.3 The crossing problem for an arbitrary boundary

Now consider the problem of the asymptotics of probability (3.7.1) or, which is the same, the asymptotics of the probability $\mathbf{P}(\nu_g \leqslant 1)$. We will assume in this subsection that

$$\mathbf{E}\xi = 0, \qquad \inf_{t\in[0,1]} g(t) \geqslant g_0 > 0, \qquad \alpha = \frac{x}{n} \to \alpha_0 \in (0, \alpha_+), \tag{3.7.12}$$

for $n \to \infty$, so the probability (3.7.1) converges to 0 as $n \to \infty$.

Denote by ω_g the value ω, for which the level lines $\mathfrak{l}_\omega(t)$ touch the curve $\alpha g(t)$ for the first time when ω increases from the zero value:

$$\omega_g := \max\left\{\omega : \inf_{t\in(0,1]} \big(\alpha g(t) - \mathfrak{l}_\omega(t)\big) \geqslant 0\right\}.$$

(Note that ω_g depends not only on g but also on α.) Denote by t_g the *tangency point of the curves* $\alpha g(t)$ and $\mathfrak{l}_{\omega_g}(t)$, so that

$$\alpha g(t_g) = \mathfrak{l}_{\omega_g}(t_g),$$

if the function $g(t)$ is continuous at the point t_g.

First consider the case when contact of the curves $g(t)$ and $\mathfrak{l}_{\omega_g}(t)$ is made *inside* the interval $(0, 1)$. Denote

$$\alpha_g := \frac{\alpha g(t_g)}{t_g}$$

and note that

$$\alpha_g = \frac{\mathfrak{l}_{\omega_g}(t_g)}{t_g} > \mathfrak{l}_{\omega_g}(1) = \omega_g,$$

$$t_g\Lambda(\alpha_g) = t_g\Lambda\left(\frac{\mathfrak{l}_{\omega_g}(t_g)}{t_g}\right) = \Lambda(\omega_g).$$

Theorem 3.7.3. *Let*

$$\mathbf{E}\xi = 0, \qquad \inf_{t\in[0,1]} g(t) = g_0 > 0, \qquad \alpha_g \in \mathcal{A}^* \tag{3.7.13}$$

and suppose that the following conditions are satisfied:

(i) *There exists a unique tangency point* t_g, *and*

$$t_g \in (\theta_{\omega_g}, 1).$$

(ii) *The function* $g(t)$ *is twice continuously differentiable in a neighbourhood of the point* t_g,

$$\alpha g''(t_g) > \mathfrak{l}''_{\omega_g}(t_g) \qquad \big(\alpha g'(t_g) = \mathfrak{l}'_{\omega_g}(t_g)\big). \tag{3.7.14}$$

(iii) *For any sufficiently small* $\delta > 0$, *there exists* $\varepsilon = \varepsilon(\delta) > 0$ *such that*

$$\inf_{t\notin(t_g)_\delta} \big(\alpha g(t) - \mathfrak{l}_{\omega_g}(t)\big) \geqslant \varepsilon. \tag{3.7.15}$$

Then, as $n \to \infty$,

$$\mathbf{P}(\nu_g \leqslant 1,\ \chi_g < \Delta) \sim e^{-n\Lambda(\omega_g)} \frac{I(\alpha_g, \beta, \Delta)}{\sigma(\alpha_g)\sqrt{t_g\lambda(\alpha_g)d''(t_g)}}, \tag{3.7.16}$$

where $d(t) = \alpha g(t) - \mathfrak{l}_{\omega_g}(t)$ *and the function* $I(\cdot)$ *is defined in* (3.6.5).

Under the conditions of Theorem 3.7.3, along with the inclusion $\alpha_g \in \mathcal{A}^*$ it is additionally assumed (see (3.7.12)) that α and, hence, α_g converge to some fixed value as $n \to \infty$. Therefore, the tangency point t_g also converges to some fixed value and condition (ii) of the theorem can be assumed to be satisfied in a neighbourhood of this fixed point.

If $\alpha = \text{const.}$ (i.e. it does not depend on n, so that we may assume $x = n$ without loss of generality) then t_g will be a fixed point, and in condition (ii) it is enough to require the twice continuous differentiability of $g(t)$ only at the point t_g (and not in its neighbourhood).

To prove Theorem 3.7.3 we will need

Lemma 3.7.4. *If conditions* (i)–(iii) *of Theorem* 3.7.3 *are satisfied, then conditions* **[g, z]** *of Theorem* 3.7.1 *are satisfied for* $z = t_g$.

Proof. Since the function $\mathfrak{l}_{\omega_g}$ is strictly concave on $[\theta_{\omega_g}, 1]$, for some $\varepsilon > 0$ we have

$$g'(t_g) = \mathfrak{l}_{\omega_g}(t_g) < \frac{\mathfrak{l}_{\omega_g}(t_g)}{t_g} - 2\varepsilon = \frac{g(t_g)}{t_g} - 2\varepsilon.$$

Clearly, an inequality of such a type is preserved in a sufficiently small neighbourhood of the point t_g, i.e. for all t from a sufficiently small neighbourhood of the point t_g we have

$$g'(t) \leqslant \frac{g(t)}{t} - \varepsilon.$$

This means that condition (3.7.2) is met for $z = t_g$.

Let us now check the validity of condition (3.7.3). We will check it in each of the three regions where u changes:

(1) in a neighbourhood of the point t;
(2) in a neighbourhood of zero;
(3) in the intermediate zone.

(1) For sufficiently small $\delta > 0$ we have

$$\inf_{|t-z|<\delta} \left(\frac{\mathfrak{l}_{\omega_g}(t)}{t} - \mathfrak{l}'_{\omega_g}(t) \right) =: 3b > 0.$$

Since the values $\mathfrak{l}_{\omega_g}(z)$ and $\mathfrak{l}'_{\omega_g}(z)$ coincide with $g(z)$ and $g'(z)$, respectively, and, moreover, the function $g(t)$ is continuously differentiable for $|t - z| < \delta$, for sufficiently small δ we also have the inequality

$$\inf_{|t-z|<\varepsilon} \left(\frac{g(t)}{t} - g'(t) \right) \geqslant 2b > 0. \tag{3.7.17}$$

If t, u are located in a δ-neighbourhood of the point z then, as $|t - u| \to 0$,

$$g(u) = g(t) + (u - t)g'(t) + o\big(|u - t|\big).$$

By virtue of (3.7.17),

$$g'(t) < \frac{g(t)}{t} - 2b$$

and, hence, for sufficiently small δ,

$$g(u) > g(t) + (u-t)\left(\frac{g(t)}{t} - 2b\right) - (t-u)b > \frac{ug(t)}{t} + b(t-u). \quad (3.7.18)$$

Note that owing to the smoothness of the function $g(t)$ and its proximity to $\mathfrak{l}_{\omega_g}(t)$, the quantity $b > 0$ under consideration does not tend to 0 as $\delta \to 0$. Therefore, inequality (3.7.18) implies the existence of $v > 0$, independent of δ, such that

$$g(u) > \frac{ug(t)}{t} + \frac{b}{2}(t-u) \quad \text{for} \quad u \in (t-v, t). \quad (3.7.19)$$

Condition (3.7.3) for $u \in (t-v, t)$ is proved.

(2) For sufficiently small $\delta > 0$ and $|t-z| < \delta$, it holds that

$$\frac{g(t)}{t} < \frac{2\mathfrak{l}_{\omega_g}(z)}{z}.$$

Hence, for $u \leqslant u_0$ and u_0 such that

$$\frac{2u_0\mathfrak{l}_{\omega_g}(z)}{z} = \frac{g_0}{2},$$

we have

$$g(u) \geqslant g_0 = \frac{2u_0\mathfrak{l}_{\omega_g}(z)}{z} + \frac{g_0}{2} \geqslant \frac{ug(t)}{t} + \frac{g_0}{2}(t-u).$$

Condition (3.7.3) in the u_0-neighbourhood of the point 0 is proved.

(3) It remains to consider the intermediate zone $u \in (u_0, t-v)$, where u_0 and v are fixed. B virtue of condition (iii) of Theorem 3.7.3, for $u \in (u_0, t-v)$ we have

$$g(u) \geqslant \mathfrak{l}_{\omega_g}(u) + \varepsilon > \frac{u\mathfrak{l}_{\omega_g}(t)}{t} + \varepsilon,$$

where $\varepsilon = \varepsilon(v)$ depends only on v. Moreover, for sufficiently small $\delta > 0$ and $|t-z| < \delta$,

$$\frac{\mathfrak{l}_{\omega_g}(t)}{t} > \frac{g(t)}{t} - \frac{\varepsilon}{2}.$$

Consequently,

$$g(u) > u\left(\frac{g(t)}{t} - \frac{\varepsilon}{2}\right) + \varepsilon > \frac{ug(t)}{t} + \frac{\varepsilon}{2} > \frac{ug(t)}{t} + \frac{\varepsilon}{2}(t-u).$$

Inequality (3.7.3) is thus proved on the whole segment $[0, t]$. The lemma is proved. □

Proof of Theorem 3.7.3. By virtue of Lemma 3.7.4, we can apply the local theorem 3.7.1 for m such that $t = m/n$ is in a neighbourhood of the point t_g. For

$$\alpha = \frac{x}{n}, \qquad x' = xg\left(\frac{m}{n}\right), \qquad \alpha' = \frac{x'}{m}, \qquad \beta = -g'\left(\frac{m}{n}\right)\alpha$$

we have

$$P_{n,m} := \mathbf{P}\left(\nu_g = \frac{m}{n},\ \chi_g < \Delta\right) = f_{(m)}(x')I(\alpha', \beta, \Delta)\big(1 + o(1)\big), \qquad (3.7.20)$$

uniformly in $\alpha' \in \mathcal{A}^*$, $\beta \in [-\alpha' + \varepsilon', \beta']$ for some $\varepsilon' > 0$ and any fixed $\beta' < \infty$, where (see (3.4.2))

$$f_{(m)}(x') + \frac{e^{-m\Lambda(\alpha')}}{\sigma(\alpha')\sqrt{2\pi m}}.$$

The function $I(\alpha, \beta, \Delta)$ is defined in (3.6.5). Here, for $t = m/n$,

$$m\Lambda(\alpha') = m\Lambda\left(\frac{x}{m}g\left(\frac{m}{n}\right)\right) = nt\Lambda\left(\frac{\alpha g(t)}{t}\right). \qquad (3.7.21)$$

Denote

$$d(t) := \alpha g(t) - \mathfrak{l}_{\omega_g}(t), \qquad (3.7.22)$$

so that

$$d(t_g) = 0, \quad d'(t_g) = 0, \quad d''(t_g) > 0,$$
$$d(t) = \frac{1}{2}d''(t_g)(t - t_g)^2 + o\big((t - t_g)^2\big)$$

for $|t - t_g| \to 0$.

Further,

$$\begin{aligned} t\Lambda\left(\frac{\alpha g(t)}{t}\right) &= t\Lambda\left(\frac{\mathfrak{l}_{\omega_g}(t) + d(t)}{t}\right) \\ &= t\Lambda\left(\frac{\mathfrak{l}_{\omega_g}(t)}{t}\right) + \lambda\left(\frac{\mathfrak{l}_{\omega_g}(t)}{t_g}\right)\frac{d''(t_g)(t - t_g)^2}{2} + o\big((t - t_g)^2\big) \\ &= \Lambda(\omega_g) + \frac{1}{2}\lambda\left(\frac{\alpha g(t_g)}{t_g}\right)d''(t_g)(t - t_g)^2\big(1 + o(1)\big) \end{aligned}$$

for $|t - t_g| \to 0$.

Since

$$\alpha' = \frac{\alpha g(t)}{t} \sim \frac{\alpha g(t_g)}{t_g} = \alpha_g,$$

returning to (3.7.20), we find

$$\begin{aligned} P_{n,m} = {} & \frac{e^{-n\Lambda(\omega_g)}I(\alpha_g, \beta, \Delta)}{\sqrt{2\pi n t_g}\,\sigma(\alpha_g)} \\ & \times \exp\left\{-\frac{n}{2}\lambda(\alpha_g)d''(t_g)(t - t_g)^2\big(1 + o(1)\big)\right\}\big(1 + o(1)\big) \end{aligned} \qquad (3.7.23)$$

uniformly in $t = m/n$ from a δ_n-neighbourhood of the point t_g, where $\delta_n \to 0$ sufficiently slowly.

Further,

$$\mathbf{P}(\nu_g \leqslant 1, \chi_g < \Delta) = \sum_{m\in A} P_{n,m} + \sum_{m\notin A} P_{n,m},$$

where

$$A := \{m \leqslant n : |m - nt_g| < n\delta_n\}. \tag{3.7.24}$$

From (3.7.23), it follows that

$$\sum_{m\in A} P_{n,m} \sim \frac{e^{-n\Lambda(\omega_g)} I(\alpha_g, \beta, \Delta)}{\sigma(\alpha_g)\sqrt{2\pi n t_g}} \sum_{m\in A} \exp\left\{-\frac{n}{2}\lambda(\alpha_g)d''(t_g)\left(\frac{m}{n} - t_g\right)^2\right\}, \tag{3.7.25}$$

where $\sum_{m\in A}$ on the right-hand side of (3.7.25) is asymptotically equivalent to the integral

$$\int_{|u-nt_g|<n\delta_n} \exp\left\{-\frac{1}{2n}\lambda(\alpha_g)d''(t_g)(u - nt_g)^2\right\} du. \tag{3.7.26}$$

After making the change of variables

$$(u - nt_g)\sqrt{\frac{\lambda(\alpha_g)d''(t_g)}{n}} = v,$$

for (3.7.26), we find the value

$$\sqrt{\frac{n}{\lambda(\alpha_g)d''(t_g)}} \int_{|v|<\sqrt{n}\,\delta_n'} e^{-v^2/2} dv \sim \sqrt{\frac{2\pi n}{\lambda(\alpha_g)d''(t_g)}}, \tag{3.7.27}$$

where δ_n' has the same order of infinitesimality as δ_n, and $\delta_n \to 0$ as $n \to \infty$ slowly enough that, $\sqrt{n}\,\delta_n \to \infty$. This means that, for the chosen δ_n,

$$\sum_{m\in A} P_{n,m} \sim e^{-n\Lambda(\omega_g)} \frac{I(\alpha_g, \beta, \Delta)}{\sigma(\alpha_g)\sqrt{t_g\lambda(\alpha_g)d''(t_g)}}.$$

Let us now estimate $\sum_{m\notin A} P_{n,m}$. Assume for a moment that $\delta_n = \delta$ is fixed. Then for some $\varepsilon > 0$ no term of this sum exceeds

$$\begin{aligned} P_{n,m} &\leqslant \mathbf{P}\left(S_m \geqslant xg\left(\frac{m}{n}\right)\right) = \mathbf{P}\left(S_m \geqslant n\alpha g\left(\frac{m}{n}\right)\right) \\ &\leqslant \mathbf{P}\left(S_m \geqslant n\left[\mathfrak{l}_{\omega_g}\left(\frac{m}{n}\right) + \varepsilon\right]\right) \leqslant \exp\left\{-m\Lambda\left(\frac{n}{m}\left[\mathfrak{l}_{\omega_g}\left(\frac{m}{n}\right) + \varepsilon\right]\right)\right\} \\ &= \exp\left\{-nt\Lambda\left(\frac{\mathfrak{l}_{\omega_g}(t) + \varepsilon}{t}\right)\right\}, \end{aligned}$$

where $t = m/n$, and, owing to the convexity of the function Λ,

$$\Lambda\left(\frac{\mathfrak{l}_{\omega_g}(t) + \varepsilon}{t}\right) \geqslant \Lambda\left(\frac{\mathfrak{l}_{\omega_g}(t)}{t}\right) + \lambda\left(\frac{\mathfrak{l}_{\omega_g}(t)}{t}\right)\frac{\varepsilon}{t}.$$

Therefore,

$$P_{n,m} \leqslant \exp\left\{-n\Lambda(\omega_g) - n\varepsilon\lambda\left(\frac{\mathfrak{l}_{\omega_g}(t)}{t}\right)\right\}.$$

Thus, $\sum_{m\notin A} P_{n,m}$, up to a bounded factor, will not exceed

$$n\exp\left\{-n\Lambda(\omega_g) - n\varepsilon\lambda\left(\frac{\mathfrak{l}_{\omega_g}(t)}{t}\right)\right\},$$

which is infinitesimally small compared with $\sum_{m\in a} P_{n,m}$.

The last conclusion obviously remains valid in the case when

$$\delta = \delta_n \to 0, \qquad \varepsilon = \varepsilon_n \to 0$$

sufficiently slowly as $n \to \infty$ (e.g. so that $n\varepsilon_n \gg \ln n$ holds true). The theorem is proved. □

Now consider the case where contact of the curves $g(t)$ and $\mathfrak{l}_{\omega_g}(t)$ occurs at the point $t_g = 1$. In this case $\alpha_g = \omega_g = \alpha g(1)$.

Theorem 3.7.5. *Suppose that conditions* (3.7.13) *are satisfied and* $t_g = 1$.

(i) *Assume from now on that the function* $g(t)$ *is twice continuously differentiable from the left at the point* $t = 1$, *relations* (3.7.14) *and* (3.7.15) *are satisfied for* $t_g = 1$ *and*

$$\alpha g'(1) = \mathfrak{l}'_{\omega_g}(1).$$

Then the probability $\mathbf{P}(\nu_g \leqslant 1, \chi_g < \Delta)$ *is equal to one-half of the right-hand side of* (3.7.15), *in which the value* t_g *is replaced by* 1 *and* ω_g *by* α_g.

(ii) *Suppose that the function* $g(t)$ *is continuously differentiable from the left at the point* $t = 1$,

$$-d'(1) = \mathfrak{l}'_{\omega_g}(1) - \alpha g'(1) \geqslant d_0 > 0$$

and the relations (3.7.15) *are satisfied for* $t_g = 1$. *Then, as* $n \to \infty$,

$$\mathbf{P}(\nu_g \leqslant 1, \chi_g < \Delta) \sim \frac{e^{-n\Lambda(\alpha_g)} I(\alpha_g, \beta, \Delta)}{\sigma(\alpha_g)\sqrt{2\pi n}(1-q)}, \tag{3.7.28}$$

where $q = e^{\lambda(\alpha_g)d'(1)}$. *The statement* (3.7.28) *can be supplemented with the following relations:*

$$\mathbf{P}(\nu_g \leqslant 1, \chi_g < \Delta) \sim \frac{f_{(n)}\big(xg(1)\big)I(\alpha_g, \beta, \Delta)}{1-q} \sim \frac{\mathbf{P}(\nu_g = 1, \chi_g < \Delta)}{1-q}.$$

Proof. (i) The proof of the first statement of the theorem is carried out in the same way as the proof of Theorem 3.7.3, the only difference being that now $t_g = 1$ and the set A is of the form

$$A = \big\{m \leqslant n :\ m \geqslant n(1-\delta_n)\big\}$$

(cf. (3.7.24)). This difference implies that the integral in (3.7.27) should be replaced by an integral over only the interval $(-\delta_n\sqrt{n}, 0)$, which is asymptotically equivalent to $\sqrt{\pi/2}$. This proves statement (i).

(ii) As in the proof of Theorem 3.7.3, we can verify that the conditions of Theorem 3.7.5(ii) imply the validity of conditions **[g, z]** for $z = 1$. Hence, according to Theorem 3.7.1 for $m = n - k$, $k = o(n)$, we have

$$P_{n,m} = \mathbf{P}\left(\nu_g = \frac{m}{n},\ \chi_g < \Delta\right)$$
$$= f_{(m)}\left(xg\left(\frac{m}{n}\right)\right) I(\alpha_g, \beta, \Delta)\big(1 + o(1)\big) \tag{3.7.29}$$

uniformly in $\alpha_g = \alpha g(1) \in \mathcal{A}^*$ and $\beta \in [-\alpha_g + \varepsilon', \beta']$ (see Theorem 3.7.1). Here, for $t = m/n$, $n \to \infty$ (see (3.4.2), (3.7.21) and (3.7.22))

$$\begin{aligned} f_{(m)}\left(xg\left(\frac{m}{n}\right)\right) &\sim \frac{e^{-m\Lambda((x/m)g(m/n))}}{\sigma(\alpha_g)\sqrt{2\pi n}}, \\ \frac{m}{n}\Lambda\left(\frac{x}{m}g\left(\frac{m}{n}\right)\right) &= t\Lambda\left(\frac{\alpha g(t)}{t}\right) = t\Lambda\left(\frac{\mathfrak{l}_{\omega_g}(t) + d(t)}{t}\right), \end{aligned} \tag{3.7.30}$$

where, as $t \uparrow 1$,

$$d(t) \equiv \alpha g(t) - \mathfrak{l}_{\omega_g}(t) = d'(1)(t-1) + o(1-t).$$

Therefore,

$$\begin{aligned} t\Lambda\left(\frac{\alpha g(t)}{t}\right) &= t\Lambda\left(\frac{\mathfrak{l}_{\omega_g}(t)}{t}\right) + t\lambda\left(\frac{\mathfrak{l}_{\omega_g}(t)}{t}\right)d'(1)(t-1) + o(1-t) \\ &= \Lambda(\alpha_g) + \lambda\big(\alpha g(1)\big)d'(1)(t-1)\big(1 + o(1)\big). \end{aligned}$$

Returning to (3.7.29) and (3.7.30), we find

$$\begin{aligned} P_{n,m} = &\frac{e^{-n\Lambda(\alpha_g)} I(\alpha_g, \beta, \Delta)}{\sigma(\alpha_g)\sqrt{2\pi n}} \\ &\times \exp\Big\{ - n\lambda(\alpha_g)d'(1)(t-1)\big(1 + o(1)\big)\Big\}\big(1 + o(1)\big) \end{aligned} \tag{3.7.31}$$

uniformly in $t = m/n$ from a δ_n-neighbourhood of the point $t = 1$, where $\delta_n \to 0$ sufficiently slowly as $n \to \infty$. Denoting by A the range of values of m,

$$A = \big\{m \leqslant n : m \geqslant n(1 - \delta_n)\big\},$$

we get

$$\mathbf{P}(\nu_g \leqslant 1, \chi_g < 1) = \sum_{m \in A} P_{n,m} + \sum_{m \notin A} P_{n,m},$$

where by virtue of (3.7.31), as $n \to \infty$,

$$\sum_{m \in A} P_{n,m} \sim \frac{e^{-n\Lambda(\alpha_g)} I(\alpha_g, \beta, \Delta)}{\sigma(\alpha_g)\sqrt{2\pi n}} \sum_{m \in A} \exp\left\{-n\Lambda(\alpha_g)d'(1)\left(\frac{m}{n} - 1\right)\right\}. \tag{3.7.32}$$

Since

$$d'(1) = \alpha g'(1) - \mathfrak{l}'_{\omega_g}(1) < -d_0 < 0, \qquad \lambda(\alpha_g)d'(1) < 0,$$

and the sum on the right-hand side of (3.7.32) is asymptotically equivalent to the sum of the geometric sequence with parameter $q = e^{\lambda(\alpha_g)d'(1)} < 1$, we have, as $n \to \infty$,

$$\sum_{m\in A} P_{n,m} \sim \frac{e^{-n\Lambda(\alpha_g)} I(\alpha_g, \beta, \Delta)}{\sigma(\alpha_g)\sqrt{2\pi n}\,(1-q)}.$$

Estimation of $\sum_{m\notin A} P_{n,m}$ is carried out in exactly the same way as in the proof of Theorem 3.7.3. The theorem is proved. □

Other types of tangency of the curve $\alpha g(t)$ and the level lines $\mathfrak{l}_\omega(t)$ (along with those already obtained in Theorems 3.7.2 and 3.7.3) are considered in [17] and [18].

3.7.4 The probability that a trajectory does not cross a boundary

In this section we consider the asymptotics of the probability that a trajectory $s_n(t)$ does not cross a boundary $g(t)$ on $[0, 1]$ in the case when the ray $t\mathbf{E}\xi$, $t > 0$, does cross this boundary. More precisely, we will study the asymptotic behaviour of the probability

$$\mathbf{P}(\nu_g > 1) := \mathbf{P}\left(\max_{k\leqslant n}\left(s_n\left(\frac{k}{n}\right) - g\left(\frac{k}{n}\right)\right) < 0\right)$$

in the case when, instead of (3.7.12), it is assumed that

$$\max_{t\in[0,1]}\big(t\mathbf{E}\xi - \alpha g(t)\big) > 0, \qquad \alpha \to \alpha_0 \in \mathcal{A}^* \tag{3.7.33}$$

for $n \to \infty$. For the sake of definiteness, let $g(1) = 1$.

Theorem 3.7.6. *Let* $g(1) = 1$ *and conditions* (3.7.33) *and* **[g, 1]** *be satisfied* (see (3.7.2), (3.7.3)). *Then*

$$\mathbf{P}(\nu_g > 1) \sim f_{(n)}(x) I(\alpha, -\alpha g'(1), -\infty), \tag{3.7.34}$$

where the function $I(\cdot)$ *was defined in* (3.6.6).

Proof. For brevity, let

$$A_{n,g} := \left\{S_{n-k} < xg\left(\frac{n-k}{n}\right) \quad \text{for all} \quad k = 1, \ldots, n-1\right\}.$$

Then

$$\begin{aligned}\mathbf{P}(\nu_g > 1) &= \mathbf{P}\left(S_{n-k} < xg\left(\frac{n-k}{n}\right) \quad \text{for all} \quad k = 0, \ldots, n\right)\\ &= \mathbf{P}(A_{n,g};\ S_n < x)\\ &= \mathbf{P}\big(A_{n,g};\ S_n \in \Delta[x-\Delta)\big) + \mathbf{P}(A_{n,g};\ S_n < x - \Delta).\end{aligned} \tag{3.7.35}$$

Here

$$g_{k,n} := \alpha \left[g\left(\frac{n-k}{n}\right) - g(1) \right] \sim -\alpha k g'(1)$$

for each $k \geqslant 1$ and $n \to \infty$, so that the boundary $g(t)$ is asymptotically linear. Therefore, by virtue of Theorem 3.7.1, for the events

$$A_{n,g} = \{S_{n-k} < x + g_{k,n} \quad \text{for all} \quad k = 1, \dots, n-1\}$$

we have

$$\mathbf{P}\big(A_{n,g};\; S_n \in \Delta[x - \Delta)\big) \sim f_{(n)}(x) I(\alpha, \beta, -\Delta)$$

for $\beta = -\alpha g'(1) > -\alpha$. Furthermore, if conditions (3.7.12) and **[g, 1]** are simultaneously satisfied then necessarily

$$\alpha = \alpha g(1) < \mathbf{E}\xi, \qquad \lambda(\alpha) < 0.$$

This means that the integral $I(\alpha, \beta, -\infty)$ is finite and the ratio

$$\frac{I(\alpha, \beta, -\Delta)}{I(\alpha, \beta, -\infty)}$$

can be made arbitrarily close to 1 by appropriate choice of sufficiently large Δ. However, the second term on the right-hand side of (3.7.35) does not exceed $\mathbf{P}(S_n < x - \Delta)$, and for $\alpha < \mathbf{E}\xi$ can be made arbitrarily small compared with the first term by choosing large Δ. This proves (3.7.34). The theorem is proved. □

If condition **[g, 1]** is not satisfied then the structure of the asymptotics of $\mathbf{P}(\nu_g > 1)$ becomes quite complicated.

The results of Chapter 4 imply the following statement about the rough (logarithmic) asymptotics of $\mathbf{P}(\nu_g > 1)$. Let $\mathbf{E}\xi = 0$, $g_- := \min_{t \in [0,1]} g(t) < 0$ and condition $[\mathbf{C}_0]$ be satisfied. Suppose that $h(t)$ is a curve that can be represented as a thread which stretches between the points $(0, 0)$ and $(1, g_-)$ and is a lower envelope line of the set of points (t, v) such that $v \geqslant g(t)$. Then

$$\ln \mathbf{P}(\nu_g > 1) \sim -n \int_0^1 \Lambda\big(\alpha h'(t)\big) dt, \qquad \alpha = \frac{x}{n}.$$

It is possible to find the exact asymptotics of $\mathbf{P}(\nu_g > 1)$ only for *moderately* large deviations, when $x = o(n)$ for $\mathbf{E}\xi = 0$, $\min_{t \in [0,1]} g(t) < 0$ and other rather restrictive assumptions are valid (see [132]).

3.8 Supplement. Boundary crossing problems in the multidimensional case

3.8.1 Introduction

Let ξ be a non-degenerate vector in $\mathbb{R}^d$, i.e. a vector satisfying the following condition: *there is no plane* $L = L(\lambda, c) := \big\{x : \langle \lambda, x \rangle = c\big\} \subset \mathbb{R}^d$ *such that*

$$\mathbf{P}(\xi \in L) = 1.$$

This condition will always be assumed to be met.

Suppose that random vectors $\{\xi_i\}_{i=1}^{\infty}$ are independent and have the same distribution as a vector ξ. Denote, as before, $S_0 = 0, S_n = \xi_1 + \cdots + \xi_n$. The most typical boundary crossing problems for multidimensional walks consist in the following.

Consider a fixed set $B \subseteq \mathbb{R}^d$, the closure of which does not include the origin, and let us study the distributions related to the time and place of the first entry of a trajectory $\{S_n\}_{n=1}^{\infty}$ into the set tB for $t \to \infty$. The first entry time into the set tB is denoted by

$$\eta = \eta(tB) := \min\{n \geqslant 1 : S_n \in tB\};$$

if $S_n \notin tB$ for all $n \geqslant 1$, we define $\eta(tB) = \infty$.

Define also the random vector

$$\chi = \chi(tB) := (1 - p)\xi_\eta,$$

where $p = \inf\{u \in (0, 1] : S_{\eta-1} + u\xi_\eta \in tB\}$. The random vector χ is called the *magnitude of the first entry into the set tB* or, as in the one-dimensional case, the *magnitude of the first overshoot over the boundary* $t\Gamma$ of the set tB. The position of the trajectory of S_n at the first entry time into tB is a vector $S_{\eta(tB)}$, which we represent as the sum

$$S_{\eta(tB)} = \theta(tB) + \chi(tB),$$

where $\theta(tB) = S_{\eta-1} + p\xi(\eta)$ is the point on the boundary $t\Gamma$ of the set tB at which the walk crossed the boundary $t\Gamma$ for the first time.

Thus, the triple

$$(\eta(tB), \theta(tB), \chi(tB)) \tag{3.8.1}$$

defines the moment, place and magnitude of the first entry into the set tB by the walk $\{S_n\}$. These random variables are not defined on the set $\eta = \infty$. There is some indeterminacy in the definition of the triple (3.8.1), which is due to the implied 'linear straightness' of the motion from the point $S_{\eta-1}$ to the point S_η. Instead of (η, θ, χ), one could consider the equivalent (in some sense) triple of variables $(\eta, \hat{\theta}, \hat{\xi})$, where $\hat{\theta} \equiv S_\eta$, $\hat{\xi} \equiv \xi_\eta = S_\eta - S_{\eta-1}$, the definition of which does not involve any indeterminacy. However, the statements of results for the triple (η, θ, χ) turn out to be somewhat simpler and clearer. For that reason, and also following the tradition in the theory of boundary crossing problems, we will use the triple (3.8.1). Note also that S_η and χ are essentially dependent. For example, if Γ is a hyperplane orthogonal to the vector N then $\langle S_\eta - x, \mathrm{N}\rangle = \langle \chi, \mathrm{N}\rangle$, where x is some point on the surface of $t\Gamma$.

Limit theorems for the distribution of the $(2d + 1)$-dimensional vector

$$(\eta, \theta, \chi)$$

and its components constitute the essence of boundary crossing problems for the multidimensional random walks considered in this section. They are multidimensional analogues of the problems considered in section 3.3. Of course, boundary crossing problems may have a somewhat different form, as, for example, does the well-known problem of the first time and place when a random walk $\{x + S_n\}_{n=0}^{\infty}$ crosses the boundary of the positive orthant $\mathbb{R}^{d+} = \{\alpha \in \mathbb{R}^d : \alpha_i \geqslant 0,\ i = 1, \dots, d\}$ for x from the interior $(\mathbb{R}^{d+})$ of the orthant $\mathbb{R}^{d+}$, $|x| \to \infty$ (see e.g. [28]).

Typically, we will leave aside the proper limiting distributions of the normalised (η, θ) in the case $a = \mathbf{E}\xi = 0$; these may be found by a diffusion approximation (there is a large number of papers in this field; see e.g. [103], [111], [113], [114], [176]).

Our main aim is to study the probabilities of large deviations, i.e. the asymptotics of probabilities of the form $\mathbf{P}(\eta \in A,\ \theta \in \Theta)$, when either the set Θ is a considerable distance from the ray $\{\alpha = au, u \geqslant 0\}$, or the set A is a considerable distance from the value $m = |x|/|a|$, where x is the crossing point of the boundary $t\Gamma$ of the set tB by the ray $\{\alpha = au, u \geqslant 0\}$ $(a \neq 0)$. In this case, those probabilities will tend to zero. In such problems, typically we will assume that Cramér's condition **[C]** is satisfied. For the sake of completeness, in some cases we will also study proper limiting distributions when the above-mentioned properties of the sets A and Θ are not satisfied, $|a| \neq 0$. In that case Cramér's condition will not be necessary.

This section contains the main results obtained in [53].

3.8.2 The time and place of first entry into the set tB when $a = \mathbf{E}\xi \neq 0$ and the ray $\{ua : u \geqslant 0\}$ intersects the set B

As before, we will denote by $[B]$ and (B) the closure and the interior of the set $B \subseteq \mathbb{R}^d$. The boundary of the set B is determined by the identity $\Gamma = [B] \setminus (B)$.

Below, the symbol e will denote vectors of unit length; say, for a vector $\alpha \neq 0$ let

$$\mathrm{e}(\alpha) = \frac{\alpha}{|\alpha|}.$$

Denote by $\Gamma(z) = z \inf\{u > 0 : uz \in \Gamma\}$ the point of first intersection of the boundary Γ and the ray $\{uz : u > 0\}$, if they do intersect. Then $|\Gamma(z)|$ is the distance from the origin to the set Γ along the ray $\{uz : u > 0\}$ (if this ray does not intersect the set Γ then let, by definition, $\big|\Gamma(z)\big| = \infty$). Clearly, $v = t\Gamma(z)$ is the point of intersection of this ray and the boundary $t\Gamma$ of the set tB, $\Gamma(cz) = \Gamma(z)$.

By the symbols c and C with or without indices, we will denote various constants, which may be different if they appear in different formulas. For $\varepsilon > 0$, by $(x)_\varepsilon$ and $(A)_\varepsilon$ we will denote the ε-neighbourhoods of a point $x \in \mathbb{R}^d$ and a set $A \subseteq \mathbb{R}^d$ respectively.

We will need the following condition on the smoothness of the boundary Γ of the set B in a neighbourhood of the point $\Gamma(z)$ (for $z = a$, in this section).

$[\mathbf{D}_1(\mathbf{z})]$ *The function $|\Gamma(\mathrm{e})| > 0$ in some neighbourhood of the point $\mathrm{e}(z)$ satisfies the Lipschitz condition*

$$\big|\,|\Gamma(\mathrm{e}')| - |\Gamma(\mathrm{e}'')|\,\big| \leqslant c|\mathrm{e}' - \mathrm{e}''|$$

for $\mathrm{e}', \mathrm{e}'' \in \big(\mathrm{e}(z)\big)_{\varepsilon}$, $\varepsilon > 0$, $c < \infty$.

Note that instead of $[\mathbf{D}_1(\mathbf{z})]$ one can consider the following closely related condition.

$[\mathbf{D}_1^*(\mathbf{z})]$ *In some neighbourhood of the point* $\Gamma(\mathrm{e}(z))$ *the surface* Γ *is continuous (the function* $\Gamma(\mathrm{e})$ *is continuous); almost everywhere in this neighbourhood the surface* Γ *is differentiable, i.e. there exists a matrix* $\Gamma'(\mathrm{e})$ *with elements*

$$\gamma_{i,j}(\mathrm{e}) = \frac{\partial \Gamma_i(\mathrm{e})}{\partial \mathrm{e}_j},$$

and

$$|\gamma_{i,j}(\mathrm{e})| \leqslant C, \quad \langle z, \mathrm{N}_{\Gamma(\mathrm{e})}\rangle \geqslant \delta, \tag{3.8.2}$$

where $C < \infty$, $\delta > 0$, $\mathrm{e} \in \big(\mathrm{e}(z)\big)_\varepsilon$, $\mathrm{N}_{\Gamma(\mathrm{e})}$ *is the unit normal vector to the surface* Γ *at the point* $\Gamma(\mathrm{e})$, *directed inward to the set* B; e_j, $j = 1, \dots, d$, *are the coordinates of the vector* e. *Moreover, for almost all points* e', e'' *from the neighbourhood* $\big(\mathrm{e}(z)\big)_\varepsilon$ *the function* $g(t) = \Gamma(\mathrm{e}' + t(\mathrm{e}'' - \mathrm{e}'))$ *is absolutely continuous on the segment* $[0, 1]$.

The condition $[\mathbf{D}_1^*(\mathbf{z})]$ implies condition $[\mathbf{D}_1(\mathbf{z})]$. Indeed for almost all e', e'' we have

$$\Gamma(\mathrm{e}'') - \Gamma(\mathrm{e}') = g(1) - g(0) = \int_0^1 g'(u)\,du.$$

Hence, by virtue of the first inequality in (3.8.2), for $\mathrm{e}', \mathrm{e}'' \in \big(\mathrm{e}(z)\big)_\varepsilon$ the following inequality is true:

$$|\Gamma(\mathrm{e}'') - \Gamma(\mathrm{e}')| \leqslant C|\mathrm{e}'' - \mathrm{e}'|.$$

Together with the second inequality from (3.8.2), it implies condition $[\mathbf{D}_1(\mathbf{z})]$.

Denote by L the plane

$$L = L(a, 0) \equiv \{\alpha : \langle \alpha, a\rangle = 0\},$$

which is orthogonal to the vector a, and by $\mathcal{P}^L$ the projector on to L, so that

$$z \equiv \mathcal{P}^L z + \langle \mathrm{e}, z\rangle \mathrm{e} \quad \text{for} \quad \mathrm{e} = \mathrm{e}(a)$$

for any $z \in \mathbb{R}^d$.

For two random vectors $b^{(1)}$ and $b^{(2)}$, defined in general on different probability spaces, we will write

$$b^{(1)} \underset{d}{=} b^{(2)},$$

if the distributions of these vectors are equal.

Denote by ζ a Gaussian random vector with zero mean and covariance matrix M_ξ equal to the covariance matrix of the vector ξ.

Theorem 3.8.1. *Let* $\mathbf{E}|\xi|^2 < \infty$, $a = \mathbf{E}\xi \neq 0$ *and condition* $[\mathbf{D}_1(\mathbf{a})]$ *be satisfied. Then the following representation holds true:*

$$(\eta(tB), \theta(tB)) \underset{d}{=} (\eta^*, \theta^*),$$

$$\eta^* = \frac{t}{|a|}\left|\Gamma\left(a + \frac{1}{\sqrt{m}}\mathcal{P}^L\zeta\right)\right| - \frac{\sqrt{m}}{|a|}\langle \mathrm{e}(a), \zeta\rangle + \varepsilon_1(t)\sqrt{t}, \tag{3.8.3}$$

$$\theta^* = t\Gamma\left(a + \frac{1}{\sqrt{m}}\mathcal{P}^L\zeta\right) + \varepsilon_2(t)\sqrt{t} \in t\Gamma,$$

where $m = t|\Gamma(a)|/|a|$ *and, as* $t \to \infty$,

$$\left|\varepsilon_i(t)\right| \underset{P}{\to} 0, \quad i = 1, 2.$$

The first term in the second line of (3.8.3) determines the time spent on deterministic straight-line movement with speed $|a|$ from the origin to the point $t\Gamma\big(a + \frac{1}{\sqrt{m}}\mathcal{P}^L\zeta\big)$. The second term reflects the influence of the dispersion caused by the perturbation of this motion by a motion with random jumps $\langle \mathrm{e}(a), \xi\rangle$ along the same direction. The value $m = t|\Gamma(a)|/|a|$ is the time spent on deterministic movement from 0 to the point $t\Gamma(a)$.

Along with the condition $[\mathbf{D}_1(\mathbf{z})]$, we consider the condition

$[\mathbf{D}_{1+}(\mathbf{z})]$ *The boundary* Γ *in a neighbourhood of the point* $\Gamma(\mathrm{e}(z))$ *is continuously differentiable. By this we mean that the function* $\Gamma(\mathrm{e})$ *in a neighbourhood of the point* $\mathrm{e}(z)$ *is continuously differentiable.*

If the condition $[\mathbf{D}_{1+}(\mathbf{z})]$ is satisfied then in some neighbourhood of the point $\Gamma(\mathrm{e}(z))$ the unit normal vector $\mathrm{N}_{\Gamma(\mathrm{e}(z))}$ is well defined (we will assume it is directed inward the set B) and conditions (3.8.2) are satisfied, so condition $[\mathbf{D}_{1+}(\mathbf{z})]$ implies that conditions $[\mathbf{D}_1^*(\mathbf{z})]$and $[\mathbf{D}_1(\mathbf{z})]$ are met.

Suppose that condition $[\mathbf{D}_{1+}(\mathbf{a})]$ is met. For brevity, let $\mathrm{N}_{\Gamma(a)} = \mathrm{N}$. If the vector N is collinear with a then, for fixed z and $t \to \infty$,

$$t\Gamma\left(a + \frac{1}{\sqrt{m}}z\right) = t\Gamma(a) + o(\sqrt{m}),$$

and, as a corollary from Theorem 3.8.1, we obtain the well-known 'one-dimensional' result

$$\eta(tB) \underset{D}{=} m - \frac{\sqrt{m}}{|a|}\langle \mathrm{e}(a), \zeta\rangle + o_p(\sqrt{m}).$$

Consider now the general case when the normal vector N is not necessarily collinear with a. The tangent plane T to the surface Γ at the point $\Gamma(a)$ is defined by

$$T = \{\alpha : \langle \mathrm{N}, \alpha - \Gamma(a)\rangle = 0\}.$$

Consider the point

$$T(z) = zu_0, \quad u_0 = \inf\{u : uz \in T\},$$

of intersection of the ray $\{\alpha = uz : u > 0\}$ and the surface T. It is easy to find that

$$T(z) = z\frac{\langle \mathrm{N}, \Gamma(a)\rangle}{\langle \mathrm{N}, z\rangle},$$

$$T\left(a + \frac{1}{\sqrt{m}}z\right) = \Gamma(a) + \frac{z}{\sqrt{m}}\frac{m}{t} - \frac{\Gamma(a)}{\sqrt{m}}\frac{\langle \mathrm{N}, z\rangle}{\langle \mathrm{N}, a\rangle}.$$

Since for $m \to \infty$

$$\Gamma\left(a + \frac{1}{\sqrt{m}}\mathcal{P}^L\zeta\right) = T\left(a + \frac{1}{\sqrt{m}}\mathcal{P}^L\zeta\right) + o_p\left(\frac{1}{\sqrt{m}}\right),$$

we obtain the following statement.

For random vectors $b^{(t)}$ depending on t, we will write

$$b^{(t)} \Longrightarrow b$$

as $t \to \infty$, if the distributions of $b^{(t)}$ weakly converge to the distribution of the vector b.

Corollary 3.8.2. *Suppose that condition* $[\mathbf{D}_{1+}(\mathbf{a})]$ *is met, where* $a = \mathbf{E}\xi \neq 0$. *Then, as* $t \to \infty$,

$$\left(\frac{\eta(tB) - m}{\sqrt{m}}, \frac{\theta(tB) - x}{\sqrt{m}}\right) \Longrightarrow \left(-\frac{1}{|a|}\frac{\langle \mathrm{N}, \zeta\rangle}{\langle \mathrm{N}, \mathrm{e}(a)\rangle}, \ \zeta - \mathrm{e}(a)\frac{\langle \mathrm{N}, \zeta\rangle}{\langle \mathrm{N}, \mathrm{e}(a)\rangle}\right), \quad (3.8.4)$$

where $x = t\Gamma(a)$.

Thus, under the condition $[\mathbf{D}_{1+}(\mathbf{a})]$, the joint distribution of η and θ is *asymptotically normal*. If the condition $[\mathbf{D}_{1+}(\mathbf{a})]$ is not satisfied, this may be not true in general.

Now consider in detail the vector $(\theta(tB) - x)/\sqrt{m}$ on the left-hand side of (3.8.4). If the boundary Γ is twice continuously differentiable in a neighbourhood of the point $\Gamma(a)$, then this vector has an explicit orthogonal decomposition into two components, one of which lies in the plane $L(\mathrm{N}, 0) = T - \Gamma(a)$ and the other along the vector N. Obviously, for the component

$$\frac{\mathcal{P}^{L(\mathrm{N},0)}(\theta(tB) - x)}{\sqrt{m}}$$

lying in $L(\mathrm{N}, 0)$, the statement (3.8.4) obtained in Corollary 3.8.2 will remain true:

$$\frac{\mathcal{P}^{L(\mathrm{N},0)}(\theta(tB) - x)}{\sqrt{m}} \Longrightarrow \zeta - \mathrm{e}(a)\frac{\langle \mathrm{N}, \zeta\rangle}{\langle \mathrm{N}, \mathrm{e}(a)\rangle}.$$

In order to find the second component, note that, by virtue of (3.8.3),

$$\theta \underset{D}{=} t\Gamma\left(a + \frac{1}{\sqrt{m}}\mathcal{P}^L\zeta\right) + o_p(1)\sqrt{t} \in t\Gamma.$$

If the boundary Γ is twice continuously differentiable in a neighbourhood of the point $\Gamma(a)$ and for $z \in \mathbb{R}^d$ the vector $t\Gamma(a+\frac{1}{\sqrt{t}}\mathcal{P}^L z)+\sqrt{t}o_p(1)$ lies on the boundary $t\Gamma$, then

$$t\Gamma\left(a+\frac{1}{\sqrt{t}}\mathcal{P}^L z\right)+\sqrt{t}o_p(1)=t\Gamma\left(a+\frac{1}{\sqrt{t}}\mathcal{P}^L(z+o_p(1)\right)$$
$$=t\Gamma(a)+\sqrt{t}(1+o_p(1))(z-z^{(1)})+(1+o_p(1))\mathrm{N}(z-z^{(1)})\frac{R}{2}(z-z^{(1)})^T,$$

where

$$z^{(1)}=\mathrm{e}(a)\frac{\langle \mathrm{N},z\rangle}{\langle \mathrm{N},\mathrm{e}(a)\rangle},$$

R is the curvature matrix of the surface Γ at the point $\Gamma(a)$ and N is the unit normal vector to the surface Γ at the point $\Gamma(a)$. Since the vector $z-z^{(1)}$ is orthogonal to the vector N, for $x=t\Gamma(a)$, $t\Gamma(a+\frac{1}{\sqrt{t}}\mathcal{P}^L z)+\sqrt{t}o(1)\in t\Gamma$ the following equality is true:

$$\left\langle \mathrm{N},\left(t\Gamma(a+\frac{1}{\sqrt{t}}\mathcal{P}^L z)+\sqrt{t}o(1)-x\right)\right\rangle=(1+o(1))\mathrm{N}(z-z^{(1)})\frac{R}{2}(z-z^{(1)})^T.$$

Thus, we have proved

Corollary 3.8.3. *Suppose that condition* $[\mathbf{D}_{1+}(\mathbf{a})]$ *is met, where* $a=\mathbf{E}\xi\neq 0$, *and also that the surface* Γ *is twice continuously differentiable in some neighbourhood of the point* $\Gamma(a)$. *Then, as* $t\to\infty$,

$$\left(\frac{\eta(tB)-m}{\sqrt{m}},\ \frac{\mathcal{P}^{L(\mathrm{N},0)}(\theta(tB)-x)}{\sqrt{m}},\ \langle \mathrm{N},\theta(tB)-x\rangle\right)$$
$$\Longrightarrow\left(-\frac{1}{|a|}\langle \mathrm{e}(a),\zeta^{(1)}\rangle,\ \zeta-\zeta^{(1)},\ (\zeta-\zeta^{(1)})\frac{|\Gamma(a)|R}{2|a|}(\zeta-\zeta^{(1)})^T\right),$$

where $\zeta^{(1)}=\mathrm{e}(a)\langle \mathrm{N},\zeta\rangle/\langle \mathrm{N},\mathrm{e}(a)\rangle$.

Now we provide an estimate of the tail of the distribution of the length $|\chi|$ of the vector of first entry $\chi(tB)$ of the walk S_n into the set tB.

For a measurable set $W\subset\mathbb{R}^d$ define the renewal measure (function)

$$H(W)=\sum_{n=0}^{\infty}\mathbf{P}(S_n\in W),$$

where $S(0)=0$. Let

$$\tau(W)=\sum_{n=0}^{\infty}I_{\{S_n\in W\}}$$

denote the time spent by the walk S_n inside the set W. Obviously,

$$H(W)=\mathbf{E}\tau(W).$$

Denote, for $k = 1, 2, \ldots$,

$$A_k(W) = (W)_k \setminus (W)_{k-1},$$

where, as before, for $\varepsilon > 0$, the set $(W)_\varepsilon$ is the ε-neighbourhood of the set W, $(W)_0 = W$. Let

$$h(W) = \sup_{k \geqslant 1} H(A_k(W))$$

and denote by

$$\pi(u) = \int_u^\infty \mathbf{P}(|\xi| > v)dv$$

the 'double tail' of the distribution of the random variable $|\xi|$.

For $z \in \mathbb{R}^d$, $\varepsilon > 0$, introduce the cone

$$C(\varepsilon, z) = z + \{v \in \mathbb{R}^d : |\mathrm{e}(v) - \mathrm{e}(a)| \leqslant \varepsilon\}$$

with vertex at the point z and with 'angle' 2ε at the ray $\{v = z + ta : \ t \geqslant 0\}$.

Theorem 3.8.4. (i) *For any* $u \geqslant 0$,

$$\mathbf{P}(|\chi(tB)| > u;\ \eta(tB) < \infty) \leqslant h(tB)\pi(u). \tag{3.8.5}$$

If the boundary of the set B *is a plane* $L(\mathrm{N}, c) := \{v \in \mathbb{R}^d : \langle \mathrm{N}, v\rangle = c\}$, *and* $\langle \mathrm{N}, a\rangle > 0$, *then on the right-hand side of* (3.8.5) *the function* $\pi(u)$ *can be replaced by* $\pi_N(u) = \int_u^\infty \mathbf{P}(|\xi| > v, \langle \mathrm{N}, \xi\rangle > 0)dv$.

(ii) *Suppose the set* B *is such that, for some* $\varepsilon > 0$ *and any* $v \in B$,

$$C(\varepsilon, v) \subseteq B. \tag{3.8.6}$$

Then, for any $t \geqslant 0$,

$$\mathbf{P}(\eta(tB) < \infty) = 1, \quad h(tB) \leqslant \bar{h} := H(\overline{C}(\varepsilon, \Gamma(a))) < \infty, \tag{3.8.7}$$

where $\overline{C}(\cdot)$ *is the complementary set of* $C(\cdot)$. *Thus, if* (3.8.5) *is satisfied, for all* $u \geqslant 0$, $t \geqslant 0$ *we have*

$$\mathbf{P}(|\chi(tB)| > u) \leqslant \bar{h}\pi(u). \tag{3.8.8}$$

Suppose that the unit vector N is such that $\langle \mathrm{N}, a\rangle > 0$. Then it is obvious that the set $B = \{v \in \mathbb{R}^d : \langle \mathrm{N}, v\rangle \geqslant c\}$, bounded by the plane $L(\mathrm{N}, c)$, satisfies condition (3.8.6).

Note that in part (ii) of Theorem 3.8.4 we consider sets B which satisfy some condition 'at all points of the boundary' Γ. If we considered only conditions 'on some points of the boundary' Γ then it would be not be possible to obtain an equality that is uniform in $t > 0$,

$$\mathbf{P}(|\chi(tB)| > u) \leqslant c_1\pi(u), \quad c_1 < \infty,$$

and valid for all $u > 0$. One can see this by considering the set $B = \{\alpha : \alpha_{(1)} \geqslant 1, \ \alpha_{(2)} \geqslant -v\}$ in $\mathbb{R}^2$, $v > 0$. If $a = \mathbf{E}\xi = (1,0)$ then, obviously, with positive probability $p(t)$, the first entry into the set tB happens through the horizontal segment of the boundary. Since the vector ξ has zero mean in the direction orthogonal to a, it is possible to estimate the tail of the distribution of the overshoot over the horizontal boundary by using only the 'the triple tail of the distribution' (see [129]),

$$\bar{\pi}(u) = \int_u^\infty \pi(v)\,dv.$$

If one does not need to obtain an inequality which is uniform in $t > 0$, then it is possible to use only the following conditions on the boundary in a neighbourhood of the point $\Gamma(a)$.

Corollary 3.8.5. *Suppose that the set B satisfies the conditions of Theorem* 3.8.1. *Then, for any fixed $u \geqslant 0$ and some $\varepsilon > 0$, the following inequality holds true:*

$$\limsup_{t\to\infty} \mathbf{P}(|\chi(tB)| > u) \leqslant \bar{h}\pi(u),$$

where $\bar{h} = H(\overline{C}(\varepsilon, \Gamma(a))) < \infty$.

If the boundary Γ in a neighbourhood of the point $\Gamma(a)$ is continuously differentiable then the statement of Theorem 3.8.1 can be extended by a statement about the existence of a limiting distribution for the magnitude of the vector $\chi(tB)$ corresponding to the first entry into the set tB.

We will need the following notation. As before, let N be the unit normal vector to the boundary Γ at the point $\Gamma(a)$. For $u \geqslant 0$, define the probability

$$p(u) := \mathbf{P}(\inf_{k\geqslant 0} \langle \mathrm{N}, S_k\rangle \geqslant -u),$$

which is positive for all $u \geqslant 0$ if $\langle \mathrm{N}, a\rangle > 0$ (see e.g. [22]). Introduce the distributions

$$F(u, dw) = \mathbf{P}\left(\left(1 - \frac{u}{\langle \mathrm{N}, \xi\rangle}\right)\xi \in dw \,|\, \langle \mathrm{N}, \xi\rangle > u\right).$$

In essence, these are the distributions of the vector $\chi(W)$ of the entry into the set $W = \Pi(\mathrm{N}, u)$ of the walk S_n, $n = 0, 1, \dots,$ under the condition that the entry occurs on the first step. Here, as before, $\Pi(\mathrm{N}, u) = \{v \in \mathbb{R}^d : \langle \mathrm{N}, v\rangle > u\}$ is the open half-space bounded by the plane having normal N.

Theorem 3.8.6. *Let $a = \mathbf{E}\xi \neq 0$, let condition* $[\mathbf{D}_{1+}(\mathbf{a})]$ *be satisfied and let the random variable $\xi^{\langle v\rangle} := \langle v, \xi\rangle$ in some neighbourhood of the point* N *be non-lattice. Then the weak convergence of distributions takes place as $t \to \infty$:*

$$\mathbf{P}(\chi(tB) \in dw) \Longrightarrow \frac{1}{\langle \mathrm{N}, a\rangle} \int_{u=0}^\infty p(u) F(u, dw)\,du. \tag{3.8.9}$$

Corollary 3.8.7. *Under the conditions of Theorem 3.8.6, the following inequality holds true:*

$$\lim_{t\to\infty} \mathbf{P}\Big(\langle \mathrm{N}, \chi(tB)\rangle > s\Big) = \frac{1}{\mathbf{E}\eta_+^{(\mathrm{N})}} \int_s^\infty \mathbf{P}\big(\eta_+^{(\mathrm{N})} > t\big)\, dt,$$

where $\eta_+^{(\mathrm{N})}$ is the first positive sum in the walk $\langle \mathrm{N}, S_n\rangle$, $n = 0, 1, \ldots$

In other words, the weak limit of the distribution $\mathbf{P}(\langle \mathrm{N}, \chi(tB)\rangle \in dw)$ coincides with the distribution of the overshoot over an infinitely distant barrier by the one-dimensional random walk $\langle \mathrm{N}, S_n\rangle$, $n = 0, 1, \ldots$ (see e.g. [92]).

From Corollaries 3.8.2 and 3.8.3 and Theorem 3.8.6, we have the following.

Corollary 3.8.8. *Under the conditions of Theorem 3.8.6, as $t \to \infty$,*

$$\left(\frac{\eta(tB) - m}{\sqrt{m}},\ \frac{(\theta(tB) - x)}{\sqrt{m}},\ \chi(tB)\right) \Longrightarrow \left(-\frac{1}{|a|}\langle \mathrm{e}(a),\ \zeta^{(1)}\rangle,\ \zeta - \zeta^{(1)},\ \hat{\chi}\right),$$

where $\zeta^{(1)} = a\langle \mathrm{N}, \zeta\rangle / \langle \mathrm{N}, a\rangle$, the random vector $\hat{\chi}$ has the distribution from the right-hand side of (3.8.9), *which does not depend on the Gaussian vector ζ. If, moreover, the surface Γ is twice differentiable in some neighbourhood of the point $\Gamma(a)$ then the above statement can be supplemented by the decomposition of $\theta(tB) - x$ for $x = t\Gamma(a)$ in Corollary 3.8.3: as $t \to \infty$,*

$$\left(\frac{\eta(tB) - m}{\sqrt{m}},\ \frac{\mathcal{P}^{L(\mathrm{N})}(\theta(tB) - x)}{\sqrt{m}},\ \langle \mathrm{N}, \theta(tB) - x\rangle,\ \chi(tB)\right)$$
$$\Longrightarrow \left(-\frac{1}{|a|}\langle \mathrm{e}(a), \zeta^{(1)}\rangle,\ \zeta - \zeta^{(1)},\ (\zeta - \zeta^{(1)})\frac{|\Gamma(a)|R}{2|a|}(\zeta - \zeta^{(1)})^T,\ \hat{\chi}\right).$$

3.8.3 Local limit theorems for η, θ, χ under arbitrary mutual location of the vector $a = \mathbf{E}\xi$ and the set B

In this section we will assume that Cramér's condition **[C]** holds true. As before, let $A(\lambda) = \ln \psi(\lambda)$,

$$\mathcal{A} = \{\lambda : \psi(\lambda) < \infty\}, \qquad \mathcal{A}' = \{A'(\lambda) : \lambda \in \mathcal{A}\},$$

$$\mathbf{F}^{(\lambda)}(U) = \frac{\mathbf{E}(e^{\langle\lambda,\xi\rangle};\ \xi \in U)}{\psi(\lambda)}, \qquad {}^{\alpha}\mathbf{F}(U) = \mathbf{F}^{(\lambda(\alpha))}(U),$$

and let ${}^{\alpha}\xi$ be a random vector with distribution ${}^{\alpha}\mathbf{F}$. For $\alpha \in \mathcal{A}'$, the following relations hold true:

$$\mathbf{E}{}^{\alpha}\xi = \alpha, \qquad M(\alpha) := \mathbf{E}({}^{\alpha}\xi - \alpha)^T({}^{\alpha}\xi - \alpha) = \big(\Lambda''(\alpha)\big)^{-1}.$$

To formulate the main statements, we will need additional notation. Let x be some point of the boundary $t\Gamma$ of the set tB. Below we will study the asymptotics of the probability

$$\mathbf{P}(\eta(tB) = n,\ S_n \in x + W)$$

assuming that the surface Γ satisfies the condition $[\mathbf{D}_{1+}(\boldsymbol{\alpha})]$, $\alpha = x/n \in \mathcal{A}'$.

Let $\mathrm{N} = \mathrm{N}_{\Gamma(\alpha)}$ be the unit normal vector to the surface Γ at the point $\Gamma(\alpha) \in \Gamma$ (or to the surface $t\Gamma$ at the point x; note that $x/t \in \Gamma$, and, hence, we have $x/t \to \Gamma(\mathrm{e}_0) = \text{const.}$ as $\mathrm{e}(\alpha) \to \mathrm{e}_0$). Recall that in $\mathbb{R}^d$ we define the subspace

$$\Pi(\mathrm{N}) = \Pi(\mathrm{N}, 0) \equiv \{v : \langle \mathrm{N}, v \rangle \leqslant 0\},$$

which is located 'below' the tangent plane $L(\mathrm{N}) = L(\mathrm{N}, 0) := \{v \in \mathbb{R}^d : \langle \mathrm{N}, v \rangle = 0\}$ to the surface $t\Gamma - x$ at the point 0. 'Above' $L(\mathrm{N})$ lies the open half-space $\overline{\Pi}(\mathrm{N}) = \{v : \langle \mathrm{N}, v \rangle > 0\}$.

For $\alpha \in \mathcal{A}'$, introduce the function

$$p_\alpha = p_\alpha(z) \equiv \mathbf{P}(\inf_{n \geqslant 1} \langle \mathrm{N}, {}^\alpha S_n \rangle \geqslant \langle \mathrm{N}, z \rangle),$$

where $\mathrm{N} = \mathrm{N}_{\Gamma(\alpha)}$, ${}^\alpha S_n = {}^\alpha\xi_1 + \cdots + {}^\alpha\xi_n$, and the independent terms ${}^\alpha\xi_i$ have the common distribution ${}^\alpha\mathbf{F}$. It is known (see e.g. [39]) that if $\langle \mathrm{N}, \alpha \rangle > 0$ then the value $p_\alpha(z)$ is positive for all $z \in \Pi(\mathrm{N})$.

For $\alpha \in \mathcal{A}'$, define the σ-finite measure $Q(\alpha) = Q(\alpha, W)$ with support in $\overline{\Pi}(\mathrm{N})$, assuming, for a Borel set $W \subseteq \overline{\Pi}(\mathrm{N})$, that

$$Q(\alpha, W) = \frac{1}{\psi(\lambda(\alpha))} \int_{z \in \Pi(\mathrm{N})} e^{-\langle \lambda(\alpha), z \rangle} p_\alpha(z) \mathbf{P}(z + \xi \in W)\, dz. \qquad (3.8.10)$$

One can show (see [53]) that for $W \subseteq \overline{\Pi}(\mathrm{N})$ we have

$$Q(\alpha, W) \equiv \int_W e^{-\langle \lambda(\alpha), w \rangle} q_\alpha(w)\, dw, \quad q_\alpha(w) = \int_{\Pi(\mathrm{N})} p_\alpha(v) \mathbf{P}(w - {}^\alpha\xi \in dv), \qquad (3.8.11)$$

so that the measure $Q(\alpha)$ is absolutely continuous with respect to the Lebesgue measure in $\mathbb{R}^d$.

We will say that a vector ξ is non-lattice if all its coordinates are non-lattice.

Theorem 3.8.9. *Let ξ be a non-lattice vector and condition* **[C]** *be satisfied. Suppose that for a fixed non-zero vector $\alpha_0 \in \mathcal{A}'$ the set B satisfies the condition* $[\mathbf{D}_{0+}(\boldsymbol{\alpha}_0)]$, *$x \in t\Gamma$, and that*

$$\alpha = \frac{x}{n} \to \alpha_0$$

as $n \to \infty$. Then for any $\Delta_0 > 0$, $C < \infty$, it is possible to choose a subsequence $\Delta_n \to 0$ such that

$$\mathbf{P}\big(\eta(tB) = n,\ S_n \in x + \Delta[y)\big) = \frac{Q(\alpha, \Delta[y))}{(2\pi n)^{d/2}\sigma(\alpha)} e^{-n\Lambda(\alpha)}(1 + \varepsilon_n), \qquad (3.8.12)$$

where $\sigma^2(\alpha) = \det M(\alpha)$,

$$\lim_{n \to \infty} \sup |\varepsilon_n| = 0,$$

and the supremum is taken over all $\Delta > 0$, $\Delta_n \leqslant \Delta \leqslant \Delta_0$, $y \in \mathbb{R}^d$, $|y| \leqslant C$.

One can obtain several corollaries from Theorem 3.8.9. If W is an arbitrary set in $\overline{\Pi}(\mathrm{N})$ from the class of sets with a sufficiently 'thin' boundary then approximating

W by unions of cubes, it is possible to obtain a uniform (over this class) variant of statement (3.8.12). For that, denote by $\mathcal{F}(c)$ the class of measurable sets W lying in the ball $(0)_c$ of radius c, for which, for any $\varepsilon \in (0, 1)$, the following inequality holds true:

$$\mu((\partial W)_\varepsilon) \leqslant c\varepsilon,$$

where ∂W is the boundary of W, $\mu(.)$ in the Lebesgue measure in $\mathbb{R}^d$. Any bounded set with a smooth boundary belongs to the class $\mathcal{F}(c)$ for an appropriate c.

Corollary 3.8.10. *Suppose that the conditions of Theorem 3.8.9 are satisfied. Then there exists $\delta > 0$ such that, for $\alpha = x/n$, $x \in t\Gamma$, and any $c < \infty$,*

$$\mathbf{P}\big(\eta(t) = n,\ S_n \in x + W\big) = \frac{1}{(2\pi n^{d/2})\sigma(\alpha)} e^{-n\Lambda(\alpha)}\Big(Q(\alpha, W) + \varepsilon_n\Big), \quad (3.8.13)$$

where

$$\lim_{n\to\infty} \sup |\varepsilon_n| = 0,$$

and the sup *is taken over the class $|\alpha - \alpha_0| \leqslant \delta$, $W \in \mathcal{F}(c)$.*

Let us proceed with studying the asymptotics of the probability of the event

$$\{\eta(tB) = n\} \quad (3.8.14)$$

(this probability always tends to 0 as $n \to \infty$) without fixing the location of S_n in the case when $s := t/n \to s_0 > 0$ as $n \to \infty$. One can see that the main contribution to the probability of this event is made by the trajectories which (after being shrunk n times) cross the boundary $s\Gamma$ of the set sB for the first time in a neighbourhood of some non-random point $\alpha_0 = \alpha(s_0) \in s_0\Gamma$. This point (we will call it the *most probable point* in the set s_0B) is defined as follows.

Denote

$$\Lambda(U) = \inf_{v\in U} \Lambda(v), \quad U \subseteq \mathbb{R}^d.$$

If one takes the set sB as the set U then it is not hard to see that, for the set U, the infimum mentioned above will be attained at the boundary $s\Gamma$. Further, define the point $\alpha(s)$, on the boundary $s\Gamma$ of the set $U = sB$, where the minimum of $\Lambda(v)$ is attained,

$$\Lambda\big(\alpha(s)\big) = \Lambda(sB). \quad (3.8.15)$$

The point $\alpha(t/n)$ is the most probable point in the set $(t/n)B$. To study the asymptotics of the probability $\mathbf{P}\big(\eta(tB) = n\big)$, we will need the following additional assumption on the boundary of the set B.

[$\mathbf{D}_2$(s)] *The vector $\alpha(s)$ is unique in a neighbourhood of the point $\gamma := \Gamma(\mathrm{e}(\alpha(s)))$, and the boundary Γ is twice continuously differentiable (the function $\Gamma(\mathrm{e})$ is twice continuously differentiable).*

If the boundary Γ at the point γ is twice continuously differentiable then, as we have already noted, at this point it is possible to define not only the unit normal vector $\mathrm{N} = \mathrm{N}_\gamma$ but also the curvature matrix $R = R_\gamma$. Under the condition **[D$_2$(s)]**, the point $\alpha(s) \in s\Gamma$ is the unit tangency point of the surfaces $s\Gamma$ and $\{v : \Lambda(v) = \Lambda(\alpha(s))\}$. Recall that the unit vector N is directed into the set B; the vector $\lambda(\alpha(s))$ is directed outward from the set $\{v : \Lambda(v) \leqslant \Lambda(\alpha(s))\}$. Therefore, the unit normals to these surfaces at the point $\alpha(s)$ coincide:

$$\mathrm{N}_{\Gamma(\alpha(s))} = \mathrm{e}(\lambda(\alpha(s))).$$

Introduce the integral

$$I(s) := \int_{L(\mathrm{N},0)} l(v)\mu'(dv), \quad l(v) = l(v,s) := \exp e^{-v\frac{s\Lambda''(\alpha)+|\lambda(\alpha)|R}{2}v^T}, \tag{3.8.16}$$

where $\mu'(dv)$ is the Lebesgue measure in the plane $L(\mathrm{N},0)$, $R = R_\gamma$, $\mathrm{N} = \mathrm{N}_\gamma$, $\gamma = \Gamma(\alpha)$, $\alpha = \alpha(s)$.

Theorem 3.8.11. *Suppose that the vector ξ is non-lattice, condition* **[C]** *is met and for some $s_0 > 0$ condition* **[D$_2$(s$_0$)]** *holds for the set B, and, moreover, that $\alpha(s_0) \in \mathcal{A}'$, $I(s_0) < \infty$. Then, for $s = t/n \to s_0$,*

$$\mathbf{P}(\eta(tB) = n) = \frac{p(s)}{\sqrt{n}} e^{-n\Lambda(\alpha(s))}(1 + o(1)), \tag{3.8.17}$$

where $p(s)$ is a known continuous function. Also, weak convergence takes place as $n \to \infty$:

$$\mathbf{P}\left(\frac{\mathcal{P}^L(S_n - x)}{\sqrt{t}} \in dv \mid \eta(tB) = n\right) \Longrightarrow l(v, s_0)\mu'(dv), \tag{3.8.18}$$

where the function $l(v,s)$ was defined in (3.8.16), *$v \in L(\mathrm{N})$ and $\mu'(dv)$ is the Lebesgue measure on $L(\mathrm{N})$.*

3.8.4 Integral limit theorems for the time of reaching $\eta(tB)$

Theorems 3.8.9 and 3.8.11 allow us to find the asymptotics of the events

$$\{\eta(tB) \leqslant n\}, \quad \{n \leqslant \eta(tB) < \infty\} \tag{3.8.19}$$

for integer $n = n(t)$ such that $n/t \to u$, $0 \leqslant u < \infty$, in the case when these probabilities tend to 0 (the situation when they do not tend to 0 is described in Theorem 3.8.1).

First we provide a theorem about the *logarithmic asymptotics* of such probabilities, which was obtained in [50].

Suppose $n/t \to u > 0$. Under wide assumptions (see Theorem 3.8.11 and, for example, [50]), the following relations are true:

$$\ln \mathbf{P}\big(\eta(tB) = n\big) \sim \ln \mathbf{P}\big(S_n \in tB\big) \sim -n\Lambda\left(\frac{t}{n}B\right) \sim -tu\Lambda\left(\frac{1}{u}B\right). \tag{3.8.20}$$

From (3.8.20) it follows that the logarithmic asymptotics of the events (3.8.19) are determined by the functions

$$D_u(B) = \inf_{\alpha\in B} D_u(\alpha), \quad D_u(\alpha) = u\Lambda\left(\frac{\alpha}{u}\right).$$

Theorem 3.8.12 ([50]). *Let $0 \leqslant b < \infty$ be a fixed number and, for a measurable bounded set B, the numbers $u_+ \in [b,\infty)$, $u_- \in [0,b]$, $u_\pm = u_\pm(b)$ be such that*

$$\inf_{u\leqslant b} D_u(B) = D_{u_-}(B), \quad \inf_{u\geqslant b} D_u(B) = D_{u_+}(B).$$

Suppose the condition

$$D_{u_\pm}([B]) = D_{u_\pm}((B))$$

holds true, and the function $D_u(B)$ is continuous in u at the points $u = u_\pm$. Then for integer $n = n(t) \geqslant 2$, $n/t \to b$, the following identities hold:

$$\lim_{t\to\infty} \frac{1}{t} \ln \mathbf{P}\big(n \leqslant \eta(tB) < \infty\big) = -D_{u_+}(B),$$

$$\lim_{t\to\infty} \frac{1}{t} \ln \mathbf{P}\big(\eta(tB) < n\big) = -D_{u_-}(B).$$

From Theorem 3.8.12 it follows that, in particular, a crucial role in the description of the asymptotics of the probability

$$\mathbf{P}\big(\eta(tB) < \infty\big)$$

is played by the *second deviation function*

$$D(\alpha) = \inf_{u>0} D_u(\alpha),$$

which was introduced and studied in the paper [50] (see also section 2.9). The logarithmic asymptotics of this probability has the form

$$\lim_{t\to\infty} \frac{1}{t} \ln \mathbf{P}\big(\eta(tB) < \infty\big) = -D(B),$$

where $D(B) = \inf_{\alpha\in B} D(\alpha)$.

Next we provide theorems about the exact asymptotics of the probabilities of the events (3.8.19). Let u^* be the value of the number u at which the infimum of the function $D_u(B)$ over $u > 0$ is attained:

$$D_{u^*}(B) = \inf_{u>0} D_u(B) \equiv D(B).$$

If $u^* > 0$ then

$$D(B) = D_{u^*}(B) = u^*\Lambda\left(\alpha\left(\frac{1}{u^*}\right)\right),$$

where, according to (3.8.15), the point $\alpha(1/u^*) \in (1/u^*)\Gamma$ is the most probable point of the set $(1/u^*)B$. As was established in [50], if $\alpha(1/u^*) \in \mathcal{A}', I(1/u^*) < \infty$ then the minimum point $u^* > 0$ is unique and

$$\sigma^2 := D''_{u^*}(B) = \frac{\partial^2}{\partial u^2} D_u(B)\bigg|_{u=u^*} > 0. \tag{3.8.21}$$

As one can see from Theorem 3.8.12, an important circumstance determining the asymptotics of the probabilities of the events (3.8.19) is the location of the number $b = \lim_{t\to\infty} n/t$ relative to u^*.

Theorem 3.8.13. *Suppose that the vector ξ is non-lattice and condition* **[C]** *is met.*

(i) *Let $n/t \to b \leqslant u^*$, $(u^* - n/t)\sqrt{t} \to \infty$ and suppose that, for the set B condition* $[\mathbf{D}_2(1/b)]$ *is met and that $\alpha(1/b) \in \mathcal{A}'$, $I(1/b) < \infty$. Then*

$$\mathbf{P}\big(\eta(tB) \leqslant n\big) = \frac{c_1}{(1 - e^{-|D'_{n/t}(B)|})\sqrt{t}} e^{-tD_{n/t}(B)}\big(1 + o(1)\big),$$

where the constant $c_1 = c_1(b) > 0$ is known in an explicit form, $D'_v(B) = (\partial/\partial u)\, D_u(B)|_{u=v}$.

(ii) *Let*

$$\frac{n}{t} = \frac{1}{u^*} + y\frac{1}{\sqrt{t}}, n \geqslant 1,$$

and suppose that for the set B condition $[\mathbf{D}_2(1/b)]$ *is met and that $\alpha(1/u^*) \in \mathcal{A}'$, $I(1/u^*) < \infty$. Then there exists a function $y(t) \to \infty$ such that for $-y(t) \leqslant y \leqslant \infty$*

$$\mathbf{P}(\eta(tB) < n) = c_2\Phi(y/\sigma)e^{-tD(B)}(1 + o(1)) \tag{3.8.22}$$

and for $y \leqslant y(t)$

$$\mathbf{P}\big(\infty > \eta(tB) \geqslant n\big) = c_2(1 - \Phi(y/\sigma))e^{-tD(B)}\big(1 + o(1)\big), \tag{3.8.23}$$

where

$$\Phi(y) = \frac{1}{\sqrt{2\pi}} \int_{-\infty}^{y} e^{-u^2/2}\, du$$

is the standard normal law, the constant $\sigma^2 = D''_{u^*}(V) > 0$ is determined by equality (3.8.21) *and the constant $c_2 > 0$ is known in an explicit form.*

(iii) *Let $n/t \to b \geqslant u^*$, $(n/t - u^*)\sqrt{t} \to \infty$, and suppose that for the set B condition* $[\mathbf{D}_2(1/b)]$ *is met and that $\alpha(1/b) \in \mathcal{A}'$, $I(1/b) < \infty$. Then*

$$\mathbf{P}(\infty > \eta(tB) \geqslant n) = \frac{c_3}{(1 - e^{-|D'_{n/t}(B)|})\sqrt{t}} e^{-tD_{n/t}(B)}(1 + o(1)),$$

where the constant $c_3 = c_3(b) > 0$ is known in an explicit form.

In the papers [50] and [9] the asymptotics of the function $H(tB)$ as $t \to \infty$ was studied, where

$$H(tB) = \sum_{n=1}^{\infty} \mathbf{P}(S_n \in tB)$$

is the multidimensional renewal function. It was proved there, in particular, that if the condition $[\mathbf{D}_2(1/b)]$ is satisfied at the point s^* and $\alpha(s^*) \in \mathcal{A}'$, $I(s^*) < \infty$, then

$$H(tB) \sim c_1 e^{-tD(B)},$$

where the constant $c_1 > 0$ is known in an explicit form ([9]). From Theorem 3.8.13, it follows that under these conditions the relation

$$\mathbf{P}(\eta(tB) < \infty) \sim c_2 e^{-tD(B)}.$$

is true. This implies that

$$\mathbf{P}\big(\eta(tB) < \infty\big) \sim \frac{c_2}{c_1} H(tB).$$

This result was stated as a hypothesis in [50].

Proofs of the above statements can be found in [53]. Cramér's condition on the characteristic function $\psi(it)$, which was used in [53], is superfluous if the vector ξ is non-lattice. This follows from Theorem 2.3.2, relation (2.3.5) from which was used in [53].

3.9 Supplement. Analytic methods for boundary crossing problems with linear boundaries

3.9.1 Introduction

If one studies the asymptotics of the exiting (or not exiting) of the trajectory of a random walk $\{S_k\}$ through a linear boundary, then by a 'rotation' transformation (i.e. by passing to the random walk $\{S_k + bk\}$, $b = \text{const.}$) the problem can be reduced to studying the joint distribution of the maximum of the partial sum $S_1, \dots, S_{n-1}$ and the last sum S_n. Thus, the problem reduces to studying the asymptotics of the probability

$$u^y_{x,n} := \mathbf{P}(\overline{S}_{n-1} < x,\ S_n \geqslant x + y) \tag{3.9.1}$$

for $x > 0$ and arbitrary y. If $y > 0$, we will speak about the time $\eta(x)$ of the first crossing of the level x; if $y < 0$, the probability $u^y_{x,n} - u^0_{x,n}$ will describe the joint distribution[1] of $\overline{S}_n = \max_{k \leqslant n} S_k$ and S_n. In this section, we present analytic methods for studying the probabilities (3.9.1) which were developed in [16]. The technical level of that work is beyond the scope of the present book. Therefore, we

[1] We note that, in fact, it is sufficient to consider only the case $y < 0$, because the problem of studying $\eta(x)$ reduces to this case by means of the representation

$$\mathbf{P}\big(\eta(x) = n+1,\ \chi(x) < \Delta\big) = \int_{-\infty}^{0} \mathbf{P}(\overline{S}_{n-1} < x,\ S_n \in x + dy)\,\mathbf{P}\big(\xi \in (-y, -y+0)\big).$$

However, for the sake of completeness and simplicity of exposition, we will present results for both negative and positive y.

will explain only the essence of the method and the main results in a simplified and shortened form.

Analytic methods of studying the asymptotics of $u_{x,n}^y$ are based on the fact that the probability $u_{x,n}^y$ satisfies an integro-difference equation. Indeed, by virtue of the total probability formula with respect to the first jump, we have

$$
\begin{aligned}
u_{x,n+1}^y &= \int_{-\infty}^{x} u_{x-t,n}^y dF(t) \quad \text{for} \quad n \geqslant 1, \\
u_{x,1}^y &= 1 - F(x+y).
\end{aligned} \tag{3.9.2}
$$

The generating function $u_x^y(z) := \sum_{n=1}^{\infty} u_{x,n}^y z^n$ for $|z| < 1$ allows us to write equation (3.9.2) in the form

$$
u_x^y(z) = z \int_{-\infty}^{x} u_{x-t}^y(z) dF(t) + z\big(1 - F(x+y)\big). \tag{3.9.3}
$$

This is an integral equation of the Stieltjes type on a semi-axis. The solution of this equation consists of the following three steps.

(1) A generalisation of the well-known Wiener–Hopf solution of an equation of this type, written in terms of a Riemann integral (in (3.9.3), this can take place if there exists the density $f(t) = F'(t)$), to equations of the Stieltjes type. This method consists in finding an explicit form of the dual transform

$$
\mathfrak{u}^y(z,\lambda) := \int_0^{\infty} e^{i\lambda x} u_x^y(z) dx
$$

in terms of the components of the factorisation (see the definitions below) of the function (in the present case)

$$
\mathfrak{v}_z(\lambda) := 1 - z\varphi(\lambda), \qquad \varphi(\lambda) = \mathbf{E} e^{i\lambda\xi} = \int e^{i\lambda x} dF(x).
$$

Here and below the argument λ is complex-valued. Since we are dealing with integrals of the Stieltjes type, the index of equation (3.9.3) is, generally speaking, not defined and one has to consider a special kind of factorisation – the so-called V-factorisation (see the definition below).

(2) The second step consists in inverting the obtained transforms with respect to the variable λ and finding an asymptotic representation for $u_x^y(z)$ as $x \to \infty$. That turns out to be possible if one selects a simple pole of the function $\mathfrak{u}^y(z,\lambda)$ in the plane of the variable λ.

(3) The third step consists in inverting the transform $u_x^y(z)$ and finding the sought-for asymptotics $u_{x,n}^y$ as $x \to \infty$, $n \to \infty$, using the method of steepest descent and its modifications.

Since the components of the factorisation play an important role in the whole presentation and appear in the formulations of the main results, we need to introduce some notions and notation.

3.9.2 V-factorisation and the explicit form of the double transforms $\mathfrak{u}^y(z,\lambda)$

I. The rings V, $\mathfrak{V}$. Let V be the ring of complex-valued functions $v(t)$ $(-\infty < t < \infty)$ with bounded variation, for which multiplication is defined as convolution and the norm is $\|v\| = \mathrm{Var}_{-\infty,\infty} v(t)$, and let $\mathfrak{V}$ be the ring with respect to the standard multiplication of the Fourier–Stieltjes transforms $\mathfrak{v}(\lambda)$ of functions from V with the natural norm $\|\mathfrak{v}\| = \| \int_{\infty}^{\infty} e^{i\lambda t} dv(t)\| = \|v\|$. We define the rings $\mathfrak{V}(\mu_-)$ and $\mathfrak{V}(\mu_-, \mu_+)$ $(\mu_- \leqslant \mu_+)$, also with respect to the usual multiplication, by the relations

(1) $\mathfrak{v} \in \mathfrak{V}(\mu_-)$ if $\widetilde{\mathfrak{v}}(\lambda) := \mathfrak{v}(i\mu_- + \lambda) \in \mathfrak{V}$; $\|\mathfrak{v}\|_{\mathfrak{V}(\mu_-)} = \|\widetilde{\mathfrak{v}}\|_{\mathfrak{V}}$ (so that $\mathfrak{V}(0) = \mathfrak{V}$);

(2) $\mathfrak{v} \in \mathfrak{V}(\mu_-, \mu_+)$ if $\mathfrak{v} \in \mathfrak{V}(\mu_-)$ and the function v in the representation

$$\mathfrak{v}(\lambda) = \int_{-\infty}^{\infty} e^{(i\lambda + \mu_-)t} dv(t), \quad v \in V, \tag{3.9.4}$$

has the property

$$\int_{-\infty}^{\infty} e^{(\mu_- - \mu_+)t} \big| dv(t) \big| < \infty. \tag{3.9.5}$$

The norm in $\mathfrak{V}(\mu_-, \mu_+)$ can be defined as, for example, the norm in $\mathfrak{V}(\mu_-)$, so that $\mathfrak{V}(\mu_-, \mu_+)$ is the ring[2] of functions that can be represented for any λ, $\mu_- \leqslant -\mathrm{Im}\lambda \leqslant \mu_+$, in the form $\int_{-\infty}^{\infty} e^{i\lambda t} dv(t)$, $v \in V$.

We will say that $\mathfrak{v} \in \mathfrak{V}_+(\mu_-)$ ($\mathfrak{v} \in \mathfrak{V}_-(\mu_-)$), if $\mathfrak{v} \in \mathfrak{V}(\mu_-)$ and the function $\mathfrak{v}$ is representable in the form

$$\mathfrak{v}(i\mu_- + \lambda) = \int_0^{\infty} e^{i\lambda t} dv(t) \qquad \left(\mathfrak{v}(i\mu_- + \lambda) = \int_{-\infty}^{0} e^{i\lambda t} dv(t) \right), \qquad v \in V.$$

In what follows, we will use the following shortened notation for such relations: $\mathfrak{v} \in \mathfrak{V}_{\pm}(\mu_-)$ if $\mathfrak{v}(i\mu_- + \lambda) = \int_0^{\infty} e^{\pm i\lambda t} dv(t)$. Accordingly, $V_{\pm}$ will denote the sets of functions from V which are constant for $t \leqslant 0$. The sets $\mathfrak{V}_{\pm}(\mu_-)$, $V_{\pm}$ are subrings of the rings $\mathfrak{V}(\mu_-)$, V, respectively.

Denote by $\Pi(\mu_-, \mu_+)$ the zone $\mu_- \leqslant \mathrm{Im}\,\lambda \leqslant \mu_+$, and by $\big(\Pi(\mu_-, \mu_+)\big)$ the interior of $\Pi(\mu_-, \mu_+)$. The line $\Pi(\mu_-, \mu_-)$ is denoted by $\Pi(\mu_-)$ and the half-planes $\Pi(\mu_-, \infty)$ and $\Pi(-\infty, \mu_-)$ by $\Pi_+(\mu_-)$ and $\Pi_-(\mu_-)$, respectively.

Functions from $\mathfrak{V}(\mu_-, \mu_+)$ are analytic in the interior of $\Pi(\mu_-, \mu_+)$ and are continuous in this zone, including at its boundary points. Functions from $\mathfrak{V}_{\pm}(\mu_-)$ have the same properties in the regions $\Pi_{\pm}(\mu_-)$.

[2] The fact that $\mathfrak{V}(\mu_-, \mu_+)$ is a ring follows from the inequality

$$\int_{-\infty}^{\infty} e^{(\mu_- - \mu_+)t} \left| d \int_{-\infty}^{\infty} v_2(t-x) dv_1(x) \right| < \infty,$$

which is true in the case when v_1 and v_2 satisfy condition (3.9.5).

Everywhere below, elements of the rings $\mathfrak{V}(\cdot)$ will be denoted by Greek letters and their images in V by Latin letters. Regions will be denoted by calligraphic letters. The letter c (with or without indices) will denote some constant.

It is not hard to verify the following properties of the rings $\mathfrak{V}$ and their preimages (see [16]).

(i) *If* $\mathfrak{v} \in \mathfrak{V}(\mu_-, \mu_0)$ *and* $\mathfrak{v} \in \mathfrak{V}(\mu_0, \mu_+)$ $(\mu_- \leqslant \mu_0 \leqslant \mu_+)$ *then* $\mathfrak{v} \in \mathfrak{V}(\mu_-, \mu_+)$.

(ii) *If* $\mathfrak{v} \in \mathfrak{V}(\mu_-)$ *then the function* v *from the representation*

$$\mathfrak{v}(\lambda) = \int_{-\infty}^{\infty} e^{i\lambda t} dv(t)$$

has the property

$$\left|v(t) - v(\infty)\right| \leqslant ce^{\mu_- t}.$$

Denote by $\mathfrak{R}(\cdot)$ the subring of the ring $\mathfrak{V}(\cdot)$ which consists of elements for which v in representation (3.9.4) is an absolutely continuous function.

(iii) *If* $\mathfrak{v} \in \mathfrak{V}(\mu_-, \mu_+)$ *and* $\mathfrak{v}(i\mu_0) = 0$, $i\mu_0 \in \big(\Pi(\mu_-, \mu_+)\big)$, *then*

$$\mathfrak{r}(\lambda) = \frac{\mathfrak{v}(\lambda)}{\lambda - i\mu_0} \in \mathfrak{R}(\mu_-, \mu_+),$$

and the function r *in the representation*

$$\mathfrak{r}(\lambda) = \int_{-\infty}^{\infty} e^{i\lambda t} r(t) dt$$

has the property

$$\left|r(t)\right| = O(e^{\mu_- t}) \quad as \quad t \to +\infty.$$

Similar statements are true when $\mathfrak{v} \in \mathfrak{V}_+(\mu_-)$ *or* $\mathfrak{v} \in \mathfrak{V}_-(\mu_+)$.

II. Factorisation. Definitions. (1) A function $\mathfrak{v}(\lambda)$ which is analytic inside and continuous, including at the boundary, in the strip $\Pi(\mu_-, \mu_+)$, $\mu_- \leqslant \mu_+)$, *admits a factorisation in* $\Pi(\mu_-, \mu_+)$ *if it can be represented in the form*

$$\mathfrak{v}(\lambda) = \mathfrak{v}_+(\lambda)\mathfrak{v}_-(\lambda), \qquad \lambda \in \Pi(\mu_-, \mu_+), \tag{3.9.6}$$

where the functions $\mathfrak{v}_\pm(\lambda_\pm)$ are analytic inside and continuous, including at the boundaries, in the regions $\Pi_\pm(\mu_\pm)$, respectively. The functions $\mathfrak{v}_+(\lambda)$ are the positive components of the factorisation; the functions $\mathfrak{v}_-(\lambda)$ are the negative components.

(2) *A function* $\mathfrak{v}$ *admits a canonical factorisation (c.f.) in the strip* $\Pi(\mu_-, \mu_+)$ if it admits a factorisation and in the representation (3.9.6) it holds that

$$\sup_{\Pi(\mu_-,\mu_+)} \left|\mathfrak{v}_\pm(\lambda)\right| < \infty, \qquad \inf_{\Pi(\mu_-,\mu_+)} \left|\mathfrak{v}_\pm(\lambda)\right| > 0. \tag{3.9.7}$$

(3) By a *V-factorisation (V-f.)* of a function $\mathfrak{v} \in \mathfrak{V}(\mu_-, \mu_+)$ in the strip $\Pi(\mu_-, \mu_+)$ we mean the representation (3.9.6), where $\mathfrak{v}_\pm(\lambda) \in \mathfrak{V}(\mu_\pm)$.

(4) *The canonical V-factorisation (c.V-f.) of a function* $\mathfrak{v} \in \mathfrak{V}(\mu_-, \mu_+)$ *in the strip* $\Pi(\mu_-, \mu_+)$ is the V-f. in $\Pi(\mu_-, \mu_+)$, such that

$$\inf_{\Pi_+(\mu_-)} \left|\mathfrak{v}_+(\lambda)\right| > 0, \qquad \inf_{\Pi_-(\mu_+)} \left|\mathfrak{v}_-(\lambda)\right| > 0. \tag{3.9.8}$$

By definition, two factorisations coincide if their components differ only by constant factors.

If $\mathfrak{v}(\lambda)$ *admits the c.f.* (in particular, the c.V-f.) *then the factorisation* (3.9.6) *in the class of all expansions* (3.9.6) *having the property*

$$\inf_{\Pi_+(\mu_-)} \left|\mathfrak{v}_+(\lambda)\right| > 0 \qquad \left(or \quad \inf_{\Pi_-(\mu_+)} \left|\mathfrak{v}_-(\lambda)\right| > 0\right)$$

is unique and, hence, coincides with the c.f.

Indeed, let

$$\mathfrak{v}(\lambda) = \mathfrak{u}_+(\lambda)\mathfrak{u}_-(\lambda), \qquad \lambda \in \Pi(\mu_-, \mu_+),$$

be a factorisation different from (3.9.6) and having the property

$$\inf_{\Pi_+(\mu_-)} \left|\mathfrak{u}_+(\lambda)\right| > 0.$$

Then

$$\frac{\mathfrak{v}_+(\lambda)}{\mathfrak{u}_+(\lambda)} = \frac{\mathfrak{u}_-(\lambda)}{\mathfrak{v}_-(\lambda)}, \qquad \lambda \in \Pi(\mu_-, \mu_+), \tag{3.9.9}$$

where the two ratios are analytic and bounded in $\Pi_\pm(\mu_\pm)$, respectively, and, hence, coincide in $\Pi(\mu_-, \mu_+)$. This means that both parts of the equality (3.9.9) are analytic bounded functions in the whole plane, so they are necessarily equal to a constant. The statement is proved.

The positive component of a factorisation $\mathfrak{v}_+(\lambda)$ can be found in the form of a contour integral of the function $\ln \mathfrak{v}(\mu)$. In [39], § 12.5, and [22], § 18, several cases where the components of a factorisation can be found explicitly are considered. This can be done when $v(t)$ for $t > 0$ (or $v(t)$ for $t < 0$) is an exponential polynomial, i.e. it has the form

$$v(t) = \sum_{l=1}^{L} \sum_{k=0}^{K} c_{kl} t^k e^{-\lambda_l t}, \qquad t > 0, \quad L < \infty, \quad K < \infty, \quad \lambda_l > 0.$$

III. Properties of the function $\mathfrak{v}_z(\lambda) := 1 - z\varphi(\lambda)$. In what follows we will assume that the following Cramér's condition is satisfied:

[C$_0$]

$$\lambda_- = \inf\left\{\lambda : \psi(\lambda) < \infty\right\} < 0, \qquad \lambda_+ = \sup\left\{\lambda : \psi(\lambda) < \infty\right\} > 0.$$

In the paper [16], the results of which are presented in this section, the following condition **[F$_a$]** is also used:

[F$_a$] *The distribution* **F** *has a non-zero absolutely continuous component.*

Perhaps this condition can be relaxed to the condition

[R$^\kappa$].

$$\sup_{|t|\to\infty} \big|\varphi(t)\big| \leqslant \kappa < 1.$$

If condition **[C$_0$]** is satisfied, introduce numbers $\mu_\pm$ defined by the relations

$$\mu_\pm = \begin{cases} -\lambda_\mp & \text{if} \quad \psi(\lambda_\mp) < \infty, \\ -\lambda_\mp \mp \delta & \text{if} \quad \psi(\lambda_\mp) = \infty, \end{cases}$$

for sufficiently small $\delta > 0$ that the inequality $\mu_- < \mu_+$ holds true.

Clearly, for any z, $|z| \leqslant 1$,

$$\mathfrak{v}_z(\lambda) = 1 - z\varphi(\lambda) \in \mathfrak{V}(\mu_-, \mu_+).$$

For $|z| < 1$, one can define the function $\ln\big(1 - z\varphi(\lambda)\big)$, to be understood in the sense of a principal value. Therefore

$$\begin{aligned} \mathfrak{v}_z(\lambda) = e^{\ln(1-z\varphi(\lambda))} &= \exp\bigg\{-\sum_{n=1}^{\infty} \frac{z^n\varphi^n(\lambda)}{n}\bigg\} = \exp\bigg\{-\sum_{n=1}^{\infty} \frac{z^n}{n}\mathbf{E}e^{i\lambda S_n}\bigg\} \\ &= \exp\bigg\{-\sum_{n=1}^{\infty} \frac{z^n}{n}\mathbf{E}(e^{i\lambda S_n};\, S_n \geqslant 0)\bigg\} \exp\bigg\{\sum_{n=1}^{\infty} \frac{z^n}{n}\mathbf{E}(e^{i\lambda S_n};\, S_n < 0)\bigg\} \\ &= \mathfrak{v}_{z+}(\lambda)\mathfrak{v}_{z-}(\lambda), \end{aligned} \tag{3.9.10}$$

where

$$\mathfrak{v}_{z+}(\lambda) = \exp\bigg\{-\sum_{n=1}^{\infty} \frac{z^n}{n}\mathbf{E}(e^{i\lambda S_n};\, S_n \geqslant 0)\bigg\}. \tag{3.9.11}$$

Clearly, the exponent factor on the right-hand side of (3.9.11) is a function from $\mathfrak{V}_+(0)$ when $|z| < 1$. The exponent itself has same property, i.e. it is the function $\mathfrak{v}_{z+}(\lambda)$. A similar assertion is true for $\mathfrak{v}_{z-}(\lambda) \in \mathfrak{V}_-(0)$. Besides that,

$$\inf\big|\mathfrak{v}_{z\pm}(\lambda)\big| > 0.$$

Thus, for $|z| < 1$, the function $\mathfrak{v}_z(\lambda)$ admits the c.V-f. (3.9.10) on the axis $\Pi(0)$. Obviously, the same is true for the function $\mathfrak{v}_z^{-1}(\lambda)$.

The function $\mathfrak{v}_z(\lambda)$ plays an important role in a series of problems. The function

$$\mathfrak{v}_z^{-1}(\lambda) = \frac{1}{1 - z\varphi(\lambda)} = \sum_{n=0}^{\infty} z^n\varphi^n(\lambda)$$

is the double transform of the distribution $\mathbf{P}(S_n < x)$ (with respect to x and n). It turns out that the double transform of $\mathbf{P}(\overline{S}_n < x)$ can be found in terms of the positive component $\mathfrak{v}_{z+}(\lambda)$ of the factorisation of the function $\mathfrak{v}_z(\lambda)$. Namely, the following result is true (see [16], formulas (24) and (25)). Let $v_{z\pm}^{\pm}(t)$ be the preimages of the functions $\mathfrak{v}_{z\pm}^{\pm 1}$.

Theorem 3.9.1. *Suppose that the conditions* **[C$_0$]**, **[F$_a$]** *are met. Then, for* $y \geqslant 0$, $|z| < 1$, $\lambda \in (\Pi_+(0))$,

$$\mathfrak{u}^{-y}(z,\lambda) = -\frac{e^{i\lambda y}}{i\lambda} + \frac{\mathfrak{v}_{z+}(0)}{i\lambda \mathfrak{v}_{z+}(\lambda)} + \frac{1}{\mathfrak{v}_{z+}(\lambda)} \int_0^y e^{i\lambda t}\big[v_{z-}^{-}(0) - v_{z-}^{-}(t-y)\big]dt, \tag{3.9.12}$$

$$\mathfrak{u}^{y}(z,\lambda) = \frac{\mathfrak{v}_{+}(0) - \mathfrak{v}_{z+}(\lambda)}{i\lambda \mathfrak{v}_{z+}(\lambda)} + \frac{1}{\mathfrak{v}_{z+}(\lambda)} \int_0^\infty e^{i\lambda t}\big[v_{z+}^{+}(t+y) - v_{z+}^{+}(t)\big]dt. \tag{3.9.13}$$

In particular, from (3.9.12) and (3.9.13), for $y = 0$ we get the following representation of the double transform of $\mathbf{P}(\eta(x) = n)$:

$$\mathfrak{u}^{0}(z,\lambda) = \sum_{n=1}^{\infty} z^n \int_0^\infty e^{i\lambda x}\mathbf{P}(\eta(x) = n)dx = \frac{\mathfrak{v}_{+}(0) - \mathfrak{v}_{z+}(\lambda)}{i\lambda \mathfrak{v}_{z+}(\lambda)}.$$

Theorem 3.9.1 completes the first stage of the study of the proposed problem.

3.9.3 Inversion of double transforms

The two subsequent stages require us to investigate the analytic properties of the component $\mathfrak{v}_{z+}(\lambda)$ and, first of all, to find its zeros (the poles of the function $\mathfrak{v}_{z+}^{-1}(\lambda)$) in the plane of the variable λ. In order to do this, we will modify the function $\mathfrak{v}_z(\lambda)$ somewhat, so that the new function admits the c.V-f. for $|z| \geqslant 1$ and in a sufficiently wide strip of values λ. Note that for $|z| = 1$, and, moreover, for $|z| \geqslant 1$, the above arguments regarding the possible existence of a c.V-f. of the function $\mathfrak{v}_z(\lambda)$ fail to be valid, and the function $\mathfrak{v}_z(\lambda)$ for $z = 1$ may not admit the c.f. on the axis $\Pi(0)$ since $\mathfrak{v}_z(0) = 0$ for $z = 1$. Also, if the condition **[R$^\kappa$]** is not satisfied then

$$\liminf_{|t|\to\infty} \mathfrak{v}_1(t) = 0.$$

Thus, as well as imposing assumptions **[R$^\kappa$]** (or **[F$_a$]**) we have to eliminate the zeros of the function $\mathfrak{v}_z(\lambda)$.

Suppose, as before, that $\psi(\lambda) = \varphi(-i\lambda)$. Since $\psi''(\lambda) > 0$ on the whole segment $[\mu_-, \mu_+]$ (at the ends of the segment the derivative is understood as the right or the left derivative), there exists a unique point $\lambda_0 \in [\mu_-, \mu_+]$, at which $\min_{[\mu_-,\mu_+]} \psi(\lambda)$ is attained ($\lambda_0 = \lambda(0)$, if $\mu_\pm = \lambda_\pm$). This means that for positive $z \leqslant z_0 = 1/\psi(\lambda_0)$ the equation

$$\psi(\lambda) = \frac{1}{z}$$

has no more than two real solutions,

$$\lambda_-(z) \leqslant \lambda_+(z).$$

Let

$$z_\pm = \frac{1}{\psi(\lambda_\pm(z))}.$$

Then the values $-\lambda_+(z) \leqslant -\lambda_-(z)$ will be real zeros of the function $\mathfrak{v}_z(i\lambda) = 1 - z\psi(-\lambda)$, defined respectively for $z \in [z_-, z_0]$, $z \in [z_+, z_0]$. According to implicit function theorems (see e.g. [126]), $\lambda_\pm(z)$ will be analytic within, and also in some neighbourhoods (in the z-plane) of the segments $[z_\pm, z_0]$. In the case when

$$\lambda_- < \lambda_0 < \lambda_+,$$

the point z_0 is a second-order branching point of the functions $\lambda_\pm(z)$, which form at this point a single circular system.

Define the functions $\mu_\pm(z)$ for positive z as follows:

$$\mu_\mp(z) = \begin{cases} -\lambda_\pm, & \text{if} \quad z \leqslant z_\pm, \\ -\lambda_\pm(z), & \text{if} \quad z_0 < z < z_\pm, \\ -\lambda_0, & \text{if} \quad z \geqslant z_0. \end{cases}$$

Let

$$\mu_{\gamma-}(|z|) = \max\left[\mu_-(|z|) - \gamma, \mu_-\right], \qquad \mu_{\gamma+}(|z|) = \min\left[\mu_+(|z|) + \gamma, \mu_+\right],$$
$$\Pi_\gamma = \Pi\big(\mu_{\gamma-}(|z|), \mu_{\gamma+}(|z|)\big).$$

One of the most important statements here is the following.

Let λ_0 be an interior point of the segment $[\mu_-, \mu_+]$, so that $z_0 > z_\pm$. Consider the zone $\mathcal{K}_{\delta_1,\delta}$, defined by the inequalities $\big\{|\mathrm{Im}\, z| \leqslant \delta;\ \mathrm{Re}\, z > 0;\ \overline{z} + \delta_1 \leqslant |z| \leqslant z_0 + \delta\big\}$, where $\overline{z} = \max(z_-, z_+)$.

Choose numbers δ_1 and $\delta > 0$ such that $\mathcal{K}_{\delta_1,\delta}$ does not contain other singularities of the function $\lambda_\pm(z)$ except the point $z = z_0$. Since the algebraic second-order branching point z_0 is common for the functions $\lambda_\pm(z)$, any polynomial symmetric with respect to $\lambda_\pm(z)$ will be a single-valued analytic function in $\mathcal{K}_{\delta_1,\delta}$. This implies, in particular, that the function

$$\mathfrak{w}_z(\lambda) = \big[1 - z\varphi(\lambda)\big]\, \frac{\big[\lambda - i(\mu_- - 1)\big]\big[\lambda - i(\mu_+ + 1)\big]}{\big[\lambda + i\lambda_+(z)\big]\big[\lambda + i\lambda_-(z)\big]}$$

will be analytic in the variables z and λ inside and continuous, including at the boundary, for $z \in \mathcal{K}_{\delta_1,\delta}$, $\lambda \in \Pi(\mu_-, \mu_+)$. It is not hard to show that $\mathfrak{w}_z(\lambda) \in \mathfrak{V}(\mu_-, \mu_+)$ for any $z \in \mathcal{K}_{\delta_1,\delta}$. The following theorem holds true.

Theorem 3.9.2. *For any $\delta_1 > 0$ and sufficiently small $\delta > 0$ and $\gamma > 0$, the functions $\mathfrak{w}_z^{\pm 1}(\lambda)$ for $z \in \mathcal{K}_{\delta_1,\delta}$ admit the c.V-f. in the strip Π_γ. The components $\mathfrak{w}_{z\pm}$ $(\mathfrak{w}_{z\pm}^{-1})$ of the factorisation of the function $\mathfrak{w}_z$ $(\mathfrak{w}_z^{-1})$ can be defined as functions which are analytic in the variables z and λ inside and continuous, including at the boundary, in the regions $z \in \mathcal{K}_{\delta_1,\delta}$, $\lambda \in \Pi_\pm\big(\lambda_{\gamma\mp}(|z|)\big)$.*

Theorem 3.9.2 implies that

$$\mathfrak{v}_z(\lambda) = \frac{[\lambda + i\lambda_-(z)][\lambda + i\lambda_+(z)]}{[\lambda - i(\mu_- - 1)][\lambda - i(\mu_+ + 1)]}\,\mathfrak{w}_{z+}(\lambda)\mathfrak{w}_{z-}(\lambda),$$

and one can take

$$\mathfrak{v}_{z+}^{-1}(\lambda) = \frac{[\lambda - i(\mu_- - 1)]}{[\lambda + i\lambda_+(z)]}\,\mathfrak{w}_{z+}^{-1}(\lambda).$$

This means that the function $\mathfrak{v}_{z+}^{-1}(\lambda)$ in (3.9.12) and (3.9.13) has an isolated pole at the point $-i\lambda_+(\lambda)$. Since the 'preimage' of a pole is an exponential function, for $\lambda \in \big(\Pi_+(-\lambda_+(z))\big)$ we have

$$\frac{1}{\lambda + i\lambda_+(z)} = i\int_0^\infty e^{i\lambda t - t\lambda_+(z)}dt;$$

this allows us to find an asymptotic representation for $x \to \infty$ of the functions $u_x^y(z)$. In the next few statements, for $y \geqslant 0$ it will be more convenient to use the generating functions

$$q_x^y(z) = u_x^0(z) - u_x^y(z) \quad \text{and} \quad Q_x^y(z) = u_x^{-y}(z) - u_x^0(z),$$

which correspond to the probabilities

$$q_{x,n}^y = \mathbf{P}(\overline{S}_{n-1} < x,\, x \leqslant S_n < x + y) \quad \text{and} \quad Q_{x,n}^y = \mathbf{P}(\overline{S}_n < x, S_n \geqslant x - y)$$

for $x \to \infty$ and fixed $y > 0$.

Theorem 3.9.3. *For $x \to \infty$ and some $\gamma > 0$,*

$$q_x^y(z) = e^{-x\lambda_+(z)}\frac{W_1(z,y)}{\mathfrak{W}_{z+}\big(-i\lambda_+(z)\big)} + o\Big(\exp\big\{x\mu_{\gamma-}(|z|)\big\}\Big), \tag{3.9.14}$$

$$Q_x^y(z) = e^{-x\lambda_+(z)}\frac{W_2(z,y)}{\mathfrak{W}_{z+}\big(-i\lambda_+(z)\big)} + o\Big(\exp\big\{x\mu_{\gamma-}(|z|)\big\}\Big), \tag{3.9.15}$$

where

$$\mathfrak{W}_{z+}(\lambda) = \frac{\mathfrak{w}_{z+}(\lambda)}{\lambda - i(\mu_- - 1)}$$

is the positive component of the V-f. of the function

$$\frac{1 - z\varphi(\lambda)}{(\lambda + i\lambda_+(z))(\lambda + i\lambda_-(z))}; \tag{3.9.16}$$

$W_1(z,y)$ is the 'preimage' of the function $\mathfrak{W}_{z+}(\lambda)$ and

$$\mathfrak{W}_{z+}(\lambda) = \int_0^\infty e^{i\lambda y}dW_1(z,y),$$

satisfying the condition $W_1(z,0) = 0$. The function $W_2(z,y)$ is defined in a more complicated way (see Lemma 9 in [16]).

The last stage of inversion of the double transform consists in the inversion of the generating functions $q_x^y(z)$, $Q_x^y(z)$ using the inversion formulas (for integration of the functions $q_x^y(z)z^{-n-1}$, $Q_x^y(z)z^{-n-1}$ along the respective contours) with an application of the steepest descent method for $x \to \infty$.

3.9.4 Results of the asymptotic analysis. First passage time

I. *First consider the case* $x = o(n)$ *as* $n \to \infty$. Note that in this case the approaches that were used in section 3.5 do not give the desired results. The analytic methods considered in that chapter allow us to obtain such results only if one assumes that $x \to \infty$. As has been noted before, in order to obtain the asymptotics of the probabilities $q_{x,n}^y$, $Q_{x,n}^y$ using the relations (3.9.14), (3.9.15), one should use the inversion formula to integrate the functions $q_x^y(z)z^{-n-1}$, $Q_x^y(z)z^{-n-1}$ along the corresponding contour (see [16]). According to the steepest descent method and formulas (3.9.14), (3.9.15), this contour should pass through the point z^*, at which is attained

$$\min_z e^{-x\lambda_+(z)} z^{-n-1}. \tag{3.9.17}$$

This minimum will determine the main exponential term in the sought-for asymptotics. But $z = 1/\psi\big(\lambda_+(z)\big)$ and, for $x/n = \alpha$,

$$\begin{aligned}\min_z \big(-x\lambda_+(z) - n\ln z\big) &= \min_z \Big(-x\lambda_+(z) + n\ln\psi\big(\lambda_+(z)\big)\Big)\\ &= -n\max_\lambda \big(\alpha\lambda - \ln\psi(\lambda)\big) = -n\Lambda(\alpha).\end{aligned}$$

In addition, note that asymptotic analysis of the integral of the function

$$\frac{e^{-x\lambda_+(z)}W_j(z,y)}{z^{n+1}\mathfrak{W}_{z+}\big(i\lambda_+(z)\big)}, \qquad j = 1, 2,$$

is complicated owing to the fact that for $x = o(n)$ (when the point z^*, where the minimum in (3.9.17) is attained, is close to $z_0 = 1/\min\psi(\lambda)$) the functions $\lambda_\pm(z)$ have a common branching point at the point $z = z_0$. Therefore, the integration contour should go around the point z_0 (see [16], § 3).

Further on, we will write

$$R = \Xi_{k_1,\dots,k_l}(\varepsilon_1,\dots,\varepsilon_l),$$

if a function R admits an asymptotic expansion in powers of $\varepsilon_1,\dots,\varepsilon_l$ which do not contain terms of order $\varepsilon_1^{j_1},\dots,\varepsilon_l^{j_l}$ where, simultaneously, $j_1 < k_1,\dots,j_l < k_l$.

Theorem 3.9.4. *Let $x = o(n)$, $\alpha = x/n$ and let $\lambda_0 \in (\lambda_-, \lambda_+)$ be the point where $\min \psi(\lambda)$ is attained, $z_0 = 1/\psi(\lambda_0)$. Then*

$$q_{x,n}^{y} = \mathbf{P}\big(\eta(x) = n,\ \chi(x) < y\big) = \frac{x z_0^{1/2} e^{-n\Lambda(\alpha)}}{\sqrt{2\pi\psi''(\lambda_0)}\mathfrak{W}_{z_0+}(-i\lambda_0) n^{3/2}} \times \left\{ W_1(z_0, y) + \frac{n}{x}\, \Xi_{1,2}(n^{-1}\alpha) + O\left(\frac{n^{3/2} e^{-\gamma x}}{x}\right) \right\}, \tag{3.9.18}$$

for some $\gamma > 0$. The functions $\mathfrak{W}_{z+}(\lambda)$, $W_1(z, y)$ are defined in Theorem 3.9.3.

The exponential factor $e^{-n\Lambda(\alpha)}$, which appears in (3.9.18) and subsequent formulas, is provided in the paper [16] in a somewhat different form, since, in 1961, when [16] was written, the deviation function $\Lambda(\alpha)$ had not been yet introduced in the literature as a convenient and adequate notion (it appeared in the corresponding places in an implicit form).

II. *Now let $\alpha = x/n$ be in some neighbourhood of the point $\alpha_0 > 0$.* Assume, as before, that $\lambda(\alpha)$ is the solution of the equation

$$\frac{\psi'(\lambda)}{\psi(\lambda)} = \alpha, \qquad \alpha_+ = \frac{\psi'(\lambda_+)}{\psi(\lambda_+)}.$$

Theorem 3.9.5. *Let $\alpha_0 \in (0, \alpha_+)$, $z_\alpha = 1/\psi\big(\lambda(\alpha)\big)$. Then, for sufficiently small $\varepsilon > 0$,*

$$\mathbf{P}\big(\eta(x) = n,\ \chi(x) < y\big) = \frac{\alpha e^{-n\Lambda(\alpha)} W_1(z_\alpha, y)}{\sqrt{2\pi n}\sigma(\alpha)\mathfrak{W}_{z_\alpha+}(-i\lambda(\alpha))}\big(1 + o(1)\big) \tag{3.9.19}$$

uniformly in $\alpha \in [\alpha_0 - \varepsilon, \alpha_0 + \varepsilon]$.

Here $\lambda(\alpha)$, $\sigma(\alpha)$ have the same meaning as before (see e.g. section 2.2).

In [16], this theorem is formulated differently (see Theorem 5 in [16]), using the value of the constant factor on the right-hand side of (3.9.19) at the fixed point α_0 and replacing $o(1)$ with the asymptotic expansion of $\Xi_{1,1}(n^{-1}, \alpha - \alpha_0)$ as $\alpha \to \alpha_0$, $n \to \infty$.

Under the conditions of Theorems 3.9.4 and 3.9.5, the probability $\mathbf{P}\big(\eta(x) = n\big)$ is equal to the right-hand sides of (3.9.18) and (3.9.19), where $W_1(z_\alpha, \infty) = \mathfrak{W}_{z_\alpha+}(0)$. If $\alpha \to \alpha_0$ as $n \to \infty$ then there exists the limit

$$\lim_{n\to\infty} \mathbf{P}\big(\chi(x) < y \,|\, \eta(x) = n\big) = \mathbf{E}_{\alpha_0}(y),$$

$$\int_0^\infty e^{i\lambda y} d\mathbf{E}_\alpha(y) = \frac{\mathfrak{W}_{z_\alpha+}(\lambda)}{\mathfrak{W}_{z_\alpha+}(0)}.$$

3.9.5 The joint distribution of $\overline{S}_n$ and S_n

In [16], limit theorems for the asymptotic behaviour of

$$\mathbf{P}(\overline{S}_n < x,\ S_n \geqslant x - y), \qquad \mathbf{P}(\overline{S}_n \geqslant x,\ S_n < x - y)$$

were obtained for different combinations of the speeds of growth of x and y as $n \to \infty$. When using the steepest descent method, there arises an additional difficulty related to the fact that in some cases a pole of the integrand function appears in a neighbourhood of the saddle point (this always occurs when the probabilities in question have a non-zero limit). Because the formulations of those results are quite long, here we will present only the simplest, which are mainly related to the case $x = o(n)$ and, thus, supplement the results of section 3.4.

Theorem 3.9.6. *Suppose that* $\mathbf{E}\xi = 0$, $x = o(n)$, $S := \sup S_n$. *Then, as* $x \to \infty$,

$$\mathbf{P}(S \geqslant x) = \frac{\mathfrak{W}_{1+}(0)}{\mathfrak{W}_{1+}(-i\lambda_+(1))} e^{-x\lambda_+(1)}\big(1 + o(1)\big),$$

$$\mathbf{P}(\overline{S}_n < x) = \mathbf{P}(S < x) + \frac{xe^{-n\Lambda(\alpha)}\mathfrak{W}_{z_0+}(0)}{\sqrt{2\pi}\sigma\, n^{3/2}(z_0 - 1)\mathfrak{W}_{z_0+}(-i\lambda_0)}\big(1 + o(1)\big).$$

If y is fixed then

$$\mathbf{P}(\overline{S}_n < x,\ S_n \geqslant x - y) = \frac{xe^{-n\Lambda(\alpha)}W_2(z_0, y)}{\sqrt{2\pi}\sigma\, n^{3/2}\mathfrak{W}_{z_0+}(-i\lambda_0)}\big(1 + o(1)\big).$$

If x and y are comparable with n then in some cases the problem simplifies in a known way, since it becomes possible to find a dominant asymptotics. Consider, for example,

$$P := \mathbf{P}(\overline{S}_n < x,\ S_n \geqslant x - y)$$

for $x = \alpha n$, $y = \beta n$ ($\alpha > 0$ and $\beta > 0$ are separated from 0). Clearly, as $n \to \infty$,

$$\begin{aligned}
&P \sim \mathbf{P}(\overline{S}_n < x) \quad \text{if} \quad \mathbf{E}\xi > \alpha,\\
&P \to 1 \quad \text{if} \quad \mathbf{E}\xi \in (\alpha - \beta, \alpha),\\
&P \sim \mathbf{P}(S_n \geqslant x - y) \quad \text{if} \quad \mathbf{E}\xi < \alpha - \beta.
\end{aligned}$$

If the random variables ξ are arithmetic and bounded then a similar but more complete asymptotic analysis, including the case of fixed deviations of x (i.e. not growing with n), was conducted in [15]. A combination of the approaches of the papers [15] and [16] allows one to perform the necessary asymptotic analysis for arbitrary (unbounded) arithmetic ξ satisfying Cramér's condition $[\mathbf{C}_0]$.

3.9.6 Extension of the factorisation method of solving boundary crossing problems to other objects

I. A number of results obtained by A.A. Borovkov in [16] using the approach described above were carried over by B.A. Rogozin in [160] and [161] to processes with independent increments.

II. Consider a renewal process of the following type. Suppose that along with a sequence of sums $\{S_n\}$ there is given a sequence of sums $T_k = \sum_{j=1}^{k} \tau_j$, $T_0 = 0$, of independent non-negative identically distributed random variables which do not depend on $\{S_k\}$. Let

$$n(T) = \min\{k \geqslant 1 : T_k \geqslant T\} = 1 + \sup\{k \geqslant 0 : T_k < T\}$$

and consider the two-dimensional random walk $\{S_k, T_k\}$ for $k = 0, 1, \ldots, n(T)$. In the paper [70], A.A. Borovkov and B.A. Rogozin carried out a complete asymptotic analysis of the distribution of the first passage time of $\{S_k\}_{k \leqslant n(T)}$ through a high level and the joint distribution of $S_{n(T)}$ and $\overline{S}_{n(T)}$.

III. Many results obtained in [16] can be carried over to a sequence of sums of random variables defined on the states (or transitions) of a finite ergodic Markov chain (see section 4.11). Here, first of all, we mention the paper of E.L. Presman [154], and also the papers of G.D. Miller [128], K. Arndt [2] and [3], K.A. Borovkov [73].

IV. Factorisation methods can be used also in a number of boundary crossing problems, for example, problems with two boundaries. Let $\eta(y,x) = \inf\{n \geqslant 1 : S_n \notin (-y,x)\}$, $x > 0$, $y > 0$. In the papers of V.I. Lotov [118], [119] etc., the asymptotics of the probability $\mathbf{P}(\eta(x,y) = n,\ S_n \in A), \mathbf{P}(\eta(x,y) > n,\ S_n \in B)$ was found for sets $A \notin (x,-y)$ and $B \in (x,-y)$, for $x > \sqrt{n}$, $y > \sqrt{n}$, $x + y = o(n)$, $n \to \infty$. Some of those results in [120]–[123], [108] were extended to random walks on a finite Markov chain, compound renewal processes and processes with independent increments.

3.10 Finding the numerical values of large deviation probabilities

In connection with possible applications of the results obtained in Chapters 2 and 3 (see e.g. Chapter 6), one can consider the problem of finding numerical values for the parameters which define probabilities of large deviations. For example, to find the probability $\mathbf{P}(S_n \geqslant x)$ it is necessary to know the values of the parameters $\Lambda(\alpha)$, $\lambda(\alpha)$, $\sigma(\alpha)$ for $\alpha = x/n$ (see Theorem 2.2.3, formula (2.2.8)). To compute $\mathbf{P}(S \geqslant x)$ for $\mathbf{E}\xi < 0$, one needs to know the values of the parameters

$$\lambda_1 \quad \text{and} \quad p := \frac{I(\alpha_1, \infty)}{\alpha_1} \tag{3.10.1}$$

(see Theorem 3.4.6, formula (3.4.24)).

Below we provide a procedure that can be used to obtain numerical sequential approximations of the these parameters.

3.10.1 Sequential approximations for the values $\lambda(\alpha)$, $\Lambda(\alpha)$, $\sigma(\alpha)$

It is clear that if the functions $\lambda(\alpha)$ and $\Lambda(\alpha)$ are known in an explicit form (see the examples in section 1.1), the above-mentioned problems do not arise. We will assume that the functions $\lambda(\alpha)$ and $\Lambda(\alpha)$ are not known and will construct sequential approximations for their values at a given fixed point $\alpha \in \mathcal{A}'$.

I. *The zeroth approximation.* The key role in the sequential approximations will be played by approximations of $\lambda(\alpha)$, i.e. of the solution of the equation

$$A'(\lambda) := \frac{\psi'(\lambda)}{\psi(\lambda)} = \alpha.$$

First, compute the values of the integrals

$$a = \mathbf{E}\xi = \int t dF(t), \qquad a_2 = \mathbf{E}\xi^2 = \int t^2 dF(t).$$

To be specific, let us assume that $\alpha > a$. If there are no other methods for choosing the zeroth approximation (for example, a method using the proximity of the distribution $\mathbf{F}$ to distributions for which an explicit form of $\lambda(\alpha)$ is known), then as the zeroth approximation $\lambda_{(0)}$ of $\lambda(\alpha)$ one can take the value

$$\lambda_{(0)} = \frac{\alpha - a}{\sigma^2}, \qquad \sigma^2 = a_2 - a^2,$$

which corresponds to the normal distribution with parameters (a, σ^2). Then it is necessary to verify that $\lambda_{(0)} < \lambda_+$ and $\alpha < \alpha_+$. For that, it is sufficient to specify $\lambda_{(+)} < \lambda_+$ such that

$$\psi(\lambda_{(+)}) < \infty, \qquad \alpha < \frac{\psi'(\lambda_{(+)})}{\psi(\lambda_{(+)})}, \qquad \lambda_{(0)} \leqslant \lambda_{(+)}. \tag{3.10.2}$$

If the last inequality is not satisfied, one can take $\lambda_{(+)}$ as $\lambda_{(0)}$.

II. *Subsequent approximations.* Since the function $A'(\lambda)$ is analytic in $[\lambda_{(0)}, \lambda(\alpha)]$ and $A''(\lambda) > 0$, for values of $|\lambda_{(0)} - \lambda(\alpha)|$ that are not too large the following approximation works well:

$$A'(\lambda(\alpha)) \equiv \alpha \approx A'(\lambda_{(0)}) + (\lambda(\alpha) - \lambda_{(0)})A''(\lambda_{(0)}).$$

So, as the first approximation, $\lambda_{(1)}$, it is natural to define

$$\lambda_{(1)} := \lambda_{(0)} + \frac{\alpha - A'(\lambda_{(0)})}{A''(\lambda_{(0)})}.$$

As the second approximation, $\lambda_{(2)}$, we define

$$\lambda_{(2)} := \lambda_{(1)} + \frac{\alpha - A'(\lambda_{(1)})}{A''(\lambda_{(1)})}$$

and so on. This is the well-known Newton's sequential approximation method with fast (quadratic) speed of convergence (see e.g. [105]). Each step of this method requires one to compute

$$A'(\lambda) \quad \text{and} \quad A''(\lambda) = \frac{\psi''(\lambda)}{\psi(\lambda)} - (A'(\lambda))^2$$

at the points $\lambda = \lambda_{(j)}, j = 0, 1, \ldots$, i.e. to compute for $\lambda = \lambda_{(j)}$ the three integrals

$$\psi(\lambda) = \int e^{\lambda t} dF(t), \qquad \psi(\lambda) = \int t e^{\lambda t} dF(t), \qquad \psi''(\lambda) = \int t^2 e^{\lambda t} dF(t).$$

If we call each of these computations a computational operation (this does not pose a difficulty for modern computers) then each step of the above-mentioned procedure requires three operations (the zeroth approximation requires two operations to compute a and a_2, and, possibly, additional operations to check (3.10.2)).

III. *Approximations for* $\Lambda(\alpha)$ *and* $\sigma(\alpha)$. The jth approximation $\Lambda_{(j)}$ for $\Lambda(\alpha)$ has the form

$$\Lambda_{(j)} = \lambda_{(j)}\alpha - \ln \psi(\lambda_{(j)}).$$

Clearly, $\Lambda_{(j)}$ converges from below to $\Lambda(\alpha)$. Since the definitive element of the asymptotics of $\mathbf{P}(S_n \geqslant x)$ is the factor $e^{-n\Lambda(\alpha)}$, $\alpha = x/n$, then one should stop the computational procedure at a step j when the difference $n(\Lambda_{(j)} - \Lambda_{(j-1)})$ is small. For example, in the case when 10 per cent relative approximation precision is required, one should choose j such that

$$\left|\Lambda_{(j)} - \Lambda_{(j-1)}\right| < \frac{1}{10n}. \tag{3.10.3}$$

If n is not too large (say, $n < 30$), a three-step approximation may be sufficient to achieve this precision.

As an approximation of $\sigma(\alpha)$, one should take the value

$$\sigma_{(j)} = \frac{\psi''(\lambda_{(j)})}{\psi(\lambda_{(j)})} - \alpha^2.$$

Then the approximate value of $\mathbf{P}(S_n \geqslant x)$ will be determined by the formula (2.2.8), in which $\Lambda(\alpha)$, $\lambda(\alpha)$ and $\sigma(\alpha)$ should be replaced by, respectively, $\Lambda_{(j)}$, $\lambda_{(j)}$ and $\sigma_{(j)}$ for step j, which ensures, first of all, a sufficiently good approximation of $\Lambda(\alpha)$ (see e.g. (3.10.3)).

3.10.2 Sequential approximations for the values λ_1, α_1, p (see (3.10.1))

The value λ_1 for $a = \mathbf{E}\xi < 0$, $\lambda_+ > 1$, is the positive solution of the equation $\psi(\lambda) = 1$. Here, again, it is natural to use Newton's method. If there are no other options, the solution of the equation $\psi_{(0)}(\lambda) = 1$, where $\psi_{(0)}$ corresponds to the normal distribution with parameters (a, σ^2), can be taken as the zeroth approximation. Since $\ln \psi_{(0)}(\lambda) = \lambda a + \sigma^2\lambda^2/2$, as the zeroth approximation one should take

$$\lambda_{(0)} = -\frac{2a}{\sigma^2};$$

and verify that $\psi'(\lambda_{(0)}) > 0$. If this inequality is not satisfied (this occurs only in rare cases), then as the zeroth approximation one should take $\lambda^*_{(0)} \in (\lambda_{(0)}, \lambda_+)$ such that $\psi'(\lambda^*_{(0)}) > 0$.

Since for small values $\lambda_1 - \lambda_{(j)}, j \geqslant 0$, we have

$$1 = \psi(\lambda_1) \approx \psi(\lambda_{(j)}) + (\lambda_1 - \lambda_{(j)})\psi'(\lambda_{(j)}),$$

as the $(j+1)$th approximation $\lambda_{(j+1)}$ one should take the value

$$\lambda_{(j+1)} = \lambda_{(j)} + \frac{1 - \psi(\lambda_{(j)})}{\psi'(\lambda_{(j)})}.$$

According to (3.4.24) and (3.10.1),

$$\mathbf{P}(S \geqslant x) = p e^{-\lambda_1 x}\big(1 + o(1)\big) \tag{3.10.4}$$

as $x \to \infty$; thus for large x the approximation $\lambda_{(j)}$ is required to satisfy additional properties, for example,

$$\big|\lambda_{(j)} - \lambda_{(j-1)}\big| < \frac{1}{10x}$$

(see (3.10.3)).

The number α_1, which appears in several statements in Chapter 3, is given by (see Section 3.4.2)

$$\alpha_1 = \psi'(\lambda_1),$$

so that the jth approximation for α_1 is $\psi'(\lambda_{(j)})$.

The coefficient $p < 1$ in (3.10.4) depends on the distribution $\mathbf{F}$ in a very complicated way (see section 3.4) and it is not possible to obtain simple numerical approximation procedures for it. One can use the different interpretation of p provided in [39] to propose a method of statistical estimation for it. Let $\{S_k^{(\lambda_1)}\}$ be a random walk with jumps $\xi_k^{(\lambda_1)}$, $\mathbf{E}\xi^{(\lambda_1)} = \psi'(\lambda_1) > 0$, which have distribution $\mathbf{F}^{(\lambda_1)}$. Let $\chi^{(\lambda_1)}$ be the first overshoot of the random walk $\{S_k^{(\lambda_1)}\}$ over an infinitely distant barrier. Then Theorem 3.4.11 implies that in relation (3.10.4)

$$p = \mathbf{E}e^{-\lambda_1 \chi^{(\lambda_1)}}. \tag{3.10.5}$$

The relation (3.10.5) allows us to construct a consistent statistical estimator for p in the following way. If there is a good approximation of $\lambda_{(j)}$ for λ_1 then the distribution $\mathbf{F}^{(\lambda_1)}$ can be considered known. We can simulate the random walk $\{S_k^{(\lambda_1)}\}$ until the first positive sum $\chi_1(0)$ appears. Repeating such a simulation n times, we can obtain an empirical distribution function H_n^* for the quantity $\chi_1(0)$. But the distribution $\chi^{(\lambda_1)}$ can be explicitly represented through the distribution $\chi_1(0)$ (see Corollary 10.4.1 in [39]; there is a typographical error in the statement of the corollary):

$$\mathbf{P}(\chi^{(\lambda_1)} \geqslant v) = \frac{1}{\mathbf{E}\chi_1(0)} \int_v^\infty \mathbf{P}\big(\chi_1(0) > t\big)dt.$$

Therefore,

$$\mathbf{E}e^{-\lambda_1 \chi^{(\lambda_1)}} = \frac{1}{\lambda_1 \mathbf{E}\chi_1(0)}\left[1 - \int_0^\infty e^{-\lambda_1 v}\mathbf{P}\big(\chi_1(0) \in dv\big)\right]. \tag{3.10.6}$$

Let $\mathrm{x}_1, \dots, \mathrm{x}_n$ be empirical values of $\chi_1(0)$ (the points of jumps of the function H_n^*), $\mathrm{y}_k = e^{-\lambda_1 \mathrm{x}_k}$,

$$\overline{\mathrm{x}}_n = \frac{1}{n}\sum_{k=1}^{n} \mathrm{x}_k,\ \overline{\mathrm{y}}_n = \frac{1}{n}\sum_{k=1}^{n} \mathrm{y}_k.$$

Then, using the substitution method (see [30], § 2.13), we obtain for p the consistent estimator

$$p^* = \frac{[1-\overline{\mathrm{y}}_n]}{\lambda_1 \overline{\mathrm{x}}_n}. \tag{3.10.7}$$

One can show that the estimator p^* is asymptotically normal and, using it, one can construct a confidence interval for p. Indeed, consider the event (see (3.10.5), (3.10.6))

$$\begin{aligned} B := \left\{p - p^* < \frac{v}{\sqrt{n}}\right\} &= \left\{\frac{1-\mathbf{E}\mathrm{y}_1}{\mathbf{E}\mathrm{x}_1} - \frac{1-\overline{\mathrm{y}}_n}{\overline{\mathrm{x}}_n} < \frac{v\lambda_1}{\sqrt{n}}\right\} \\ &= \left\{(\overline{\mathrm{x}}_n - \mathbf{E}\mathrm{x}_1)(1-\mathbf{E}\mathrm{y}_1) + (\overline{\mathrm{y}}_n - \mathbf{E}\mathrm{y}_1)\mathbf{E}\mathrm{x}_1 < \frac{v\lambda_1}{\sqrt{n}}\,\overline{\mathrm{x}}_n\,\mathbf{E}\mathrm{x}_1\right\}, \end{aligned}$$

and let

$$z_k = (\mathrm{x}_k - \mathbf{E}\mathrm{x}_1)(1-\theta-\mathbf{E}\mathrm{y}_1) + (\mathrm{y}_k - \mathbf{E}\mathrm{y}_1)\mathbf{E}\mathrm{x}_1, \tag{3.10.8}$$

where $\theta = v\lambda_1/\sqrt{n}$. The random variables z_k are independent, centered and identically distributed, and the event B can be written as

$$B = \left\{\sum_{k=1}^{n} z_k < v\lambda_1\sqrt{n}(\mathbf{E}\mathrm{x}_1)^2\right\}.$$

Hence, we can use the central limit theorem:

$$\mathbf{P}\left(p - p^* < \frac{v}{\sqrt{n}}\right) \to \Phi\left(\frac{v\lambda_1(\mathbf{E}\mathrm{x}_1)^2}{\sigma_z}\right), \tag{3.10.9}$$

or, equivalently,

$$\mathbf{P}\left(\frac{(p-p^*)(\mathbf{E}\mathrm{x}_1)^2}{\sigma_z} < \frac{v}{\sqrt{n}}\right) \to \Phi(v\lambda_1), \tag{3.10.10}$$

where $\sigma_z^2 = \mathbf{D}z_1$. Relations (3.10.9) and (3.10.10) do not formally mean that the estimator p^* is asymptotically normal, since σ_z depends on θ and hence on v. However, it is not hard to see that, when computing σ_z^2, the terms containing v will sum up to $O_p(1/\sqrt{n})$. Therefore, they will be negligible and can be ignored, preserving (3.10.9) and (3.10.10). This implies the asymptotic normality of p^*.

Furthermore, it is clear that relation (3.10.10) (where θ is replaced by 0) is preserved if one replaces $\mathbf{E}\mathrm{x}_1, \mathbf{E}\mathrm{y}_1, \sigma_z^2$ with consistent estimators $\overline{\mathrm{x}}_n, \overline{\mathrm{y}}_n, (\sigma_z^2)^* = n^{-1}\sum_{k=1}^{n}(z_k^*)^2$, respectively, where

$$z_k^* = (\mathrm{x}_k - \overline{\mathrm{x}}_n)(1-\overline{\mathrm{y}}_n) + (\mathrm{y}_k - \overline{\mathrm{y}}_n)\overline{\mathrm{x}}_n.$$

The computation of $(\sigma_z^2)^*$ is quite easy and we leave it to the reader.

From relation (3.10.10) and the remarks made above, it follows that the interval with end points

$$p^* \pm \frac{\sigma_z^* v_{\delta/2}}{\lambda_1(\overline{x}_n)^2\sqrt{n}},$$

where v_δ is the normal quantile of level $1-\delta$ $(\Phi(v_\delta) = 1-\delta)$, will be an asymptotic confidence interval at significance level $1 - \delta$.

Here we also recall that, along with (3.10.4), the inequality

$$\mathbf{P}(S \geqslant x) \leqslant e^{-\lambda_1 x}$$

is always true (see Theorem 1.1.2), and also true are the exact inequalities for $\mathbf{P}(S \geqslant x)$ from Theorem 3.4.8.

4

Large deviation principles for random walk trajectories

As before, let $\xi, \xi_1, \xi_2, \dots$ be a sequence of independent identically distributed d-dimensional random vectors,

$$S_0 := 0, \qquad S_n := \sum_{i=1}^{n} \xi_i \quad \text{for} \quad n \geqslant 1. \tag{4.0.1}$$

Everywhere in what follows it will be assumed that the distribution of ξ is non-degenerate and satisfies Cramér's moment condition in some form (unless the contrary is explicitly stated). Further, let $s_n = s_n(t)$, $t \in [0, 1]$, be a trajectory of the 'normed' random walk $\{S_k\}_{k=0}^{n}$, constructed in some way (which will be specified later) on the nodes $\big(k/n, S_k/x\big)$, $k = 0, \dots, n$. The trajectories $s_n(t)$ will be considered as elements of certain metric spaces of functions on $[0, 1]$. We will study the asymptotic properties of distributions of the trajectories $s_n(t)$, $t \in [0, 1]$, in most general form: for an arbitrary measurable set B (in the corresponding measurable space of functions) we consider the asymptotic behaviour of the probability $\mathbf{P}(s_n \in B)$ as $n \to \infty$, when $x \gg \sqrt{n}$, $x = O(n)$. Certainly, this general frame implies a loss of accuracy of the results. One cannot expect to describe the asymptotic behaviour of $\mathbf{P}(s_n \in B)$ precisely if B is an arbitrary set. However, if we restrict our study to so-called crude (or logarithmic) asymptotics, i.e. asymptotics of $\ln \mathbf{P}(s_n \in B)$ as $n \to \infty$, then it is possible to get relatively general and complete results. The present chapter is devoted to such results. As we noted before, the trajectories $s_n(t)$ will be considered as elements of certain functional spaces. These spaces differ in both the nature of their elements and the metric used.

In that connection, we will need several concepts and results related to large deviation principles (l.d.p.'s) for arbitrary metric spaces.

4.1 On large deviation principles in metric spaces

Let $(\mathbb{Y}, \rho)$ be a metric space with metric ρ and σ-algebra $\mathfrak{B}(\mathbb{Y}, \rho)$ of Borel sets in it. Further, let $\{\eta_n,\ n = 1, 2, \dots\}$ be a sequence of random elements in $\langle \mathbb{Y}, \mathfrak{B}(\mathbb{Y}, \rho)\rangle$. If, for some set $B_0 \in \mathfrak{B}(\mathbb{Y}, \rho)$ and for any $\varepsilon > 0$ the condition

$$\lim_{n\to\infty} \mathbf{P}\big(\eta_n \in (B_0)_\varepsilon\big) = 1 \tag{4.1.1}$$

is met, where $(B)_\varepsilon$ is an ε-neighbourhood of the set B, then, for any set $B \in \mathfrak{B}(\mathbb{Y}, \rho)$ such that $(B_0)_\varepsilon \cap B = \varnothing$ for some $\varepsilon > 0$, the probability $\mathbf{P}(\eta_n \in B)$ will tend to 0 as $n \to \infty$ and will characterise the distribution of η_n in the large deviation zone. If η_n converges in probability to a non-random element y_0: $\mathbf{P}\big(\eta_n \in (y_0)_\varepsilon\big) \to 1$ for any $\varepsilon > 0$ as $n \to \infty$, where $(y)_\varepsilon = \big\{\widetilde{y} \in \mathbb{Y} : \rho(y, \widetilde{y}) < \varepsilon\big\}$ is an ε-neighbourhood of the point y, then (4.1.1) holds for $B_0 = \{y_0\}$.

Definition 4.1.1. We will say that *a sequence $\{\eta_n\}$ satisfies the local large deviation principle in the space $(\mathbb{Y}, \rho)$ (the l.l.d.p. in the space $(\mathbb{Y}, \rho)$, or simply the l.l.d.p.), if there exists a numerical sequence $z_n \to \infty$ as $n \to \infty$, and a function $D = D(y)$: $\mathbb{Y} \to [0, \infty]$, such that, for any $y \in \mathbb{Y}$,*

$$\lim_{\varepsilon\to 0} \overline{\lim_{n\to\infty}} \frac{1}{z_n} \ln \mathbf{P}(\eta_n \in (y)_\varepsilon) = \lim_{\varepsilon\to 0} \underline{\lim_{n\to\infty}} \frac{1}{z_n} \ln \mathbf{P}(\eta_n \in (y)_\varepsilon) = -D(y). \tag{4.1.2}$$

Definition 4.1.1 is equivalent to the following.

Definition 4.1.2. *For any $y \in \mathbb{Y}$ and any sequence ε_n converging to zero slowly enough as $n \to \infty$ the following relation holds true:*

$$\lim_{n\to\infty} \frac{1}{z_n} \ln \mathbf{P}\big(\eta_n \in (y)_{\varepsilon_n}\big) = -D(y). \tag{4.1.3}$$

Proof. of the equivalence of Definitions 4.1.1 and 4.1.2. Put

$$d_{n,\varepsilon} := \Big|\mathbf{P}^{1/z_n}\big(\eta_n \in (y)_\varepsilon\big) - e^{-D(y)}\Big|.$$

Then it is clear that (4.1.2) is equivalent to the relation

$$\lim_{\varepsilon\to 0} \overline{\lim_{n\to\infty}} d_{n,\varepsilon} = 0. \tag{4.1.4}$$

Let (4.1.4) hold true. Then

$$d_{(\varepsilon)} := \overline{\lim_{n\to\infty}} d_{n,\varepsilon} \to 0 \quad \text{as} \quad \varepsilon \to 0,$$

and hence for any $\delta > 0$ there exists an $\varepsilon(\delta)$ such that $d_{(\varepsilon(\delta))} < \delta$. This, in turn, means that there exists an $n(\delta)$ such that, for all $n \geqslant n(\delta)$, the following inequality holds:

$$d_{n,\varepsilon} \leqslant 2\delta. \tag{4.1.5}$$

Without loss of generality, one can assume that the functions $\varepsilon(\delta)$ and $n(\delta)$ are monotone. Then, denoting by $\delta(\varepsilon)$ and ε_n generalised solutions to the equations $\varepsilon(\delta) = \varepsilon$ and $n(\delta(\varepsilon)) = n$, respectively, we can rewrite relation (4.1.5) as

$$d_{n,\varepsilon_n} < 2\delta(\varepsilon_n)$$

for all $n > n(\delta(\varepsilon))$. Since $\delta(\varepsilon_n) \to 0$ as $n \to \infty$, we obtain (4.1.3). However, the function $n(\delta)$ can be chosen to grow at an arbitrarily fast rate. Hence the 'inverse' function ε_n can decay at an arbitrarily slow rate. This proves (4.1.3).

Conversely, now let (4.1.3), i.e.

$$d_{n,\varepsilon_n} \to 0 \quad \text{as} \quad n \to \infty, \tag{4.1.6}$$

for all ε_n such that $\varepsilon_n \downarrow 0$ slowly enough as $n \uparrow \infty$. Assume the contrary to what we want to show: that relation (4.1.4) is false, and therefore there exists a sequence $\{\varepsilon(k)\}$, $\varepsilon(k) \downarrow 0$ as $k \uparrow \infty$ such that $d_{(\varepsilon(k))} > 2c = \text{const.}$ for all k. This means that for every $k \geqslant 1$ there exists a sequence $n(l,k)$, $l = 1, 2, \ldots$, that increases unboundedly in l and is such that

$$d_{n(l,k),\varepsilon(k)} > c \quad \text{for all} \quad l \geqslant 1, \;\; k \geqslant 1.$$

It is clear that one can always choose a sequence $(n,\ \varepsilon_n)$, from the array $\{n(l,k), \varepsilon(k);\ l \geqslant 1,\ k \geqslant 1\}$, such that $\varepsilon_n \downarrow 0$ arbitrarily slowly as $n \to \infty$. This contradicts relation (4.1.6). We have proved the desired equivalence.

Definitions 4.1.1 and 4.1.2 elucidate the probabilistic meaning of $z_n D(y)$: *for small ε the values $-\ln \mathbf{P}(\eta_n \in (y)_\varepsilon)$ grow with n approximately as $z_n D(y)$*. It follows from (4.1.2) that the factors z_n and $D(y)$ in that product are defined up to a constant factor.

Note that from Definition 4.1.1 it necessarily follows that *the function $D(y)$ is lower semicontinuous*; i.e., for any $y \in \mathbb{Y}$ one has

$$\varliminf_{\rho(y_k,y)\to 0} D(y_k) \geqslant D(y). \tag{4.1.7}$$

More precisely, if the second relation in (4.1.2) holds true then (4.1.7) is also true.

Indeed, for all large enough k one has

$$(y_k)_\varepsilon \subset (y_k)_{\varepsilon+\rho(y_k,y)} \subset (y_k)_{2\varepsilon}.$$

Therefore, for such k, by virtue of the second relation in (4.1.2) one has

$$\begin{aligned} -D(y_k) &= \lim_{\varepsilon\to\infty} \varliminf_{n\to\infty} \frac{1}{z_n} \ln \mathbf{P}\big(\eta_n \in (y_k)_\varepsilon\big) \leqslant \varliminf_{n\to\infty} \frac{1}{z_n} \ln \mathbf{P}\big(\eta_n \in (y_k)_\varepsilon\big) \\ &\leqslant \varliminf_{n\to\infty} \frac{1}{z_n} \ln \mathbf{P}(\eta_n \in (y_k)_{\varepsilon+\rho(y_k,y)}) \leqslant \varliminf_{n\to\infty} \frac{1}{z_n} \ln \mathbf{P}(\eta_n \in (y_k)_{2\varepsilon}). \end{aligned}$$

From this, again by virtue of the second relation in (4.1.2), we get

$$\varlimsup_{k\to\infty} (-D(y_k)) = -\varliminf_{k\to\infty} D(y_k) \leqslant \lim_{\varepsilon\to 0} \varliminf_{n\to\infty} \frac{1}{z_n} \ln \mathbf{P}(\eta_n \in (y)_{2\varepsilon}) = -D(y).$$

This proves (4.1.7).

There is one more definition equivalent to Definitions 4.1.1 and 4.1.2.

Definition 4.1.3. This definition is obtained from Definition 4.1.1 if we add to the latter the assumption that *the function $D(y)$ is lower semicontinuous* and replace property (4.1.2) with the following:

For any $y \in \mathbb{Y}$ there exist functions $\delta(\varepsilon) \to 0$ and $\beta(\varepsilon) \to 0$ as $\varepsilon \to 0$, and one has

$$\overline{\lim_{n\to\infty}} \frac{1}{z_n} \ln \mathbf{P}\big(\eta_n \in (y)_\varepsilon\big) \leqslant -D\big((y)_{\delta(\varepsilon)}\big) + \beta(\varepsilon), \tag{4.1.8}$$

where $D(B) := \inf_{y\in B} D(y)$. *Moreover, for any* $y \in \mathbb{Y}$ *and* $\varepsilon > 0$,

$$\underline{\lim_{n\to\infty}} \frac{1}{z_n} \ln \mathbf{P}(\eta_n \in (y)_\varepsilon) \geqslant -D(y). \tag{4.1.9}$$

Later we will see that in (4.1.8) one necessarily has $\delta(\varepsilon) \geqslant \varepsilon$ (see Theorem 4.1.6).

Proof of the equivalence of Definitions 4.1.1 and 4.1.3. (i) Let the left-hand side of (4.1.2) be equal to the right-hand side. Then

$$\overline{\lim_{n\to\infty}} \frac{1}{z_n} \ln \mathbf{P}(\eta_n \in (y)_\varepsilon) \leqslant -D(y) + \beta(\varepsilon) \leqslant -D\big((y)_{\delta(\varepsilon)}\big) + \beta(\varepsilon).$$

If the middle expression in (4.1.2), is equal to the right-hand expression, then

$$\underline{\lim_{n\to\infty}} \frac{1}{z_n} \ln \mathbf{P}\big(\eta_n \in (y)_\varepsilon\big) \geqslant -D(y) - \beta(\varepsilon).$$

Since the left-hand side of this inequality decreases monotonically as $\varepsilon \downarrow 0$, one can apply the inequality with ε/k in place of ε to obtain

$$\underline{\lim_{n\to\infty}} \frac{1}{z_n} \ln \mathbf{P}\big(\eta_n \in (y)_\varepsilon\big) \geqslant \underline{\lim_{n\to\infty}} \frac{1}{z_n} \ln \mathbf{P}\big(\eta_n \in (y)_{\varepsilon/k}\big) \geqslant -D(y) - \beta(\varepsilon/k).$$

Since k is arbitrary, relation (4.1.9) holds true. The lower semicontinuity (4.1.7) of the function D from Definition 4.1.1 was established earlier.

(ii) Conversely, let (4.1.8) hold true. Then, for any sequence $\varepsilon_k \to 0$ as $k \to \infty$, there exists a sequence $y_k \in (y)_{\delta(\varepsilon_k)}$ such that $D(y_k) \leqslant D\big((y)_{\delta(\varepsilon_k)}\big) + 1/k$. Therefore, for any k, by virtue of (4.1.8) one has

$$\overline{\lim_{n\to\infty}} \frac{1}{z_n} \ln \mathbf{P}\big(\eta_n \in (y)_{\varepsilon_k}\big) \leqslant -D(y_k) + \frac{1}{k} + \beta(\varepsilon_k),$$

$$\lim_{k\to\infty} \overline{\lim_{n\to\infty}} \frac{1}{z_n} \ln \mathbf{P}\big(\eta_n \in (y)_{\varepsilon_k}\big) \leqslant -\underline{\lim_{k\to\infty}} D(y_k).$$

However, $y_k \to y$ as $k \to \infty$ and hence, by virtue of the lower semicontinuity of the function $D(y)$, we get

$$\lim_{k\to\infty} \overline{\lim_{n\to\infty}} \frac{1}{z_n} \ln \mathbf{P}\big(\eta_n \in (y)_{\varepsilon_k}\big) \leqslant -D(y).$$

Together with (4.1.9) this means that (4.1.2) holds true. The desired equivalence is proved.

Note that the l.l.d.p. is defined differently in [60]. The definition in [60] contains excessive requirements, in particular the requirement of uniformity which is not peculiar to the notion of locality. This narrows the area of application of the definition.

Denote by (B) the interior of the sets $B \in \mathfrak{B}(\mathbb{Y}, \rho)$ (i.e. the totality of all points that are contained in B together with a neighbourhood thereof), and by $[B] = \mathbb{Y} \setminus (\overline{B})$ denote the closure of B, where $\overline{B}$ is the complement of B.

Definition 4.1.4. We will say that *the sequence* $\{\eta_n\}$ *satisfies the extended large deviation principle in the space* $(\mathbb{Y}, \rho)$ (*the e.l.d.p. in the space* $(\mathbb{Y}, \rho)$ *or simply the e.l.d.p.*) if there exists a numerical sequence $z_n \to \infty$ as $n \to \infty$ and a *lower semicontinuous function* $D(y)$ such that, for any $B \in \mathfrak{B}(\mathbb{Y}, \rho)$, one has the inequalities

$$L^*(B) := \overline{\lim_{n\to\infty}} \frac{1}{z_n} \ln \mathbf{P}(\eta_n \in B) \leqslant -D(B+), \tag{4.1.10}$$

$$L_*(B) := \underline{\lim_{n\to\infty}} \frac{1}{z_n} \ln \mathbf{P}(\eta_n \in B) \geqslant -D\big((B)\big), \tag{4.1.11}$$

where

$$D(B) := \inf_{y\in B} D(y), \qquad D(B+) := \lim_{\delta\to 0} D((B)_\delta). \tag{4.1.12}$$

In (4.1.12) we assume that $D(\varnothing) = \infty$, i.e. the lower bound of the function on the empty set is equal to ∞.

To provide somewhat fuller characterisations of the concepts introduced in Definitions 4.1.1–4.1.4 we will sometimes be write 'the l.l.d.p. (or e.l.d.p.) with parameters (z_n, D) or $(z_n, D(y))$'.

If $D(B+) = D((B))$ then one can replace the right-hand sides of (4.1.10) and (4.1.11) with $D(B)$; then these equations become equalities instead of inequalities. The meaning of these equalities is that the asymptotics of $\ln \mathbf{P}(\eta_n \in B)$ *admits factorisation*, i.e. it can be represented as a product of two factors, z_n and $D(B)$, of which the first depends only on n and the second only on B.

Recall that in the literature by the *large deviation principle (l.d.p.)* for a sequence of random elements $\{\eta_n\}$ in a metric space $(\mathbb{Y}, \rho)$ one understands the following (see e.g. [155], [156], [178], [179]):

(1) The space $(\mathbb{Y}, \rho)$ is assumed to be *complete and separable*.
(2) There is a so-called *deviation function* (rate function) $D = D(y)$, $y \in \mathbb{Y}$, taking values in $[0, \infty]$, that has the following two properties:
(2a) it is *lower semicontinuous*;
(2b) *the set* $D_v := \{y : D(y) \leqslant v\}$ *is compact in* $(\mathbb{Y}, \rho)$ *for any* $v \geqslant 0$.
(3) Relations (4.1.10) and (4.1.11) hold true with *the right-hand side of* (4.1.10) *replaced by* $D([B])$ *and* z_n *replaced by* n.

Some of the listed assumptions of the l.d.p. (and especially (1), (2b) and the condition $z_n = n$) are rather restrictive and do not hold in many important problems related to random walks (see below and also [38], [60]). It is because of this that we introduced Definition 4.1.4. There we used the adjective 'extended' and also included a reference to a metric space (that can be different even in closely related problems; see below) to stress the difference between the e.l.d.p. and the 'usual' l.d.p. considered in the literature (see e.g. [155], [156], [178], [179]). The difference is shown in the following:

(1) In the e.l.d.p. the space $(\mathbb{Y},\rho)$ is not assumed to be complete and separable.
(2) In the e.l.d.p. there is no assumption that, for any $v \geqslant 0$, the set D_v is compact (see below and also [38], [60]).
(3) In the e.l.d.p., any growth rate z_n is admissible for $-\ln \mathbf{P}(\eta_n \in B)$ (as we have already pointed out, $z_n = n$ in the 'usual' l.d.p.; see e.g. [179]).
(4) In the 'usual' l.d.p., instead of $D(B+)$ we have $D([B])$ on the right-hand side of (4.1.10). As $[B] \subseteq (B)_\varepsilon$ for any $\varepsilon > 0$, one has $D([B]) \geqslant D(B+)$. In Lemma 4.1.5 below we give conditions sufficient for $D([B]) = D(B+)$.

Note also that, in the definition of the e.l.d.p., we state the metric space $(\mathbb{Y},\rho)$ to which the elements η_n belong, and we sometimes indicate the parameters (z_n, D). In what follows, for the sake of expositional uniformity the same information will be provided for the l.d.p. as well; i.e., we will write 'l.d.p. in $(\mathbb{Y},\rho)$ with parameters (z_n, D)'. This information is important, as the parameters can be different even when studying very close objects. In the literature devoted to the l.d.p. this information, as a rule, is omitted since the space $\mathbb{Y}$ and the choice $z_n = n$ are fixed.

It follows from the above that the l.d.p. in $(\mathbb{Y},\rho)$ with parameters (z_n, D) always implies the e.l.d.p. in $(\mathbb{Y},\rho)$ with the same parameters (z_n, D), but not vice versa.

5. Because of the differences listed in items 1–4 above and also owing to the broader conditions ensuring the validity of the e.l.d.p. (compared to the conditions for the l.d.p.; see below), using the e.l.d.p. instead of the l.d.p. enables one to substantially broaden both the class of objects for which assertions of the l.d.p.-type hold true and also the conditions under which inequalities (4.1.10), (4.1.11) will hold for objects already studied.

Further, we will present conditions sufficient for the equality

$$D\big([B]\big) = D(B+) \tag{4.1.13}$$

to hold.

Lemma 4.1.5. ([60]) *If the set D_v is compact for any $v \geqslant 0$ then* (4.1.13) *holds true.*

Thus, for compact D_v the right-hand sides in inequalities (4.1.10) and (4.1.11) for the e.l.d.p. and l.d.p. coincide.

The l.d.p. for sums S_n of random vectors, i.e. for random elements

$$\eta_n = \frac{S_n}{n} \in \mathbb{Y} = \mathbb{R}^d,$$

was proved in section 2.7. The role of the function $D(y)$ in (4.1.2) was played there by the deviation function $\Lambda(y)$, $y \in \mathbb{R}^d$, studied in detail in sections 1.1 and 1.2 (see also [39], Chapter 9). In this chapter we will study the l.d.p. for scaled trajectories $s_n(t)$, say, of the form $S_{[nt]}/x$ or their continuous modifications in a suitable metric space of functions on $[0, 1]$, e.g. in the space of functions without discontinuities of the second kind or in the space of continuous functions.

If the Cramér moment condition is satisfied then one can take $z_n = n$ (the usual l.d.p.) or $z_n = x^2/n$ (the moderately large deviation principle) or $z_n \gg n$ (for super-large deviations; for more details, see [60]).

When studying large deviations in functional spaces, the function D is sometimes referred to as the *action functional.* To make the terminology more consistent, and since the meaning of the latter term when applied to random walks is somewhat unclear, we will call D the *(large) deviation function* when dealing with the state space of the process, and call it the *(large) deviation functional* when dealing with the trajectory space of the process.

Now consider the conditions that enable one to obtain the e.l.d.p. from the l.l.d.p. Recall that a set $T \subset \mathbb{Y}$ is called *totally bounded* in $\mathbb{Y}$ (see e.g. [110]), if, for any $\varepsilon > 0$, there exists a finite ε-covering of the set T (i.e., there exists a collection of points $y_1, \ldots, y_R$ in $\mathbb{Y}$, $R = R(\varepsilon) < \infty$, such that $\bigcup_{j=1}^{R}(y_j)_\varepsilon \supset T$). As it is well known, if a totally bounded subset of a complete metric space is closed then it is compact (see e.g. [110]). A compact set is always totally bounded.

We will need the following condition on the distribution of $\{\eta_n\}$:

[K] *There is a family of imbedded compact sets K_v, $v > 0$, such that, for any $N > 0$, there exists a $v > 0$ such that, for any $\varepsilon > 0$, one has*

$$\overline{\lim_{n\to\infty}} \frac{1}{z_n} \ln \mathbf{P}(\eta_n \notin (K_v)_\varepsilon) \leqslant -N. \tag{4.1.14}$$

The condition **[K$_0$]** which is obtained from **[K]** by replacing $(K_v)_\varepsilon$ in (4.1.14) with K_v, is sufficient for **[K]** to hold.

Now consider a more general condition.

[TB] *There exists a family of imbedded totally bounded sets T_v, $v > 0$, such that, for any $N > 0$, there exists a $v > 0$ such that, for any $\varepsilon > 0$, one has*

$$\overline{\lim_{n\to\infty}} \frac{1}{z_n} \ln \mathbf{P}(\eta_n \notin (T_v)_\varepsilon) \leqslant -N. \tag{4.1.15}$$

The condition **[TB$_0$]**, which is obtained from **[TB]** by replacing $(T_v)_\varepsilon$ in (4.1.15) with T_v, is sufficient for **[TB]** .

In what follows, conditions **[K]** or **[TB]** will be met for the sets $T_v = D_v := \{y : D(y) \leqslant v\}$.

There are at least two ways to prove the l.d.p. (or the e.l.d.p.) using the l.l.d.p. One is based on using the l.l.d.p. and condition **[K]**. The other uses the l.l.d.p. and condition **[TB]** but, in that case, in the l.l.d.p. (see Definition 4.1.3), one has to assume the *uniformity of the upper bounds on the sets T_v* (see the remark preceding Definition 4.1.1).

If one can find 'highly likely' compacts then, of course, one should use the first way. If not then one should be looking for 'highly likely' totally bounded sets (which exist under broader conditions) and then proving the uniformity of upper bounds on these sets in the l.l.d.p.

The above-mentioned two ways are followed in the next assertions elucidating the relationship between the l.l.d.p and the e.l.d.p.

Theorem 4.1.6. (i) *Let the l.l.d.p. and condition* **[K]** *hold. Then relation* (4.1.10) *(from the e.l.d.p.) holds true.*

(ii) *Conversely, if relation* (4.1.10) *holds true then* (4.1.8) *holds for* $\delta(\varepsilon) \geqslant \varepsilon$.

(iii) *Inequalities* (4.1.9) *(from the l.l.d.p.) and* (4.1.11) *(from the e.l.d.p.) are equivalent.*

Thus, under condition **[K]** the l.l.d.p. and the e.l.d.p. are equivalent.

Theorem 4.1.7. (i) *Let the condition* **[TB]** *be satisfied and, for any* $v > 0$, *the upper bounds in the l.l.d.p. be uniform on the set* T_v *(the functions* $\delta(\varepsilon)$ *and* $\beta(\varepsilon)$ *in* (4.1.8) *do not depend on* y *for* $y \in T_v$*). Then relation* (4.1.10) *(from the e.l.d.p.) holds true.*

(ii) *Moreover, assertions* (ii), (iii) *from Theorem 4.1.6 hold true.*

Thus, if the condition **[TB]** is met and uniformity takes place in (4.1.8) then the l.l.d.p. and the e.l.d.p. are equivalent. In that case, one necessarily has $\delta(\varepsilon) \geqslant \varepsilon$.

Corollary 4.1.8. *If relations* (4.1.10), (4.1.11) *and*

$$D\big((B)\big) = D(B+)$$

hold true, then there exists the limit

$$\lim_{n\to\infty} \frac{1}{z_n} \ln \mathbf{P}(\eta_n \in B) = -D(B).$$

Remark 4.1.9. (i) Condition **[K$_0$]** is also used to establish the 'usual' l.d.p. In that case, however, one assumes that the space $(\mathbb{Y}, \rho)$ is complete and separable, and one requires in addition that the sets $D_v = \big\{y : D(y) \leqslant v\big\}$ are compact.

(ii) All the assertions of the present chapter will also remain true when the space $(\mathbb{Y}, \rho)$ is 'pseudometric', i.e. when ρ is a pseudometric. (In that case, the equality $\rho(y, \widetilde{y}) = 0$ will not, generally speaking, imply that $y = \widetilde{y}$.)

Proof of Theorem 4.1.6. (i) Fix an arbitrary $M < \infty$ and set $N := \min\{D(B+),\ M\}$ (the number M is introduced for the case where $D(B+) = \infty$). For that N, by virtue of condition **[K]** there exists a compact set $K = K_v$ for which (4.1.14) holds true. Fix $\varepsilon > 0$ and $\delta > 0$. For each $y \in K$, by virtue of (4.1.2) one can choose $\varepsilon_y \in (0, \varepsilon)$ such that

$$\varlimsup_{n\to\infty} \frac{1}{z_n} \ln \mathbf{P}(\eta_n \in (y)_{2\varepsilon_y}) \leqslant -\min\{D(y) - \delta, M\}. \tag{4.1.16}$$

From the open cover $(y)_{\varepsilon_y}$ of the compact K, choose a finite subcover

$$(y_1)_{\beta_1},\ldots,(y_R)_{\beta_R}, \quad \text{where} \quad \beta_k := \varepsilon_{y_k}, \quad k=1,\ldots,R.$$

For $\varepsilon_0 := \min\{\beta_1,\ldots,\beta_R\}$, one has

$$K \subset \bigcup_{k=1}^{R}(y_k)_{\beta_k}, \quad (K)_{\varepsilon_0} \subset \bigcup_{k=1}^{R}(y_k)_{2\beta_k}.$$

Hence,

$$B \subset B\bigcap(K)_{\varepsilon_0} + \overline{(K)}_{\varepsilon_0} \subset U_B + \overline{(K)}_{\varepsilon_0}, \qquad \bigcup_{k\in\mathcal{R}}\{y_k\} \subset (B)_{2\varepsilon}, \tag{4.1.17}$$

where

$$U_B := \bigcup_{k\in\mathcal{R}}(y_k)_{2\beta_k}, \quad \mathcal{R} := \Big\{k\in\{1,\ldots,R\} : B\bigcap(y_k)_{2\beta_k} \neq \emptyset\Big\},$$

$$\mathbf{P}(\eta_n\in B) \leqslant \sum_{k\in\mathcal{R}}\mathbf{P}\big(\eta_n\in(y_k)_{2\beta_k}\big) + \mathbf{P}\big(\eta_n\notin\overline{(K)}_{\varepsilon_0}\big). \tag{4.1.18}$$

By virtue of (4.1.16) and the second relation in (4.1.17), one has, for all $k\in\mathcal{R}$, the inequalities

$$\begin{aligned}\overline{\lim_{n\to\infty}}\frac{1}{z_n}\ln\mathbf{P}(\eta_n\in(y_k)_{2\beta_k}) &\leqslant -\min\big\{D(y_k)-\delta, M\big\}\\ &\leqslant -\min\big\{D((B)_{2\varepsilon})-\delta, M\big\} =: -D_{\varepsilon,\delta,M}.\end{aligned}$$

Using (4.1.14), we now obtain that

$$\overline{\lim_{n\to\infty}}\frac{1}{z_n}\ln\mathbf{P}(\eta_n\notin\overline{(K)_{\varepsilon_0}}) \leqslant -\min\{D(B+),M\} \leqslant -D_{\varepsilon,\delta,M}.$$

Since $k\in\mathcal{R}$ assumes only a finite number $R<\infty$, values, there exists a sequence $\theta_n\to 0$ as $n\to\infty$ that does not depend on k and is such that, for all $k\in\mathcal{R}$ and all large enough n, one has

$$\begin{aligned}\mathbf{P}\big(\eta_n\in(y_k)_{2\beta_k}\big) &\leqslant \exp\big\{-z_n[D_{\varepsilon,\delta,M}-\theta_n]\big\},\\ \mathbf{P}\big(\eta_n\notin\overline{(K)}_{\varepsilon_0}\big) &\leqslant \exp\big\{-z_n[D_{\varepsilon,\delta,M}-\theta_n]\big\}.\end{aligned}$$

The right-hand sides of these inequalities do not depend on k. Hence, by virtue of (4.1.18),

$$\begin{aligned}\mathbf{P}(\eta_n\in B) &\leqslant (R+1)\exp\big\{-z_n[D_{\varepsilon,\delta,M}-\theta_n]\big\},\\ \overline{\lim_{n\to\infty}}\frac{1}{z_n}\ln\mathbf{P}(\eta_n\in B) &\leqslant -D_{\varepsilon,\delta,M} = -\min\Big\{D\big((B)_{2\varepsilon}\big)-\delta, M\Big\}.\end{aligned}$$

Since ε,δ,M are all arbitrary, relation (4.1.10) follows.

(ii) Conversely, if (4.1.10) holds true then, setting $B=(y)_\varepsilon$, one obtains that

$$\overline{\lim_{n\to\infty}}\frac{1}{z_n}\ln\mathbf{P}(\eta_n\in(y)_\varepsilon) \leqslant -D\big((y)_\varepsilon+\big) \leqslant -D\big((y)_{\delta(\varepsilon)}\big)$$

for any function $\delta(\varepsilon) \to 0$ as $\varepsilon \to 0$, $\delta(\varepsilon) \geqslant \varepsilon$. Therefore, it follows from the lower semicontinuity of the function $D(y)$ that

$$\lim_{\varepsilon\to 0}\varlimsup_{n\to\infty}\frac{1}{z_n}\ln\mathbf{P}\big(\eta_n\in(y)_\varepsilon\big)\leqslant -\lim_{\varepsilon\to 0}D\big((y)_{\delta(\varepsilon)}\big)=-D(y).$$

(iii) Let (4.1.9) hold true. For any $y \in (B)$ and all sufficiently small $\varepsilon > 0$, one has $(y)_\varepsilon \subset (B) \subset B$. Therefore,

$$\varliminf_{n\to\infty}\frac{1}{z_n}\ln P_n(B)\geqslant \varliminf_{n\to\infty}\frac{1}{z_n}\ln P_n\big((y)_\varepsilon\big)\geqslant -D(y).$$

Since this relation holds true for any $y \in (B)$, one can put $-D\big((B)\big)$ on the right-hand side of the inequality. This proves (4.1.11).

Conversely, assume that (4.1.11) holds true. For $B = (y)_\varepsilon$ one has that the right-hand side of (4.1.11) is not less than $-D\big((y)_\varepsilon\big) \geqslant -D(y)$. This proves (4.1.9). Theorem 4.1.6 is proved. □

Proof of Theorem 4.1.7. (i). Let M be fixed and v be such that relation (4.1.15) holds true for $N = \min\{M, D(B+)\}$. Then, for all large enough n,

$$\mathbf{P}(\eta_n\in B)\leqslant e^{-z_nN}+\mathbf{P}\big(\eta_n\in B\cap(T_v)_\varepsilon\big). \tag{4.1.19}$$

Cover the set T_v with an ε-net with centres at the points $y_k \in T_v$, $k = 1,\dots,R(\varepsilon)$. Then

$$B\cap(T_v)_\varepsilon\subset\bigcup_{k\in\mathcal{R}}(y_k)_{2\varepsilon},\quad \mathcal{R}:=\Big\{k\in\big\{1,\dots,R(\varepsilon)\big\}:\ B(y_k)_{2\varepsilon}\neq\varnothing\Big\},$$

$$\mathbf{P}\big(\eta_n\in B\cap(T_v)_\varepsilon\big)\leqslant\sum_{k\in\mathcal{R}}\mathbf{P}\big(\eta_n\in(y_k)_{2\varepsilon}\big). \tag{4.1.20}$$

By virtue of inequalities (4.1.8) and the uniformity of these upper bounds for $k \in \mathcal{R}$, one has

$$\varlimsup_{n\to\infty}\frac{1}{z_n}\ln\mathbf{P}(\eta_n\in(y_k)_{2\varepsilon})\leqslant -D((y_k)_{\delta(2\varepsilon)})+\beta(2\varepsilon)\leqslant -D(U_B)+\beta(2\varepsilon),$$

where $\delta(2\varepsilon)$ and $\beta(2\varepsilon)$ are the same for all $y_k \in T_v$ and

$$U_B:=\bigcup_{k\in\mathcal{R}}(y_k)_{\delta(2\varepsilon)}.$$

Note that, as the set U_B is part of $(B)_{3\varepsilon+\delta(2\varepsilon)}$, one has $D(U_B) \geqslant D((B)_{3\varepsilon+\delta(2\varepsilon)})$. Since k assumes only finitely many values, there exists a sequence $\theta_n \to 0$ as $n \to \infty$ that does not depend on k and is such that, for all $n \geqslant n(\varepsilon)$, $k \in \mathcal{R}$, one has

$$\mathbf{P}(\eta_n\in(y_k)_{2\varepsilon})\leqslant\exp\{-z_n[D((B)_{3\varepsilon+\delta(2\varepsilon)})-\beta(2\varepsilon)-\theta_n]\}.$$

Because the right-hand side here does not depend on k, it follows from (4.1.20) that

$$\mathbf{P}(\eta_n \in B(T_v)_\varepsilon) \leqslant R(\varepsilon)\exp\{-z_n[D((B)_{3\varepsilon+\delta(2\varepsilon)}) - \beta(2\varepsilon) - \theta_n]\},$$

$$\overline{\lim_{n\to\infty}} \frac{1}{z_n} \ln \mathbf{P}(\eta_n \in B(T_v)_\varepsilon) \leqslant -D((B)_{3\varepsilon+\delta(2\varepsilon)}) + \beta(2\varepsilon).$$

From here and (4.1.19) one obtains that

$$\overline{\lim_{n\to\infty}} \frac{1}{z_n} \ln \mathbf{P}(\eta_n \in B) \leqslant -\min\{D((B)_{3\varepsilon+\delta(2\varepsilon)}) + \beta(2\varepsilon), \min\{M, D(B+)\}\}.$$

Passing to the limit as $M \to \infty$, $\varepsilon \to 0$, we establish (4.1.10).

(ii) Assertion (ii) is proved in exactly the same way as in Theorem 4.1.6. Theorem 4.1.7 is proved. □

We will now make some remarks concerning the concepts we are dealing with here.

Definition 4.1.10. (1) A sequence $y_k \in B$ is referred to as a $D(B)$*-sequence* if $D(y_k) \to D(B)$ as $k \to \infty$.

(2) A point y_B is called the (asymptotically) ρ*-most probable* for B if there exists a $D(B)$-sequence y_k that converges to y_B as $k \to \infty$.

It is clear that a $D(B)$-sequence always exists. On the one hand, if, for a set B, there exists a $D(B)$-sequence y_k that converges to y_B as $k \to \infty$, and the point y_B belongs to the set B, then $D(y_B) \geqslant D(B)$. On the other hand, by virtue of the lower semicontinuity of the function $D(y)$, one has

$$D(B) = \lim_{k\to\infty} D(y_k) \geqslant D(y_B).$$

Hence $D(y_B) = D(B)$.

If the sets D_v are compact for all $v \geqslant 0$ then, for any measurable set B, there exists a ρ-most probable point $y_B \in [B]$. Indeed, let $\{y_n\}$ be a $D(B)$-sequence. For any given ε and all large enough k, one has $y_k \in D_{v+\varepsilon}$ for $v = D(B)$. Since $D_{v+\varepsilon}$ is a compact set, one can assume without loss of generality that $y_k \to y_B \in [B]$.

As Example 4.3.1 below shows, in the general case the most probable points do not necessarily exist. When dealing with random processes, it is natural to refer to the most probable points as the *most probable trajectories*.

The following lemma will be useful in what follows.

Lemma 4.1.11. *Assume that two sequences, η_n and $\bar\eta_n$, are given on a common probability space in such a way that, for any $h > 0$, one has*

$$\overline{\lim_{n\to\infty}} \frac{1}{z_n} \ln \mathbf{P}\big(\rho(\eta_n, \bar\eta_n) > h\big) = -\infty.$$

Then the following assertions hold true.

(i) *If the sequence η_n satisfies upper bounds in the l.l.d.p. with functions $\delta(\varepsilon)$ and $\beta(\varepsilon)$ then the sequence $\bar\eta_n$ satisfies upper bounds in the l.l.d.p. with the same parameters and functions $\bar\delta(\varepsilon) = \delta(\varepsilon(1+\tau))$ and $\bar\beta(\varepsilon) = \beta(\varepsilon(1+\tau))$ for any $\tau > 0$ (see Definition 4.1.3).*

(ii) *If the sequence* η_n *satisfies upper bounds in the e.l.d.p. then the sequence* $\bar{\eta}_n$ *satisfies upper bounds in the e.l.d.p. with the same parameters.*

(iii) *If the sequence* η_n *satisfies lower bounds in the l.l.d.p. (e.l.d.p.) then the sequence* $\bar{\eta}_n$ *satisfies lower bounds in the l.l.d.p. (e.l.d.p.) with the same parameters.*

Proof of the lemma is almost obvious and we leave it to the reader.

Remark 4.1.12. It is not always possible to prove a particular version of the l.d.p. by following Definitions 4.1.1–4.1.4 exactly. In a number of cases one can prove relations (4.1.2) ((4.1.10), (4.1.11)) not for any $y \in \mathbb{Y}$ ($B \in \mathfrak{B}$) but only for elements y (sets B) from some subclasses $\mathbb{Y}_0$ of elements of $\mathbb{Y}$ (subclasses $\mathfrak{B}_0$ of sets from $\mathfrak{B}$). For example, in Theorem 4.2.9 and Corollary 4.2.4 below, relations (4.1.10) are established only for the classes of convex or bounded sets in the respective functional metric spaces. In such cases, we will say that *the l.l.d.p. (e.l.d.p.) holds in the space* $(\mathbb{Y}, \rho)$ *with parameters* (z, D) *from the subclass of elements* $\mathbb{Y}_0$ (*the subclass of sets* $\mathfrak{B}_0$).

Remark 4.1.13. The concept of the l.d.p. is not necessarily connected with a sequence of random elements. The concept can be extended to arbitrary σ-finite measures on $(\mathbb{Y}, \rho)$. For instance, if $\mathbb{Y} = \mathbb{R}^d$, $\xi \in \mathbb{R}^d$, $\mathbf{E}\xi \neq 0$, one can take the renewal function

$$H(B) = \sum_{n=0}^{\infty} \mathbf{P}(S_n \in B)$$

as a measure, where the S_n are defined in (4.0.1). The l.d.p. describes the 'crude' asymptotic behaviour of $H(TB)$ for sets TB, $0 \notin B$, that 'run off' as $T \to \infty$ (i.e. the distance between 0 and the set TB tends to infinity as T tends to infinity) (i.e. the distance between 0 and the set TB tends to infinity as T tends to infinity). The l.l.d.p. describes these asymptotics for 'running-off' neighbourhoods $T(\alpha)_\varepsilon$ as $\varepsilon \to 0$ sufficiently slowly while $T \to \infty$, $\alpha \in \mathbb{R}^d$. The asymptotics have the form

$$\lim_{T \to \infty} \frac{1}{T} \ln H\big(T(\alpha)_\varepsilon\big) = -D(\alpha),$$

where $D(\alpha)$ is a deviation function. It is convex and semicontinuous as described above. In section 2.9 it was stated that the l.l.d.p. for the renewal functions holds true under weak assumptions.

The meaning of the l.d.p. remains as before: it means that the logarithm of a measure *TB* has an asymptotic factorisation (as $T \to \infty$), i.e. it is asymptotically equivalent to the product of two factors. In this case one factor depends on T only while the second factor depends on B only, and both factors are relatively 'regular' functions.

It is possible to consider the l.d.p. for the distribution of a random variable $\xi \in \mathbb{R}^d$. If $P(B) = \mathbf{P}(\xi \in B)$ then the l.d.p. has the form

$$\lim_{T\to\infty} \frac{1}{T} \ln P\big(T(\alpha)_\varepsilon\big) = -\Lambda(\alpha). \tag{4.1.21}$$

Here the 'collectiveness' of the principle is absent, and the implementation of (4.1.21) depends on the smoothness of the distribution of ξ. For instance, the limit relation (4.1.21) in the one-dimensional case for $\alpha > 0$ holds true when the condition of 'crude' smoothness for the right tail of the distribution of ξ is met:

$$\ln \mathbf{P}(\xi \geqslant T) \sim -\Lambda(T) \sim \lambda_+ T$$

as $T \to \infty$. It is easy to see that in this case the distribution of ξ satisfies the l.d.p. with deviation function $D(\alpha) = \alpha\lambda_+$, with $\alpha \geqslant 0$.

In these examples one can consider also a 'triangular' setup, when the measure H depends on the parameter $T \to \infty$.

4.2 Deviation functional (or integral) for random walk trajectories and its properties

In this section we will find the form of the deviation functional $D(\cdot)$ in the l.d.p. for random walk trajectories and study its properties.

First we consider the space $\mathbb{C} = \mathbb{C}^d[0,1]$ and a random continuous polygonal line $s_n = s_n(t)$ in the space, with nodes $(k/n, S_k/n)$, $k = 0, 1, \dots, n$. Denote by $I(f)$ the Lebesgue integral

$$I(f) := \int_0^1 \Lambda(f'(t))dt, \quad \text{where} \quad f'(t) = \big(f'_{(1)}(t), \dots, f'_{(d)}(t)\big), \tag{4.2.1}$$

defined for the functions f from the space $\mathbb{C}_a \subset \mathbb{C}$ of absolutely continuous functions on the segment [0, 1]. If the condition $[\mathbf{C}_\infty]$ is met (see subsection 1.1.1) then the l.l.d.p. holds true for random walk trajectories $s_n = s_n(\cdot)$:

$$\lim_{\varepsilon\to 0}\lim_{n\to\infty} \frac{1}{n} \ln \mathbf{P}\big(s_n \in (f)_\varepsilon\big) = -\begin{cases} I(f) & \text{if } f \in \mathbb{C}_a, \ f(0) = 0, \\ \infty & \text{otherwise,} \end{cases} \tag{4.2.2}$$

where $(f)_\varepsilon$ is a ε-neighbourhood of f in the uniform metric. Therefore, in this case the functional $D(f)$ in relations 4.1.1–4.1.3 coincides with $I(f)$. Statement (4.2.2) follows from the results in [19] (in the one-dimensional case, $d = 1$) and [131] (in the multidimensional case, $d > 1$), where the l.d.p. was proved for the trajectories $s_n(t)$ provided that condition $[\mathbf{C}_\infty]$ is met.

If condition $[\mathbf{C}_\infty]$ is not satisfied then in general the relation (4.2.2) is not true.

If only one of the two conditions $[\mathbf{C}]$ and $[\mathbf{C}_0]$ is met, in order to state the local and extended large deviation principles, we have to extend the space $\mathbb{C}_a$. On the extension we define a functional $J(f)$ (which coincides with $I(f)$ when condition $[\mathbf{C}_\infty]$ is met and $f \in \mathbb{C}_a$) such that the left-hand side of (4.2.2) for f from the extended space (with $f(0) = 0$) is described by the functional $J(f)$ when only one of the conditions **[C]** and **[C$_0$]** is met. This functional will also be an integral (like $I(f)$), but a more general one.

4.2.1 Deviation integral in the one-dimensional case, $d = 1$

I. *The space* $\mathbb{D}$ *of functions without discontinuities of the second kind and the metric* ρ. As the space of trajectories for random walks, we take the extension $\mathbb{D}$ of the well-known space $\mathbb{D}[0,1]$ of functions without discontinuities of the second kind. This space consists of the functions $f = f(t)$: $[0,1] \to \mathbb{R}$, possessing both one-sided limits at every point $t \in (0,1)$ and such that at every discontinuity point $t \in (0,1)$ the value $f(t)$ lies in the segment $[f(t-0), f(t+0)]$, while the values $f(0)$ and $f(1)$ need not coincide with $f(+0)$ and $f(1-0)$, respectively. Note that every function f in the extended space $\mathbb{D}$ retains *separability*: for every countable-everywhere dense subset U of $[0,1]$ containing 0 and 1, one has

$$\sup_{(a,b)} f(t) = \sup_{(a,b)\cap U} f(t).$$

As the metric we take $\rho = \rho_{\mathbb{F}}$, introduced in [21] for the space $\mathbb{F}$ which includes $\mathbb{D}$. The topology generated by $\rho_{\mathbb{F}}$ coincides with the Skorokhod topology M_2 described in [169]. We can also consider ρ as an extension of the Lévy metric on the space of non-decreasing functions (see e.g. [96]) to the larger space $\mathbb{D}$.

The metric ρ is defined as follows. Consider the graph of a given function $f \in \mathbb{D}$ as the simply connected set Γf in $[0,1] \times \mathbb{R}$ whose cross-section at t coincides with the segment $[f(t-0), f(t+0)]$. This set coincides with the curve $\big(t, f(t)\big)$ everywhere except at the discontinuity points of t. At a discontinuity point t the points $\big(t, f(t-0)\big)$ and $\big(t, f(t+0)\big)$ are joined by a straight line segment. At every point of the set Γf we construct the open ball (in the Euclidean metric) of radius ε in the two-dimensional space $[0,1] \times \mathbb{R}$. Denote the domain obtained as the intersection of the strip $0 \leqslant t \leqslant 1$ with the union of these balls by $(\Gamma f)_\varepsilon$ and call it the ε-neighbourhood of the graph Γf for the function f.

We will write $\rho(f,g) < \varepsilon$ when $\Gamma f \in (\Gamma g)_\varepsilon$ and $\Gamma g \in (\Gamma f)_\varepsilon$ simultaneously. In other words,

$$\rho(f,g) := \max\big\{r(f,g), r(g,f)\big\}, \qquad r(f,g) := \max_{v\in\Gamma f} \min_{u\in\Gamma g} |u - v|,$$

where $u = (u_1, u_2)$, $v = (v_1, v_2)$, $u_1, v_1 \in [0,1]$, $u_2, v_2 \in \mathbb{R}$, and $|\cdot|$ means the Euclidean norm on $\mathbb{R}^2$.

It is easy to see that $\Gamma f \in (\Gamma g)_{\varepsilon_1}$ and $\Gamma g \in (\Gamma h)_{\varepsilon_2}$ imply $\Gamma f \in (\Gamma h)_{\varepsilon_1+\varepsilon_2}$, while $\Gamma h \in (\Gamma g)_{\varepsilon_2}$ and $\Gamma g \in (\Gamma f)_{\varepsilon_1}$ imply $\Gamma h \in (\Gamma f)_{\varepsilon_1+\varepsilon_2}$. Hence ρ satisfies the triangle inequality $\rho(f,h) \leqslant \rho(f,g) + \rho(g,h)$. Clearly,

$$\rho(f,g) \leqslant \rho_{\mathbb{C}}(f,g), \quad \text{where} \quad \rho_{\mathbb{C}}(f,g) := \sup_{0\leqslant t\leqslant 1} \big|f(t) - g(t)\big|. \tag{4.2.3}$$

It would be more correct to call ρ a *pseudometric*, since in general $\rho(f,g)=0$ fails to imply that $f = g$ (if the functions $f,\ g \in \mathbb{D}$ differ only at discontinuity points then their graphs coincide and $\rho(f,g)=0$).

In what follows, we refer to ρ as the *metric*, and this should never lead to confusion.

As becomes clear below (see assertion (iii) of Theorem 4.2.3), we should be interested mainly in functions from the space $\mathbb{V}$ of functions of bounded variation. In this case there always exists a continuous parametrisation of the curve Γf in $\mathbb{R}^2$. As a parameter we can take, for instance, the length of the curve or a monotone continuous transformation of the length. In these cases along with ρ we can use the metric $\rho_{\mathbb{V}}$ (in a certain sense it is more adapted to the problems under study), which is determined as follows.

Let $\mathbf{f}(s) \in \mathbb{R}^2$, $s \in [0,1]$, be a continuous parametrisation of the curve Γf in $\mathbb{R}^2$. We will assume that $\rho_{\mathbb{V}}(f,g) < \varepsilon$ if there exists a monotone continuous mapping $r(s) : [0,1] \to [0,1]$ such that

$$\sup_{s\in[0,1]} \left|\mathbf{f}(s) - \mathbf{g}\big(r(s)\big)\right| < \varepsilon.$$

It is not difficult to see that the metric $\rho_{\mathbb{V}}$ is 'weaker' than the uniform metric and the Skorokhod metric, but 'stronger' than ρ. For the details see section 4.7.

II. *The deviation integral.* Consider the partitions

$$\mathbf{t}_K = \{t_0, t_1, \ldots, t_K\}, \quad t_0 = 0 < t_1 < \cdots < t_{K-1} < t_K = 1,$$

of the segment $[0,1]$ into disjoint half-closed intervals $\Delta_k[t_k) := [t_k, t_k + \Delta_k)$, $k = 0, \ldots, K-2$, and the segment $\Delta_{K-1}[t_{K-1}] := [t_{K-1}, t_K]$, where $\Delta_k := t_{k+1} - t_k$, $k = 0, \ldots, K-1$. We refer to a sequence $\{\mathbf{t}_K\}$ of these partitions as *dense* whenever

$$\max_{0\leqslant k\leqslant K-1} \Delta_k \to 0 \quad \text{as} \quad K \to \infty.$$

Definition 4.2.1. We will say that *the deviation integral* $J(f)$ of a function $f \in \mathbb{D}$ exists if the limit

$$\lim_{K\to\infty} \sum_{k=0}^{K-1} \Delta_k \Lambda\left(\frac{f(t_{k+1}) - f(t_k)}{\Delta_k}\right) =: J(f) \tag{4.2.4}$$

exists for every dense sequence of partitions $\mathbf{t}_K$ and this limit does not depend on the choice of sequence.

The Riemann-type integral sum in (4.2.4) appears when we approximate the logarithm of the probability that the trajectory $s_n(t)$ (or another version of it) lies in some neighbourhood of the curve $f(t)$ (see below).

If f is *absolutely continuous* then it is natural to express $J(f)$ as follows:

$$J(f) = \int_0^1 \Lambda(g(t))\, dt, \quad \text{where} \quad g(t) = f'(t).$$

By definition, $J(f)$ is the limit of special Riemann-type integral sums constructed from the *average values* of g on the intervals $\Delta_k[t_k)$:

$$\frac{1}{\Delta_k}\int_{t_k}^{t_k+\Delta_k} g(u)\,du = \frac{f(t_{k+1})-f(t_k)}{\Delta_k}.$$

If $f(t)$ is absolutely continuous then $f'(t)$ is a measurable function such that

$$f(t)-f(t_0) = \int_{t_0}^{t} f'(u)\,du, \quad t \geqslant t_0.$$

Hence the function $\Lambda(f'(t))$ is also measurable, and its Lebesgue integral $I(f)$ in (4.2.1) always exists. We have to prove the existence of the limit (4.2.4) for $f \in \mathbb{D}$.

As already noted, the construction of the deviation integral is forced by the nature of the phenomena under study, while $J(f)$ itself is the *Riemann–Darboux integral*

$$\int F(t,u) := \lim_{K\to\infty} \sum_{k<K} F(t_k, t_{k+1}) \tag{4.2.5}$$

of the interval function

$$F(t,u) = (u-t)\Lambda\left(\frac{f(u)-f(t)}{u-t}\right). \tag{4.2.6}$$

Riemann and Darboux studied the integral (4.2.5) (see e.g. [157], pp. 27–36). In some cases the existence of $J(f)$ follows from Darboux's theorem (see [157], p. 33, and the remark after Theorem 4.2.2).

Definition 4.2.1 remains valid if we replace the deviation function $\Lambda(\alpha)$ by an arbitrary lower semi-continuous convex function $M = M(\alpha)$ mapping $\mathbb{R}$ into the extended semi-axis $[0,\infty]$. In this case we denote the corresponding functional by J^M. In particular, for $M(\alpha) = |\alpha|$, we obtain the total variation of $f \in \mathbb{D}$ as $J^M(f)$:

$$J^M(f) = \operatorname{Var} f.$$

III. *Existence and main properties of the deviation integral.* Given a partition $\mathbf{t}_K$ of the segment $[0,1]$ and a function $f = f(t)$, denote by $f^{\mathbf{t}_K} \in \mathbb{C}_a$ the continuous polygonal line with nodes

$$\big(0,f(0)\big),\ldots,\big((t_k,f(t_k)\big),\ldots,\big(1,f(1)\big).$$

Then the sum on the left-hand side of (4.2.4) can be written as $I(f^{\mathbf{t}_K})$. One of the main assertions of this subsection is the following theorem.

Theorem 4.2.2. *If $f \in \mathbb{D}$ then the deviation integral $J(f)$ (which can be finite or infinite) always exists. Moreover,*

$$J(f) = \sup I(f^{\mathbf{t}_K}), \tag{4.2.7}$$

where the supremum is taken over all partitions $\mathbf{t}_K$ of $[0,1]$.

If $\rho(f,g) = 0$ *then*

$$J(f) = J(g). \tag{4.2.8}$$

Observe that, since Λ is a convex function (see subsection 1.1.2), the interval function $F(t,u)$ defined in (4.2.6), is *semi-additive*, i.e. it satisfies, for $t < u < v$, the inequality

$$F(t,v) \leqslant F(t,u) + F(u,v).$$

For functions of this type the integral sums in (4.2.5) fail to decrease as the partitions $\mathbf{t}_K$ becomes finer (i.e. new points are added). Suppose henceforth that the function $F(t,u)$ is continuous at every point v, in the sense that $F(v-\varepsilon_1, v+\varepsilon_2) \to 0$ as $\varepsilon_1 + \varepsilon_2 \to 0$, $\varepsilon_1 > 0$, $\varepsilon_2 > 0$; and, moreover, suppose that the supremum $\mathbf{S}(F)$ of the integral sums in (4.2.5) over all partitions $\mathbf{t}_K$ is finite. Under these assumptions the Darboux theorem establishes the existence of the integral $\int F(t,u)$. Furthermore, it holds that $\int F(t,u) = \mathbf{S}(F)$ (cf. (4.2.7) in Theorem 4.2.2). Theorem 4.2.2 generalises Darboux's theorem to functions $F(t,u)$ of the particular form (4.2.6), which, in general, are not continuous, while $\mathbf{S}(F)$ is not necessarily finite.

Now, we list the main properties of the functional $J(f)$.

Theorem 4.2.3. (i) *The functional* $J(f)$ *is convex:*

$$J(pf_1 + (1-p)f_2) \leqslant pJ(f_1) + (1-p)J(f_2), \quad p \in [0,1],\ f_i \in \mathbb{D},\ i = 1,2.$$

(ii) $J(f)$ *is lower semi-continuous with respect to the metric* ρ*:*

$$\varliminf_{\rho(f_n,f)\to 0} J(f_n) \geqslant J(f). \tag{4.2.9}$$

(iii) *If condition* **[C]** *is met and* $\mathrm{Var} f = \infty$ *then* $J(f) = \infty$. *If condition* $[\mathbf{C}_0]$ *is met then there exist constants* $c_1 > 0$, $c_2 > 0$ *such that*

$$J(f) \geqslant c_1 \mathrm{Var} f - c_2.$$

(iv) *If* $f \in \mathbb{D} \setminus \mathbb{C}_a$ *and condition* $[\mathbf{C}_\infty]$ *is met then* $J(f) = \infty$.

(v) *For every function* $f \in \mathbb{C}_a$ *one has* $J(f) = I(f)$.

(vi) *For every set B, one has*

$$J(B) \leqslant I_a(B), \tag{4.2.10}$$

where

$$J(B) := \inf_{f\in B} J(f), \qquad I_a(B) := \inf_{f\in B\cap\mathbb{C}_a} I(f).$$

For every open subset B in $(\mathbb{D},\rho)$ *one has*

$$J(B) = I_a(B). \tag{4.2.11}$$

Assertion (4.2.11) *remains true for every set* $B \subset \mathbb{C}$ *which is open with respect to the metric* $\rho_{\mathbb{C}}$.

(vii) *If condition* $[\mathbf{C}_\infty]$ *is met then, for any* $v \geqslant 0$, *the set* $J_v := \{f \in \mathbb{C} : f(0) = 0, J(f) \leqslant v\}$ *is compact in* $(\mathbb{C}, \rho_{\mathbb{C}})$.

Denote by $(f)_{\mathbb{C},\varepsilon}$ and $(f)_\varepsilon$ the ε-neighbourhoods of a point f in the uniform metric $\rho_{\mathbb{C}}$ and the metric ρ, respectively.

Corollary 4.2.4. *The relations*

$$J(f) = \lim_{\varepsilon\to 0} I_a((f)_{\mathbb{C},\varepsilon}) \; for f \in \mathbb{C}, \quad J(f) = \lim_{\varepsilon\to 0} I_a((f)_\varepsilon) \; for f \in \mathbb{D} \tag{4.2.12}$$

hold true.

Proof of Corollary 4.2.4. (a) We will verify the second equality in (4.2.12). Since by assertion (vi) of Theorem 4.2.3 one has $J((f)_\varepsilon) = I_a((f)_\varepsilon)$, it suffices to show that

$$J(f) = \lim_{\varepsilon\to 0} J((f)_\varepsilon) \quad \text{for} \quad f \in \mathbb{D}. \tag{4.2.13}$$

Since $J(f) \geqslant J((f)_\varepsilon)$, we have

$$\overline{\lim_{\varepsilon\to 0}}\, J((f)_\varepsilon) \leqslant J(f). \tag{4.2.14}$$

However, for arbitrary $\delta > 0$, $N < \infty$ and for every k there exists a function $f_k \in (f)_{1/k}$ such that $J((f)_{1/k}) \geqslant \min\{J(f_k) - \delta, N\}$ (we introduce the number N for the case $J(f) = \infty$). Assertion (ii) of Theorem 4.2.3 yields

$$\underline{\lim}_{k\to\infty}\, J(f_k) \geqslant J(f);$$

hence

$$\underline{\lim}_{\varepsilon\to 0}\, J((f)_\varepsilon) = \underline{\lim}_{k\to\infty}\, J((f)_{1/k}) \geqslant \min\{J(f) - \delta, N\}.$$

Since δ and N are arbitrary, this implies the inequality

$$\underline{\lim}_{\varepsilon\to 0}\, J((f)_\varepsilon) \geqslant J(f),$$

which along with (4.2.14) establishes (4.2.13).

(b) Now verify the first equality in (4.2.12). Since $(f)_{\mathbb{C},\varepsilon} \subset (f)_\varepsilon$, one has

$$I_a((f)_\varepsilon) \leqslant I_a((f)_{\mathbb{C},\varepsilon}).$$

Therefore, assertion (vi) of Theorem 4.2.3 yields $J((f)_\varepsilon) = I_a((f)_\varepsilon) \leqslant I_a((f)_{\mathbb{C},\varepsilon})$ and hence, in view of the already established second equality in (4.2.12), we get on the one hand

$$J(f) = \lim_{\varepsilon\to 0} I_a((f)_\varepsilon) \leqslant \overline{\lim_{\varepsilon\to 0}}\, I_a((f)_{\mathbb{C},\varepsilon}). \tag{4.2.15}$$

On the other hand, for $\varepsilon > 0, f \in \mathbb{C}$ and a fixed dense sequence of partitions $\mathbf{t}_K$, there exists a K large enough that $f^{\mathbf{t}_K} \in (f)_{\mathbb{C},\varepsilon}$. Therefore, $I_a((f)_{\mathbb{C},\varepsilon}) \leqslant I(f^{\mathbf{t}_K})$ and, by assertion (4.2.7) of Theorem 4.2.2, for any $\varepsilon > 0$ one has

$$J(f) = \sup I(f^{\mathbf{t}_K}) \geqslant I_a((f)_{\mathbb{C},\varepsilon}).$$

This inequality along with (4.2.15) establishes the first inequality of (4.2.12). Corollary 4.2.4 is proved. □

Note that (4.2.12) and (4.2.7) can serve as definitions of the deviation integral in the spaces $(\mathbb{C}, \rho_{\mathbb{C}})$ and $(\mathbb{D}, \rho)$ respectively (see the definition of the functional $I^{\lim}(f)$ in section 4.3 and in [62]).

As before, put

$$\lambda_+ := \sup\{\lambda : \psi(\lambda) < \infty\} \geqslant 0, \quad \lambda_- := \inf\{\lambda : \psi(\lambda) < \infty\} \leqslant 0.$$

If the condition $[\mathbf{C}_+] := \{\lambda_+ > 0\} \subset [\mathbf{C}]$ is met then there exist finite constants $c_1 > 0$, $c_2 > 0$ such that

$$\Lambda(\alpha) \geqslant c_1\alpha - c_2 \quad \text{for} \quad \alpha \geqslant 0. \tag{4.2.16}$$

(see property (Λ6) in subsection 1.1.2). A similar relation holds provided that $[\mathbf{C}_-] := \{\lambda_- < 0\} \subset [\mathbf{C}]$.

Assertion (iii) of Theorem 4.2.3 shows that, while studying J, our main interest is in the functions $f \in \mathbb{D}$ in the space $\mathbb{V}$ of functions of bounded variation. It turns out that it is possible to find the Riemann–Darboux integral for these functions explicitly. Recall that, by the Lebesgue theorem, each function f of bounded variation admits a unique decomposition in the form

$$f = f_a + f_s + f_\partial, \quad f_a(0) = f(0), \quad f_s(0) = f_\partial(0) = 0, \tag{4.2.17}$$

where f_a, f_s, f_∂ are the absolutely continuous, singular (continuous) and discrete components of f, respectively.

One has the unique representation, for $f \in \mathbb{V}$,

$$f(t) = f_+(t) - f_-(t), \; t \in [0,1], \quad f(0) = f_+(0), \quad f_-(0) = 0$$

as the difference of two non-decreasing functions $f_\pm$ such that $\mathrm{Var} f = \mathrm{Var} f_+ + \mathrm{Var} f_-$. Put $\mathrm{Var}^\pm f := \mathrm{Var} f_\pm = f_\pm(1) - f_\pm(0)$.

Theorem 4.2.5. *For every function $f = f_a + f_s + f_\partial \in \mathbb{V}$, the representation*

$$J(f) = I(f_a) + \lambda_+ \mathrm{Var}^+(f_s + f_\partial) - \lambda_- \mathrm{Var}^-(f_s + f_\partial). \tag{4.2.18}$$

holds.

According to (4.2.18),

$$J(f_\partial) = \Lambda(0) + \lambda_+ \mathrm{Var}^+ f_\partial - \lambda_- \mathrm{Var}^- f_\partial,$$

and one has a similar expression for the components of f_s. Hence, in the case $\Lambda(0) = 0$ (which always holds, provided that $\mathbf{E}\xi = 0$), the integral $J(f)$, while failing to be a linear functional of f, turns out to be additive with respect to the representation (4.2.17):

$$J(f) = J(f_a) + J(f_s) + J(f_\partial).$$

Also note that in the important particular case $f_s \equiv 0$, one has the representation

$$J(f) = I(f_a) + \lambda_+ \sum_k f_{\partial,k+} - \lambda_- \sum_k f_{\partial,k-}, \tag{4.2.19}$$

where $f_{\partial,k+}$ ($f_{\partial,k-}$), $k = 1, 2, \ldots$, are enumerated positive (negative) jumps of the function f_∂.

From the assertions presented above it is clear that the value of $J(f)$ does not depend on the location of the jumps of f or on the sets where the changes in the components $f_{s\pm}$ are concentrated. Moreover, $J(f)$ is a non-decreasing linear function of the variables $\mathrm{Var} f_{\partial\pm}$, $\mathrm{Var} f_{s\pm}$.

4.2.2 Proofs

I. *Proof of Theorem 4.2.2.* We need the following properties of the deviation function, appearing in subsection 1.2.2 (here we return to the general case $d \geqslant 1$).

(1) *The function* $\Lambda(\alpha)$ is *convex*: for $\alpha, \beta \in \mathbb{R}^d$, $p \in [0, 1]$, one has

$$\Lambda(p\alpha + (1-p)\beta) \leqslant p\Lambda(\alpha) + (1-p)\Lambda(\beta).$$

(2) $\Lambda(\alpha)$ is *lower semicontinuous*:

$$\varliminf_{\alpha_n \to \alpha} \Lambda(\alpha_n) \geqslant \Lambda(\alpha), \quad \alpha \in \mathbb{R}^d.$$

In what follows, the behaviour of the functions $A(\lambda) = \ln \psi(\lambda)$ and $\Lambda(\alpha)$ at infinity plays a substantial role. If condition $[\mathbf{C}_0]$ is met then, as we can observe in (4.2.16), *there exist finite constants* $c_1 > 0$, $c_2 > 0$ *such that*

$$\Lambda(\alpha) \geqslant c_1|\alpha| - c_2 \quad \text{for} \quad \alpha \in \mathbb{R}^d. \tag{4.2.20}$$

However, if condition $[\mathbf{C}_\infty]$ is met then $\Lambda(\alpha)$ tends to infinity faster than any linear function: *for some* $c \geqslant 0$ *and some continuous function* $u(t)$ *tending to infinity as* $t \to \infty$, *one has*

$$\Lambda(\alpha) \geqslant u(|\alpha|)|\alpha| - c \quad \text{for} \quad \alpha \in \mathbb{R}^d \tag{4.2.21}$$

(see subsection 1.2.2).

Now we resume proving the theorem. We have to verify that for *all dense sequences of partitions* $\mathbf{t}_K$ *the common limit* $\lim_{K\to\infty} I(f^{\mathbf{t}_K})$ *exists*. We say that *a partition* $\mathbf{u}_L$ *is included in a partition* $\mathbf{t}_K$, and write $\mathbf{u}_L \subset \mathbf{t}_K$, whenever $K \geqslant L$ and the set $\{u_0, \ldots, u_L\}$ is a subset of $\{t_0, \ldots, t_K\}$. Since the deviation function $\Lambda(\alpha)$ is convex, one has

$$I(f^{\mathbf{u}_L}) \leqslant I(f^{\mathbf{t}_K}), \quad \text{if } \mathbf{u}_L \subset \mathbf{t}_K. \tag{4.2.22}$$

Furthermore, for a fixed partition $\mathbf{t}_K = (t_0, \ldots, t_K)$, consider a sequence of partitions $\mathbf{t}_K^{(n)} = \left(t_0^{(n)}, \ldots, t_K^{(n)}\right)$ 'converging' to $\mathbf{t}_K$ as $n \to \infty$, i.e. such that

$$\lim_{n\to\infty} \max_{1\leqslant k\leqslant K-1} \left|t_k - t_k^{(n)}\right| = 0.$$

Since the deviation function $\Lambda(\alpha)$ is lower semicontinuous, one has

$$\underline{\lim}_{n\to\infty} I(f^{\mathbf{t}_K^{(n)}}) \geqslant I(f^{\mathbf{t}_K}). \tag{4.2.23}$$

Consider two dense partitions $\mathbf{u}_{L_n}^{(n)}$, $\mathbf{v}_{M_n}^{(n)}$ ($L_n, M_n \to \infty$ as $n \to \infty$). Fix a partition $\mathbf{u}_{L_{n_0}}^{(n_0)}$, and (removing the superfluous elements from $\mathbf{v}_{M_n}^{(n)}$) construct a sequence of partitions $\mathbf{w}_{L_{n_0}}^{(n)}$ satisfying two relations: (a) $\mathbf{w}_{L_{n_0}}^{(n)} \subset \mathbf{v}_{M_n}^{(n)}$ for all large enough n; (b) $\lim_{n\to\infty} \max_{1\leqslant k\leqslant L_{n_0}-1} \left|w_k^{(n)} - u_k^{(n_0)}\right| = 0$. By (4.2.22), (4.2.23) and relations (a), (b), one has

$$\underline{\lim}_{n\to\infty} I(f^{\mathbf{v}_{K_n}^{(n)}}) \geqslant \underline{\lim}_{n\to\infty} I(f^{\mathbf{w}_{L_{n_0}}^{(n)}}) \geqslant I(f^{\mathbf{u}_{L_{n_0}}^{(n_0)}}).$$

Since on the right-hand side of the last inequality one has an arbitrary partition from the sequence $\mathbf{u}_{L_n}^{(n)}$, the inequality also holds if we replace this right-hand side by the upper limit $\overline{\lim}_{n\to\infty} I(f^{\mathbf{u}_{L_n}^{(n)}})$. Therefore

$$\underline{\lim}_{n\to\infty} I(f^{\mathbf{v}_{M_n}^{(n)}}) \geqslant \overline{\lim}_{n\to\infty} I(f^{\mathbf{u}_{L_n}^{(n)}}). \tag{4.2.24}$$

Similarly, we get

$$\underline{\lim}_{n\to\infty} I(f^{\mathbf{u}_{L_n}^{(n)}}) \geqslant \overline{\lim}_{n\to\infty} I(f^{\mathbf{v}_{M_n}^{(n)}}). \tag{4.2.25}$$

Inequalities (4.2.24) and (4.2.25) yield the first assertion of Theorem 4.2.2.

The second assertion follows clearly from the first: if $\{\mathbf{t}_K\}$ is a dense sequence of partitions with $I(f^{\mathbf{t}_K}) \to \sup I(f^{\mathbf{t}_K})$ as $K \to \infty$ then

$$J(f) = \lim_{K\to\infty} I(f^{\mathbf{t}_K}) = \sup I(f^{\mathbf{t}_K}).$$

Finally, let us verify (4.2.8). If $\rho(f,g) = 0$ then $f(t) = g(t)$ everywhere, possibly except at discontinuity points. Since $J(f^{\mathbf{t}_K}) = J(g^{\mathbf{t}_K})$ for the partitions $\mathbf{t}_K$ which avoid the discontinuity points of the function f, the first assertion of the theorem implies that $J(f) = J(g)$. The theorem is proved. □

II. *Proof of Theorem 4.2.3* (i) Since the deviation function $\Lambda(\alpha)$ is convex, the functional $I(f)$ on the class of absolutely continuous functions is convex as well. By Theorem 4.2.2 this property is preserved for the functional $J(f)$ on $\mathbb{D}$.

(ii) The lower semicontinuity of $J(f)$ follows from next four remarks.

(1) For all $\varepsilon > 0$ and $N < \infty$ there is a partition $\mathbf{t}_K$ such that

$$J(f^{\mathbf{t}_K}) \geqslant \min\{J(f) - \varepsilon, N\} \tag{4.2.26}$$

(the number N is introduced for the case when $J(f) = \infty$).

(2) If $\rho(f,g) = 0$ and $f_n \to f$ then $J(f) = J(g)$ and $f_n \to g$. Therefore, without loss of generality, we may assume that f is left-continuous (or right-continuous) at all interior discontinuity points. Since $\rho(f_n,f) \to 0$, there is a collection of points $\mathbf{t}_{K,n} = \{t_k^{(n)}\}_{k=0}^K$ such that

$$\lim_{n\to\infty} \max_{k\leqslant K}\{|t_k^{(n)} - t_k| + |f_n(t_k^{(n)}) - f(t_k)|\} = 0. \tag{4.2.27}$$

(3) Construct the polygonal line $f_n^{\mathbf{t}_{K,n}}$ from the collection of points $(t_k^{(n)}, f_n(t_k^{(n)}))$, $0 \leqslant k \leqslant K$. Since $\Lambda(\alpha)$ is lower semicontinuous,

$$\begin{aligned}\varliminf_{n\to\infty} I(f_n^{\mathbf{t}_{K,n}}) &= \varliminf_{n\to\infty} \sum_{k<K} (t_{k+1}^{(n)} - t_k^{(n)})\Lambda\left(\frac{f_n(t_{k+1}^{(n)}) - f_n(t_k^{(n)})}{t_{k+1}^{(n)} - t_k^{(n)}}\right)\\ &\geqslant \sum_{k<K}(t_{k+1} - t_k)\Lambda\left(\frac{f(t_{k+1}) - f(t_k)}{t_{k+1} - t_k}\right) = J(f^{\mathbf{t}_K}).\end{aligned} \tag{4.2.28}$$

(4) Finally,

$$J(f_n) \geqslant I(f_n^{\mathbf{t}_{K,n}}). \tag{4.2.29}$$

Combining (4.2.26), (4.2.28) and (4.2.29), we obtain

$$\varliminf_{n\to\infty} J(f_n) \geqslant J(f^{\mathbf{t}_K}) \geqslant \min\{J(f) - \varepsilon, N\}.$$

Since $\varepsilon > 0$ and $N < \infty$ are arbitrary, (4.2.9) is proved.

(iii) Suppose that $\mathrm{Var} f = \infty$. This means that for every dense sequence of partitions $\mathbf{t}_K$ one has

$$\lim_{K\to\infty} \sum_{k=0}^{K-1} |f(t_{k+1}) - f(t_k)| = \infty.$$

Since, furthermore,

$$\left|\sum_{k=0}^{K-1} (f(t_{k+1}) - f(t_k))\right| = |f(1) - f(0)| < \infty,$$

we have simultaneously

$$\lim_{K\to\infty} \sum_{k\in\mathcal{K}_+} (f(t_{k+1}) - f(t_k)) = \infty, \quad \lim_{K\to\infty} \sum_{k\in\mathcal{K}_-} |f(t_{k+1}) - f(t_k))| = \infty, \tag{4.2.30}$$

where $\mathcal{K}_+ := \{k \leqslant K-1 : f(t_{k+1}) - f(t_k) > 0\}$, $\mathcal{K}_- := \{k \leqslant K-1 : f(t_{k+1}) - f(t_k) < 0\}$. Since condition [**C**] is fulfilled, it follows that either take over $\lambda_+ > 0$ or $\lambda_- < 0$. For definiteness, we assume that $\lambda_+ > 0$. Then (4.2.16) implies

$$I(f^{\mathbf{t}_K}) \geqslant \sum_{k\in\mathcal{K}_+} \Delta_k \Lambda\left(\frac{f(t_{k+1})-f(t_k)}{\Delta_k}\right) \geqslant c_1 \sum_{k\in\mathcal{K}_+} (f(t_{k+1})-f(t_k)) - c_2.$$

By the first inequality in (4.2.30) we get

$$J(f) = \lim_{K\to\infty} I(f^{\mathbf{t}_K}) = \infty.$$

The second claim of (iii) follows from (4.2.20) and

$$I(f^{\mathbf{t}_K}) \geqslant c_1 \sum_{k=0}^{K-1} \Delta_k \left|\frac{f(t_{k+1})-f(t_k)}{\Delta_k}\right| - c_2.$$

Statement (iii) is proved.

(iv) Suppose that condition $[\mathbf{C}_\infty]$ is met and $f \in \mathbb{D}$ is not absolutely continuous. Then (see e.g. [157], p. 58) for some $m > 0$ and every $\delta > 0$ there is a collection of disjoint intervals (r_i, s_i), $i = 1,\ldots,N$, which depends on δ, with

$$\sum_{i\leqslant N}(s_i - r_i) < \delta, \qquad \left|\sum_{i\leqslant N}(f(s_i) - f(r_i))\right| \geqslant m.$$

Since by Theorem 4.2.2 we have

$$J(f) \geqslant \delta \sum_{i\leqslant N} \frac{(s_i - r_i)}{\delta} \Lambda\left(\frac{f(s_i) - f(r_i)}{s_i - r_i}\right),$$

by the convexity of Λ it follows that

$$J(f) \geqslant \delta\Lambda\left(\frac{\sum_{i\leqslant N}(f(s_i) - f(r_i))}{\delta}\right).$$

Using (4.2.21), we obtain

$$J(f) \geqslant \delta\left(u\left(\frac{m}{\delta}\right)\frac{m}{\delta} - c\right) = mu\left(\frac{m}{\delta}\right) - \delta c.$$

Since $u(t) \to \infty$ as $t \to \infty$, we can make the right-hand side of this inequality arbitrarily large by choosing δ appropriately. This means that $J(f) = \infty$.

(v) Since the polygonal line $f^{\mathbf{t}_K}$ 'rectifies' $f \in \mathbb{C}_a$ and Λ is convex,

$$I(f^{\mathbf{t}_K}) = \int_0^1 \Lambda(f'_{\mathbf{t}_K}(t))\,dt \leqslant \int_0^1 \Lambda(f'(t))\,dt = I(f).$$

Therefore,

$$J(f) := \lim_{K\to\infty} I(f^{\mathbf{t}_K}) \leqslant I(f). \tag{4.2.31}$$

Now we verify the converse inequality

$$J(f) \geqslant I(f). \tag{4.2.32}$$

Denote by $g^{(K)} = g^{(K)}(t)$ the derivative of the polygonal line $f^{\mathbf{t}_K}$. Then $g^{(K)}(t)$ is a step function which is constant on every interval defined by the partition $\mathbf{t}_K$. Meanwhile, on every interval (u, v) of the step function, the derivative equals

$$\frac{f^{\mathbf{t}_K}(v) - f^{\mathbf{t}_K}(u)}{v-u} = \frac{f(v) - f(u)}{v-u}.$$

It is known (see e.g. [157], p. 86) that, for a dense sequence of partitions and an absolutely continuous function $f(t)$, for almost all (with respect to Lebesgue measure) points $t \in [0, 1]$, the derivative $f'(t)$ is the limit of the sequence of step functions $g^{(K)}$:

$$f'(t) = \lim_{K\to\infty} g^{(K)}(t).$$

This means, by the lower semi-continuity of $\Lambda(\alpha)$, that for almost all t in $[0, 1]$ one has

$$\varliminf_{K\to\infty} \Lambda(g^{(K)}(t)) \geqslant \Lambda(f'(t)).$$

By Fatou's lemma,

$$J(f) = \lim_{K\to\infty} I(f^{\mathbf{t}_K}) = \varliminf_{K\to\infty} I(f^{\mathbf{t}_K}) = \varliminf_{K\to\infty} \int_0^1 \Lambda(g^{(K)}(t))\,dt$$
$$\geqslant \int_0^1 \varliminf_{K\to\infty} \Lambda(g^{(K)}(t))\,dt \geqslant \int_0^1 \Lambda(f'(t))\,dt = I(f).$$

Inequality (4.2.32) is proved. Along with (4.2.31) this establishes statement (v).

(vi) We need

Lemma 4.2.6. *For any function $f \in \mathbb{D}$ and arbitrary dense sequence $\mathbf{t}_K$ of partitions of* $[0, 1]$ *we have*

$$\lim_{K\to\infty} \rho(f, f^{\mathbf{t}_K}) = 0. \qquad (4.2.33)$$

Proof. Fix $f \in \mathbb{D}$ and a dense sequence $\mathbf{t}_K = \{t_0, \dots, t_K\}$ of partitions of $[0, 1]$.

Obviously, for every $\varepsilon > 0$ the number of discontinuity points of f with norm greater than ε is finite; otherwise, a countable set of these points would have the limit point t_0 on $[0, 1]$. At this point the function f would have a discontinuity of the second kind. Denote by f_∂ a step function formed by the jumps of f, with norm greater than ε. Put $f_1 = f - f_\partial$. Then there is a $\Delta > 0$ such that

$$\sup_{0\leqslant u\leqslant v\leqslant 1, v-u\leqslant\Delta} |f_1(v) - f_1(u)| \leqslant 2\varepsilon$$

(otherwise, we would find a point $t_0 \in [0, 1]$ at which f would have a discontinuity of the second kind). Therefore, for $t_j - t_{j-1} \leqslant \Delta$,

$$\sup_{t_{j-1}\leqslant t\leqslant t_j} \left|f_1(t) - f_1^{\mathbf{t}_K}(t)\right| \leqslant \sup_{t_{j-1}\leqslant t\leqslant t_j} |f_1(t) - f_1(t_j)| + |f_1(t_{j-1}) - f_1(t_j)| \leqslant 4\varepsilon.$$

Hence, for all sufficiently large K with

$$\max_{j\leqslant K}(t_j - t_{j-1}) \leqslant \Delta,$$

one has

$$\rho_{\mathbb{C}}\big(f_1, f_1^{\mathbf{t}_K}\big) \leqslant 4\varepsilon. \tag{4.2.34}$$

It is not difficult to see that

$$\overline{\lim_{K\to\infty}}\, \rho\big(f_1 + f_\partial, f_1 + f_\partial^{\mathbf{t}_K}\big) = 0. \tag{4.2.35}$$

It follows from the triangle inequality and (4.2.3) that

$$\begin{aligned}\rho\big(f_1 + f_\partial, f_1^{\mathbf{t}_K} + f_\partial^{\mathbf{t}_K}\big) &\leqslant \rho\big(f_1 + f_\partial, f_1 + f_\partial^{\mathbf{t}_K}\big) + \rho\big(f_1 + f_\partial^{\mathbf{t}_K}, f_1^{\mathbf{t}_K} + f_\partial^{\mathbf{t}_K}\big)\\ &\leqslant \rho\big(f_1 + f_\partial, f_1 + f_\partial^{\mathbf{t}_K}\big) + \rho_{\mathbb{C}}\big(f_1, f_1^{\mathbf{t}_K}\big).\end{aligned}$$

Therefore, (4.2.34) and (4.2.35) yield $\overline{\lim}_{K\to\infty}\, \rho(f, f^{\mathbf{t}_K}) \leqslant 4\varepsilon$. Since $\varepsilon > 0$ is arbitrary, this proves (4.2.33). The lemma is proved. □

We resume proving statement (vi). It is obvious that

$$J(B) = \inf_{g\in B} J(g) \leqslant \inf_{g\in B\cap\mathbb{C}_a} J(g) = \inf_{g\in B\cap\mathbb{C}_a} I(g) = I_a(B). \tag{4.2.36}$$

Therefore, (4.2.10) holds true.

If $J(B) = \infty$ then

$$J(B) \geqslant I_a(B). \tag{4.2.37}$$

If $J(B) < \infty$ then for any ε there exists a function $f \in B \subset \mathbb{D}$ with $J(B) + \varepsilon \geqslant J(f)$. For this function and a dense sequence $\mathbf{t}_K$ of partitions, Lemma 4.2.6 yields $\rho(f^{\mathbf{t}_K}, f) \to 0$ as $K \to \infty$. Therefore, for all sufficiently large K, the polygonal line $f^{\mathbf{t}_K}$ lies in the open set B along with f. Moreover, $J(B) + \varepsilon \geqslant J(f) \geqslant I(f^{\mathbf{t}_K}) \geqslant I_a(B)$. Therefore, (4.2.37) holds true as well. The second claim of (vi) follows from (4.2.37) and (4.2.36).

The third claim of (vi) is verified in the same way.

(vii) We verify the compactness of J_v provided that condition $[\mathbf{C}_\infty]$ is met. For a function f in J_v and the function $u(t)$ in (4.2.21) one has

$$\begin{aligned}|f(t) - f(t+\Delta)| &\leqslant \int_t^{t+\Delta} |f'(s)|\, ds\\ &\leqslant \int_t^{t+\Delta} |f'(s)|\mathbf{1}_{\{|f'(s)|<1/\sqrt{\Delta}\}}\, ds + \int_t^{t+\Delta} |f'(s)|\mathbf{1}_{\{|f'(s)|\geqslant 1/\sqrt{\Delta}\}}\, ds\\ &\leqslant \sqrt{\Delta} + \frac{1}{u(1/\sqrt{\Delta})} \int_t^{t+\Delta} u(|f'(s)|)|f'(s)|\, ds\\ &\leqslant \sqrt{\Delta} + \frac{1}{u(1/\sqrt{\Delta})} \int_t^{t+\Delta} \Lambda(f'(u))\, du \leqslant \sqrt{\Delta} + \frac{v}{u(1/\sqrt{\Delta})}.\end{aligned}$$

Thus, for $f \in J_v$ its continuity modulus $\omega_f(\Delta)$ admits a bound which is uniform in J_v:

$$\omega_f(\Delta) \leqslant \sqrt{\Delta} + \frac{v}{u(1/\sqrt{\Delta})} \to 0 \quad \text{as} \quad \Delta \to 0.$$

By Arzela's criterion, means that J_v is a totally bounded subset of $(\mathbb{C}, \rho_{\mathbb{C}})$. Since $J(f)$ is a lower semicontinuous functional (see statement (ii) of Theorem 4.2.3), the totally bounded set J_v is closed, and so it is compact. This proves statement (vii).

Theorem 4.2.3 is proved. □

III. *Proof of Theorem 4.2.5.* Observe now that the limit

$$\lim_{t\to\infty} \frac{1}{t}\Lambda(\alpha t) = \Lambda_\infty(\alpha) := \begin{cases} \lambda_+\alpha & \text{if } \alpha > 0, \\ -\lambda_-\alpha & \text{if } \alpha < 0 \end{cases} \tag{4.2.38}$$

always exists. This follows from the representation

$$\Lambda(\alpha) = \Lambda(0) + \int_0^\alpha \lambda(t)\,dt, \quad \lambda(t) := \Lambda'(t),$$

and $\lim_{t\to\pm\infty} \lambda(t) = \lambda_\pm$ (see e.g. [39]).

First, let us prove Theorem 4.2.5 in the case $f_s \equiv 0$.

Lemma 4.2.7. *If $f_s \equiv 0$ then*

$$\begin{aligned} J(f) &= I(f_a) + \sum_{k=1}^{\infty} \Lambda_\infty(f_{\partial,k}) \\ &= I(f_a) + \lambda_+ \sum f_{\partial,k+} - \lambda_- \sum f_{\partial,k-} \\ &= I(f_a) + \lambda_+ \operatorname{Var}^+ f_\partial - \lambda_- \operatorname{Var}^- f_\partial, \end{aligned} \tag{4.2.39}$$

where $f_{\partial,k\pm}$ are defined in (4.2.19).

Proof. At first assume for simplicity that f has finitely many jumps, say N. Consider a dense sequence $\mathbf{t}_K$ of partitions for sufficiently large $K = K_N$ that every interval of the partition contains at most one jump. Split the deviation integral $I(f^{\mathbf{t}_K})$ into two parts: $I(f^{\mathbf{t}_K}) = I_1(f^{\mathbf{t}_K}) + I_2(f^{\mathbf{t}_K})$, where I_1 comprises the terms of the sum in (4.2.4) over the intervals Δ_k without jumps while I_2 comprises those terms over the intervals containing jumps. The total length L_K of the intervals in I_2 vanishes as $K \to \infty$. Therefore, it is clear that

$$I_1(f^{\mathbf{t}_K}) \to I(f_a) \quad \text{as} \quad K \to \infty. \tag{4.2.40}$$

If an interval $\Delta_{j,K} = (t_{j,K}^-, t_{j,K}^+)$ of $\mathbf{t}_K$ contains a jump with index j and of size $f_{\partial,j}$ then

$$|\Delta_{j,K}| := t_{j,K}^+ - t_{j,K}^- \to 0, \quad f(t_{j,K}^+) - f(t_{j,K}^-) \to f_{\partial,j} \quad \text{as} \quad K \to \infty.$$

Therefore, by (4.2.38),

$$|\Delta_{j,K}|\Lambda\left(\frac{f(t_{j,K}^{+}) - f(t_{j,K}^{-})}{|\Delta_{j,K}|}\right) \to \Lambda_\infty(f_{\partial,j}), \quad I_2(f^{\mathbf{t}_K}) \to \sum_{j=1}^{N} \Lambda_\infty(f_{\partial,j})$$

as $K \to \infty$. Along with (4.2.30) this proves (4.2.39) for $N < \infty$.

If there are infinitely many jumps then order them so that their absolute values decrease, and denote by f_N the function f with all jumps having indices greater than N removed. Then

$$\mathrm{Var}(f - f_N) \to 0, \quad \rho_{\mathbb{C}}(f, f_N) \to 0, \quad \rho(f, f_N) \to 0 \quad \text{as} \quad N \to \infty.$$

Therefore, since $J(f)$ is lower semicontinuous,

$$\varliminf_{N\to\infty} J(f_N) \geqslant J(f). \tag{4.2.41}$$

Furthermore, express the prelimit integral $I(f^{\mathbf{t}_K})$ as the sum

$$I(f^{\mathbf{t}_K}) = I_1(f^{\mathbf{t}_K}) + I_2(f^{\mathbf{t}_K}),$$

where I_1 comprises the terms of the sum in (4.2.4) over the intervals Δ_k where there are no jumps of f_N while I_2 comprises terms over the intervals containing jumps. By the argument above,

$$\lim_{K\to\infty} I_2(f^{\mathbf{t}_K}) = \sum_{k=1}^{N} \Lambda_\infty(f_{\partial,k}). \tag{4.2.42}$$

Denote by $g_K(t)$ the derivative of the polygonal line $f^{\mathbf{t}_K}(t)$ with respect to t. Then

$$I_1(f^{\mathbf{t}_K}) = \int_0^1 \Lambda(g_K(t))\mathbf{I}_{t\notin U_K}\, dt,$$

where U_K is the union of finitely many intervals of $\mathbf{t}_K$ containing the jumps of f_N.

Obviously,

$$\lim_{K\to\infty} \mathbf{I}_{t\notin U_K} = 1 \quad \text{for all } t, \text{ except for finitely many points.}$$

It is known (see e.g. [157]) that for a dense sequence $\mathbf{t}_K$ of partitions we have the convergence $g_K(t) \to f_a'(t)$ as $K \to \infty$ for almost all (with respect to Lebesgue measure) $t \in [0,1]$. Since the deviation function $\Lambda(\alpha)$ is lower semicontinuous,

$$\varliminf_{K\to\infty} \Lambda(g_K(t))\mathbf{I}_{t\notin U_K} \geqslant \Lambda(f_a'(t)) \quad \text{for almost all } t \in [0,1].$$

Therefore, by Fatou's lemma,

$$\begin{aligned}\varliminf_{K\to\infty} I_1(f^{\mathbf{t}_K}) &= \varliminf_{K\to\infty} \int_0^1 \Lambda(g_K(t))\mathbf{I}_{t\notin U_K}\, dt \\ &\geqslant \int_0^1 \varliminf_{K\to\infty} \Lambda(g_K(t))\mathbf{I}_{t\notin U_K}\, dt \geqslant \int_0^1 \Lambda(f_a'(t))\, dt = I(f_a).\end{aligned}$$

So, by virtue of (4.2.42),

$$J(f) = \lim_{K\to\infty} I(f^{\mathbf{t}_K}) \geqslant I(f_a) + \sum_{k=1}^{N} \Lambda_\infty(f_{\partial,k}) = J(f_N).$$

Thus, by (4.2.41) we get $J(f) = \lim_{N\to\infty} J(f_N)$. Lemma 4.2.7 is proved. □

We now resume proving Theorem 4.2.5. Write $f_s = f_s^+ - f_s^-$, where $f_s^\pm$ are non-decreasing functions with $\operatorname{Var} f_s = \operatorname{Var} f_s^+ + \operatorname{Var} f_s^-$. We give a proof under the simplifying assumption[1] that for sufficiently small $\delta > 0$ there exists a collection of disjoint intervals $\Delta_i^\pm$, $i = 1, 2, \ldots$, of total length δ such that we can embed the negligible set on which the changes in $f_s^\pm$ are concentrated into the union $\bigcup \Delta_i^\pm$ of these intervals. For instance, Cantor's staircase satisfies this assumption. Meanwhile, we may assume without loss of generality that the $\Delta_i^\pm$ do not contain the discontinuity points of f (if, say, Δ_i does contain a discontinuity point then we can always replace it by the union of two intervals obtained by subdividing Δ_i into two parts at the discontinuity point).

First, assume for simplicity that $f_s = f_s^+$. Then we can approximate f_s in the uniform metric by a sequence of jump-like (discrete) non-decreasing functions $g_{(\delta)}$, with jumps, say, at the midpoints of the intervals $\Delta_i = \Delta_i^+ = (t_i^-, t_i^+)$ and of sizes equal to the increments of f_s on these intervals. Since $\operatorname{Var} g_{(\delta)} = \operatorname{Var} f_s$, Lemma 4.2.7 yields

$$J(f_a + f_\partial + g_{(\delta)}) = I(f_a) + \lambda_+ \operatorname{Var}^+(f_\partial + g_{(\delta)}) - \lambda_- \operatorname{Var}^- f_\partial.$$

But $\rho(f_s, g_{(\delta)}) \to 0$, and $\rho(f, f_a + f_\partial + g_{(\delta)}) \to 0$ as $\delta \to 0$. Thus, since J is lower semicontinuous,

$$\varliminf_{\delta\to 0} J(f_a + f_\partial + g_{(\delta)}) \geqslant I(f_a) + \lambda_+ \operatorname{Var}^+(f_\partial + f_s) - \lambda_- \operatorname{Var}^- f_\partial = J(f).$$

Now find a lower bound for $J(f)$. Construct an absolutely continuous function $\hat{f}$ as follows: it coincides with $f_a + f_s$ everywhere except for the intervals Δ_i, where we replace $f_a + f_s$ by a linear function possessing on Δ_i the same increment as $f_a + f_s$. The function $\hat{f} + f_\partial$ is a rectification of f and, since Λ is convex, we have

$$J(\hat{f} + f_\partial) \leqslant J(f). \tag{4.2.43}$$

But $J(\hat{f} + f_\partial) = I(\hat{f}) + \lambda_+ \operatorname{Var}^+ f_\partial - \lambda_- \operatorname{Var}^- f_\partial$. Denote by $|\Delta_i|$ the length of Δ_i. Then, for every fixed N,

$$J(\hat{f}) = I_1 + I_2 + I_3,$$

[1] The proof of Theorem 4.2.5 in the general case turns out to be more cumbersome. See [135].

where

$$I_1 := \int_{A_\delta} \Lambda(f_a'(t))\,dt, \quad A_\delta := [0,1] \setminus \bigcup_i \Delta_i,$$

$$I_2 := \sum_i |\Delta_i| \Lambda(R_i) \mathbf{I}_{R_i \leqslant N}, \quad R_i := \frac{\hat{f}(t_i^+) - \hat{f}(t_i^-)}{|\Delta_i|},$$

$$I_3 := \sum_i |\Delta_i| \Lambda(R_i) \mathbf{I}_{R_i > N}.$$

Since $\sum_i |\Delta_i| = \delta \to 0$, we have $I_1 \to I(f_a)$ as $\delta \to 0$.

If $\lambda_+ < \infty$ then $\Lambda(\alpha) \leqslant c < \infty$ when $\alpha \in [0,N]$, $I_2 \leqslant c\sum_i |\Delta_i| = c\delta \to 0$ as $\delta \to \infty$.

Since $\Lambda(\alpha) = \alpha\lambda_+(1+\theta(\alpha))$ and $|\theta(\alpha)| \leqslant \theta_N \to 0$ for $\alpha > N$ as $N \to \infty$, we have

$$I_3 = \lambda_+ \sum_i \big(\hat{f}(t_i^+) - \hat{f}(t_i^-)\big)(1+\theta_{i,N}), \quad |\theta_{i,N}| \leqslant \theta_N. \tag{4.2.44}$$

If $\sum_i |\Delta_i| = \delta \to 0$ then the right-hand side of the last relation converges to $\lambda_+ \operatorname{Var} f_s + O(\theta_N)$. Since $I_2 + I_3$ is independent of N, the sum in (4.2.44) converges to $\lambda_+ \operatorname{Var} f_s$ as $N \to \infty$. Hence, by (4.2.43),

$$\begin{aligned} J(f) \geqslant \varliminf_{\delta\to 0} J(\hat{f} + f_\partial) &= \varliminf_{\delta \to 0} J(\hat{f}) + \lambda_+ \operatorname{Var}^+ f_\partial - \lambda_- \operatorname{Var}^- f_\partial \\ &= I(f_a) + \lambda_+ \operatorname{Var}^+(f_\partial + f_s) - \lambda_- \operatorname{Var}^- f_\partial. \end{aligned}$$

Together with (4.2.43) this proves (4.2.18).

If $\lambda_+ = \infty$ then it is not difficult to see that $I_3 \to \infty$ as $\delta \to 0$, $f_s \not\equiv 0$, and so $J(f) = \infty$.

The cases $f_s = -f_s^-$ and $f_s = f_s^+ - f_s^-$ can be treated similarly.

The proof of Theorem 4.2.5 is complete. □

4.2.3 The deviation integral in the case $d \geqslant 1$

Given two points $\alpha, \beta \in \mathbb{R}^d$ with $d > 1$, define $[\alpha, \beta] \subset \mathbb{R}^d$ as the straight line segment connecting these points. The definition of the space $\mathbb{D}$ of functions $f = f(t) : [0,1] \to \mathbb{R}^d$ without discontinuities of the second kind repeats, in the general case $d \geqslant 1$, the definition in the case $d = 1$, if we assume that for every $t \in (0,1)$ the value $f(t)$ lies on the segment $[f(t-0), f(t+0)]$. The values $f(0)$ and $f(1)$ need not coincide (as in the case $d = 1$) with $f(+0)$ and $f(1-0)$ respectively. Note that every coordinate $f_{(i)} = f_{(i)}(t)$ of $f = (f_{(1)}, \ldots, f_{(d)}) \in \mathbb{D}$ belongs to the corresponding 'one-dimensional' space $\mathbb{D}$ (in the case $d = 1$).[2] Therefore, the

[2] The converse is false in general: a function $f = (f_{(1)}, \ldots, f_{(d)})$ whose coordinates lie in $\mathbb{D}$ (with $d = 1$), can itself lie outside $\mathbb{D}$ since the value $f(t)$ at a discontinuity point may lie outside $[f(t-0), f(t+0)]$.

functions f in $\mathbb{D}$ in the general case $d \geqslant 1$ retain separability: for every countable everywhere dense subset U of $[0,1]$ containing 0 and 1, we have

$$\sup_{(a,b)} f(t) := (\sup_{(a,b)} f_{(1)}(t), \dots, \sup_{(a,b)} f_{(d)}(t))$$
$$= (\sup_{(a,b)\cap U} f_{(1)}(t), \dots, \sup_{(a,b)\cap U} f_{(d)}(t)) =: \sup_{(a,b)\cap U} f(t).$$

The definition of the metric ρ in the general case $d \geqslant 1$ repeats the definition for the case $d = 1$. Lemma 4.2.6 remains valid in the case $d \geqslant 1$.

Definition 4.2.1 of the deviation integral $J(f)$ for a function $f \in \mathbb{D}$ is also the same in the case $d \geqslant 1$, with the obvious replacement of the scalar differences $f(t_{k+1}) - f(t_k)$ by vector differences.

As before, when, instead of the deviation function $\Lambda(\alpha)$, we use in Definition 4.2.1 an arbitrary convex lower semicontinuous function $M = M(\alpha)$, mapping $\mathbb{R}^d$ into $[0, \infty]$, we denote the corresponding integral by $J^M = J^M(f)$.

By the *variation* of $f \in \mathbb{D}$ in the case $d \geqslant 1$ we refer to to the functional $\operatorname{Var} f := J^M(f)$, where $M = M(\alpha) := |\alpha|$, $\alpha \in \mathbb{R}^d$. By Theorem 4.2.8 (see below), the variation $\operatorname{Var} f$, finite or infinite, of every function $f \in \mathbb{D}$ is always defined. Meanwhile, since the vector $\alpha = (\alpha_{(1)}, \dots, \alpha_{(d)})$ satisfies

$$\frac{1}{\sqrt{d}} \sum_{i=1}^{d} |\alpha_{(i)}| \leqslant |\alpha| \leqslant \sum_{i=1}^{d} |\alpha_{(i)}|,$$

for $f = (f_{(1)}, \dots, f_{(d)}) \in \mathbb{D}$ we have

$$\frac{1}{\sqrt{d}} \sum_{i=1}^{d} \operatorname{Var} f_{(i)} \leqslant \operatorname{Var} f \leqslant \sum_{i=1}^{d} \operatorname{Var} f_{(i)}. \tag{4.2.45}$$

Theorem 4.2.8. *If $f \in \mathbb{D}$ then the (finite or infinite) deviation integral $J(f)$ always exists. Moreover,*

$$J(f) = \sup I(f^{\mathbf{t}_K}),$$

where the supremum is taken over all partitions $\mathbf{t}_K$ of $[0,1]$.

Theorem 4.2.9. *Statements* (i)–(vii) *of Theorem 4.2.3 remain valid for $d \geqslant 1$.*

Proofs of Theorems 4.2.8, 4.2.9. These proofs repeat those of of Theorems 4.2.2 and 4.2.3 in the case $d = 1$, with the exception of the proof of claim (iii) of Theorem 4.2.3.

Proof of statement (iii) of Theorem 4.2.9. First, note that, by property $(\overrightarrow{\Lambda}7)$ in subsection 1.2.2, for any $\lambda \in \mathbb{R}^d$, $\lambda \neq 0$, one has

$$\Lambda^{(\langle \lambda, \xi \rangle)}(t) = \inf_{\alpha : \langle \lambda, \alpha \rangle = t} \Lambda(\alpha), \quad t \in \mathbb{R},$$

where $\Lambda^{(\zeta)}$ is the deviation function of ζ. Therefore, for any $\lambda, \alpha \in \mathbb{R}^d$ the inequality $\Lambda^{(\langle \lambda, \xi \rangle)}(\langle \lambda, \alpha \rangle) \leqslant \Lambda(\alpha)$ holds, and for any function $f \in \mathbb{D}$ we have

$$J^{\Lambda^{(\langle\lambda,\xi\rangle)}}(\langle\lambda,f\rangle) \leqslant J(f). \tag{4.2.46}$$

Assume that $\mathrm{Var} f = \infty$. By condition **[C]**, fulfilled for the random vector ξ, the function $\psi(\lambda) = \mathbf{E}e^{\langle\lambda,\xi\rangle}$ is finite in the neighbourhood $(\mu)_\varepsilon$ of a point μ for some $\mu \in \mathbb{R}^d$ and $\varepsilon > 0$. Using (4.2.45), we can show that there exists a nonzero vector $\lambda = \lambda_f \in (\mu)_\varepsilon$, such that $\mathrm{Var}\langle\lambda,f\rangle = \infty$. Since **[C]** is fulfilled for the random variable $\langle\lambda,\xi\rangle$, claim (iii) of Theorem 4.2.3 implies that $J^{\Lambda^{(\langle\lambda,\xi\rangle)}}(\langle\lambda,f\rangle) = \infty$. Hence, by (4.2.46) we obtain $J(f) = \infty$. Statement (iii) of Theorem 4.2.9 is proved.

Corollary 4.2.10. *Corollary 4.2.4 remains valid for $d \geqslant 1$.*

Proof of Corollary 4.2.10. This repeats the proof of Corollary 4.2.4. □

Consider now the decomposition of the deviation integral in the case $d \geqslant 1$. By statement (iii) of Theorem 4.2.9, the functions f in the class $\mathbb{V}$ of functions of bounded variation are of major interest. It is clear from (4.2.45) that if a function $f = (f_{(1)},\ldots,f_{(d)}) \in \mathbb{D}$ has bounded variation then so does every coordinate $f_{(i)}$ and by (4.2.17),

$$f_{(i)} = f_{(i)a} + f_{(i)s} + f_{(i)\partial}.$$

Therefore, for every $f \in \mathbb{V}$ we have an analogue of (4.2.17):

$$f = f_a + f_s + f_\partial, \qquad f_a(0) = f(0), \qquad f_s(0) = f_\partial(0) = 0, \tag{4.2.47}$$

where

$$\begin{aligned} f_a &:= (f_{(1)a},\ldots,f_{(d)a}), \\ f_s &:= (f_{(1)s},\ldots,f_{(d)s}), \\ f_\partial &:= (f_{(1)\partial},\ldots,f_{(d)\partial}) \end{aligned}$$

are absolutely continuous, singular (continuous) and discrete components respectively.

Along with the deviation function $\Lambda(\alpha)$ define the functions

$$\Lambda_\infty(\alpha) := \sup_{\lambda\in\mathcal{A}} \langle\lambda,\alpha\rangle,\ \alpha \in \mathbb{R}^d, \quad \mathcal{A} := \{\lambda : \psi(\lambda) < \infty\},$$

where, as before, $\psi(\lambda) = \mathbf{E}e^{\langle\lambda,\xi\rangle}$ is the Laplace transform of the distribution of the random vector ξ. The function $\Lambda_\infty(\alpha)$ is convex lower semicontinuous and linear along every ray $L_{\mathrm{e}} := \{\alpha = t\mathrm{e} : t \geqslant 0\}$, $\mathrm{e} \in \mathbb{R}^d$, $|\mathrm{e}| = 1$ (see e.g. [159]). These properties of Λ_∞ allow us to define for any function $f \in \mathbb{D}$ the integral $J^{\Lambda_\infty}(f)$ along with the 'main' deviation integral $J(f) = J^{\Lambda}(f)$.

The following statement generalises Theorem 4.2.5 to the case $d \geqslant 1$.

Theorem 4.2.11. *For every function $f = f_a + f_s + f_\partial \in \mathbb{V}$ we have*

$$J(f) = I(f_a) + J^{\Lambda_\infty}(f_s) + J^{\Lambda_\infty}(f_\partial). \tag{4.2.48}$$

Since by (4.2.48) we have $J(f_\partial) = \Lambda(0) + J^{\Lambda_\infty}(f_\partial)$ and the same expression holds for the component f_s, in the case $\Lambda(0) = 0$ (which always holds provided that $\mathbf{E}\xi = 0$) the integral $J(f)$, while failing to be a linear functional of f, turns out to be additive with respect to the decomposition (4.2.47):

$$J(f) = J(f_a) + J(f_s) + J(f_\partial).$$

Note also that in the important particular case $f_s \equiv 0$ we have

$$J(f) = I(f_a) + \sum_{k=1}^{\infty} \Lambda_\infty(f_{\partial,k}), \tag{4.2.49}$$

where $f_{\partial,k}$, $k = 1, 2, \ldots$, are the enumerated sizes of the jumps of f_∂.

The proof of Theorem 4.2.11 in the case $f_s \equiv 0$ (i.e. (4.2.49)) repeats that of Lemma 4.2.7. We only have to observe that

$$\lim_{t\to\infty} \frac{\Lambda(\alpha t)}{t} = \Lambda_\infty(\alpha) \quad \text{for} \quad \alpha \in \mathbb{R}^d,\ |\alpha| \neq 0.$$

The proof of Theorem 4.2.11 in the case $f_s \not\equiv 0$, and under the same simplifying assumptions as those stated while proving Theorem 4.2.5, is similar to the proof of Theorem 4.2.5. See the proof of Theorem 4.2.11 in general case in [135].

The results of this section were obtained in [61].

4.3 Chebyshev-type exponential inequalities for trajectories of random walks

In this section we obtain upper exponential bounds for $\mathbf{P}(s_n \in B)$, which enable us to get the desired bounds in l.d.p. (cf. (4.1.8), (4.1.10)). These inequalities are of independent interest.

Consider continuous random polygonal lines $s_n = s_n(t), t \in [0, 1]$, with vertices at the points $(k/n,\ S_k/n)$, $k = 0, 1, \ldots, n$, where $S_0 = 0$. The trajectories $s_n(t)$ will be considered as elements of the metric space $(\mathbb{C}, \rho_{\mathbb{C}})$, where $\mathbb{C} = \mathbb{C}^d[0, 1]$ is the space of continuous functions $f = f(t)$ on the segment $[0, 1]$ with values in $\mathbb{R}^d$, endowed with the uniform metric

$$\rho_{\mathbb{C}}(f, g) := \max_{0\leqslant t\leqslant 1} |f(t) - g(t)|.$$

We will be interested in finding upper bounds for the probabilities $\mathbf{P}(s_n \in B)$, with arbitrary measurable sets B from $\mathbb{C}$. An important role will be played here by the Lebesgue integral

$$I(f) := \int_0^1 \Lambda(f'(t))dt$$

(see (4.2.1), defined on the space of continuous functions $\mathbb{C}_a$ and called the deviation integral (or action functional); see section 4.2. For any set B from $\mathbb{C}$ set

$$I_a(B) := \inf_{f\in B\cap\mathbb{C}_a,\, f(0)=0} I(f),$$

where the infimum over an empty set is defined to be equal to ∞. As before (see section 1.4), denote by $\Lambda^{(\gamma)}$ the deviation function of a random variable $\gamma = \Lambda(\xi)$.

The main assertion of this subsection is the following theorem.

Theorem 4.3.1. (i) *For any open convex set $B \subset \mathbb{C}$ one has the inequality*

$$\mathbf{P}(s_n \in B) \leqslant e^{-nI_a(B)}. \tag{4.3.1}$$

(ii) *For any Borel set $B \subset \mathbb{C}$,*

$$\mathbf{P}(s_n \in B) \leqslant e^{-nI_a([B^{\mathrm{con}}])}, \tag{4.3.2}$$

where B^{con} is the convex envelope of B.

(iii) *If B is a Borel set, $I_a(B) \geqslant \mathbf{E}\gamma$, $\gamma = \Lambda(\xi)$, then*

$$\mathbf{P}(s_n \in B) \leqslant e^{-n\Lambda^{(\gamma)}(I_a(B))}. \tag{4.3.3}$$

The next lemma turns out to be useful for both Theorem 4.3.1 and our further assertions.

Lemma 4.3.2. *If $B \subset \mathbb{C}$ is a convex set, $I_a((B)) < \infty$, then*

$$I_a((B)) = I_a(B) = I_a([B]). \tag{4.3.4}$$

Proof. Consider the auxiliary functional

$$I^{\lim}(f) := \lim_{\varepsilon \to 0} I_a((f)_\varepsilon),$$

defined on all elements of $f \in \mathbb{C}$. The functional $I^{\lim}$ coincides with the deviation integral J introduced in section 4.2 (see Corollary 4.2.4), although these functionals are defined in a different way. In terms of the notation (4.1.12), the functional $I^{\lim}(f)$ could be written in the form $I_a(f+)$.

We will need the following lemma.

Lemma 4.3.3. *The functional $I^{\lim}(f)$ is convex and lower semicontinuous.*

Proof. (i) First we verify that the functional $I^{\lim}$ is convex. For $p \geqslant 0$, $q \geqslant 0$, $p+q = 1$ and $f, g \in \mathbb{C}$, there exist sequences $f_n \to f$ and $g_n \to g$ as $n \to \infty$ such that

$$pI^{\lim}(f) + qI^{\lim}(g) = \lim_{n\to\infty} (pI(f_n) + qI(g_n)).$$

From the convexity of $I(f)$ one has $pI(f_n) + qI(g_n) \geqslant I(pf_n + qg_n)$. Therefore,

$$pI^{\lim}(f) + qI^{\lim}(g) \geqslant \overline{\lim_{n\to\infty}}\, I(pf_n + qg_n).$$

Further, for any $\varepsilon > 0$ and all large enough n we have $pf_n + qg_n \in (pf + qg)_\varepsilon$. Hence

$$I(pf_n + qg_n) \geqslant I((pf + qg)_\varepsilon), \quad pI^{\lim}(f) + qI^{\lim}(g) \geqslant I((pf + qg)_\varepsilon).$$

Passing to the limit on the right-hand side of the last inequality as $\varepsilon \to 0$, we obtain

$$pI^{\lim}(f) + qI^{\lim}(g) \geqslant I^{\lim}(pf + qg).$$

This establishes the convexity of the functional $I^{\lim}(f)$.

(ii) Now we prove the lower semicontinuity of $I^{\lim}(f)$. Take an arbitrary sequence f_n that converges to f as $n \to \infty$. For any fixed $\varepsilon > 0$ there exists an $N = N_\varepsilon < \infty$ such that, for any $\delta \in (0, \varepsilon)$, $n \geqslant N$, one has

$$(f_n)_\delta \subset (f)_{2\varepsilon}, \quad I_a((f_n)_\delta) \geqslant I_a((f)_{2\varepsilon}).$$

Hence, for $n \geqslant N$,

$$I^{\lim}(f_n) = \lim_{\delta\to 0} I_a\big((f_n)_\delta\big) \geqslant I_a\big((f)_{2\varepsilon}\big), \qquad \varliminf_{n\to\infty} I^{\lim}(f_n) \geqslant I_a\big((f)_{2\varepsilon}\big).$$

Passing to the limit as $\varepsilon \to 0$, we get

$$\varliminf_{n\to\infty} I^{\lim}(f_n) \geqslant I^{\lim}(f).$$

This proves the lower semicontinuity of the functional $I^{\lim}(f)$ and hence Lemma 4.3.3 as well. □

Now we return to the proof of Lemma 4.3.2. By virtue of Corollary 1.2.3, for the convex and lower semicontinuous functional

$$J(f) := I^{\lim}(f)$$

on $\mathbb{C}$ one has the equalities

$$J((B)) = J(B) = J([B]), \tag{4.3.5}$$

in which the value $J(B)$ for an arbitrary set $B \subset \mathbb{C}$ is defined as

$$J(B) := \inf_{f\in B} J(f).$$

Now we will prove (4.3.4). Since for $f \in \mathbb{C}_a$ one has $I(f) \geqslant J(f)$ and, for any $B \subset \mathbb{C}$,

$$I_a(B) = \inf_{f\in B\cap\mathbb{C}_a} I(f),$$

we obtain

$$I_a((B)) \geqslant I_a(B) \geqslant I_a([B]) \geqslant J([B]). \tag{4.3.6}$$

Let us show that

$$J((B)) \geqslant I_a((B)). \tag{4.3.7}$$

For any $f \in (B)$ and any small enough $\varepsilon > 0$ one has

$$(f)_\varepsilon \subset (B), \qquad I_a((f)_\varepsilon) \geqslant I_a((B)).$$

Hence

$$J(f) = I^{\lim}(f) = \lim_{\varepsilon \to 0} I_a((f)_\varepsilon) \geqslant I_a((B)), \qquad f \in (B).$$

This implies (4.3.7). Relations (4.3.5), (4.3.6) and (4.3.7) imply (4.3.4). Lemma 4.3.2 is proved. □

Note that if condition $[\mathbf{C}_\infty]$ is met then, as was shown in Theorem 4.2.3, one has $I^{\lim}(f) = \infty$ for $f \in \mathbb{C} \setminus \mathbb{C}_a$. Since $I^{\lim}(f) = I(f)$ for $f \in \mathbb{C}_a$, we see that, in this case, the required properties of convexity and lower semicontinuity of the functional $I^{\lim}(f)$ on $\mathbb{C}$ are an obvious consequence of the properties of I on $\mathbb{C}_a$, and hence the assertion of Lemma 4.3.2 follows immediately from Corollary 1.2.3.

The next assertion follows from Theorem 4.3.1 and Lemma 4.3.2.

Corollary 4.3.4. *Let $B \subset \mathbb{C}$ be a convex set. Then*

(i)

$$\mathbf{P}(s_n \in B) \leqslant e^{-nI_a([B])}. \tag{4.3.8}$$

(ii) *If, in addition, at least one of the following conditions is satisfied,*
  (a) *the set B is open,*
  (b) *the set B is closed,*
  (c) $I_a((B)) < \infty$,
  then, on the right-hand side of (4.3.8), *one can replace $[B]$ by B.*

We will show below that if B satisfies conditions (a) or (c) of assertion (ii), then there exists the limit

$$\lim_{n\to\infty} \frac{1}{n} \ln \mathbf{P}(s_n \in B) = -I_a(B). \tag{4.3.9}$$

Proof of Theorem 4.3.1. (i) Denote by $\mathbb{L}_n$ the subclass of the space $\mathbb{C}$ that consists of piecewise linear continuous functions $l = l(t)$ with $l(0) = 0$ which are linear on each of the intervals $((i-1)/n,\ i/n)$, $i = 1, \dots, n$. For example, the random polygonal line s_n is an element of the space $\mathbb{L}_n$.

Consider the mapping $H : \mathbb{L}_n \to \mathbb{R}^{dn}$, which maps a polygonal line $l \in \mathbb{L}_n$ to the vector $Hl = \vec{h} := (h_1, \dots, h_n)$, where $h_i := n\big(l(i/n) - l((i-1)/n)\big)$, $i = 1, \dots, n$. The inverse mapping $H^{(-1)}$ reconstructs a polygonal line in $\mathbb{L}_n$ with vertices at the points $(i/n, H_i/n)$, where $H_i = h_1 + \dots + h_i$ for $i = 1, \dots, n$, from the vector of increments $\vec{h} \in \mathbb{R}^{dn}$. Clearly, H is a one-to-one continuous (under the 'natural' uniform metric $\rho_{\mathbb{C}}$ on $\mathbb{L}_n$ and the Euclidean norm on $\mathbb{R}^{dn}$) linear mapping of the space $\mathbb{L}_n$ onto $\mathbb{R}^{dn}$. This mapping establishes a correspondence between the random vector $Hs_n = \vec{\xi} := (\xi_1, \dots, \xi_n)$ and the random polygonal line s_n. Note also that, for any open convex set $B \subset \mathbb{C}$, the image $H(B \cap \mathbb{L}_n)$ is open and convex in the space $\mathbb{R}^{dn}$. Therefore, by virtue of Theorem 1.3.1, we have in the dn-dimensional space the relations

$$\mathbf{P}(s_n \in B) = \mathbf{P}(s_n \in B \cap \mathbb{L}_n) = \mathbf{P}(\vec{\xi} \in H(B \cap \mathbb{L}_n)) \leqslant e^{-\Lambda^{(\vec{\xi})}(H(B\cap\mathbb{L}_n))}. \tag{4.3.10}$$

Further, observe that for any vector $\vec{h} \in \mathbb{R}^{dn}$, the deviation function $\Lambda^{(\vec{\xi})}(\vec{h})$ of a random vector $\vec{\xi}$ due to the independence of its components has the form

$$\Lambda^{(\vec{\xi})}(\vec{h}) = \sum_{k=1}^{n} \Lambda(h_i) = n \int_0^1 \Lambda(l'(t))dt = nI(l),$$

where $l = H^{(-1)}\vec{h}$ and $\Lambda(\alpha) = \Lambda^{(\xi)}(\alpha)$. In other words, for $l \in \mathbb{L}_n$, $\vec{h} = Hl$, one has the equality $\Lambda^{(\vec{\xi})}(\vec{h}) = \Lambda^{(\vec{\xi})}(Hl) = nI(l)$. Therefore, using an obvious notation,

$$\Lambda^{(\vec{\xi})}(H(B \cap \mathbb{L}_n)) = nI_a(B \cap \mathbb{L}_n) \geqslant nI_a(B).$$

This means that the right-hand side of (4.3.10) does not exceed $e^{-nI_a(B)}$. Assertion (i) of Theorem 4.3.1 is proved.

Assertion (ii) can be proved in a similar way to Theorem 1.3.2.

To prove (iii) we basically need to repeat the proof of Theorem 1.3.4. Since $B \cap \mathbb{C}_a \subset \{f : I(f) \geqslant I_a(B)\}$, for $I_a(B) \geqslant \mathbf{E}\gamma$ one has

$$\mathbf{P}(s_n \in B) = \mathbf{P}(s_n \in B \cap \mathbb{C}_a) \leqslant \mathbf{P}(I(s_n) \geqslant I_a(B)) = \mathbf{P}\left(\int_0^1 s_n'(t)dt \geqslant I_a(B)\right) =$$

$$= \mathbf{P}\left(\sum_{i=1}^{n} \Lambda(\xi_k) \geqslant nI_a(B)\right) = \mathbf{P}\left(\sum_{i=1}^{n} \gamma_k \geqslant nI_a(B)\right) \leqslant e^{-n\Lambda^{(\gamma)}(I_a(B))}. \quad (4.3.11)$$

This proves (4.3.3). The theorem is proved. □

Note that there are no 'crude' bounds in the proof of inequality (4.3.3) (see the proof of Corollary 1.3.7), and therefore the bound (4.3.3) for the probability of hitting the set B_v, which is determined by the level surface at level v, is exponentially unimprovable in the infinite-dimensional case (unlike inequality (1.3.14) in the finite-dimensional case). The fact that the argument of the exponent in (4.3.3) cannot be improved also follows from the form of the logarithmic asymptotics for the probability $\mathbf{P}(s_n \in B_v) = \mathbf{P}(\sum_{k=1}^n \gamma_k \geqslant nv)$, for which (4.3.11) gives a bound in the case of sets $B_v = \{f : I(f) \geqslant v\}$, $v = I_a(B)$. According to the large deviation principle (see subsection 2.7.2),

$$\ln \mathbf{P}\left(\sum_{k=1}^{n} \gamma_k \geqslant nv\right) \sim -n\Lambda^{(\gamma)}(v) \quad \text{as} \quad n \to \infty.$$

One can also add that $\mathbf{P}(s_n \in B_v) \to 1$ for any $v < \mathbf{E}\gamma$, while for values $v \in [\mathbf{E}\gamma, v_+)$, where v_+ is known, one can give a closed-form expression for the exact asymptotics for that probability (see section 2.2).

Since $\Lambda^{(\gamma)}(v) < v - \mathbf{E}\gamma$ for $v > \mathbf{E}\gamma$ (see (1.4.8)), a comparison of (4.3.11) with (4.3.1) shows that, in the infinite-dimensional case, the probability of s_n hitting B_v is exponentially greater than the probability that it will hit an arbitrary convex set $B \subset B_v$ that touches the set $\overline{B}_v$ from the outside. In the finite-dimensional case, the two probabilities differ from each other just by a power function factor (see subsection 1.3.4).

4.4 Large deviation principles for continuous random walk trajectories. Strong versions

In what follows we will assume that the random vector ξ has a *non-degenerate* distribution in $\mathbb{R}^d$, i.e. it satisfies the following condition:

For any $\lambda \in \mathbb{R}^d$, $|\lambda| \neq 0$, $u \in \mathbb{R}$, *one has* $\mathbf{P}(\langle \lambda, \xi \rangle = u) < 1$.

As a rule, it will be assumed in what follows that the jumps ξ_j of the random walk $\{S_n\}$ (see (4.0.1)) satisfy the *Cramér moment condition*.

4.4.1 A strong version of the 'usual' l.d.p. for trajectories of random walks

Consider continuous random polygonal lines $s_n = \{s_n(t);\ t \in [0,1]\}$, with nodes at the points $(k/n, S_k/x)$, $k = 0, 1, \ldots, n$:

$$s_n = s_n(t) := \frac{1}{x}\big(S_{[nt]} + \{nt\}\xi_{[nt]+1}\big), \qquad 0 \leqslant t \leqslant 1,$$

where $[t]$ and $\{t\}$ are the integer and fractional parts of the real number t, respectively. Here and elsewhere in this section we will consider values $x \sim \alpha_0 n$ as $n \to \infty$, $\alpha_0 = \text{const.} > 0$. In what follows we will assume, without loss of generality, that $\alpha_0 = 1$. Thus, in this and later sections we will be dealing with 'proper' large deviations of order n, which lead to the values $z_n = n$ in the notation of section 4.1. Moderately large deviations satisfying $x = o(n)$, $z_n = o(n)$, $x \gg \sqrt{n}$ as $n \to \infty$ will be dealt with in Chapter 5.

We will denote by $\mathfrak{A}(\mathbb{C})$ the σ-algebra of subsets of $\mathbb{C}$ that is generated by cylinder sets. It is well known (see e.g. [94], p. 580), that the σ-algebra $\mathfrak{A}(\mathbb{C})$ coincides with the σ-algebra $\mathfrak{B}(\mathbb{C}, \rho_{\mathbb{C}})$ of Borel subsets of $\mathbb{C}$ generated by the uniform metric $\rho_{\mathbb{C}}$.

The triplet $\langle (\mathbb{R}^d)^n, (\mathfrak{B}^d)^n, \mathbf{P}_n \rangle$, where $\mathfrak{B}^d$ is the σ-algebra of Borel sets in $\mathbb{R}^d$ and $\mathbf{P}_n$ is the distribution of the sequence $\{0, S_1, \ldots, S_n\}$, will be taken as our *original probability space*. The function s_n is a measurable mapping of the original probability space into the probability space

$$\langle \mathbb{C}, \mathfrak{A}(\mathbb{C}), \mathbf{P} \rangle, \tag{4.4.1}$$

where $\mathbf{P}$ is the distribution on $\mathfrak{A}(\mathbb{C})$ corresponding to the process s_n. In the present subsection, the probability space (4.4.1) will be considered as basic.

In what follows, an important role will be played by the deviation functional (integral) $J(f)$ introduced and studied in section 4.2.

For an arbitrary set $B \in \mathfrak{B}(\mathbb{C}, \rho_{\mathbb{C}})$ and function $f \in \mathbb{C}$, denote by $(B)_{\mathbb{C}}$, $[B]_{\mathbb{C}}$, $(f)_{\mathbb{C},\varepsilon}$ the interior and closure of B and the ε-neighbourhood of f in the metric $\rho_{\mathbb{C}}$. The following assertion establishes a relationship between the functionals J and I.

Lemma 4.4.1. (i) *For any* $f \in \mathbb{C}$,

$$J(f) = \lim_{\varepsilon\to 0} I_a\big((f)_{\mathbb{C},\varepsilon}\big) = \lim_{\varepsilon\to 0} J\big((f)_{\mathbb{C},\varepsilon}\big). \tag{4.4.2}$$

(ii) *If a set* B *is convex and* $I_a((B)_{\mathbb{C}}) < \infty$ *then one has*

$$I_a(B) = I_a([B]_{\mathbb{C}}) = I_a((B)_{\mathbb{C}}) = J(B) = J([B]_{\mathbb{C}}) = J((B)_{\mathbb{C}}). \tag{4.4.3}$$

Proof. (i) The first equality in (4.4.2) was proved in Corollary 4.2.4. The second equality in (4.4.2) follows from assertion (vi) of Theorem 4.2.3 (see (4.2.11)).

(ii) The first two equalities in (4.4.3) follow from Lemma 4.3.2. By virtue of assertion (vi) of Theorem 4.2.3, one has $I_a((B)_{\mathbb{C}}) = J((B)_{\mathbb{C}})$ and hence $J((B)_{\mathbb{C}}) < \infty$. The last two equalities in (4.4.3) follow from Corollary 1.2.3. The lemma is proved. □

I. *The local l.d.p.* Everywhere in what follows it will be assumed that $\alpha = x/n \to 1$ as $n \to \infty$. It is not hard to see that the more general case when $\alpha = x/n \to \alpha_0$, $\alpha_0 > 0$, as $n \to \infty$, can be reduced to the case $\alpha_0 = 1$ by scaling both random variables and the sets B.

Theorem 4.4.2 (A strong version of the l.l.d.p.). *(i) For any* $f \in \mathbb{C}$, $\varepsilon > 0$, $\tau > 0$.

$$\overline{\lim_{n\to\infty}} \frac{1}{n} \ln \mathbf{P}\big(s_n \in (f)_{\mathbb{C},\varepsilon}\big) \leqslant -J((f)_{\mathbb{C},\varepsilon(1+\tau)}). \tag{4.4.4}$$

(ii) *For any* $f \in \mathbb{C}$, $\varepsilon > 0$,

$$\underline{\lim}_{n\to\infty} \frac{1}{n} \ln \mathbf{P}(s_n \in (f)_{\mathbb{C},\varepsilon}) \geqslant -J((f)_{\mathbb{C},\varepsilon}). \tag{4.4.5}$$

(iii) *If* $x = n$ *then*

$$\mathbf{P}\big(s_n \in (f)_{\mathbb{C},\varepsilon}\big) \leqslant e^{-nJ((f)_{\mathbb{C},\varepsilon})} \tag{4.4.6}$$

and there exist

$$\lim_{n\to\infty} \frac{1}{n} \ln \mathbf{P}\big(s_n \in (f)_{\mathbb{C},\varepsilon}\big) = -J\big((f)_{\mathbb{C},\varepsilon}\big),$$

$$\lim_{\varepsilon\to 0}\lim_{n\to\infty} \frac{1}{n} \ln \mathbf{P}\big(s_n \in (f)_{\mathbb{C},\varepsilon}\big) = -J(f).$$

In other words, the l.l.d.p. (in its strong form) holds in $(\mathbb{C}, \rho_{\mathbb{C}})$ with parameters (n, J) and functions $\delta(\varepsilon) = \varepsilon(1+\tau)$ and $\beta(\varepsilon) = 0$ (see section 4.1). Observe that the l.l.d.p. always holds in $(\mathbb{C}, \rho_{\mathbb{C}})$ without any restrictions.

Proof. (i) *The upper bound.* We will make use of the following assertion for convex sets, which follows from Corollary 4.3.4 and assertion (vi) in Theorem 4.2.3.

Lemma 4.4.3. *Let a convex set* B *possess at least one of the following properties:*

(a) *B is open;*
(b) *B is closed;*

(c)

$$I_a\left(\left(\frac{x}{n}B\right)_{\mathbb{C}}\right) < \infty \qquad \left(\text{or } J\left(\left(\frac{x}{n}B\right)_{\mathbb{C}}\right) < \infty\right).$$

Then, for all n, one has

$$\mathbf{P}(s_n \in B) \leqslant e^{-nI_a(xB/n)} \leqslant e^{-nJ(\frac{x}{n}B)}. \tag{4.4.7}$$

In case (a) *the second inequality in* (4.4.7) *can be replaced by an equality.*

In what follows we take advantage of the following useful assertion, which allows us to restrict our considerations, when proving the l.l.d.p., to large deviations of the form $x = n$ and, in situations where $a = \mathbf{E}\xi$ is finite, to assume that $\mathbf{E}\xi = 0$.

Lemma 4.4.4. (1) *If, for $x = n$, the process s_n satisfies the l.l.d.p. in $(\mathbb{C}, \rho_{\mathbb{C}})$ with functions $\delta(\varepsilon)$, $\beta(\varepsilon)$* (see Definition 4.1.3) *then it satisfies the same principle for $x \sim n$, $n \to \infty$, with functions $\delta(\varepsilon(1+\tau))$, $\beta(\varepsilon(1+\tau))$ for any $\tau > 0$.*
(2) *Let $x = n$ and let b be an arbitrary vector. If the l.l.d.p. in $(\mathbb{C}, \rho_{\mathbb{C}})$ with parameters (n, J) and functions $\delta(\varepsilon)$, $\beta(\varepsilon)$ holds for the process s_n constructed out of random vectors ξ_k then a process s_n, of the same kind but constructed out of random vectors $\xi_k + b$, satisfies the l.l.d.p. with parameters (n, J_b), where $J_b(f) := J(f - b\mathrm{e})$, $\mathrm{e} = \mathrm{e}(t) := t$, and with the same functions $\delta(\varepsilon)$, $\beta(\varepsilon)$.*

Proof. (1) *The upper bound.* Let the process s_n be constructed in the case where $\alpha := x/n \to 1$, $n \to \infty$. Put $s_n^* = xs_n/n$, so that the process s_n^* is constructed for $x = n$ and therefore satisfies the l.l.d.p. with parameters (z_n, J). Further, for $g \in \mathbb{C}$ one has $|g| := \sup_{0\leqslant t\leqslant 1} |g(t)| < \infty$,

$$\alpha g = g + (\alpha - 1)g \in (g)_{\mathbb{C},\theta},$$

where $\theta \leqslant |\alpha - 1||g|$. Hence, for any $\tau > 0$ and all large enough n one obtains $\theta < \tau\varepsilon$. Therefore,

$$\mathbf{P}\big(s_n \in (f)_{\mathbb{C},\varepsilon}\big) = \mathbf{P}\big(s_n^* \in \alpha(f)_{\mathbb{C},\varepsilon}\big) \leqslant \mathbf{P}\big(s_n^* \in (f)_{\mathbb{C},\varepsilon(1+\tau)}\big), \tag{4.4.8}$$

where $\mathbf{P}\big(s_n^* \in (f)_{\mathbb{C},\varepsilon(1+\tau)}\big)$ satisfies the upper bound in the l.l.d.p. (see Definition 4.1.3). Hence, this bound, with ε replaced by $\varepsilon(1+\tau)$, also holds for the left-hand side of inequality (4.4.8).

The lower bound. Similarly, for large enough n,

$$\mathbf{P}(s_n \in (f)_{\mathbb{C},\varepsilon}) \geqslant \mathbf{P}(s_n^* \in (f)_{\mathbb{C},\varepsilon/2}),$$

and the left-hand side of that inequality satisfies the same bounds (4.1.9) as the right-hand side.

(2) The process $s_n^{(b)} = s_n + b\mathrm{e}$, $\mathrm{e} = \mathrm{e}(t) := t$, $0 \leqslant t \leqslant 1$, is again a process of the same kind as s_n, for which $s_n^{(b)}(k/n) = (S_k + kb)/n$. Therefore, it remains to make use of the equality $\mathbf{P}\big(s_n^{(b)} \in (f)_{\mathbb{C},\varepsilon}\big) = \mathbf{P}\big(s_n \in (f - b\mathrm{e})_{\mathbb{C},\varepsilon}\big)$ and the l.l.d.p. for s_n. Lemma 4.4.4 is proved. □

Thus, in what follows, when proving the l.l.d.p. one can always assume that $x = n$ and $\mathbf{E}\xi = 0$ (if the mean $\mathbf{E}\xi$ is finite).

Let us continue the proof of Theorem 4.4.2. Since the set $(f)_{\mathbb{C},\varepsilon}$ is convex and open, the required upper bounds (4.4.4), (4.4.6) follow from Lemmas 4.4.3 and 4.4.4.

(ii) *The lower bound.* To prove the lower bound we first state the weaker version of (4.4.5), i.e. we prove relation (4.4.5) but with the right-hand side replaced by $-J(f)$. The proof of this weaker assertion will be split into three steps.

(1) First assume that $d = 1$, the variable ξ is bounded and f is a continuous polygonal line with nodes at the points $(t_k, f(t_k))$, $t_k = k\Delta$, $k = 0, \ldots, R+1$, $R+1$, where $R+1 = 1/\Delta$ is an integer.

If $|\xi| \leqslant N$ then it suffices to consider only those functions f for which $|f'(t)| \leqslant N$ for all $t \neq t_k, k = 0, \ldots, R$. Indeed, assuming, for example, that $f'(t) > N$ for $t \in (t_k, t_{k+1})$, one has $\Lambda(f'(t)) = \infty$ on that interval and $J(f) \geqslant \int_{t_k}^{t_{k+1}} \Lambda(f'(t))dt = \infty$, and so inequality (4.4.5) holds true.

For simplicity assume that Δ is a multiple of $1/n$. For given $\varepsilon > 0$, $\delta > 0$, consider the events

$$A_\delta := \bigcap_{k=1}^{R} \left\{ \left| s_n(t_k) - f(t_k) \right| < k\delta \right\}, \qquad B_\varepsilon := \left\{ \max_{0 \leqslant t \leqslant 1} \left| s_n(t) - f(t) \right| < \varepsilon \right\}.$$

For a fixed polygonal line f and numbers $N < \infty$, $\varepsilon > 0$ first choose $\Delta = 1/(R+1)$ and then $\delta > 0$ such that

$$\Delta < \frac{\varepsilon}{4N}, \qquad R\delta < \frac{\varepsilon}{2}. \tag{4.4.9}$$

If

$$s_n(t_k) \in \left[f(t_k) - k\delta, f(t_k) + k\delta \right],$$

and hence

$$s_n(t_k) \in \left[f(t_k) - \frac{\varepsilon}{2}, f(t_k) + \frac{\varepsilon}{2} \right],$$

then, by virtue of (4.4.9), at the time t_{k+1} the trajectory of $s_n(t)$ will fall below the level

$$f(t_k) + \frac{\varepsilon}{2} + N\Delta \leqslant f(t_{k+1}) + 2N\Delta + \frac{\varepsilon}{2} \leqslant f(t_{k+1}) + \varepsilon.$$

Similarly, one has

$$s_n(t_{k+1}) > f(t_k) - \frac{\varepsilon}{2} - N\Delta \geqslant f(t_{k+1}) - 2N\Delta - \frac{\varepsilon}{2} \geqslant f(t_{k+1}) - \varepsilon.$$

This means that, for the chosen Δ and δ, the following holds:

$$A_\delta \subset B_\varepsilon, \qquad \mathbf{P}(B_\varepsilon) \geqslant \mathbf{P}(A_\delta). \tag{4.4.10}$$

For brevity, put $f(t_{k+1}) - f(t_k) =: \Delta f(t_k)$ Then, for $m = n\Delta$, one has

$$\bigcap_{k=0}^{R} \left\{ \Delta s_n(t_k) \in \big(\Delta f(t_k)\big)_{\mathbb{C},\delta} \right\} \subset A_\delta,$$

$$\mathbf{P}(A_\delta) \geqslant \prod_{k=0}^{R} \mathbf{P}\Big(s_n(\Delta) \in \big(\Delta f(t_k)\big)_{\mathbb{C},\delta}\Big) = \prod_{k=0}^{R} \mathbf{P}\Big(S_m \in n\big(\Delta f(t_k)\big)_{\mathbb{C},\delta}\Big).$$

According to the l.l.d.p., for the sums S_m we have

$$\begin{aligned} \varliminf_{n\to\infty} \frac{1}{n} \ln \mathbf{P}(A_\delta) &\geqslant \sum_{k=0}^{R} \varliminf_{n\to\infty} \frac{1}{n} \ln \mathbf{P}\left(\frac{S_m}{m} \in \frac{1}{\Delta} \big(\Delta f(t_k)\big)_{\mathbb{C},\delta} \right) \\ &\geqslant -\sum_{k=0}^{R} \Delta \Lambda \left(\frac{\Delta f(t_k)}{\Delta} \right) = -I(f). \end{aligned} \tag{4.4.11}$$

The desired weaker version of bound (4.4.5) follows from here and (4.4.10).

If Δ is not a multiple of $1/n$ then one should take as partition points $t_{k,n} := [n\Delta k]/n \to t_k := \Delta k$ as $n \to \infty$. In this case, the increments of f on the intervals $(t_{k,n}, t_{k+1,n})$ differ from $\Delta f(t_k)$ by quantities not exceeding $2N/n$, while the number of independent increments of $s_n(t)$ on the intervals $(t_{k,n}, t_{k+1,n})$ stays within the limits $\Delta n \pm 1$. Therefore, after some small obvious changes in the preceding argument, we again obtain the limiting relation (4.4.11).

(2) Now we remove the assumption that ξ is bounded. We truncate the random variables ξ_i at levels $\pm N$, $N > 0$. One has

$$\begin{aligned} \mathbf{P}\big(s_n \in (f)_{\mathbb{C},\varepsilon}\big) &\geqslant \mathbf{P}\big(s_n \in (f)_{\mathbb{C},\varepsilon},\ \max_{1\leqslant i\leqslant n} |\xi_i| \leqslant N\big) \\ &= \mathbf{P}\Big(s_n \in (f)_{\mathbb{C},\varepsilon} \,\Big|\, \max_{1\leqslant i\leqslant n} |\xi_i| \leqslant N\Big) \mathbf{P}^n\big(|\xi| \leqslant N\big), \end{aligned}$$

$$\begin{aligned} &\varliminf_{n\to\infty} \frac{1}{n} \ln \mathbf{P}\big(s_n \in (f)_{\mathbb{C},\varepsilon}\big) \\ &\quad \geqslant \varliminf_{n\to\infty} \frac{1}{n} \ln \mathbf{P}\Big(s_n \in (f)_{\mathbb{C},\varepsilon} \,\Big|\, \max_{1\leqslant i\leqslant n} |\xi_i| \leqslant N\Big) + \ln \mathbf{P}\big(|\xi| \leqslant N\big). \end{aligned} \tag{4.4.12}$$

Denote by ${}^{(N)}\xi$ a random variable with the distribution

$$\mathbf{P}({}^{(N)}\xi \in B) := \mathbf{P}\big(\xi \in B \,\big|\, |\xi| \leqslant N\big).$$

By ${}^{(N)}s_n$ we will denote the random polygonal line constructed using the sums ${}^{(N)}S_j = {}^{(N)}\xi_1 + \cdots + {}^{(N)}\xi_j$, where the ${}^{(N)}\xi_i$ are independent copies of ${}^{(N)}\xi$. Clearly,

$$\mathbf{P}\big(s_n \in (f)_{\mathbb{C},\varepsilon} |\ \max_{1\leqslant i\leqslant n} |\xi_i| \leqslant N\big) = \mathbf{P}\big({}^{(N)}s_n \in (f)_{\mathbb{C},\varepsilon}\big),$$

so the inequality (4.4.12) can be rewritten as

$$\varliminf_{n\to\infty} \frac{1}{n} \ln \mathbf{P}\big(s_n \in (f)_{\mathbb{C},\varepsilon}\big) \geqslant \varliminf_{n\to\infty} \frac{1}{n} \ln \mathbf{P}\big({}^{(N)}s_n \in (f)_{\mathbb{C},\varepsilon}\big) + \ln \mathbf{P}\big(|\xi| \leqslant N\big). \tag{4.4.13}$$

As before, we will keep for ${}^{(N)}\xi$ the notation introduced for all characteristics of ξ adding the left upper index (N) only.

By virtue of the already proven weaker version of inequality (4.4.5) for bounded ξ, one has

$$\varliminf_{n\to\infty} \frac{1}{n} \ln \mathbf{P}\big({}^{(N)}s_n \in (f)_{\mathbb{C},\varepsilon}\big) \geqslant -{}^{(N)}I(f), \tag{4.4.14}$$

where (see (4.4.11))

$${}^{(N)}I(f) = \Delta \sum_{k=0}^{R-1} {}^{(N)}\Lambda\left(\frac{\Delta f(t_k)}{\Delta}\right).$$

Now take any other polygonal line g in $(f)_{\mathbb{C},\varepsilon}$ (with breaks at the same points t_k, $g(0) = 0$). Then there exists a $\delta < \varepsilon$ such that $(g)_{\mathbb{C},\delta} \subset (f)_{\mathbb{C},\varepsilon}$ and, hence, the left-hand side of (4.4.14) will be larger than

$$\varliminf_{n\to\infty} \frac{1}{n} \ln \mathbf{P}\big({}^{(N)}s_n \in (g)_{\mathbb{C},\delta}\big) \geqslant -{}^{(N)}I(g). \tag{4.4.15}$$

Since

$$\big|\Delta g(t_0) - \Delta f(t_0)\big| < \varepsilon, \qquad \big|\Delta g(t_k) - \Delta f(t_k)\big| < 2\varepsilon \quad \text{for} \quad k \geqslant 1,$$

applying the preceding arguments and in (4.4.15) taking the supremum over $g \in (f)_{\mathbb{C},\varepsilon}$, we can replace the right-hand side of (4.4.14) (which is independent of g) by

$$-\Delta\left[{}^{(N)}\Lambda\left(\left(\frac{\Delta f(t_0)}{\Delta}\right)_{\varepsilon/\Delta}\right) + \sum_{k=1}^{R-1} {}^{(N)}\Lambda\left(\left(\frac{\Delta f(t_k)}{\Delta}\right)_{2\varepsilon/\Delta}\right)\right]. \tag{4.4.16}$$

To bound this expression we use the following assertion on the convergence of the deviation functions ${}^{(N)}\Lambda(\alpha)$ as $N \to \infty$. We formulate and prove the assertion in the general multidimensional case $d \geqslant 1$, in order to make the proofs in the cases $d = 1$ and $d > 1$ similar and to avoid repetition.

As before, let S be a convex envelope of the support of distribution of ξ.

Lemma 4.4.5. *If $\alpha \in (\mathrm{S})$ then*

$${}^{(N)}\Lambda(\alpha) \to \Lambda(\alpha) \quad as \quad N \to \infty.$$

If $\alpha \in \partial\mathrm{S}$ then, for any $\varepsilon > 0$, one has

$${}^{(N)}\Lambda\big((\alpha)_\varepsilon\big) \to \Lambda\big((\alpha)_\varepsilon\big) \quad as \quad N \to \infty.$$

If $\alpha \notin [\mathrm{S}]$ then

$${}^{(N)}\Lambda(\alpha) = \Lambda(\alpha) = \infty.$$

The lemma implies

Corollary 4.4.6. *For any α there exists $\varepsilon = \varepsilon(\alpha)$ such that*

$$\lim_{N\to\infty} {}^{(N)}\Lambda\big((\alpha)_\varepsilon\big) = \Lambda\big((\alpha)_\varepsilon\big) \leqslant \Lambda(\alpha).$$

It is easy to verify, that in the one-dimensional case $d = 1$ the convergence ${}^{(N)}\Lambda(\alpha) \to \Lambda(\alpha)$ as $N \to \infty$ holds true for all α. However, in the case $d > 1$, only the convergence of the values ${}^{(N)}\Lambda\big((\alpha)_\varepsilon\big)$ takes place, for all α.

Proof of Lemma 4.4.5. Denote

$$\psi_N(\lambda) = \mathbf{E}\big(e^{\langle\lambda,\xi\rangle},\ |\xi| < N\big), \qquad \delta_N = \ln \mathbf{P}\big(|\xi| < N\big).$$

Then

$$\delta_N \uparrow 0, \qquad \psi_N(\lambda) \uparrow \psi(\lambda), \qquad A_N(\lambda) := \ln \psi_N(\lambda) \uparrow A(\lambda)$$

as $N \uparrow \infty$. Next,

$$\begin{aligned} {}^{(N)}A(\lambda) &= A_N(\lambda) - \delta_N, \\ {}^{(N)}\Lambda(\alpha) &= \sup\big(\langle\alpha,\lambda\rangle - {}^{(N)}A(\lambda)\big) = \Lambda_N(\alpha) + \delta_N, \end{aligned}$$

where

$$\Lambda_N(\alpha) := \sup\big(\langle\alpha,\lambda\rangle - A_N(\lambda)\big) \downarrow \quad \text{as} \quad N \uparrow \infty.$$

Since $\langle\alpha,\lambda\rangle - A_N(\lambda)$ and $\langle\alpha,\lambda\rangle - A(\lambda)$ are concave functions, it is easy to see that for $\alpha \in (\mathrm{S}) = (\Lambda_{<\infty})$, where $\Lambda_{<\infty} = \{\alpha : \Lambda(\alpha) < \infty\}$, one has

$$\Lambda_N(\alpha) \downarrow \Lambda(\alpha), \qquad {}^{(N)}\Lambda(\alpha) \to \Lambda(\alpha)$$

as $N \uparrow \infty$. This can also be deduced from convergence theorems in convex analysis.

Now let $\alpha \in \partial\mathrm{S}$. For $\beta \in (\alpha)_\varepsilon \cap (\mathrm{S})$, by virtue of the first statement of the lemma,

$$\overline{\lim_{N\to\infty}}\, {}^{(N)}\Lambda\big((\alpha)_\varepsilon\big) \leqslant \overline{\lim_{N\to\infty}}\, {}^{(N)}\Lambda(\beta) = \Lambda(\beta).$$

It follows that the left-hand side of this inequality (which does not depend on β) is not greater than $\inf\big\{\Lambda(\beta) : \beta \in (\alpha)_\varepsilon \cap (\mathrm{S})\big\}$. Since the function $\Lambda(\beta)$ is continuous on $\partial\mathrm{S}$ from within the set S, the infimum can be taken over the set $\beta \in (\alpha)_\varepsilon \cap [\mathrm{S}]$. This proves the upper bound in the second statement in the lemma, since $\Lambda(\beta) = \infty$ for $\beta \notin [\mathrm{S}]$. The lower bound is obvious, since ${}^{(N)}\Lambda(\alpha) = \Lambda_N(\alpha) + \delta_N \geqslant \Lambda(\alpha) + \delta_N$.

The last statement in the lemma follows from the inclusion ${}^{(N)}\Lambda_{<\infty} \subset \Lambda_{<\infty}$. The lemma is proved. □

We now return to part (3) the proof of Theorem 4.4.2. By virtue of (4.4.13)–(4.4.16) and Corollary 4.4.6, letting $N \to \infty$ we get

$$\underline{\lim_{n\to\infty}}\, \frac{1}{n} \ln \mathbf{P}\big(s_n \in (f)_{\mathrm{C},\varepsilon}\big) \geqslant -\Delta \sum_{k=0}^{R-1} \Lambda\left(\frac{\Delta f(t_k)}{\Delta}\right) = -I(f) = -J(f).$$

This proves relation (4.4.5) with $-J(f)$ on the right-hand side.

(3) *Now let f be an arbitrary continuous function, $f(0) = 0$.* For an integer $R \geqslant 1$ consider a partition $\mathbf{t}_R = (t_0, \ldots, t_{R+1})$, where $t_k = k/(R+1)$ for $k = 0, \ldots, R+1$. Choose R large enough that

$$(f^{\mathbf{t}_R})_{\mathbb{C},\varepsilon/2} \subset (f)_{\mathbb{C},\varepsilon}$$

(the polygonal line $f^{\mathbf{t}_R}$ was defined in subsection 4.2.1). Then

$$\mathbf{P}\big(s_n \in (f)_{\mathbb{C},\varepsilon}\big) \geqslant \mathbf{P}\big(s_n \in (f^{\mathbf{t}_R})_{\mathbb{C},\varepsilon/2}\big).$$

Applying to the right-hand side of this inequality the bound that we found in step (2) for the continuous polygonal lines $f^{\mathbf{t}_R}$, we obtain

$$\varliminf_{n\to\infty} \frac{1}{n} \ln \mathbf{P}\big(s_n \in (f)_{\mathbb{C},\varepsilon}\big) \geqslant -I(f^{\mathbf{t}_R}).$$

It remains to make use of the inequality $I(f^{\mathbf{t}_R}) \leqslant J(f)$ (see (4.2.7)). Then the lower bound (4.4.5) is established.

The transition to the multivariate case $d \geqslant 2$ does not require any changes in the proof.

In order to obtain (4.4.5) from its weaker version, we have to repeat arguments from the proof of the lower bound in Theorem 1.1.4 (see the proof of (2.7.7) using (2.7.1)). Thus, (4.4.5) and its weaker version are equivalent.

The assertions (i) and (ii) of Theorem 4.4.2 are proved. Assertion (iii) follows in an obvious way from (4.4.5), (4.4.7) and Lemma 4.4.1. Theorem 4.4.2 is proved. □

II. *The strong version of the l.d.p.* For trajectories s_n we have the l.d.p. in $(\mathbb{C}, \rho_{\mathbb{C}})$ in the following form.

Theorem 4.4.7. *Let $\alpha = x/n \to 1$ as $n \to \infty$. Then*

(i) *If condition $[\mathbf{C}_\infty]$ is met, for any $B \in \mathfrak{B}(\mathbb{C}, \rho_{\mathbb{C}})$ we have*

$$\varlimsup_{n\to\infty} \frac{1}{n} \ln \mathbf{P}(s_n \in B) \leqslant -J([B]_{\mathbb{C}}). \tag{4.4.17}$$

(ii) *For any $B \in \mathfrak{B}(\mathbb{C}, \rho_{\mathbb{C}})$ one always has the inequality*

$$\varliminf_{n\to\infty} \frac{1}{n} \ln \mathbf{P}(s_n \in B) \geqslant -J((B)_{\mathbb{C}}). \tag{4.4.18}$$

Theorem 4.4.7 is a strong version of the 'usual' l.d.p. in $(\mathbb{C}, \rho_{\mathbb{C}})$ with parameters (z_n, D), where $z_n = n$, $D(f) = J(f)$, which was proved in [19] in the case $d = 1$ under the condition $[\mathbf{C}_\infty]$ (for both assertions, (i) and (ii)). In [131] that result was extended to multivariate random walks.

Note that, in the case where the right-hand sides of inequalities (4.4.17), (4.4.18) do not coincide, the upper bound (4.4.17) may be 'bad'. For instance, this bound makes no sense for the set $B = B_v := \{f : J(f) \geqslant v\}$ since the closure $[B_v]_{\mathbb{C}}$ coincides with $\mathbb{C}$, and hence $J([B]_{\mathbb{C}}) = 0$. The fact that $[B_v]_{\mathbb{C}}$ coincides with $\mathbb{C}$ follows since, for any u, ε and $f \in \mathbb{C}$ one can easily construct a 'fast oscillating' function $g \in \mathbb{C}_a$, such that $\rho(f, g) < \varepsilon$ and $\operatorname{Var} g := \int_0^1 |g'(t)|dt \geqslant u$. Since, under the condition $[\mathbf{C}_0]$ for some $c_1 > 0$, $c_2 > 0$ and all $\alpha \in \mathbb{R}^d$ one has the inequality (see e.g. (Λ6))

$$\Lambda(\alpha) \geqslant c_1|\alpha| - c_2, \tag{4.4.19}$$

we conclude that $J(g) \geqslant c_1 u - c_2$. Thus, we can construct a function g that is arbitrarily close to f and is such that $J(g) \geqslant v$ ($g \in B_v$ for $u \geqslant (v + c_2)/c_1$).

The lower bound in assertion (ii) of Theorem 4.4.7 holds without any moment conditions. As for the upper bounds, in some cases one can also obtain them without moment conditions. This can be done for the class of convex sets. Simple exponentially unimprovable upper bounds for $\mathbf{P}(s_n \in B)$ for such sets were obtained in section 4.3 without any *moment conditions* or restrictions on the *order of deviation*. This substantially simplifies the formulations and makes the assertions quite general (see Theorem 4.4.9 below).

Proof of Theorem 4.4.7. (i) To prove Theorem 4.4.7 we will need to verify condition $[\mathbf{K}_0]$ (see section 4.1).

$[\mathbf{K}_0]$. *There is a family of imbedded compacts K_v; $v > 0$, in $(\mathbb{C}, \rho_C)$ such that, for any $N > 0$, there exists a $v > 0$ for which*

$$\overline{\lim_{n\to\infty}} \frac{1}{n} \ln \mathbf{P}(s_n \notin K_v) \leqslant -N. \tag{4.4.20}$$

We will need the following lemma.

Lemma 4.4.8. *The property $[\mathbf{K}_0]$ is equivalent to the condition $[\mathbf{C}_\infty]$.*

Proof. (i) First we show that $[\mathbf{K}_0]$ implies $[\mathbf{C}_\infty]$. The property $[\mathbf{K}_0]$ means that for any $N < \infty$ there exists a compact set K_v and a number $n_v < \infty$ such that

$$\mathbf{P}(s_n \notin K_v) \leqslant e^{-Nn} \quad \text{for} \quad n \geqslant n_v.$$

Put $\Omega_v(\Delta) := \sup_{f\in K_v} \omega_f(\Delta)$ for $\Delta \geqslant 0$, where

$$\omega_f(\Delta) := \max\Big\{\big|f(t) - f(u)\big| : t, u \in [0,1],\ |t-u| \leqslant \Delta\Big\}$$

is the continuity modulus of f. Clearly, the function $\Omega_v(\Delta)$ is continuous and non-decreasing with respect to Δ $\Omega_v(0) = 0$. For $n \geqslant n_v$ one has

$$\mathbf{P}\left(\frac{1}{n}|\xi_1| > \Omega_v\left(\frac{1}{n}\right)\right) = \mathbf{P}\left(\left|s_n\left(\frac{1}{n}\right)\right| > \Omega_v\left(\frac{1}{n}\right)\right) \leqslant \mathbf{P}(s_n \notin K_v) \leqslant e^{-Nn}.$$

If necessary, we can increase the value of n_v so that, for $n \geqslant n_v$, one would have $\Omega_v(1/n) \leqslant 1$. Then, for large enough n,

$$\mathbf{P}\big(|\xi_1| \geqslant n\big) \leqslant \mathbf{P}\left(\frac{1}{n}\,|\xi_1| > \Omega_v\left(\frac{1}{n}\right)\right) \leqslant e^{-nN}$$

holds, which clearly implies $[\mathbf{C}_\infty]$.

(ii) It suffices to establish the desired implication $[\mathbf{C}_\infty] \Longrightarrow [\mathbf{K}_0]$ in the one-dimensional case $d = 1$. Indeed, if condition $[\mathbf{C}_\infty]$ is met for the vector $\xi = (\xi_{(1)}, \dots, \xi_{(d)})$ then it is also met for each component $\xi_{(i)}$. By virtue of the implication $[\mathbf{C}_\infty] \Longrightarrow [\mathbf{K}_0]$ in the case $d = 1$, for each $i = 1, \dots, d$ one can choose a family of compacts $K_{i,v}$ in the one-dimensional space $(\mathbb{C}, \rho_\mathbb{C})$ such that, for a given $N < \infty$ and some $v_i = v_{i,N}$, the compact set K_{i,v_i}, satisfies (4.4.20). Then the direct product $K_v := K_{i,1} \times \cdots \times K_{i,d}$ will give us the desired compact in the d-dimensional space $(\mathbb{C}, \rho_\mathbb{C})$ since, for any $N < \infty$ and $v := \max\{v_1, \dots, v_d\}$, the compact K_v satisfies (4.4.20).

So, assume that $d = 1$. By virtue of assertions (iv), (vii) of Theorem 4.2.3, $K_v := \{f \in \mathbb{C}_a : I(f) \leqslant v\}$ is a family of imbedded compact sets.

Further, we will prove that for any $N < \infty$ there exists $v = v_N$ such that the compact K_v satisfies (4.4.20). To this end, along with ξ consider the random variable $\gamma := \Lambda(\xi)$, and construct the deviation function for it:

$$\Lambda^{(\gamma)}(t) := \sup_\lambda\{\lambda t - \ln \mathbf{E}e^{\lambda\gamma}\}, \quad t \in \mathbb{R}.$$

We will make use of Theorem 1.4.1. It implies that if $[\mathbf{C}]$ is met for ξ then $[\mathbf{C}_0]$ is satisfied for $\gamma = \Lambda(\xi)$ and $\Lambda^{(\gamma)}(t) \geqslant t - \mathbf{E}\gamma$ for all t (see (1.4.8)).

From this and Chebyshev's exponential inequality with $v \geqslant \mathbf{E}\gamma$, one has

$$\mathbf{P}(s_n \notin K_v) = \mathbf{P}\big(I(s_n) > v\big) = \mathbf{P}\left(\frac{1}{n}\sum_{i=1}^{n} \Lambda(\xi_i) > v\right)$$
$$\leqslant e^{-n\Lambda^{(\gamma)}(v)} \leqslant e^{-n(v - \mathbf{E}\gamma)}.$$

Hence, (4.4.20) holds for $v = N + \mathbf{E}\gamma$. The lemma is proved. □

We now continue the proof of assertion (i) of Theorem 4.4.7. By Theorem 4.4.2, one has an upper bound in the l.l.d.p. for s_n (see Definition 4.1.1). It follows from Lemma 4.4.8 that the condition $[\mathbf{K}]_0$ is satisfied. Hence, by virtue of Theorem 4.1.6, one has

$$\overline{\lim_{n\to\infty}}\ \frac{1}{n} \ln \mathbf{P}(s_n \in B) \leqslant -J(B+), \tag{4.4.21}$$

i.e. the upper bound from the e.l.d.p. in the space $(\mathbb{C}, \rho_\mathbb{C})$ holds true for parameters (n, J). It follows from assertion (vii) of Theorem 4.2.3 that $J_v = \{g \in \mathbb{C} : J(g) \leqslant v\}$ is a compact subset of $(\mathbb{C}, \rho_\mathbb{C})$. Therefore, by Lemma 4.4.1, one has the equality $J(B+) = J([B]_\mathbb{C})$, i.e. the right-hand side of (4.4.21) can be replaced by $-J([B]_\mathbb{C})$. Assertion (i) of Theorem 4.4.7 is proved.

(ii) By virtue of Theorem 4.1.6, assertion (ii) of Theorem 4.4.7 follows from assertion (ii) of Theorem 4.4.2. Theorem 4.4.7 is proved. □

III. *The l.d.p. for convex sets.* As we have seen, in the l.d.p. for continuous trajectories s_n condition $[\mathbf{C}_\infty]$ is required only for the upper bounds. For convex sets this moment condition can be removed or significantly weakened, while the assertion of the l.d.p. itself can be made stronger.

We will say that set B satisfies the condition **[B]** given below if at least one of the conditions (a), (c) in Lemma 4.4.3 is met:

[B]. (a). *B is open*

and/or

(c)

$$I_a\left(\frac{x}{n}(B)_{\mathbb{C}}\right) < \infty \qquad \left(or \quad J\left(\frac{x}{n}(B)_{\mathbb{C}}\right) < \infty\right).$$

In addition to assertion (ii) of Theorem 4.4.7 one has the following upper bounds in the l.d.p.

Theorem 4.4.9. *Let $B \subset \mathbb{C}$ be a convex set. Then the following assertions hold true.*

(i) *For any $n \geqslant 1$,*

$$\mathbf{P}(s_n \in B) \leqslant e^{-nI_a((x/n)[B]_{\mathbb{C}})} \leqslant e^{-nJ((x/n)[B]_{\mathbb{C}})}. \tag{4.4.22}$$

If, in addition, condition **[B]** *is satisfied then one can replace $[B]_{\mathbb{C}}$ in* (4.4.22) *by B.*

(ii) *If at least one of the following two conditions is met,*
 (d) *the set B is bounded,*
 (e) *condition* $[\mathbf{C}_0]$ *is satisfied,*
 then, for $\alpha = x/n \to 1$ as $n \to \infty$, one has

$$\varlimsup_{n\to\infty} \frac{1}{n} \ln \mathbf{P}(s_n \in B) \leqslant -J(B+). \tag{4.4.23}$$

(iii) *If $x = n$ then*

$$\varlimsup_{n\to\infty} \frac{1}{n} \ln \mathbf{P}(s_n \in B) \leqslant -J([B]_{\mathbb{C}}). \tag{4.4.24}$$

If, in addition, condition **[B]** *is met then one can replace $[B]_{\mathbb{C}}$ on the right-hand side of* (4.4.24) *by B.*

Corollary 4.4.10. *If the set B is convex and satisfies condition* **[B]** *then, for $x = n$, there exists a limit:*

$$\lim_{n\to\infty} \frac{1}{n} \ln \mathbf{P}(s_n \in B) = -J(B). \tag{4.4.25}$$

Proof of Theorem 4.4.9. (i) The first inequality in (4.4.22) follows from assertion (ii) of Theorem 4.3.1; while the second follows from assertion (vi) of Theorem 4.2.3.

Now, in addition to the assumptions of the corollary, let, the set B be open. Then, by virtue of Lemma 4.4.3 and assertion (vi) of Theorem 4.2.3, one has

$$\mathbf{P}(s_n \in B) \leqslant e^{-nJ((x/n)B)},$$

where $J((x/n)B) = I_a((x/n)B)$.

If the set B satisfies the condition $I_a((x/n)(B)_{\mathbb{C}}) < \infty$ or $J((x/n)(B)_{\mathbb{C}}) < \infty$, then, by virtue of Lemma 4.4.1, one has

$$I_a\left(\frac{x}{n}[B]_{\mathbb{C}}\right) = I_a\left(\frac{x}{n}B\right) = J\left(\frac{x}{n}B\right) = J\left(\frac{x}{n}[B]\right).$$

Assertion (i) is proved.

(ii) Let B be a bounded convex set. Then, for a given $\varepsilon > 0$ and all large enough n, one has $(x/n)[B]_{\mathbb{C}} \subset (B)_\varepsilon$. Hence, by virtue of assertion (i),

$$\varlimsup_{n\to\infty} \frac{1}{n} \ln \mathbf{P}(s_n \in B) \leqslant -J\left(\frac{x}{n}[B]_{\mathbb{C}}\right) \leqslant -J((B)_\varepsilon).$$

From this (4.4.23) follows.

If condition $[\mathbf{C}_0]$ is met, we will introduce the set $U_v = \{f : \rho_{\mathbb{C}}(0,f) \leqslant v\}$. We have

$$\mathbf{P}(s_n \in B) \leqslant \mathbf{P}(s_n \in BU_v) + \mathbf{P}\big(\rho_{\mathbb{C}}(0, s_n) > v\big), \tag{4.4.26}$$

where one can apply to the first term on the right-hand side the same argument as was used above. Assume that, for any $N < \infty$, there exist $u < \infty$ and $C < \infty$ such that

$$\mathbf{P}(\rho_{\mathbb{C}}(0, s_n) > u) \leqslant Ce^{-nN}, \quad n \geqslant 1. \tag{4.4.27}$$

Then assertion (ii) follows from (4.4.26) and (4.4.27). Thus, it remains to prove (4.4.27).

First let $d = 1$. One can assume without loss of generality that $\mathbf{E}\xi = 0$. Then, for all large enough n (see e.g. (1.1.24))

$$\mathbf{P}(\rho_{\mathbb{C}}(s_n, 0) \geqslant u) = \mathbf{P}(\max_{1\leqslant i\leqslant n} |S_i| \geqslant xu) \leqslant e^{-n\Lambda(xu/n)} + e^{-n\Lambda(-xu/n)}.$$

Under the condition $[\mathbf{C}_0]$, for some $c_1 > 0$ and all large enough n and u, by virtue of (4.4.19) one has

$$\mathbf{P}\big(\rho_{\mathbb{C}}(s_n, 0) \geqslant u\big) \leqslant 2e^{-nuc_1/2}.$$

This implies (4.4.27). In the case $d \geqslant 2$, one should use the above inequalities for each component of the trajectory s_n. Assertion (ii) is proved.

(iii) This follows from assertion (i) and Lemma 4.4.3.

Theorem 4.4.9 is proved. □

Proof of Corollary 4.4.10. If condition $[\mathbf{B}]$ is satisfied and $x = n$ then, by virtue of assertion (iii) of Theorem 4.4.9, one has the relation (4.4.24), on the right-hand side of which $-J([B]_{\mathbb{C}})$ can be replaced by $-J(B)$. Using the lower bound from Theorem 4.4.7 where we replace $J((B)_{\mathbb{C}})$ by $J(B)$, we obtain (4.4.25). Corollary 4.4.10 is proved. □

The assertions of this section show that the condition **[C$_\infty$]** is not, generally speaking, necessary for the upper bound (4.4.17) in Theorem 4.4.7. At the same time, condition **[C$_\infty$]** was used in an essential way in the proof of (4.4.17), and we do not see how to avoid that. On the other hand, we are not aware of examples where, say, only condition **[C$_0$]** is met and inequality (4.4.17) does not hold. So the question how essential is condition **[C$_\infty$]** in Theorem 4.4.7 remains open.

IV. *The l.d.p. in the triangular array setup.* An assertion similar to Theorem 4.4.7 was obtained in [177] for a different setup. In that paper, it was assumed that the distribution $\mathbf{F} = \mathbf{F}_n$ of the summand $\xi \in \mathbb{R}$ in the random walk $\{S_k;\ 0 \leqslant k \leqslant n\}$ depends on n (the so-called triangular array scheme), and the probability density function of that distribution has the following special form:

$$\mathbf{F}_n(dy) = c_n \exp\big\{-N_n h(y)\big\}dy,$$

where c_n is a scaling factor, $N_n \to \infty$ as $n \to \infty$ and $h(y)$ is an arbitrary non-negative function on $\mathbb{R}$, $h(0) = 0$, that grows at infinity faster than any linear function: $h(t) \gg |t|$ as $|t| \to \infty$. The assumption that $N_n \to \infty$ means that as $n \to \infty$ the distribution $\mathbf{F}_n$ concentrates around zero. A functional $Q(f)$, which is an analogue of the functional $I(f)$, has been defined as

$$Q(f) = Q_h(f) := \begin{cases} \int_0^1 h(f'(t))dt & \text{if } f \in \mathbb{C}_a,\quad f(0) = 0,\\ \infty & \text{otherwise,} \end{cases}$$

so that $I(f) = Q_\Lambda(f)$. Then, for $x = n$ and any set $B \in \mathfrak{B}(\mathbb{C}, \rho_{\mathbb{C}})$ satisfying the equality $Q((B)) = Q([B])$, one has the following limiting relation (see Theorem 2.1 in [177]):

$$\lim_{n\to\infty} \frac{1}{nN_n} \ln \mathbf{P}(s_n \in B) = -Q(B), \tag{4.4.28}$$

i.e. the trajectories s_n satisfy the l.d.p. in the space $(\mathbb{C}, \rho_{\mathbb{C}})$ with parameters (z_n, Q) and $z_n = nN_n$.

As we have already noted, the distribution of ξ concentrates around zero as $N_n \to \infty$. Therefore, (4.4.28) is actually a *superlarge deviation principle* (say, relative to $\mathbf{E}|\xi|$). This is also indicated by the rate at which the probability

$$\mathbf{P}(s_n \in B) = \exp\Big\{ - nN_n Q(B)\big(1 + o(1)\big)\Big\},$$

vanishes: it is greater than exponential (see Theorem 4.4.7).

4.5 An extended problem setup

4.5.1 An example leading to the extended large deviation principle

Consider an example showing that, in the case where only condition **[C]** or **[C$_0$]** (but not **[C$_\infty$]**) is satisfied, the most probable trajectory (see Definition 4.1.10) may turn out to be discontinuous (and hence the $\rho_{\mathbb{C}}$-most probable trajectory in $\mathbb{C}$ may not exist). In such a case, to preserve the l.d.p. (in one form or another) and the concept of the most probable trajectory, one has to introduce a metric ρ weaker than $\rho_{\mathbb{C}}$ in a trajectory space $\mathbb{D}$ that is broader than $\mathbb{C}$ (see section 4.2). The definition of the deviation integral $J(f)$ will be extended to that broader space in section 4.2.

Note that using $\mathbb{D}$ as the trajectory space is further justified by the fact that, along with the continuous polygonal lines $s_n(t)$, it is also natural to study the discontinuous step functions $\overline{s}_n(t) := (1/x)S_{[nt]} \in \mathbb{D}$ and some others. In what follows, by $\overline{s}_n = \overline{s}_n(t) \in \mathbb{D}$ we will also denote trajectories of a more general form, in which the value of $\overline{s}_n(t)$ on the segment $t \in [k/n,\ (k+1)/n]$, $k = 0, \dots, n-1$, can be removed from the interval $(s_n((k)/n),\ s_n((k+1)/n))$ (for more details, see subsection 4.5.2 below).

Now we return back to our continuous polygonal lines $s_n \in \mathbb{C}$ in the case $d = 1$.

Example 4.5.1. Let $g = g(t) > 0$ be a fixed continuous positive function on $[0, 1]$. Introduce the set

$$B_g := \Big\{f \in \mathbb{C} : f(0) = 0,\ \sup_{0 \leqslant t \leqslant 1} \{f(t) - g(t)\} > 0\Big\}$$

of trajectories that start at zero and exceed the boundary $g(t)$ at least at one point of the unit interval. The asymptotic behaviour of $\mathbf{P}(s_n \in B_g)$ was studied in section 3.7. Now consider that asymptotic behaviour in the following example.

Assume that $\mathbf{E}\xi = 0$ and, for fixed constants $\lambda_+ > 0$, $c > 0$ the distribution of ξ satisfies the relation $\mathbf{P}(\xi \geqslant t) \sim (c/t^3)e^{-\lambda_+ t}$ as $t \to \infty$. It is clear that condition **[C]** is met but condition **[C$_\infty$]** is not (generally speaking, it may happen that condition **[C$_0$]** also will not be met, if the left tail $\mathbf{P}(\xi < -t)$ does not satisfy Cramér's condition). In our case, $\psi(\lambda_+) < \infty$ (but $\psi(\lambda_+ + 0) = \infty$), $\psi'(\lambda_+ - 0) < \infty$ and therefore

$$\alpha_+ := \lim_{\lambda \uparrow \lambda_+} \frac{\psi'(\lambda)}{\psi(\lambda)} = \frac{\psi'(\lambda_+ - 0)}{\psi(\lambda_+)} \in (0, \infty).$$

The function $\lambda(\alpha)$ is analytic on $(0, \alpha_+)$. For $\alpha \geqslant \alpha_+$ we have $\lambda(\alpha) = \lambda_+$, so that, for $\alpha \geqslant \alpha_+$ (see also $(\Lambda 1)$) we get

$$\Lambda(\alpha) = \int_0^\alpha \lambda(v)dv = \Lambda(\alpha_+) + \lambda_+(\alpha - \alpha_+). \tag{4.5.1}$$

Assume that $x \sim n$,

$$A_{k,n} := \left\{S_k > xg\left(\frac{k}{n}\right)\right\} = \left\{s_n\left(\frac{k}{n}\right) > g\left(\frac{k}{n}\right)\right\}, \quad 1 \leqslant k \leqslant n, \quad A_n := \{s_n \in B_g\}.$$

First, let the function g be such that

$$A_n = \bigcup_{k=1}^{n} A_{k,n}. \tag{4.5.2}$$

This equality holds if, say, for all $t \in [0,1]$ there exists a second derivative $g''(t) \leqslant 0$. In particular, (4.5.2) holds true if $g(t) = g(0) + bt$, where necessarily $g(0) > 0$, $g(0) + b > 0$. For such boundaries of g and any $k = 1, \dots, n$ we have the inequalities

$$\mathbf{P}(A_{k,n}) \leqslant \mathbf{P}(A_n) \leqslant \sum_{k=1}^{n} \mathbf{P}(A_{k,n}). \tag{4.5.3}$$

By virtue of the known theorems on large and superlarge deviations (see e.g. section 2.2), for $1 \leqslant k \leqslant n$, $n \to \infty$, we have

$$\mathbf{P}(A_{k,n}) = \exp e^{-k\Lambda(\frac{x}{k}g(\frac{k}{n}))(1+o(1))}, \qquad \mathbf{P}(A_{k,n}) \leqslant e^{-k\Lambda(\frac{x}{k}g(\frac{k}{n}))}. \tag{4.5.4}$$

It follows from (4.5.3) and (4.5.4) that

$$\mathbf{P}(A_n) = \exp e^{-nH(B_g)(1+o(1))}, \tag{4.5.5}$$

where

$$H(B_g) := \lim_{n\to\infty} \inf_{0<u\leqslant 1} u\Lambda\left(\frac{x}{n}\frac{g(u)}{u}\right) = \inf_{0<u\leqslant 1} H(u), \quad H(u) := u\Lambda\left(\frac{g(u)}{u}\right). \tag{4.5.6}$$

It is not hard to see from Theorem 4.4.9 and convexity of the functional $I(f)$ that, in the class of trajectories connecting the points $(0,0)$ and $(u, g(u))$, the most probable one is the trajectory

$$f_{[u]}(t) := \begin{cases} tg(u)/u, & \text{if } \ 0 \leqslant t \leqslant u, \\ g(u), & \text{if } \ u \leqslant t \leqslant 1. \end{cases} \tag{4.5.7}$$

Since $I(f_{[u]}) = H(u)$, we see that $H(B_g)$ coincides with $I(B_g) = \inf_{0<u\leqslant 1} I(f_{[u]})$, and so formula (4.5.5) can be rewritten as

$$\mathbf{P}(A_n) = e^{-nI(B_g)(1+o(1))},$$

where

$$I(B_g) = I\big((B_g)\big) = I\big([B_g]\big) = I(\partial B_g) = \inf_{0<u\leqslant 1} I(f_{[u]}).$$

We show now that, in our case, for a broad class of functions g the ρ-most probable trajectory f_{B_g} for the set B_g (when choosing a suitable metric ρ in its definition) has the form

$$f_{B_g}(t) = \lim_{u\downarrow 0} f_{[u]}(t) := \begin{cases} 0, & \text{if } \ t = 0 \\ g(0), & \text{if } \ 0 < t \leqslant 1, \end{cases} \tag{4.5.8}$$

i.e. it is is a discontinuous function.

Indeed, if, for all $t \in (0,1]$, one has

$$\frac{g(t)}{t} \geqslant \alpha_+ \tag{4.5.9}$$

then, by virtue of (4.5.1) and (4.5.6) it holds in our case that

$$H(t) = \lambda_+ g(t) + t(\Lambda(\alpha_+) - \lambda_+\alpha_+), \quad H'(t) = \lambda_+ g'(t) + \Lambda(\alpha_+) - \lambda_+\alpha_+.$$

So, if for all $t \in (0, 1]$ we have

$$g'(t) \geqslant \alpha_+ - \frac{\Lambda(\alpha_+)}{\lambda_+}, \tag{4.5.10}$$

then the inequality $H'(t) \geqslant 0$ holds true for all $t \in (0, 1]$, the function $H(t)$ is non-decreasing and

$$\inf_{0<t\leqslant 1} H(t) = H(0) = \lambda_+ g(0).$$

Inequalities (4.5.9) and (4.5.10) describe a broad enough class of functions g. If, for instance, $g(t) = g(0) + bt$ then it is sufficient to choose $g(0) \geqslant \alpha_+, b \geqslant \alpha_+$ to ensure (4.5.9), (4.5.10).

Thus, the most probable (with respect to pointwise convergence) trajectory $f_{B_g}(t)$ is discontinuous in our example (see (4.5.8)), while the $J(B_g)$-sequence of functions $f_k = f_{[1/k]}$ (see Definition 4.1.10; the functions $f_{[u]}$ were defined in (4.5.7)) cannot converge to f_{B_g} as $k \to \infty$ in the uniform metric $\rho_{\mathbb{C}}$, because

$$\rho_{\mathbb{C}}(f_{[u]}, f_{[u/2]}) \geqslant c > 0, \qquad c = \frac{1}{2} \inf_{0\leqslant t\leqslant 1} g(t) \tag{4.5.11}$$

for all $u \in (0, 1)$. Since $I(f_{[u]}) \leqslant v$ for $v = 2I(B_g)$ and small enough $u > 0$, it follows from (4.5.11) that the set $I_v := \{f \in \mathbb{C}_a : f(0) = 0, \ \ I(f) \leqslant v\}$ cannot be compact in $\mathbb{C}$ with respect to the metric $\rho_{\mathbb{C}}$ (the sequence f_k does not contain subsequences that would converge in that metric). The convergence of f_k to f_{B_g} will hold in the metric ρ, and the set I_v will be totally bounded in the metric space $(\mathbb{D}, \ \rho)$, introduced in section 4.2. Statements (4.4.17), (4.4.18) of Theorem 4.4.7 will remain true for the set B_g provided that on the right-hand side of (4.4.17) we replace the quantity $J([B_g]) = I([B_g])$ with $\lim_{\varepsilon\to 0} I((B_g)_\varepsilon)$, so that the extended l.d.p. (e.l.d.p.) with $z_n = n$ will hold for the set B_g in $(\mathbb{D}, \rho)$. Observe also that it is natural to define here the value $J(f_{B_g})$ of the functional J for the discontinuous function f_{B_g} in (4.5.8) not as ∞ but, rather, as

$$J(f_{B_g}) := \lim_{u\to 0} I(f_{[u]}) = \lim_{u\to 0} \int_0^u \Lambda\left(\frac{g(u)}{u}\right) dt = \lim_{u\to 0} u\Lambda\left(\frac{g(u)}{u}\right) = \lambda_+ g(0).$$

We will make two remarks in connection with Example 4.5.1.

Remark 4.5.2. If the condition $g''(t) \leqslant 0$ is not met in Example 4.5.1 then our argument becomes somewhat more complicated. Instead of (4.5.4) we will have the relation

$$A_n \subseteq \bigcup_{n=1}^{n} A^*_{k,n},$$

where

$$A^*_{k,n} = \left\{S_k > x\left(g\left(\frac{k}{n}\right) - \omega_g\left(\frac{1}{n}\right)\right)\right\},$$

with $\omega_g(\Delta)$ *being* the continuity modulus of the function g. Since $\omega_g(1/n) \to 0$ as $n \to \infty$, all the required assertions remain true, but proving this requires additional effort. On the other hand, in many applied problems one is interested not in the asymptotic behaviour of $\mathbf{P}(s_n \in B_g)$ but rather in that of

$$\mathbf{P}\bigg(\bigcup_{1\leqslant k\leqslant n} A_{k,n}\bigg) = \mathbf{P}\left(\max_{1\leqslant k\leqslant n}\left\{S_k - xg\left(\frac{k}{n}\right)\right\} > 0\right).$$

Introducing here functional spaces (in which the trajectories s_n or step-function analogues thereof would lie) seems artificial and can be inconvenient, as it was when applying in Example 4.5.1 the insignificant condition $g''(t) \leqslant 0$. Introducing metric spaces is mostly justified by the fact that it enables one to find sufficient conditions for the existence of the limit

$$\lim_{n\to\infty} \frac{1}{n} \ln \mathbf{P}(\overline{s}_n \in B)$$

for sets B of a more general nature, where $\overline{s}_n$ can be constructed from the vector $(S_1, S_2, \ldots, S_n)$ in any reasonable way.

Remark 4.5.3. If $H(u) > H(0)$ for all $u > 0$ in Example 4.5.1 then $\ln \mathbf{P}(A_n) \sim -n\lambda_+ g(0)$ as $n \to \infty$, and one will have the same asymptotics for $\ln \mathbf{P}\big(S_k > ng(0)\big) \sim -n\lambda_+ g(0)$ for any fixed k or for any $k \to \infty$ such that $k = o(n)$ (see (4.5.4)). Therefore, in this case the most likely ways to reach the boundary $g(t)$ are not only by *one* large jump, as in the case when the random variable ξ has a distribution tail regularly varying at infinity (see e.g. [42]), but also by *many large jumps* (of sizes comparable with $ng(0)/k$ when the number of large jumps is k).

4.5.2 The trajectories $\overline{s}_n$

We will consider the trajectory $\overline{s}_n = \overline{s}_n(t)$ as a random process in the space $\mathbb{D}$, i.e. as a process given on the measurable space $\langle \mathbb{D}, \mathfrak{A}(\mathbb{D})\rangle$, where $\mathfrak{A}(\mathbb{D})$ is the σ-algebra generated by the cylindrical subsets of $\mathbb{D}$. The process $\overline{s}_n(t)$ is given on the same probability space as s_n, as a trajectory in $\mathbb{D}$ (i.e., as a separable process) with the following properties: $\overline{s}_n(0) = 0$ and, for any fixed $h > 0$,

$$\frac{1}{n} \ln \mathbf{P}\big(\rho(\overline{s}_n, s_n) > h\big) \to -\infty \quad \text{as} \quad n \to \infty. \tag{4.5.12}$$

This property is clearly equivalent to the existence of a sequence $h_n \to 0$ as $n \to \infty$ such that

$$\frac{1}{n} \ln \mathbf{P}\big(\rho(\overline{s}_n, s_n) > h_n\big) \to -\infty \quad \text{as} \quad n \to \infty. \tag{4.5.13}$$

As the simplest example of such a process $\overline{s}_n$ one can take the random walk

$$\overline{s}_n(t) := \frac{1}{x} S_{[nt]}, \quad t \in [0,1], \quad x \sim n \quad \text{as} \quad n \to \infty.$$

Clearly, $\rho(\overline{s}_n, s_n) \leqslant 1/n$ and property (4.5.12) holds. We can obtain another example by considering processes $\overline{s}_n$ of the form

$$\overline{s}_n(t) = s_n(t) + \theta_n(t),$$

where $\theta_n(t)$ is an arbitrary random process (which may depend on s_n), such that

$$\frac{1}{n} \ln \mathbf{P}\Big(\sup_{0 \leqslant t \leqslant 1} \big|\theta_n(t)\big| > h \Big) \to -\infty \tag{4.5.14}$$

or, equivalently,

$$\frac{1}{n} \ln \mathbf{P}\big(\rho_{\mathbb{C}}(\overline{s}_n, s_n) > h\big) \to -\infty \quad \text{as} \quad n \to \infty.$$

For instance, one can consider an arbitrary process $S(t)$ on $[0,n]$ with independent increments. Then the variables $\xi_k := S(k) - S(k-1)$ will be independent and distributed as $S(1)$. Set

$$S_k := S(k), \quad \overline{s}_n(t) := \frac{1}{x} S(nt), \quad t \in [0,1],$$

and construct, as earlier, a continuous process $s_n(t)$ from the sums S_k. Then, for the process

$$\theta_n(t) = \overline{s}_n(t) - s_n(t),$$

the inequality

$$\mathbf{P}\Big(\sup_{0 \leqslant t \leqslant 1} \big|\theta_n(t)\big| > h \Big) \leqslant n\mathbf{P}\Big(\max\big\{S^*(1), S^*(-1)\big\} > h \Big)$$

holds, where $S^*(v) := \max_{u \in [0,v]} |S(u)|$; for $t < 0$, the process $S(t)$ is defined in a natural way. It is not hard to verify that if $S(1)$ satisfies $[\mathbf{C}_\infty]$ then condition (4.5.14) will be met for the process $\theta_n(t)$.

Thus the processes $s_n, \overline{s}_n$ are given on the probability space

$$\langle \mathbb{D}, \mathfrak{A}(\mathbb{D}), \mathbf{P} \rangle,$$

which *will be considered here and in what follows as the basic underlying probability space*. Note that the Borel σ-algebra $\mathfrak{B}(\mathbb{D}, \rho)$ is embedded in the σ-algebra $\mathfrak{A}(\mathbb{D})$, which follows from the next assertion.

In what follows we use the metrics $\rho_{\mathbb{C}}$ and ρ in the spaces $\mathbb{C}$ and $\mathbb{D}$ respectively. In order to distinguish between ε-neighbourhoods and open and closed sets with respect to the metrics we will use as before the index $\mathbb{C}$ when dealing with the metric $\rho_{\mathbb{C}}$. For example, $(B)_{\mathbb{C}}$ will denote an open inner part of a set B the in metric $\rho_{\mathbb{C}}$ and

$$(f)_{\mathbb{C},\varepsilon} := \{g \in \mathbb{D} : \ \rho_{\mathbb{C}}(g,f) < \varepsilon\}.$$

We will skip the index when using the metric ρ, e.g.

$$(f)_\varepsilon := \{g \in \mathbb{D} : \rho(g,f) < \varepsilon\}$$

(the conventions that we have used already).

Lemma 4.5.4. *For any $f \in \mathbb{D}$, $\varepsilon > 0$, the sets $(f)_{\mathbb{C},\varepsilon}$, $(f)_\varepsilon$ belong to the σ-algebra $\mathfrak{A}(\mathbb{D})$ generated by the cylinder sets in $\mathbb{D}$.*

Proof. The statement that the ε-neighbourhoods belong to the σ-algebra $\mathfrak{A}(\mathbb{D})$ is an obvious consequence of the separability of the trajectories $f \in \mathbb{D}$. Denote by U a fixed countable set that is everywhere dense in $[0,1]$ and contains the points $t = 0$ and $t = 1$.

(i) The statement that the ε-neighbourhood $(f)_{\mathbb{C},\varepsilon}$ in the metric $\rho_\mathbb{C}$ belongs to the σ-algebra $\mathfrak{A}(\mathbb{D})$ follows from the representation

$$(f)_{\mathbb{C},\varepsilon} = \bigcup_{k\geqslant 1}\bigcap_{t\in U}\left\{g \in \mathbb{D} : \left|g(t) - f(t)\right| < \varepsilon - \frac{1}{k}\right\}.$$

Since $\{g \in \mathbb{D} : |g(t) - f(t)| < \varepsilon - 1/k\} \in \mathfrak{A}(\mathbb{D})$, we also have $(f)_{\mathbb{C},\varepsilon} \in \mathfrak{A}(\mathbb{D})$.

(ii) Now we prove that $(f)_\varepsilon \in \mathfrak{A}(\mathbb{D})$. Recall that Γf denotes the graph of the function $f \in \mathbb{D}$ (see the definition of the metric ρ in section 4.2). For $f \in \mathbb{D}$, $t \in [0,1]$, consider the t-sections of the sets $(\Gamma f)_\varepsilon$ and Γf, respectively:

$$(\Gamma f)_\varepsilon|_t := \left\{\alpha \in \mathbb{R}^d : (t,\alpha) \in (\Gamma f)_\varepsilon\right\}, \quad \Gamma f|_t := \left\{\alpha \in \mathbb{R}^d : (t,\alpha) \in \Gamma f\right\}.$$

We now want to represent the set $(f)_\varepsilon$ and the intersection of the two sets

$$\Gamma_1 := \left\{g \in \mathbb{D} : \Gamma g \in (\Gamma f)_\varepsilon\right\} \quad \text{and} \quad \Gamma_2 := \left\{g \in \mathbb{D} : \Gamma f \in (\Gamma g)_\varepsilon\right\}.$$

The set Γ_1 can be represented as

$$\Gamma_1 = \bigcup_{k\geqslant 1}\bigcap_{t\in U}\Gamma_1\left(f, \varepsilon - \frac{1}{k}, t\right), \quad \text{where} \quad \Gamma_1(f,\varepsilon,t) := \left\{g \in \mathbb{D} : g(t) \in (\Gamma f)_\varepsilon|_t\right\}.$$

In turn, the set $\Gamma_1(f,\varepsilon,t)$ can be represented as

$$\Gamma_1(f,\varepsilon,t) = \bigcup_{t+u\in U,\ |u|\leqslant\varepsilon} \Gamma_1(f,\varepsilon,u,t),$$

$$\Gamma_1(f,\varepsilon,u,t) := \left\{g : g(t) \in (\Gamma f|_{t+u})_{\sqrt{\varepsilon^2-u^2}}\right\}.$$

Since the t-section of the graph Γf is the segment $[f_-(t), f_+(t)] \subset \mathbb{R}^d$, where

$$f_-(t) := \lim_{u\to t,\ u\in U,\ u<t} f(u), \quad f_+(t) := \lim_{u\to t,\ u\in U,\ u>t} f(u), \quad t \in (0,1);$$

$$f_-(0) := f(0), \quad f_+(0) := \lim_{u\to 0,\ u\in U,\ u>0} f(u);$$

$$f_-(1) := \lim_{u\to 1,\ u\in U,\ u<1} f(u), \quad f_+(1) := f(1),$$

we clearly have that $\Gamma_1(f,\varepsilon,u,t) \in \mathfrak{A}(\mathbb{D})$. Therefore, $\Gamma_1(f,\varepsilon,t) \in \mathfrak{A}(\mathbb{D})$ and hence $\Gamma_1 \in \mathfrak{A}(\mathbb{D})$.

Similarly, the set Γ_2 can be represented as

$$\Gamma_2 = \bigcup_{k\geqslant 1} \bigcap_{t\in U} \Gamma_2\left(f, \varepsilon - \frac{1}{k}, t\right), \quad \text{where} \quad \Gamma_2(f,\varepsilon,t) := \big\{g : f(t) \in (\Gamma g)_\varepsilon|_t\big\}.$$

In turn, the set $\Gamma_2(f,\varepsilon,t)$ can be represented as

$$\Gamma_2(f,\varepsilon,t) = \bigcup_{t+u\in U,\ |u|\leqslant\varepsilon} \Gamma_2(f,\varepsilon,u,t),$$

$$\Gamma_2(f,\varepsilon,u,t) := \left\{g : f(t) \in (\Gamma g|_{t+u})_{\sqrt{\varepsilon^2-u^2}}\right\}.$$

Taking into account that the t-section of the graph Γg is the segment $[g_-(t), g_+(t)]$, we come to the conclusion that $\Gamma_2(f,\varepsilon,u,t) \in \mathfrak{A}(\mathbb{D})$, $\Gamma_2(f,\varepsilon,t) \in \mathfrak{A}(\mathbb{D})$, $\Gamma_2 \in \mathfrak{A}(\mathbb{D})$. The lemma is proved. □

We now show that if (4.5.12) holds true then the behaviour of the processes s_n and $\overline{s}_n$ is the same in some sense 'from the point of view' of the l.d.p. Namely, by Lemma 4.1.11 we have

Lemma 4.5.5. (i) *Let s_n satisfy the l.l.d.p. in $(\mathbb{D},\rho)$ with parameters (n,J), i.e., for $f \in \mathbb{D}$ the relations* (see Definition 4.1.3)

$$\overline{\lim_{n\to\infty}} \frac{1}{n} \ln \mathbf{P}\big(\overline{s}_n \in (f)_\varepsilon\big) \leqslant -J\big((f)_{\delta(\varepsilon)}\big) + \beta(\varepsilon)$$

are met, where $\delta(\varepsilon) \to 0$, $\beta(\varepsilon) \to 0$ as $\varepsilon \to$ and, moreover, for any $\varepsilon > 0$ we have

$$\underline{\lim_{n\to\infty}} \frac{1}{n} \ln \mathbf{P}\big(\overline{s}_n \in (f)_\varepsilon\big) \geqslant -J(f).$$

Then $\overline{s}_n$ satisfies the same relations but with $\overline{\delta}(\varepsilon) = \delta\big(\varepsilon(1+\tau)\big)$ and $\overline{\beta}_n(\varepsilon) = \beta\big(\varepsilon(1+\tau)\big)$ for any $\tau > 0$.

(ii) *Let s_n satisfy the e.l.d.p. in $(\mathbb{D},\rho)$ with parameters (n,J)* (see (4.1.10), (4.1.11)). *Then $\overline{s}_n$ satisfies the e.l.d.p. in $(\mathbb{D},\rho)$ as well, with parameters (n,J), i.e. for any $B \in \mathfrak{B}(\mathbb{D},\rho)$ one has*

$$\overline{\lim_{n\to\infty}} \frac{1}{n} \ln \mathbf{P}(\overline{s}_n \in B) \leqslant -J(B+), \tag{4.5.15}$$

$$\underline{\lim_{n\to\infty}} \frac{1}{n} \ln \mathbf{P}(\overline{s}_n \in B) \geqslant -J\big((B)\big). \tag{4.5.16}$$

4.6 Large deviation principles in the space of functions without discontinuities of the second kind

4.6.1 Statement of the main results

In this section we will deal with the one-dimensional case $d = 1$. As before, we will assume that $\alpha := x/n \to 1$ as $n \to \infty$. This does not restrict generality, compared with the assumption $\alpha \to \alpha_0 > 0$ as $n \to \infty$. The following analogue of Theorem 4.4.2 holds true for the process $\overline{s}_n$, defined in subsection 4.5.2:

Theorem 4.6.1. (i) *For any* $f \in \mathbb{D}$ *and* $\varepsilon > 0$,

$$\overline{\lim_{n\to\infty}} \frac{1}{n} \ln \mathbf{P}(\overline{s}_n \in (f)_\varepsilon) \leqslant -J((f)_{\delta(\varepsilon)})$$

for any $\delta(\varepsilon) > \sqrt{2}\varepsilon$.

(ii) *For any* $f \in \mathbb{D}$ *and* $\varepsilon > 0$

$$\underline{\lim_{n\to\infty}} \frac{1}{n} \ln \mathbf{P}(\overline{s}_n \in (f)_\varepsilon) \geqslant -J(f). \tag{4.6.1}$$

In other words, in $(\mathbb{D}, \rho)$ the l.l.d.p. holds with parameters (n, J) with $\delta(\varepsilon) > \sqrt{2}\varepsilon$ and $\beta(\varepsilon) = 0$ (see Definition 4.1.3). Note that the l.l.d.p. in $(\mathbb{D}, \rho)$, like the l.l.d.p. in $(\mathbb{C}, \rho_{\mathbb{C}})$, always holds without any restrictions. According to the results of section 4.1, the l.l.d.p. can also be written in the form (4.1.2)

$$\lim_{\varepsilon\to 0} \overline{\lim_{n\to\infty}} \frac{1}{n} \ln \mathbf{P}(\overline{s}_n \in (f)_\varepsilon) = \lim_{\varepsilon\to 0} \underline{\lim_{n\to\infty}} \frac{1}{n} \ln \mathbf{P}(\overline{s}_n \in (f)_\varepsilon) = -J(f), \tag{4.6.2}$$

or in the form (4.1.3). Observe that, since $J(f+) := \lim_{\delta\to 0} J((f)_\delta) = J(f)$, relation (4.6.2) can also be derived directly from Theorem 4.6.1, passing to the limit as $\varepsilon \to 0$. The following theorem is an analogue of Theorem 4.4.7.

Theorem 4.6.2. (i) *If condition* **[C$_0$]** *is met, for any measurable set B one has*

$$\overline{\lim_{n\to\infty}} \frac{1}{n} \ln \mathbf{P}(\overline{s}_n \in B) \leqslant -J(B+), \tag{4.6.3}$$

where $J(B+) = \lim_{\varepsilon\to 0} J((B)_\varepsilon)$.

(ii) *For any measurable set B, one always has*

$$\underline{\lim_{n\to\infty}} \frac{1}{n} \ln \mathbf{P}(\overline{s}_n \in B) \geqslant -J((B)).$$

Following Definition 4.1.4, we call the above assertion the *extended l.d.p. (e.l.d.p.) with parameters* (n, J). If $J(B+) = J((B))$ then the e.l.d.p. can be written as

$$\lim_{n\to\infty} \frac{1}{n} \ln \mathbf{P}(\overline{s}_n \in B) = -J(B).$$

If the set B in Theorem 4.4.7 is bounded or convex then the assertion of Theorem 4.4.7 can be made stronger by removing condition **[C$_0$]**:

Theorem 4.6.3. *If the set B in Theorem 4.2.8 is bounded or convex then the inequality* (4.6.3) *always holds.*

4.6.2 Proof of Theorem 4.6.1

As was the case when proving Theorem 4.4.2, we will find useful the following analogue of Lemma 4.4.4, which allows us to restrict our considerations to the case $x = n$ only. This assertion has nothing to do with dimensionality, so we will state it in the multivariate case.

Lemma 4.6.4. (1) *If the process $\overline{s}_n$ with $x = n$ satisfies the l.l.d.p. in $(\mathbb{D}, \rho)$ with functions $\delta(\varepsilon)$, $\beta(\varepsilon)$ (see Definition 4.1.3) then it also satisfies the same principle with $x \sim n$, $n \to \infty$ and functions $\delta(\varepsilon(1+\tau))$, $\beta(\varepsilon(1+\tau))$ for any $\tau > 0$.*

(2) *Let $x = n$. If, for the process $\overline{s}_n$ constructed from random vectors ξ_k, the l.l.d.p. in $(\mathbb{D}, \rho)$ holds with parameters (n, J) and functions $\delta(\varepsilon)$, $\beta(\varepsilon)$ then, for a process of the same form $\overline{s}_n$ but constructed from random vectors $\xi_k + b$, one has the l.l.d.p. with parameters (n, J_b), where $J_b(f) := J(f - b\mathrm{e})$, $\mathrm{e} = \mathrm{e}(t) := t$ and with the same functions $\delta(\varepsilon)$, $\beta(\varepsilon)$.*

Proof of Lemma 4.6.4. This proof repeats that of Lemma 4.4.4 up to obvious modifications related to changing the space $(\mathbb{C}, \rho_{\mathbb{C}})$ to $(\mathbb{D}, \rho)$ and replacing the processes s_n with $\overline{s}_n$. □

The assertion of Theorem 4.6.1 with $x = n$ will follow from bounds to be presented in the following two subsections.

4.6.3 Upper bounds

First we will obtain exact upper bounds for the probability $\mathbf{P}\big(s_n \in (f)_\varepsilon\big)$ for continuous trajectories s_n and absolutely continuous f. Recall that here we are dealing with ε-neighbourhoods $(f)_\varepsilon$ with respect to the metric ρ in the space $\mathbb{D}$ and, in general, the open sets $(f)_\varepsilon$ are not convex.

Theorem 4.6.5. *Let $f \in \mathbb{C}_a$, $r \geqslant 2$ be an arbitrary integer. Then, for all $n \geqslant 4r^2/\varepsilon$, one has*

$$\mathbf{P}\big(s_n \in (f)_\varepsilon\big) \leqslant b_n^{2(r/\varepsilon+1)} \exp\big\{-nI_a((f)_\delta)\big\},$$

where

$$b_n = \varepsilon\left(2 + \frac{1}{r}\right)n + 2r + 2, \quad \delta = \varepsilon\sqrt{2}\left(1 + \frac{2}{r}\right), I_a(B) = J(B\mathbb{C}_a).$$

This is an analogue of inequalities (4.3.1), (4.4.7) for $B = (f)_\varepsilon$.

Proof. The proof is based on the following elements: embedding the neighbourhood $(f)_\varepsilon$ into a certain union of convex sets (see (4.6.6) below) using

Chebyshev-type exponential inequalities for the probability that s_n hits such sets (see (4.4.7) or Corollary 4.3.4) and embedding these convex sets back into a somewhat bigger ρ-neighbourhood of the point f (see (4.6.10) below).

(i) First, let ε, r and n be such that $\theta := \varepsilon/r$ is a multiple of $1/n$ and $R = 1/\theta$ is an integer. We divide the segment $[0,1]$ into segments $\Delta_k := \big[(k-1)\theta, k\theta\big]$, $k = 1,\dots,R$. Set

$$B_1(\varepsilon) := \big\{g \in \mathbb{C}_a : \Gamma g \in (\Gamma f)_\varepsilon\big\}, \qquad B_2(\varepsilon) := \big\{g \in \mathbb{C}_a : \Gamma f \in (\Gamma g)_\varepsilon\big\},$$

where Γg is the graph of the function g (see subsection 4.2.1). We have to bound $\mathbf{P}(A_{n,\varepsilon})$, where

$$A_{n,\varepsilon} := \big\{s_n \in (f)_\varepsilon\big\} = \big\{\rho(s_n, f) < \varepsilon\big\} = \big\{s_n \in B_1(\varepsilon),\ s_n \in B_2(\varepsilon)\big\}.$$

Let

$$m_k^- := \min_{t\in\Delta_k} f(t), \qquad m_k^+ := \max_{t\in\Delta_k} f(t), \quad k = 1,\dots,R.$$

Further, introduce sets of integers L_k:

$$L_k := \big\{(\theta(k-1)-\varepsilon)n, \theta(k-1)-\varepsilon)n+1,\dots,(\theta k+\varepsilon)n\big\},$$
$$k = r+1,\dots,R-r.$$

The set L_k contains $(2\varepsilon+\theta)n+1$ elements. For $k \leqslant r$ and $k > R-r$ we will keep the definition L_k but take the left-hand points in L_k to be 0 for $k \leqslant r$ and take the right-hand points of L_k to be n when $k > R-r$. For all segments Δ_k, we have

$$A_{n,\varepsilon} \subset \big\{s_n \in B_1(\varepsilon)\big\} \bigcap \Bigg\{\Bigg[\bigcup_{i_k\in L_k}\Big\{s_n\Big(\frac{i_k}{n}\Big) < m_k^- + \varepsilon\Big\}\Bigg] \bigcap \Bigg[\bigcup_{j_k\in L_k}\Big\{s_n\Big(\frac{j_k}{n}\Big) > m_k^+ - \varepsilon\Big\}\Bigg]\Bigg\} \tag{4.6.4}$$

(if, for instance, $s_n(i_k/n) \geqslant m_k^- + \varepsilon$ for all $i_k \in L_k$ then the event $\big\{s_n \in B_2(\varepsilon)\big\}$ is impossible). It follows that

$$A_{n,\varepsilon} \subset \big\{s_n \in B_1(\varepsilon)\big\} \bigcap \Bigg\{\Bigg[\bigcap_{k=1}^R \bigcup_{i_k\in L_k}\Big\{s_n\Big(\frac{i_k}{n}\Big) < m_k^- + \varepsilon\Big\}\Bigg] \bigcap \Bigg[\bigcap_{k=1}^R \bigcup_{j_k\in L_k}\Big\{s_n\Big(\frac{j_k}{n}\Big) > m_k^+ - \varepsilon\Big\}\Bigg]\Bigg\}. \tag{4.6.5}$$

Denote by $\Pi^- := \Pi^-_{i_1,\dots,i_R}$ the following set of functions:

$$\Pi^-_{i_1,\dots,i_R} := \Bigg\{g \in \mathbb{C}_a : g\Big(\frac{i_1}{n}\Big) < m_1^- + \varepsilon, \dots, g\Big(\frac{i_R}{n}\Big) < m_R^- + \varepsilon\Bigg\}.$$

Similarly,

$$\Pi^+_{j_1,\dots,j_R} := \left\{ g \in \mathbb{C}_a : \; g\left(\frac{j_1}{n}\right) > m_1^+ - \varepsilon, \dots, g\left(\frac{j_R}{n}\right) > m_R^+ - \varepsilon \right\}.$$

Then the intersections of the events in the square brackets on the right-hand side of (4.6.5) can be written as

$$\bigcup_{\mathbf{R}} \{s_n \in \Pi^-_{i_1,\dots,i_R}\} \quad \text{and} \quad \bigcup_{\mathbf{R}} \{s_n \in \Pi^+_{j_1,\dots,j_R}\},$$

respectively, where $\mathbf{R}$ is the set of all tuples $(i_1, \dots, i_R)$ for $i_k \in L_k$, $k = 1, \dots, R$ (tuples $(j_1, \dots, j_R)$ for $j_k \in L_k$, $k = 1, \dots, R$). Therefore,

$$A_{n,\varepsilon} \subset \big\{s_n \in B_1(\varepsilon)\big\} \cap \Big[\bigcup_{\mathbf{R}\times\mathbf{R}} \{s_n \in \Pi^-_{i_1,\dots,i_R}, \;\; s_n \in \Pi^+_{j_1,\dots,j_R}\} \Big], \tag{4.6.6}$$

where the union is taken over the set $\mathbf{R} \times \mathbf{R}$ of all pairs of tuples $(i_1, \dots, i_R)$, $(j_1, \dots, j_R)$. Note that on the right-hand side of (4.6.6) some intersections can be empty. It follows from (4.6.6) that

$$\mathbf{P}(A_{n,\varepsilon}) \leqslant \sum_{\mathbf{R}\times\mathbf{R}} \mathbf{P}\big(s_n \in B_1(\varepsilon), \;\; s_n \in \Pi^-_{i_1,\dots,i_R}, \;\; s_n \in \Pi^+_{j_1,\dots,j_R}\big).$$

Now note that the set $B_1(\varepsilon)$ and sets of the form $\big\{g\colon\; g(i/n) < c\big\}$ or $\big\{g\colon\; g(i/n) > c\big\}$ are convex and open (in the uniform metric) in $\mathbb{C}_a$. Hence the intersections of convex open sets

$$B_1(\varepsilon)\Pi^-_{i_1,\dots,i_R}\Pi^+_{i_1,\dots,i_R}$$

are also convex and open (as we have already pointed out, some of them can be empty). Applying Chebyshev-type exponential inequalities for convex open sets (see inequality (4.4.7) in Lemma 4.4.3 or Corollary 4.3.4), we obtain

$$\mathbf{P}(A_{n,\varepsilon}) \leqslant \sum_{\mathbf{R}\times\mathbf{R}} \mathbf{P}\big(s_n \in B_1(\varepsilon)\Pi^-\Pi^+\big) \leqslant \sum_{\mathbf{R}\times\mathbf{R}} \exp\{-nI_a(B_1(\varepsilon)\Pi^-\Pi^+)\}. \tag{4.6.7}$$

Denote by $(\Gamma g)^-_{(\delta)}(t)$ $\big($respectively, $(\Gamma g)^+_{(\delta)}(t)\big)$, $t \in [0, 1]$, the lower (respectively, upper) boundary of the set $(\Gamma g)_\delta$ and consider, on the 'inner' segment Δ_k (for $r < k \leqslant R - r$), the set of functions g

$$\left\{ g\left(\frac{i_k}{n}\right) < m_k^- + \varepsilon \right\} \quad \text{for some} \quad i_k \in L_k. \tag{4.6.8}$$

On this set, one has

$$(\Gamma g)^-_{(\delta)}(t) < f(t) \quad \text{for all} \quad t \in \Delta_k, \tag{4.6.9}$$

where, for $r \geqslant 2$,

$$\delta = \sqrt{(\varepsilon + \theta)^2 + \varepsilon^2} = \varepsilon\sqrt{\left(1 + \frac{1}{r}\right)^2 + 1}$$
$$\leqslant \sqrt{2}\varepsilon\left(1 + \frac{1}{r} + \frac{1}{2r^2}\right)^{1/2} \leqslant \sqrt{2}\varepsilon\left(1 + \frac{1}{r}\right)$$

is the distance from the point $(\theta(k-1)-\varepsilon,\ m_k^- + \varepsilon)$ to $(\theta k, m_k^-)$. Indeed, if i_k in (4.6.8) is equal to $(\theta(k-1)-\varepsilon)n$ (the left-hand point of L_k) then the lower boundary of the ball of radius δ whose centre is at the point

$$\Big(\theta(k-1)-\varepsilon,\ g\big(\theta(k-1)-\varepsilon\big)\Big)$$

lies, for $t \in \Delta_k$, below the level m_k^- and hence below $f(t)$, $t \in \Delta_k$. Moreover this is true for balls of the same radius δ with centres in the 'inner' points $i_k \in L_k$. As all the above-mentioned balls lie inside $(\Gamma g)_\delta$, we obtain (4.6.9).

The above argument also applies to the terminal segments Δ_k for $k \leqslant r$ or $k > R - r$ (with the same ball radius δ), to which correspond shorter ranges of i_k-values.

Similarly, the fact that function g belongs to the set $\{g(j_k/n) > m_k^+ - \varepsilon\}$ means that

$$(\Gamma g)^+_{(\delta)}(t) > f(t) \quad \text{for all} \quad t \in \Delta_k.$$

Hence we have the inclusion

$$\begin{aligned} B_1(\varepsilon)\Pi^-\Pi^+ &\subset B_1(\varepsilon)\Big\{g:\ (\Gamma g)^-_{(\delta)}(t) < f(t) < (\Gamma g)^+_{(\delta)}(t) \quad \text{for all } t \in [0,1]\Big\} \\ &= B_1(\varepsilon)B_2(\delta) \subset (f)_\delta. \end{aligned} \qquad (4.6.10)$$

Therefore

$$I_a\big(B_1(\varepsilon)\Pi^-\Pi^+\big) \geqslant I_a\big((f)_\delta\big),$$

and, by virtue of (4.6.7),

$$\mathbf{P}(A_{n,\varepsilon}) \leqslant \exp\Big\{-nI_a\big((f)_\delta\big)\Big\} \sum_{\mathbf{R}\times\mathbf{R}} 1.$$

The maximum possible number of elements in L_k (for $r < k \leqslant R - r$) is equal to

$$b := (\theta + 2\varepsilon)n + 1.$$

Hence the number of all tuples $(i_1, \dots, i_R)$, $i_k \in L_k$, does not exceed b^R. Since the number of elements in the set $\mathbf{R}\times\mathbf{R}$ does not exceed b^{2R}, one has

$$\mathbf{P}(A_{n,\varepsilon}) \leqslant b^{2R} \exp\Big\{-nI_a\big((f)_\delta\big)\Big\}, \quad \delta \leqslant \varepsilon\sqrt{2}\Big(1+\frac{1}{r}\Big). \qquad (4.6.11)$$

(ii) Now we will remove the assumption that ε and θ are multiples of $1/n$ (we will keep the relation $\varepsilon = r\theta$; the number $r \geqslant 2$ is fixed in the conditions of the theorem).

If θ is not a multiple of $1/n$ then we consider the smallest value $\theta_n > \theta$ that is a multiple of $1/n$. Clearly, $\theta_n \in (\theta, \theta + 1/n)$. Consequently, the smallest $\varepsilon_n > \varepsilon$ that is a multiple of θ and $1/n$ has the property that $\varepsilon_n \in (\varepsilon, \varepsilon + r/n)$. Now, if $1/\theta_n$ is not an integer then we will form, as before, segments $\Delta_{k,n} = [(k-1)\theta_n, k\theta_n]$; in doing so, we will get $[1/\theta_n]$ segments of length θ_n and one shorter segment, for which all the bounds presented in part (i) of the proof remain true. In this construction, the number of segments is equal to $[1/\theta_n]+1 \leqslant 1/\theta_n+1 \leqslant 1/\theta+1$. Applying inequality (4.6.11) with chosen θ_n, ε_n (replacing $R=1/\theta$ with $R+1$), we obtain

$$\mathbf{P}(A_{n,\varepsilon}) \leqslant b_n^{2(R+1)} \exp\Big\{-nI_a\big((f)_\delta\big)\Big\},$$

where

$$b_n = (\theta_n+2\varepsilon_n)n+1 \leqslant (\theta+2\varepsilon)n+2r+2$$

and $\delta \leqslant \varepsilon_n\sqrt{2}(1+1/r) \leqslant (\varepsilon+r/n)\sqrt{2}(1+1/r) \leqslant \varepsilon\sqrt{2}(1+2/r)$ for $n \geqslant 4r^2/\varepsilon$. Theorem 4.6.5 is proved. □

We now return to the process $\overline{s}_n$ defined in (4.5.12). Applying Theorem 4.6.5 we can obtain an upper bound for the probability $\mathbf{P}\big(\overline{s}_n \in (f)_\varepsilon\big)$.

Corollary 4.6.6. *For any function $f \in \mathbb{D}$ and any integer $r \geqslant 2$, for all large enough n one has*

$$\mathbf{P}\big(\overline{s}_n \in (f)_\varepsilon\big) \leqslant (\overline{b}_n)^{2(r/\varepsilon+1)} \exp\Big\{-nJ\big((f)_{\overline{\delta}}\big)\Big\},$$

where

$$\overline{b}_n = n\varepsilon\Big(\frac{r+2}{r+1}\Big)\Big(2+\frac{1}{r}\Big)+2r+2, \quad \overline{\delta} = \varepsilon\sqrt{2}\Big(1+\frac{3}{r}\Big).$$

Proof. First we will obtain the required bound for $\mathbf{P}\big(\overline{s}_n \in (f)_\varepsilon;\ B_n\big)$, where $B_n := \big\{\rho(s_n,\overline{s}_n) \leqslant h\big\}$, $h = \varepsilon/(\sqrt{2}(r+1))$. For a given function $f \in \mathbb{D}$ there exist functions $f_k \in \mathbb{C}_a$ such that $\rho(f,f_k) < 1/k$. By the triangle inequality, on the set B_n one has

$$\rho(s_n,f_k) \leqslant h + \frac{1}{k} + \rho(\overline{s}_n,f).$$

Therefore, the event $\big\{\rho(\overline{s}_n,f) < \varepsilon;\ B_n\big\}$ implies the event $\big\{\rho(s_n,f_k) < \varepsilon+h+1/k;\ B_n\big\}$,while the inequality $\rho(f,f_k) < 1/k$ implies that $(f_k)_\gamma \subset (f)_{\gamma+1/k}$. Hence, by virtue of Theorem 4.6.5,

$$\begin{aligned}\mathbf{P}\big(\overline{s}_n \in (f)_\varepsilon;\ B_n\big) &\leqslant \mathbf{P}\big(s_n \in (f_k)_{\varepsilon+h+1/k}\big)\\ &\leqslant \overline{b}_n^{\,2(r/\varepsilon+1)} \exp\Big\{-nI_a\big((f)_{\overline{\delta}}\big)\Big\},\end{aligned}$$

where

$$\begin{aligned}\overline{b}_n &= n\Big(\varepsilon+h+\frac{1}{k}\Big)\Big(2+\frac{1}{r}\Big)+2r+2 \leqslant n\varepsilon\Big(\frac{r+2}{r+1}\Big)\Big(2+\frac{1}{r}\Big)+2r+2,\\ \overline{\delta} &= \Big(\varepsilon+h+\frac{1}{k}\Big)\sqrt{2}\Big(1+\frac{2}{r}\Big)+\frac{1}{k} \leqslant \varepsilon\sqrt{2}\Big(1+\frac{3}{r}\Big)\end{aligned}$$

for $k = N/\varepsilon$ and large enough N. Returning to bounding the probability $\mathbf{P}(\overline{s}_n \in (f)_\varepsilon)$, we find that

$$\mathbf{P}(\overline{s}_n \in (f)_\varepsilon) \leqslant \mathbf{P}(\overline{s}_n \in (f)_\varepsilon,\ B_n) + \mathbf{P}(\overline{B}_n),$$

where, for any $N > 0$ and all large enough n, one has $\mathbf{P}(\overline{B}_n) \leqslant e^{-nN}$. Clearly, for all large enough n, the term $\mathbf{P}(\overline{B}_n)$ will not exceed the right-hand side of the above bound for $\mathbf{P}(\overline{s}_n \in (f)_\varepsilon,\ B_n)$. It remains to notice that, for any open set B, one has (see property (vi) in Theorem 4.2.3)

$$I_a(B) = J(B).$$

Corollary 4.6.6 is proved. □

Corollary 4.6.7. *For $f \in \mathbb{D}$ and any integer $r \geqslant 2$,*

$$\overline{\lim_{n\to\infty}} \frac{1}{n} \ln \mathbf{P}(\overline{s}_n \in (f)_\varepsilon) \leqslant -J\big((f)_{\varepsilon\sqrt{2}(1+2/r)}\big).$$

Corollary 4.6.7 is an obvious consequence of Corollary 4.6.6.

From Corollary 4.6.7 it follows that, for any $\delta(\varepsilon) > \varepsilon\sqrt{2}$, one has

$$\overline{\lim_{n\to\infty}} \frac{1}{n} \ln \mathbf{P}(\overline{s}_n \in (f)_\varepsilon) \leqslant -J\big((f)_{\delta(\varepsilon)}\big).$$

Thus we have established a uniform upper bound in the l.l.d.p. (see Definition 4.1.3).

The proofs of Theorem 4.6.5 and Corollaries 4.6.6, 4.6.7 do not contain crude bounds. Therefore, it is hardly possible to improve upon the inequalities in Theorem 4.6.5 and Corollaries 4.6.6, 4.6.7.

Note that the proof of Theorem 4.6.5 enables one to describe the shape of the most probable trajectory f_B for sets $B = (f)_\varepsilon$ (and thus to find the value of $J((f)_\varepsilon) = I(f_B)$) for 'simple enough' functions f and small enough ε. Namely, we will assume that f has finitely many extrema and ε is sufficiently small that the distance between neighbouring minimum (maximum) points exceeds 2ε. Then, similarly to our considerations in Example 4.5.1, the trajectory f_B can be described as a stretched thread. Label those minima points t_k of f in whose ε-neighbourhoods there is a 'break point' of the function f^+ (the upper boundary of $(\Gamma f)_\varepsilon$; in the neighbourhoods of these points, the parts of $(f)_\varepsilon$ that lie, roughly speaking, on the left and right of the points t_k, are 'glued together'). Draw a circle of radius ε with centre at the point $(t_k, f(t_k))$ and take the (upper) arc that connects the neighbouring branches of $f(t)$. Further, take a thread, fix one of its ends at the point $(0,0)$ and then run the thread between the boundaries $f^\pm$ of the domain $(\Gamma f)_\varepsilon$ in such a way that it would pass between the arcs that we have drawn and the extrema points (i.e. it would be within a distance ε from the points $(t_k, f(t_k))$). Pass the other end of the thread through the point $(1, f_1)$ with $f_1 \in [f^-(1), f^+(1)]$. Then stretch the thread and, in the case $\mathbf{E}\xi = 0$, choose the point f_1 such that the length of the thread between the points $(0,0)$ and $(1, f_1)$ is minimal. It is not hard to see that this is the shortest trajectory from $(f)_\varepsilon$, and it is this line that is the most probable trajectory f_B. If $\mathbf{E}\xi \neq 0$ then the thread should be stretched in such a way that its last straight segment has slope coefficient $\mathbf{E}\xi$.

4.6.4 Lower bound

Theorem 4.6.8. *For any $\varepsilon > 0$, $f \in \mathbb{D}$ one has the inequality*

$$\varliminf_{n\to\infty} \frac{1}{n} \ln \mathbf{P}\big(\overline{s}_n \in (f)_\varepsilon\big) \geqslant -J(f).$$

Proof. For any $f \in \mathbb{D}$ there exists a sequence of polygonal lines $f_k \in \mathbb{C}_a$ such that $\rho(f,f_k) < 1/k$, $J(f_k) = I(f_k) \to J(f)$ as $k \to \infty$. We have

$$\rho(\overline{s}_n,f) \leqslant \rho(\overline{s}_n, s_n) + \rho(s_n,f_k) + \rho(f_k,f).$$

Let $B_n = \{\rho(\overline{s}_n, s_n) \leqslant h_n\}$, where $h_n \to 0$ as $n \to \infty$ (h_n is from (4.5.13)). The event $\{\rho(s_n,f_k) < \varepsilon;\ B_n\}$ implies the event $\{\rho(\overline{s}_n,f) < \varepsilon + h_n + 1/k;\ B_n\}$. Hence, for n and k such that $h_n < \varepsilon/2,\ 1/k < \varepsilon/2,$ one will have

$$\begin{aligned} P := \mathbf{P}(\rho(\overline{s}_n,f) < 2\varepsilon;\ B_n) &\geqslant \mathbf{P}\left(\rho(\overline{s}_n,f) < \varepsilon + h_n + \frac{1}{k};\ B_n\right) \\ &\geqslant \mathbf{P}(\rho(s_n,f_k) < \varepsilon;\ B_n) \geqslant \mathbf{P}(\rho_{\mathbb{C}}(s_n,f_k) < \varepsilon;\ B_n) \\ &\geqslant \mathbf{P}(\rho_{\mathbb{C}}(s_n,f_k) < \varepsilon) - \mathbf{P}(\overline{B}_n). \end{aligned}$$

Since $(1/n)\ln \mathbf{P}(\overline{B}_n) \to -\infty$, we obtain, by the l.l.d.p. in $(\mathbb{C}, \rho_{\mathbb{C}})$ (see Theorem 4.4.2) that

$$\varliminf_{n\to\infty} \frac{1}{n} \ln P \geqslant -I(f_k),$$

where the right-hand side can be made arbitrarily close to $J(f)$. It remains to make use of the inequality

$$\mathbf{P}(\rho(\overline{s}_n,f) < 2\varepsilon) \geqslant P.$$

Theorem 4.6.8 is proved. Combining this with Corollary 4.6.7, we get Theorem 4.6.1 as well. □

4.6.5 Proof of Theorem 4.6.2

We will show that Theorem 4.6.2 is a corollary of Theorem 4.6.1. To this end, we need the following lemma.

Lemma 4.6.9. *For any $z \geqslant 0$, the set*

$$M_z := \{f \in \mathbb{D} :\ \rho(f,0) \leqslant z\} = \{f \in \mathbb{D} :\ \sup_{1\leqslant t\leqslant 1} |f(t)| \leqslant z\}$$

is totally bounded in the space $(\mathbb{D}, \rho)$.

Proof. For a fixed $\varepsilon > 0$ choose an integer R such that $2/R \leqslant \varepsilon$. As an ε-net for M_z we will use the set $\mathcal{N}_\varepsilon$ of all functions g which are right-continuous at points $t \in [0, 1)$ and left-continuous at $t = 1$, and which are constant on the intervals

$$\left(\frac{k-1}{2R}, \frac{k}{2R}\right), \ k = 0, \ldots, 2R,$$

and assume the values $j/2R$, $-2zR \leqslant j \leqslant 2zR$, where, without loss of generality, one can assume that z is integer. The number $|\mathcal{N}_\varepsilon|$ of such functions equals $(4zR+1)^{2R}$. Take any function f from M_z and construct for it a simple function $g = g^{(f)} \in \mathcal{N}_\varepsilon$, following the next rule: for $i = 1, \ldots, R$, denote by

$$f_i^+ = \sup_{(i-1)/R \leqslant t \leqslant i/R} f(t), \ \ f_i^- = \inf_{(i-1)/R \leqslant t \leqslant i/R} f(t)$$

the maximum and minimum of $f(t)$ on the segment

$$\left[\frac{i-1}{R}, \frac{i}{R}\right],$$

respectively. Then the simple function $g^{(f)}(t)$ assumes on the intervals

$$\left(\frac{2i-2}{2R}, \frac{2i-1}{2R}\right), \left(\frac{2i-1}{2R}, \frac{2i}{2R}\right)$$

the values

$$\frac{[2Rf_i^+]+1}{2R}, \frac{[2Rf_i^-]-1}{2R},$$

respectively. It is obvious from the construction that one has

$$\Gamma f \subseteq (\Gamma g^{(f)})_\varepsilon, \ \ \Gamma g^{(f)} \subseteq (\Gamma f)_\varepsilon,$$

hence $\rho(f, g^{(f)}) < \varepsilon$. This proves that the set M_z is totally bounded. The lemma is proved. □

Thus, *in the space* $(\mathbb{D}, \rho)$ *the concepts of total boundedness and boundedness coincide.*

Lemma 4.6.10. *If condition* $[\mathbf{C}_0]$ *is met then the condition* $[\mathbf{TB}_0]$ (see section 4.1) *is also satisfied.*

Proof. By Lemma 4.6.9, we need only to bound $\mathbf{P}(\vec{s}_n \notin M_z)$. For any $h > 0$, one has

$$\mathbf{P}(\vec{s}_n \notin M_z) \leqslant \mathbf{P}\big(\rho(\vec{s}_n, s_n) > h\big) + \mathbf{P}\big(s_n \notin M_{z-h}\big). \tag{4.6.12}$$

By virtue of (4.5.12), it suffices to bound

$$\mathbf{P}(s_n \notin M_z) = \mathbf{P}\Big(\sup_{0 \leqslant t \leqslant 1} \big|s_n(t)\big| \geqslant z\Big)$$
$$\leqslant \mathbf{P}\big(\max_{1 \leqslant k \leqslant n} S_k \geqslant nz\big) + \mathbf{P}\big(\min_{1 \leqslant k \leqslant n} S_k \leqslant -nz\big).$$

Under condition $[\mathbf{C}_0]$, it follows from (4.4.19) that

$$\mathbf{P}(s_n \notin M_z) \leqslant e^{-n\Lambda(z)} + e^{-n\Lambda(-z)} \leqslant 2e^{-n(c_1 z - c_2)}.$$

For sufficiently large z, the required bound follows from this and (4.6.12). Lemma 4.6.10 is proved. □

To prove Theorem 4.6.2 all that remains is to make use of Theorems 4.1.7 and 4.6.1. Thus Theorem 4.6.2 is proved.

4.6.6 Proof of Theorem 4.6.3

If a set B is bounded then it is also totally bounded (see Lemma 4.6.9) and it is not hard to see from the proof of Theorem 4.6.2 that condition $[\mathbf{TB}_0]$ and hence also condition $[\mathbf{C}_0]$become superfluous. If a set B is convex then, by Theorem 4.4.9,

$$\mathbf{P}(s_n \in B) \leqslant e^{-nI_a([B]_{\mathbb{C}})},$$

where $I_a([B]_{\mathbb{C}}) \geqslant I_a([B]) \geqslant I_a((B)_\varepsilon) \geqslant J((B)_\varepsilon)$ for any $\varepsilon > 0$. Hence one has $I_a([B]_{\mathbb{C}}) \geqslant J(B+)$, and so

$$\overline{\lim_{n\to\infty}} \frac{1}{n} \ln \mathbf{P}(s_n \in B) \leqslant -J(B+). \tag{4.6.13}$$

We now make use of the argument that was employed above and this allows us to obtain (4.6.3) from (4.6.13). Theorem 4.6.3 is proved. □

Theorem 4.6.3 shows that condition $[\mathbf{C}_0]$ is not necessary for the upper bound (4.6.3). In this connection, as in the problem of how essential condition $[\mathbf{C}_\infty]$ is in Theorem 4.4.7 (see the remark at the end of section 4.4.2), the question arises of whether condition $[\mathbf{C}_0]$ is essential for the assertion of Theorem 4.4.7.

4.6.7 Supplement. Large deviation principles on the space $(\mathbb{D}, \rho)$ in the case $d \geqslant 1$

The deviation integral in the case $J(f)$ for $d \geqslant 1$ was defined and studied in subsection 4.2.3.

Theorem 4.6.11. (i) *If condition* $[\mathbf{C}_0]$ *is met then, for any* $f \in \mathbb{D}$, $\varepsilon > 0$, *one has the inequality*

$$\overline{\lim_{n\to\infty}} \frac{1}{n} \ln \mathbf{P}\big(\overline{s}_n \in (f)_\varepsilon\big) \leqslant -J((f)_{3\varepsilon}).$$

(ii) *For any* $f \in \mathbb{D}$ *and* $\varepsilon > 0$,

$$\underline{\lim_{n\to\infty}} \frac{1}{n} \ln \mathbf{P}\big(\overline{s}_n \in (f)_\varepsilon\big) \geqslant -J(f).$$

In other words, the l.l.d.p. holds in $(\mathbb{D}, \rho)$ with parameters (n, J) for $\delta(\varepsilon) = 3\varepsilon$ and $\beta(\varepsilon) = 0$ (see Definition 4.1.3). Along with that version, one can also state two other versions of the l.l.d.p., according to the equivalent Definitions 4.1.1–4.1.3 (cf. (4.6.2)).

Remark 4.6.12. The presence of condition **[C$_0$]** in assertion (i) of Theorem 4.6.11 is probably due to the method of proof. Since for all the other considered spaces and metrics the upper bound in the l.l.d.p. holds without any additional conditions (see e.g. Theorems 4.4.2, 4.6.1 and 4.8.2), one can suggest that condition **[C$_0$]** in part (i) of Theorem 4.6.11 is superfluous.

Theorem 4.6.13. *All the assertions of Theorem 4.6.2 remain true in the case* $d \geqslant 1$.

If the set B in Theorem 4.6.13 is convex or belongs to the class $V_z := \{f \in \mathbb{D} : f(0) = 0,\ \mathrm{Var} f \leqslant z\}$ for some $z < \infty$, then the assertion of Theorem 4.6.13 can be strengthened by removing condition **[C$_0$]**. In other words, we have the following result.

Theorem 4.6.14. *If the set* B *is convex or* $B \subset V_z$ *for some* $z < \infty$ *then the upper bound* (4.6.3) *holds true.*

As in the multivariate case $d > 1$ the set $B_1(\varepsilon)$ in (4.6.4) is, generally speaking, not convex, it is necessary to modify the argument that we used to establish the l.l.d.p. in the one-dimensional case. Meanwhile, the proofs of Theorems 4.6.11 and 4.6.13 become more complicated and cumbersome. For details see [64].

4.7 Supplement. Large deviation principles in the space $(\mathbb{V}, \rho_{\mathbb{V}})$

4.7.1 The space $\mathbb{V}$ and metric $\rho_{\mathbb{V}}$

If one restricts the functional space $\mathbb{D}$ to the subspace $\mathbb{V} \subset \mathbb{D}$ of functions of bounded variation then, instead of the metric ρ, one can use a metric $\rho_{\mathbb{V}}$ (see below) which is, in a sense, more adequate to the nature of the problems one is dealing with. In the space $\mathbb{V}$ this metric is weaker than the Skorokhod metric $\rho_{\mathbb{D}}$ but stronger than the metric ρ; in the space $\mathbb{C}_a \subset \mathbb{V}$, the metrics $\rho_{\mathbb{V}}$ and $\rho_{\mathbb{D}}$ are equivalent (see Lemma 4.7.1 below). The metric $\rho_{\mathbb{V}}$ is defined as follows (it is actually a pseudometric, like ρ; see our remarks in subsection 4.2.1 following the definition of the metric ρ.

Consider a parametric representation for the graph Γf (see subsection 4.2.1) of a function f from the set $V_z = \{f \in \mathbb{V} : f(0) = 0,\ \mathrm{Var} f \leqslant z\}$, choosing the trajectory length as a parameter. If $\mathrm{Var} f \leqslant z$ then the length l_f of the trajectory Γf is in the segment $[1, 1+z]$. Let $\Gamma f(l) \in \mathbb{R}^{d+1}$ be the point on the graph Γf from which the distance *along the trajectory* Γf from the origin is equal to l. Then the first component $\Gamma f(l)_{(1)}$ of the vector $\Gamma f(l)$ is the 'time epoch' $t(l)$ at which the length of the trajectory Γf reaches the value l. The enumeration of the other components of $\Gamma f(l)$ for $i \geqslant 2$ differs from that of the components of f, in that it is shifted by one.

In particular, for $f \in \mathbb{C}_a \subset \mathbb{V}$ the length of the trajectory Γf on the segment $[0, t]$ is given by the integral

$$L_f(t) := \int_0^t \sqrt{|f'(u)|^2 + 1}\, du.$$

This is an increasing continuous function assuming values from the segment $[0, l_f]$. Therefore there exists a function $t_f(l)$, $0 \leqslant l \leqslant l_f$, which is inverse to $L_f(t)$ and so the above-mentioned parametric representation of the trajectory Γf can be written as

$$\Gamma f(l) = \Big(t_f(l), f\big(t_f(l)\big)\Big), \quad 0 \leqslant l \leqslant l_f. \tag{4.7.1}$$

Now return to the general case.

We will write

$$\rho_{\mathbb{V}}(f, g) < \varepsilon$$

if there exists a continuous monotone function $q = q(t) : \ [0, l_f] \to [0, l_g]$ *such that*

$$\sup_{l \in [0, l_f]} \Big| \Gamma f(l) - \Gamma g\big(q(l)\big) \Big| < \varepsilon.$$

In other words,

$$\rho_{\mathbb{V}}(f, g) := \inf_{q \in \mathcal{Q}(l_f, l_g)} \sup_{l \in [0, l_f]} \Big| \Gamma f(l) - \Gamma g\big(q(l)\big) \Big|,$$

where $\mathcal{Q}(l_f, l_g)$ is the class of all continuous monotone mappings of [$[0, l_f]$ into $[0, l_g]$. Introducing the normalised lengths l/l_f and l/l_g of the curves Γf and Γg, respectively, one can write

$$\rho_{\mathbb{V}}(f, g) = \inf_{q \in \mathcal{Q}} \sup_{0 \leqslant v \leqslant 1} \Big| \Gamma f(v l_f) - \Gamma g\big(q(v) l_g\big) \Big|, \tag{4.7.2}$$

where $\mathcal{Q} = \mathcal{Q}(1, 1)$ is the class of continuous monotone mappings of $[0, 1]$ into $[0, 1]$. This class is now independent of f, g.

In order to understand what place the metric $\rho_{\mathbb{V}}$ occupies among other metrics we define a metric $\rho_{\mathbb{D}}$ which is stronger than ρ and moreover, equivalent to the Skorokhod metric:

$$\rho_{\mathbb{D}}(f, g) := \inf_{q \in \mathcal{Q}} \max\{\rho_{\mathbb{C}}(f, g * q), \rho_{\mathbb{C}}(q, \mathrm{e})\}, \tag{4.7.3}$$

where $g * q = g * q(t) := g(q(t))$, $\mathrm{e} = \mathrm{e}(t) = t$ for $0 \leqslant t \leqslant 1$ and by $\mathcal{Q}$ we denote the class of continuous increasing functions $q = q(t)$ mapping $[0, 1]$ into itself in a one-to-one manner.

We will verify that the metric ρ is weaker than $\rho_{\mathbb{V}}$, while $\rho_{\mathbb{V}}$ is weaker than the metric $\rho_{\mathbb{D}}$. More precisely, the following lemma holds true.

Lemma 4.7.1. *For any* $f, g \in \mathbb{V}$ *one has*

$$\rho(f, g) \leqslant \rho_{\mathbb{V}}(f, g) \leqslant \sqrt{2}\rho_{\mathbb{D}}(f, g). \tag{4.7.4}$$

For any $f, g \in \mathbb{C}_a$

$$\rho_{\mathbb{V}}(f,g) \geqslant \rho_{\mathbb{D}}(f,g). \tag{4.7.5}$$

Proof. First let $f, g \in \mathbb{C}_a$. Then, in view of the representation (4.7.1) and definition (4.7.2), we have

$$\rho_{\mathbb{V}}(f,g) = \inf_{q\in\mathcal{Q}} \sup_{0\leqslant v\leqslant 1} \sqrt{|t_f(vl_f) - t_g(q(v)l_g)|^2 + |f(t_f(vl_f)) - g(t_g(q(v)l_g))|^2}.$$

If, for the function $u = u(v) := t_f(vl_f)$ and for an arbitrary function $q_0 = q_0(u) \in \mathcal{Q}$, we set

$$q(v) := \frac{1}{l_g} t_g^{(-1)}(q_0(t_f(vl_f))) = \frac{1}{l_g} t_g^{(-1)}(q_0(u(v))) \in \mathcal{Q} \tag{4.7.6}$$

then we have

$$\begin{aligned} t_g(q(v)l_g) &= q_0(u(v)),\\ \rho_{\mathbb{V}}(f,g) &\leqslant \sup_{0\leqslant v\leqslant 1} \sqrt{|u(v) - q_0(u(v))|^2 + |f(u(v)) - g(q_0(u(v)))|^2}\\ &= \sup_{0\leqslant u\leqslant 1} \sqrt{|u - q_0(u)|^2 + |f(u) - g(q_0(u))|^2}\\ &\leqslant \sqrt{2}\max\{\sup_{0\leqslant u\leqslant 1} |u - q_0(u)|, \sup_{0\leqslant u\leqslant 1} |f(u) - g(q_0(u))|\}. \end{aligned}$$

Minimizing the right-hand side of the last inequality over $q_0 \in \mathcal{Q}$, we get $\sqrt{2}\rho_{\mathbb{D}}(f,g)$. The second inequality in (4.7.4) is proved.

Further, note that

$$\begin{aligned} \rho_{\mathbb{D}}(f,g) &= \inf_{q_0\in\mathcal{Q}} \max\{\sup_{0\leqslant u\leqslant 1} |u - q_0(u)|, \sup_{0\leqslant u\leqslant 1} |f(u) - g(q_0(u))|\}\\ &\leqslant \inf_{q_0\in\mathcal{Q}} \sup_{0\leqslant u\leqslant 1} \sqrt{|u - q_0(u)|^2 + |f(u) - g(q_0(u))|^2}. \end{aligned}$$

For the function $u = u(v) := t_g(vl_g)$ and for an arbitrary function $q = q(v) \in \mathcal{Q}$, set (see (4.7.6))

$$q_0(u) := t_g\Big(q\big(v(u)\big)l_g\Big) \in \mathcal{Q}, \quad \text{where} \quad v(u) := \frac{1}{l_g} t_g^{(-1)}(u).$$

Then

$$\begin{aligned} \rho_{\mathbb{D}}(f,g) &\leqslant \sup_{0\leqslant u\leqslant 1} \sqrt{|u - q_0(u)|^2 + |f(u) - g(q_0(u))|^2}\\ &= \sup_{0\leqslant v\leqslant 1} \sqrt{|u(v) - q_0(u(v))|^2 + |f(u(v)) - g(q_0(u(v)))|^2}\\ &= \sup_{0\leqslant v\leqslant 1} \sqrt{|t_f(vl_f) - t_g(q(v)l_g)|^2 + |f(t_f(vl_f)) - g(t_g(q(v)l_g)))|^2}. \end{aligned}$$

Minimizing over $q \in \mathcal{Q}$ the right-hand side of the last inequality, we obtain (4.7.5).

Now we show that the right-hand inequality in (4.7.4) holds for any $f, g \in \mathbb{V}$. To this end, it suffices to note that there always exists a dense sequence $\{\mathbf{t}_R\}$ of partitions of the segment [0, 1] (see subsection 4.2.1), such that simultaneously

$$\sqrt{2}\rho_{\mathbb{D}}(f,g) \geqslant \varlimsup_{R\to\infty} \sqrt{2}\rho_{\mathbb{D}}(f^{\mathbf{t}_R}, g^{\mathbf{t}_R}), \qquad \lim_{R\to\infty} \rho_{\mathbb{V}}(f^{\mathbf{t}_R}, g^{\mathbf{t}_R}) = \rho_{\mathbb{V}}(f,g), \quad (4.7.7)$$

where $f^{\mathbf{t}_R}$, $g^{\mathbf{t}_R}$ are continuous polygonal lines approximating the functions f, g, respectively, with nodes having abscissas at the points of the partition $\mathbf{t}_R$. Since the right-hand side of the first relation in (4.7.7) is greater than or equal to the left-hand side of the second relation in (4.7.7), we obtain the right-hand inequality in (4.7.4). The left-hand inequality in (4.7.4) is obvious, which means that Lemma 4.7.1 is proved. □

It follows from Lemma 4.7.1 that in the space $\mathbb{V}$ one has the following relations:

$$\mathbb{V} \cap (f)_{\mathbb{D}, \frac{1}{\sqrt{2}}\varepsilon} \subset (f)_{\mathbb{V},\varepsilon} \subset (f)_{\varepsilon},$$

where $(f)_{\mathbb{D},\varepsilon}$, $(f)_{\mathbb{V},\varepsilon}$ are ε-neighbourhoods of the point f in the metrics $\rho_{\mathbb{D}}$ and $\rho_{\mathbb{V}}$, respectively. Hence, convergence in the Skorokhod metric implies convergence in the metric $\rho_{\mathbb{V}}$, while convergence in the metric $\rho_{\mathbb{V}}$ implies convergence in the metric ρ. Therefore, all the properties of the functional $J(f)$ from Theorem 4.2.3 still hold if we replace the metric ρ with $\rho_{\mathbb{V}}$.

4.7.2 Totally bounded sets in the space $(\mathbb{V}, \rho_{\mathbb{V}})$

Lemma 4.7.2. *For any $z \geqslant 0$, the set*

$$V_z := \{f \in \mathbb{V} : f(0) = 0, \ \mathrm{Var} f \leqslant z\}$$

is totally bounded in the space $(\mathbb{V}, \rho_{\mathbb{V}})$.

Proof. The function $\Gamma_f(ul_f)$, $u \in [0, 1]$, has the following properties:

(a) The first component $t(ul_f)$ is non-decreasing in u.
(b) All its components are continuous functions of u.
(c) The absolute value of the rate of change of each component does not exceed $l_f \leqslant \mathrm{Var} f + 1$.

Therefore, the function class $\{\Gamma_f(ul_f),\ 0 \leqslant u \leqslant 1 : f \in V_z\}$ satisfies the Lipschitz condition, with the constant $z+1$ in each component. Since that condition ensures that the set is totally bounded in the space of continuous functions with uniform norm, for any $\varepsilon > 0$ there exist N functions $f_1, \ldots, f_N$ from V_z such that, for any function $f \in V_z$,

$$\begin{aligned}\min_{1\leqslant i\leqslant N} \rho_{\mathbb{V}}(f, f_i) &= \min_{1\leqslant i\leqslant N} \inf_{q\in\mathcal{Q}} \sup_{0\leqslant u\leqslant 1} \left|\Gamma_f(ul_f) - \Gamma_{f_i}\big(q(u)l_{f_i}\big)\right| \\ &\leqslant \min_{1\leqslant i\leqslant N} \sup_{0\leqslant u\leqslant 1} |\Gamma_f(ul_f) - \Gamma_{f_i}(ul_{f_i})| < \varepsilon.\end{aligned}$$

This means that the set V_z is totally bounded in the space $(\mathbb{V}, \rho_{\mathbb{V}})$. The lemma is proved. □

4.7.3 The local and extended large deviation principles in the space $(\mathbb{V}, \rho_{\mathbb{V}})$

Let the process $\overline{s}_n$ take values from the space $(\mathbb{V}, \rho_{\mathbb{V}})$ and, as before, for any $h > 0$ relation (see (4.5.12)), let

$$\lim_{n\to\infty} \frac{1}{n} \ln \mathbf{P}(\rho_{\mathbb{V}}(\overline{s}_n, s_n) > h) = -\infty \tag{4.7.8}$$

hold. All the main assertions from sections 4.6 and 4.7 for the processes $\overline{s}_n$ in (D, c) that are related to the l.l.d.p. and e.l.d.p. remain true in the space $(\mathbb{V}, \rho_{\mathbb{V}})$. More precisely, the following assertions hold true.

Theorem 4.7.3. *The l.l.d.p. with parameters (n, J) always holds true for a process $\overline{s}_n$ in $(\mathbb{V}, \rho_{\mathbb{V}})$.*

Theorem 4.7.4. (i) *If condition* $[\mathbf{C_0}]$ *is satisfied then, for any measurable set B,*

$$\overline{\lim_{n\to\infty}} \frac{1}{n} \ln \mathbf{P}(\overline{s}_n \in B) \leqslant -J(B_{\mathbb{V}}+), \tag{4.7.9}$$

where $J(B_{\mathbb{V}}+) := \lim_{\delta\to 0} J((B)_{\mathbb{V},\delta})$.

(ii) *The following inequality always holds:*

$$\underline{\lim}_{n\to\infty} \frac{1}{n} \ln \mathbf{P}(\overline{s}_n \in B) \geqslant -J\big((B)_{\mathbb{V}}\big), \tag{4.7.10}$$

where $(B)_{\mathbb{V}}$ is the open interior of the set B in the metric $\rho_{\mathbb{V}}$.

Theorem 4.7.4 means that under condition $[\mathbf{C_0}]$ the e.l.d.p. with parameters (n, J) holds for $\overline{s}_n$ in the space $(\mathbb{V}, \rho_{\mathbb{V}})$.

The following theorem is true as well.

Theorem 4.7.5. *If the set B is convex or $B \subset V_z$ for some $z < \infty$ then condition* $[\mathbf{C_0}]$ *is superfluous in the assertion of Theorem 4.7.4.*

In a sense, the assertion of the e.l.d.p. in $(\mathbb{V}, \rho_{\mathbb{V}})$ is stronger than a similar assertion in $(\mathbb{D}, \rho)$, since the difference between the upper and lower limits in (4.7.9) and (4.7.10), which is equal to $J([B]_{\mathbb{V}}+) - J((B)_{\mathbb{V}})$, does not exceed (and can actually be less than) the respective difference for the e.l.d.p. in $(\mathbb{D}, \rho)$. It may be considered as a deficiency of switching to the space $(\mathbb{V}, \rho_{\mathbb{V}})$ that we now consider a space that is smaller than $\mathbb{D}$. This narrowing of the space seems, however, to be insignificant, since $J(f) = \infty$ for $f \notin \mathbb{V}$, and the exact lower bounds $J(B) = \inf_{f\in B} J(f)$ have the property that

$$J(B) = J(B\mathbb{V}).$$

The difference between the metrics ρ and $\rho_{\mathbb{V}}$ can be illustrated by the following example. The function sequence

$$f_n(t) := \begin{cases} 0, & \text{if } 0 \leqslant t \leqslant 1/2, \\ n(t - 1/2), & \text{if } 1/2 < t \leqslant 1/2 + 1/n, \\ 1, & \text{if } 1/2 + 1/n < t \leqslant 1, \end{cases}$$

converges to the function

$$f(t) := \begin{cases} 0, & \text{if } 0 \leqslant t \leqslant 1/2, \\ 1, & \text{if } 1/2 < t \leqslant 1, \end{cases}$$

in both the metrics ρ and $\rho_{\mathbb{V}}$. However, the sequence of functions f_n^* that have on the segment $[1/2, 1/2 + 1/n]$ not one but three switches between the levels 0 and 1 (two in the upward direction and one downward) will converge to f in the metric ρ only, whereas $\rho_{\mathbb{V}}(f_n^*, f) \geqslant 1$. Such a difference between the metrics may affect the approximations of the values

$$\frac{1}{n} \ln \mathbf{P}\big(\overline{s}_n \in (f)_{\mathbb{V},\varepsilon}\big) \quad \text{and} \quad \frac{1}{n} \ln \mathbf{P}\big(\overline{s}_n \in (f)_{\varepsilon}\big)$$

with the help of the l.l.d.p. Moreover, if we take f as $f = f_n^*$, defined above for $2/n < \varepsilon$ and small ε, this difference can be essential.

In the above example, the sequence f_n^* is in V_z for $z = 3$ (if the oscillations are 'rectangular'). This is a Cauchy sequence in the space $(\mathbb{V}, \rho_{\mathbb{V}})$, but it does not converge to anything in $(\mathbb{V}, \rho_{\mathbb{V}})$ as $n \to \infty$. This shows that the space $(\mathbb{V}, \rho_{\mathbb{V}})$ is incomplete, while the set V_z is not closed and cannot be compact. The last observation means that the approach of obtaining the e.l.d.p. from the l.l.d.p., mentioned in section 4.1, is inapplicable here. That is why we needed the second approach (see section 4.1) and uniform upper bounds in the l.l.d.p. for establishing the e.l.d.p.

Proofs of Theorems 4.7.3–4.7.5 can be found in [66].

Remark 4.7.6. Since the metrics $\rho_{\mathbb{V}}$ and $\rho_{\mathbb{D}}$ are equivalent when restricted to the class $\mathbb{C}_a$ and the set $\mathbb{C}_a$ is everywhere dense in the space $(\mathbb{C}, \rho_{\mathbb{D}})$, the assertions of Theorems 4.7.3–4.7.5 will remain true if we replace the space $(\mathbb{V}, \rho_{\mathbb{V}})$ by the space $(\mathbb{C}, \rho_{\mathbb{D}})$ in the formulations of these theorems.

4.8 Conditional large deviation principles in the space $(\mathbb{D}, \rho)$

4.8.1 Conditional l.d.p. in the space $(\mathbb{D}, \rho)$

Note that Theorem 4.6.2 and the formula for conditional probabilities imply the next assertion.

Corollary 4.8.1. *Let the condition* $[\mathbf{C}_0]$ *be met and* $x \sim n$ *as* $n \to \infty$*. Then for all measurable sets* B_1, B_2 *one has*

$$\overline{\lim_{n\to\infty}} \frac{1}{n} \ln \mathbf{P}\big(\overline{s}_n \in B_1 \,|\, \overline{s}_n \in B_2\big) \leqslant -J(B_1B_2+) + J\big((B_2)\big),$$

$$\underline{\lim_{n\to\infty}} \frac{1}{n} \ln \mathbf{P}\big(\overline{s}_n \in B_1 \,|\, \overline{s}_n \in B_2\big) \geqslant -J\big((B_1B_2)\big) + J(B_2+).$$

Using this assertion in the one-dimensional case with $\mathbf{E}\xi \leqslant 0$, one can easily find in an explicit form the value of

$$\lim_{\varepsilon\to 0}\lim_{n\to\infty} \frac{1}{n}\mathbf{P}\big(\max_{k\leqslant n} S_k > x \,|\, \min_{1\leqslant k\leqslant n} S_k > -\varepsilon x\big) = \Lambda(1) + \Lambda(0)$$

corresponding to the sets

$$B_1 = \big\{f \in \mathbb{D} : \max_{t\in[0,1]} f(t) > 1\big\},$$

$$B_2 = B_2(\varepsilon) = \big\{f \in \mathbb{D} : f(t) > -\varepsilon \;\text{ for } t \in (0,1]\big\}.$$

Corollary 4.8.1 cannot be used to find the logarithmic asymptotics of the probability $\mathbf{P}\big(\overline{s}_n \in B_1 \,|\, \overline{s}_n \in B_2(\varepsilon)\big)$ for $\varepsilon = 0$, since the open set $\big(B_2(0)\big)$ does not contain functions f such that $f(0) = 0$, and, therefore, $J\big((B_2(0))\big) = \infty$.

The rest of this subsection will be devoted to a more detailed examination of the case when condition $\{\overline{s}_n \in B_2\}$ consists of fixing the value of $\overline{s}_n(1)$ in some 'small zone'. In order to simplify the formulations and proofs, we will restrict ourselves to considering the one-dimensional case $d = 1$. We will consider analogues of the l.d.p. for the conditional probabilities

$$\mathbf{P}\big(\overline{s}_n \in B \,|\, \overline{s}_n(1) \in (b)_\varepsilon\big),$$

given that the endpoint of the trajectory $\overline{s}_n$ is localized in the ε-neighbourhood $(b-\varepsilon, b+\varepsilon)$ of the point b.

We will start with the l.l.d.p. In what follows, we will be assuming that b is such that $\Lambda(b) < \infty$, so that, by virtue of the l.l.d.p., the probability of this condition,

$$\mathbf{P}\big(\overline{s}_n(1) \in (b)_\varepsilon\big),$$

is positive for all $\varepsilon > 0$ and all large enough n.

The analogue of the l.l.d.p. has the following form.

Theorem 4.8.2. *Let $x/n \to 1$ as $n \to \infty$. Then, for any function $f \in \mathbb{D}$ such that $f(1) = b$, one has*

$$\lim_{\varepsilon\to 0}\overline{\lim_{n\to\infty}} \frac{1}{n} \ln \mathbf{P}\big(\overline{s}_n \in (f)_\varepsilon \,|\, \overline{s}_n(1) \in (b)_\varepsilon\big)$$

$$= \lim_{\varepsilon\to 0}\underline{\lim_{n\to\infty}} \frac{1}{n} \ln \mathbf{P}\big(\overline{s}_n \in (f)_\varepsilon \,|\, \overline{s}_n(1) \in (b)_\varepsilon\big) = -J(f) + \Lambda(b). \quad (4.8.1)$$

Relation (4.8.1) can be called the *conditional l.l.d.p. for a trajectory $\overline{s}_n$ with localized endpoint.*

Proof. Upper bound. For $f \in \mathbb{D}$ such that $f(1) = b$, one has

$$\mathbf{P}\big(\overline{s}_n \in (f)_\varepsilon \,|\, \overline{s}_n(1) \in (b)_\varepsilon\big) \leqslant \frac{\mathbf{P}\big(\overline{s}_n \in (f)_\varepsilon\big)}{\mathbf{P}\big(\overline{s}_n(1) \in (b)_\varepsilon\big)}.$$

Moreover, by virtue of the l.l.d.p. for $\overline{s}_n(1)$ as $x \sim n$, one has

$$\lim_{\varepsilon\to 0}\lim_{n\to\infty} \frac{1}{n} \ln \mathbf{P}\big(\overline{s}_n(1) \in (b)_\varepsilon\big) = -\Lambda(b). \tag{4.8.2}$$

Hence the desired upper bound follows directly from the l.l.d.p. for $\overline{s}_n$ (Theorem 4.6.1).

The lower bound can be obtained by basically repeating the argument for the lower bound in Theorem 4.6.1. Let f_k, $k = 1, 2, \ldots$, be a sequence of continuous polygonal lines with nodes at the points $\big(j/k, f(j/k)\big)$, $j = 0, 1, \ldots, k$. Then

$$\rho(\overline{s}_n, f) \leqslant \rho(\overline{s}_n, s_n) + \rho(s_n, f_k) + \rho(f_k, f).$$

Choose k_ε such that for all $k > k_\varepsilon$ one has the inequality $\rho(f_k, f) < \varepsilon/3$, and choose n_0 large enough that, for $n \geqslant n_0$ and for the set

$$A_\varepsilon := \left\{\rho(\overline{s}_n, s_n) < \frac{\varepsilon}{3}\right\}$$

and any function f such that $J(f) < \infty$, one would have, according to (4.5.12), the inequality

$$\mathbf{P}(\overline{A_\varepsilon}) \leqslant e^{-2nJ(f)}. \tag{4.8.3}$$

Then, given that $\rho_{\mathbb{C}}(s_n, f_k) < \varepsilon/3$, one would also have $\rho(s_n, f_k) < \varepsilon/3$, and the relations $\big\{\rho(\overline{s}_n, f) < \varepsilon,\ \overline{s}_n(1) \in (b)_\varepsilon\big\}$ will hold on the set A_ε. Therefore

$$\begin{aligned}\mathbf{P}\big(\rho(\overline{s}_n, f) < \varepsilon,\ \overline{s}_n(1) \in (b)_\varepsilon\big) &\geqslant \mathbf{P}\big(\rho(\overline{s}_n, f) < \varepsilon,\ \overline{s}_n(1) \in (b)_\varepsilon,\ A_\varepsilon\big)\\ &\geqslant \mathbf{P}\big(\rho_{\mathbb{C}}(s_n, f_k) < \varepsilon/3,\ A_\varepsilon\big)\\ &\geqslant \mathbf{P}\big(\rho_{\mathbb{C}}(s_n, f_k) < \varepsilon/3\big) - \mathbf{P}(\overline{A_\varepsilon}).\end{aligned} \tag{4.8.4}$$

From the upper bound in the l.l.d.p. for continuous trajectories s_n and from the relation (4.8.3) we obtain that

$$\underline{\lim}_{n\to\infty} \frac{1}{n} \ln \mathbf{P}(\rho\big(\overline{s}_n, f) < \varepsilon,\ \overline{s}_n(1) \in (b)_\varepsilon\big) \geqslant -J(f_k),$$

where the value $J(f_k)$ can be made arbitrarily close to that of $J(f)$ by choosing an appropriate $k \geqslant k_\varepsilon$. The theorem is proved. □

Remark 4.8.3. The argument employed in the above proof will be used below in the proof of Theorem 4.8.5. Otherwise, one could have given the following simpler and shorter proof for the lower bound.

Since the set $G_\varepsilon := \{g \in \mathbb{D} : g \in (f)_\varepsilon, \ g(1) \in (b)_\varepsilon\}$ is open in $(\mathbb{D}, \rho)$, and $f \in G_\varepsilon$, one has from Theorem 4.6.2 that

$$\varliminf_{n\to\infty} \frac{1}{n} \ln \mathbf{P}\big(\rho(\overline{s}_n, f) < \varepsilon, \ \overline{s}_n(1) \in (b)_\varepsilon\big) \geqslant -J(G_\varepsilon) \geqslant -J(f).$$

The theorem is proved.

In order to somewhat simplify the proofs and formulations in the assertions that appear below, we will introduce an assumption that holds true in all examples known to us that would be of any interest in applied problems. Let $\mathbb{D}[0,1)$ be the constriction of the function space to the half-open interval $[0,1)$, endowed with the uniform metric. For given $B \in \mathfrak{B}$, $b \in \mathbb{R}$ and $\varepsilon > 0$, denote by B_- the set of functions $f_- \in \mathbb{D}[0,1)$ such that, for some $a \in (b)_\varepsilon$, one has

$$f_- \times \{f(1) = a\} \in B.$$

Clearly, the set B_- depends on b and ε. For given b, the condition on the set B has the following form.

[B, b] *For all small enough* $\varepsilon > 0$ *the following inclusion holds true:*

$$B_- \times \{f(1) = b\} \subset B \cap E^{(b)} =: B^{(b)}, \tag{4.8.5}$$

where $E^{(b)} := \{f \in \Delta : f(1) = b\}$.

Since, for any function f from B such that $f(1) \in (b)_\varepsilon$, there exists a function f_- from B_- which coincides with f on the half-open interval $[0,1)$, the following inclusion always holds true:

$$B \cap \{f(1) \in (b)_\varepsilon\} \subset B_- \times \{f(1) \in (b)_\varepsilon\}. \tag{4.8.6}$$

Therefore, the following inequality is valid:

$$\mathbf{P}\Big(\overline{s}_n \in B \cap \{f(1) \in (b)_\varepsilon\}\Big) \leqslant \mathbf{P}\Big(\overline{s}_n \in B_- \times \{f(1) \in (b)_\varepsilon\}\Big). \tag{4.8.7}$$

The condition **[B, b]** is met, if, roughly speaking, the 'end point', i.e. the point $f(1)$, is not fixed by the fact that $f \in B$, so the set $B^{(b)}$ is not empty. Condition **[B, b]** is satisfied, for instance, for the sets

$$B := \Big\{f : \sup_{t\in[0,1]} (f(t) - g(t)) > 0\Big\},$$

where the function g has the properties: $\inf_{t\in[0,1]} g(t) > 0, \ \ g(1) > b$. It is satisfied, as well, for the sets

$$B := \Big\{f : \int_0^1 G(f(t))dt > v\Big\},$$

where G is an arbitrary measurable function.

In the former example,

$$B_- = \Big\{f_- \in \mathbb{D}[0,1) : \sup_{0\leqslant t<1} (f(t) - g(t)) > 0\Big\}.$$

So, for $\varepsilon < g(1) - b$, the relation (4.8.5) will clearly hold.

In the latter example,

$$B_- = \Big\{f_- \in \mathbb{D}[0,1) : \int_0^{1-0} G(f(t))dt > v\Big\}.$$

Since, for any $f(1)$, one has

$$\int_0^{1-0} G(f(t))dt = \int_0^1 G(f(t))dt,$$

it is also clear that relation (4.8.5) holds true in that case as well.

However, the condition **[B, b]** is not always satisfied, as illustrated by the following example. Let $B = B_1 \cup B_2$, where

$$\begin{aligned} B_1 &= \{f \in \mathbb{D} : \rho_{\mathbb{C}}(f, 0) < \delta, f(1) \neq 0\}, \\ B_2 &= \{f \in \mathbb{D} : \rho_{\mathbb{C}}(f, g) < \delta, f(1) = 0\} \end{aligned} \tag{4.8.8}$$

for a given function $g \not\equiv 0$, $g(1) = 0$. Here

$$B_- = \big\{f \in \mathbb{D} : \rho_{\mathbb{C}}(f, 0) < \delta\big\} \cup \big\{f \in \mathbb{D} : \rho_{\mathbb{C}}(f, g) < \delta\big\}.$$

Therefore, for small enough $\delta > 0$, the set $B_- \times \{f(1) = 0\}$ is not a subset of $B \cap E^{(0)} = B_2$. Hence condition **[B, b]** with $b = 0$ is not satisfied.

Now we will state an analogue of the e.l.d.p.

Theorem 4.8.4. *Let $x/n \to 1$ as $n \to \infty$, B be a measurable set and the set $B^{(b)}$ be defined as in* (4.8.5).

(i) *If the conditions* $[\mathbf{C}_0]$, **[B, b]** *are met then*

$$\lim_{\varepsilon \to 0} \overline{\lim_{n \to \infty}} \frac{1}{n} \ln \mathbf{P}\big(\overline{s}_n \in B |\, \overline{s}_n(1) \in (b)_\varepsilon\big) \leqslant -J(B^{(b)}+) + \Lambda(b). \tag{4.8.9}$$

(ii) *One always has*

$$\lim_{\varepsilon \to 0} \underline{\lim_{n \to \infty}} \frac{1}{n} \ln \mathbf{P}\big(\overline{s}_n \in B |\, \overline{s}_n(1) \in (b)_\varepsilon\big) \geqslant -J((B)^{(b)}) + \Lambda(b).$$

(iii) *If the conditions* $[\mathbf{C}_0]$, **[B, b]** *are met then, for a set B that satisfies the 'continuity condition'*

$$J(B^{(b)}+) = J((B)^{(b)}) = J(B^{(b)}),$$

one has the relation

$$\lim_{\varepsilon \to 0} \underline{\lim_{n \to \infty}} \frac{1}{n} \ln \mathbf{P}(\overline{s}_n \in B |\, \overline{s}_n(1) \in (b)_\varepsilon) = -J(B^{(b)}) + \Lambda(b).$$

Observe that if $\mathbf{E}\xi = b = 0$ then, in the above example (4.8.8), the left-hand side of inequality (4.8.9) equals zero by the strong law of large numbers. Since $B^{(0)} = B_2$ in that example, the right-hand side of (4.8.9) is equal to $-J(B_2+) < 0$ for all small enough $\delta > 0$. That means that the assertion (4.8.9) fails in that example, so that condition **[B, b]** is essential for the assertion of Theorem 4.8.4 to hold true.

If condition **[B, b]** is not satisfied, one may be able to obtain other, more cumbersome upper bounds for the left-hand side of (4.8.9).

Proof of Theorem 4.8.4. The upper bound. If the first part of condition **[B, b]** is met then

$$\mathbf{P}\Big(\overline{s}_n \in B \cap \big\{f(1) \in (b)_\varepsilon\big\}\Big) \leqslant \mathbf{P}\Big(\overline{s}_n \in B_- \times \big\{f(1) \in (b)_\varepsilon\big\}\Big).$$

Since the sets B_- and $\{f(1) \in (b)_\varepsilon\}$ are specified by independent coordinates, one has the following relations:

$$B_- \times \big\{f(1) \in (b)_\varepsilon\big\} \subset (B_-)_{\mathbb{C},\varepsilon} \times \big\{f(1) \in (b)_\varepsilon\big\} = \Big(B_- \times \big\{f(1) = b\big\}\Big)_{\mathbb{C},\varepsilon}.$$

By virtue of (4.8.5), the right-hand side of that relation is a subset of $(B^{(b)})_{\mathbb{C},\varepsilon} \subset (B^{(b)})_\varepsilon$. Hence

$$\mathbf{P}\Big(\overline{s}_n \in B \cap \big\{f(1) \in (b)_\varepsilon\big\}\Big) \leqslant \mathbf{P}\big(\overline{s}_n \in (B^{(b)})_\varepsilon\big).$$

All the above considerations for $\mathbf{P}(\overline{s}_n \in B)$ are applicable to the right-hand side of this inequality, so that, by virtue of Theorem 4.6.2 (giving the upper bounds in the e.l.d.p.), one has

$$\overline{\lim_{n\to\infty}} \frac{1}{n} \ln \mathbf{P}\big(\overline{s}_n \in (B^{(b)})_\varepsilon\big) \leqslant -J((B^{(b)})_\varepsilon +) \leqslant -J\big((B^{(b)})_{2\varepsilon}\big),$$

$$\lim_{\varepsilon\to 0} \overline{\lim_{n\to\infty}} \frac{1}{n} \ln \mathbf{P}\big(\overline{s}_n \in (B^{(b)})_\varepsilon\big) \leqslant -J(B^{(b)}+).$$

Together with (4.8.2), this implies (4.8.9).

The lower bound. The set $(B) \cap \{g \in \mathbb{D} :\ g(1) \in (b)_\varepsilon\}$ is open in $(\mathbb{D}, \rho)$ and is a subset of $B \cap \{g \in \mathbb{D} :\ g(1) \in (b)_\varepsilon\}$. Hence, by Theorem 4.6.2,

$$\begin{aligned}
&\underline{\lim}_{n\to\infty} \frac{1}{n} \ln \mathbf{P}\big(\overline{s}_n \in B,\ \overline{s}_n(1) \in (b)_\varepsilon\big) \\
&\quad \geqslant \underline{\lim}_{n\to\infty} \frac{1}{n} \ln \mathbf{P}\big(\overline{s}_n \in (B),\ \overline{s}_n(1) \in (b)_\varepsilon\big) \\
&\quad \geqslant -J\Big((B) \cap \big\{g \in \mathbb{D} :\ g(1) \in (b)_\varepsilon\big\}\Big) \geqslant -J\big((B)^{(b)}\big).
\end{aligned}$$

The theorem is proved. □

4.8.2 Conditional l.d.p.s with the trajectory end point localized in a narrower zone

Consider now processes $\overline{s}_n = \overline{s}_n(t)$ of a more special form, requiring in addition that $x = n$ and

$$\overline{s}_n\left(\frac{k}{n}\right) = \frac{S_k}{n}, \qquad k = 1, \ldots, n. \tag{4.8.10}$$

For example, the step functions

$$\overline{s}_n(t) := \frac{1}{n} S_{[nt]}, \quad 0 \leqslant t \leqslant 1,$$

or the continuous trajectories $s_n(t)$ considered in section 4.4 will be processes of that kind. In a number of problems (say, when studying the empirical distribution functions), it is of interest to obtain conditional l.d.p.s under more precise localisation of the trajectory end point, for instance, for probabilities of the form

$$\mathbf{P}\big(\overline{s}_n \in B \,|\, \overline{s}_n(1) \in (b)_{\Delta/n}\big),$$

where $\Delta > 0$ is fixed. For a more precise problem formulation, we need to distinguish here between the non-lattice and arithmetic cases.

In the *non-lattice* case we will understand by Δ any value from the interval (Δ_1, Δ_2), where $\Delta_1 = \Delta_{1,n} \to 0$ slowly enough as $n \to \infty$, $\Delta_2 = \Delta_{2,n} = o(n)$.

In the *arithmetic* case, we will be dealing with the probability

$$\mathbf{P}\big(\overline{s}_n \in B \,|\, \overline{s}_n(1) = b\big) = \mathbf{P}\big(\overline{s}_n \in B \,|\, S_n = bn\big),$$

where bn is an integer.

If $\Delta > 1$ then the answers in both the non-lattice and arithmetic cases will have the same form.

First consider the non-lattice case. One has the following conditional analogues of the l.d.p. with a narrow localisation of the trajectory end point.

Theorem 4.8.5. *Assume that the random variable ξ is non-lattice, $x = n$ and condition* (4.8.10) *is met. Then, for any function $f \in \mathbb{D}$, $f(1) = b$, one has*

$$\begin{aligned}
&\lim_{\varepsilon \to 0} \overline{\lim_{n \to \infty}} \frac{1}{n} \ln \mathbf{P}\big(\overline{s}_n \in (f)_\varepsilon \,|\, \overline{s}_n(1) \in (b)_{\Delta/n}\big) \\
&\qquad = \lim_{\varepsilon \to 0} \underline{\lim_{n \to \infty}} \frac{1}{n} \ln \mathbf{P}(\overline{s}_n \in (f)_\varepsilon \,|\, \overline{s}_n(1) \in (b)_{\Delta/n}) = -J(f) + \Lambda(b).
\end{aligned}$$

The conditional analogue of the e.l.d.p. with the trajectory end point localised inside $(b - \Delta/n, b + \Delta/n)$ has the following form.

Theorem 4.8.6. *Let the random variable ξ be non-lattice and let $x = n$, the condition* (4.8.10) *be met and B be an arbitrary measurable set.*

(i) *If conditions* $[\mathbf{C}_0]$, $[\mathbf{B}, \mathbf{b}]$ *are met then*

$$\overline{\lim_{n \to \infty}} \frac{1}{n} \ln \mathbf{P}(\overline{s}_n \in B \,|\, \overline{s}_n(1) \in (b)_{\Delta/n}) \leqslant -J(B^{(b)}+) + \Lambda(b).$$

(ii) *One always has the inequality*

$$\underline{\lim_{n \to \infty}} \frac{1}{n} \ln \mathbf{P}\big(\overline{s}_n \in B \,|\, \overline{s}_n(1) \in (b)_{\Delta/n}\big) \geqslant -J\big((B)^{(b)}\big) + \Lambda(b).$$

(iii) *If conditions* **[C₀]**, **[B, b]** *are satisfied then, for a set* $B^{(b)}$ *satisfying the continuity condition*

$$J\big(B^{(b)}+\big) = J\big((B)^{(b)}\big) = J\big(B^{(b)}\big),$$

one has

$$\lim_{n\to\infty} \frac{1}{n} \ln \mathbf{P}\big(\overline{s}_n \in B |\, \overline{s}_n(1) \in (b)_{\Delta/n}\big) = -J\big(B^{(b)}\big) + \Lambda(b).$$

Proof of Theorem 4.8.5. The upper bound. To prove Theorems 4.8.5 and 4.8.6, we will need the following complement to the l.l.d.p. for S_n, which follows from Theorem 2.8.13.

Let ξ be a non-lattice random variable and $\Delta \in (\Delta_1, \Delta_2)$. Then

$$\lim_{n\to\infty} \frac{1}{n} \ln \mathbf{P}\big(\overline{s}_n(1) \in (b)_{\Delta/n}\big) = -\Lambda(b). \tag{4.8.11}$$

Since

$$\mathbf{P}\big(\overline{s}_n \in (f)_\varepsilon,\ \overline{s}_n(1) \in (b)_{\Delta/n}\big) \leqslant \mathbf{P}\big(\overline{s}_n \in (f)_\varepsilon\big),$$

the desired upper bound follows from (4.8.11) and also the l.l.d.p. for $\overline{s}_n$.

The lower bound for $\mathbf{P}\big(\overline{s}_n \in (f)_\varepsilon,\ \overline{s}_n(1) \in (b)_{\Delta/n}\big)$ is obtained by mostly repeating the argument used to prove the lower bound in Theorem 4.8.2. This time, however, instead of (4.8.4) we will have the inequality

$$\begin{aligned}\mathbf{P}\big(\rho(\overline{s}_n, f) < \varepsilon,\ s_n(1) \in (b)_{\Delta/n}\big)&\\ \geqslant \mathbf{P}\big(\rho(s_n, f) < \varepsilon/3,\ s_n(1) \in (b)_{\Delta/n}\big) - \mathbf{P}(\overline{A}_\varepsilon).&\end{aligned}$$

Here, to bound the first probability on the right-hand side, one should just repeat the argument in the proof of Theorem 4.4.2, the only difference being that now one has to bound the probabilities for the increments $S_{kn\theta} - S_{(k-1)n\theta}$ (on the partition intervals $\{k\theta,\ k \leqslant 1/\theta\}$) to be in the respective intervals of the width $2\theta\Delta$. To do that, one should use (4.8.11). The theorem is proved. □

Proof of Theorem 4.8.6. The upper bound. Since, for any $\varepsilon > 0$ and large enough n, one has

$$\mathbf{P}\big(\overline{s}_n \in B,\ \overline{s}_n(1) \in (b)_{\Delta/n}\big) \leqslant \mathbf{P}\big(\overline{s}_n \in B,\ \overline{s}_n(1) \in (b)_\varepsilon\big),$$

we see that the proof of the upper bound will just repeat the argument establishing the desired bound in Theorem 4.8.4.

The lower bound. For any function $f \in (B), f(1) = b$, and all small enough $\varepsilon > 0$ we have

$$\mathbf{P}\big(\overline{s}_n \in B,\ \overline{s}_n(1) \in (b)_{\Delta/n}\big) \geqslant \mathbf{P}\big(\overline{s}_n \in (f)_\varepsilon,\ \overline{s}_n(1) \in (b)_{\Delta/n}\big).$$

Using the lower bound from Theorem 4.8.5, one obtains

$$\varliminf_{n\to\infty} \frac{1}{n} \ln \mathbf{P}\big(\overline{s}_n \in B,\ \overline{s}_n(1) \in (b)_{\Delta/n}\big) \geqslant -J(f) + \Lambda(b).$$

Repeating the argument from the proof of Theorem 4.1.6 that enabled us to obtain the lower bound in the e.l.d.p. from that in the l.l.d.p., we conclude that

$$\varliminf_{n\to\infty} \frac{1}{n} \ln \mathbf{P}\big(\overline{s}_n \in B,\ \overline{s}_n(1) \in (b)_{\Delta/n}\big) \geqslant -J\big((B)^{(b)}\big) + \Lambda(b).$$

The theorem is proved. □

Now consider the *arithmetic* case. The following analogues of the l.d.p. for a fixed trajectory end point hold true.

Theorem 4.8.7. *Let a random variable* ξ *have an arithmetic distribution,* $x = n$ *and the condition* (4.8.10) *be met. Then, for any function* $f \in \mathbb{D}$ *such that* $f(1) = b$, *one has*

$$\lim_{\varepsilon\to 0}\varlimsup_{n\to\infty} \frac{1}{n} \ln \mathbf{P}\left(\overline{s}_n \in (f)_\varepsilon \,\middle|\, \overline{s}_n(1) = \frac{[bn]}{n}\right)$$
$$= \lim_{\varepsilon\to 0}\varliminf_{n\to\infty} \frac{1}{n} \ln \mathbf{P}\left(\overline{s}_n \in (f)_\varepsilon \,\middle|\, \overline{s}_n(1) = \frac{[bn]}{n}\right) = -J(f) + \Lambda(b).$$

The conditional analogue of the e.l.d.p. with fixed trajectory end point takes the following form.

Theorem 4.8.8. *Let a random variable* ξ *have an arithmetic distribution,* $x = n$ *and the condition* (4.8.10) *be met, with* B *be a measurable set.*

(i) *If conditions* $[\mathbf{C_0}]$, $[\mathbf{B}, \mathbf{b}]$ *are satisfied then*

$$\varlimsup_{n\to\infty} \frac{1}{n} \ln \mathbf{P}\left(\overline{s}_n \in B \,\middle|\, \overline{s}_n(1) = \frac{[bn]}{n}\right) \leqslant -J(B^{(b)}+) + \Lambda(b).$$

(ii) *One always has the inequality*

$$\varliminf_{n\to\infty} \frac{1}{n} \ln \mathbf{P}\left(\overline{s}_n \in B \,\middle|\, \overline{s}_n(1) = \frac{[bn]}{n}\right) \geqslant -J\big((B)^{(b)}\big) + \Lambda(b).$$

(iii) *If conditions* $[\mathbf{C_0}]$, $[\mathbf{B}, \mathbf{b}]$ *are met then, for a set* $B^{(b)}$ *satisfying the continuity condition*

$$J(B^{(b)}+) = J\big((B)^{(b)}\big) = J(B^{(b)}),$$

one has

$$\lim_{n\to\infty} \frac{1}{n} \ln \mathbf{P}\left(\overline{s}_n \in B \,\middle|\, \overline{s}_n(1) = \frac{[bn]}{n}\right) = -J\big(B^{(b)}\big) + \Lambda(b).$$

Proofs of Theorems 4.8.7 and 4.8.8. These proofs use arguments similar to those used above to demonstrate Theorems 4.8.5 and 4.8.6. One just has to and in corresponding places the next assertion, which follows from Theorem 2.8.13, instead of relation (4.8.11):

Let ξ be an arithmetic random variable. Then

$$\lim_{n\to\infty} \frac{1}{n} \ln \mathbf{P}\left(\overline{s}_n(1) = \frac{[nb]}{n}\right) = -\Lambda(b).$$ □

4.9 Extension of results to processes with independent increments

Let $S(t)$, $t \geqslant 0$, be a homogeneous process with independent increments, i.e. a process whose characteristic function is

$$\mathbf{E}e^{ivS(t)} = e^{tr(v)},$$

where, according to the Lévy–Khintchine representation,

$$r(v) = \beta(v; q, \sigma, \mathcal{B}) := iqv - \frac{v^2\sigma^2}{2} + \int_{-\infty}^{\infty} \left(e^{ivx} - 1 - \frac{ivx}{1+x^2} \right) \frac{1+x^2}{x^2}\, d\mathcal{B}(x)$$

and $\mathcal{B} = \mathcal{B}(x)$ is a non-decreasing function of bounded variation which is continuous at $x = 0$.

We will need Cramér's moment condition **[C]** for an distribution of the random variable $S(1)$:

[C]$\psi(\lambda) := \mathbf{E}e^{\lambda S(1)} < \infty$ *for some real* $\lambda \neq 0$.

By analogy with previous considerations, denote

$$\lambda_+ := \sup\{\lambda : \psi(\lambda) < \infty\}, \qquad \lambda_- := \inf\{\lambda : \psi(\lambda) < \infty\},$$

$$b_0(\lambda) := \int_{-1}^{1} \left(e^{\lambda x} - 1 - \frac{\lambda x}{1+x^2} \right) \frac{1+x^2}{x^2}\, d\mathcal{B}(x),$$

$$b_1(\lambda) := \int_{|x|>1} \left(e^{\lambda x} - 1 - \frac{\lambda x}{1+x^2} \right) \frac{1+x^2}{x^2}\, d\mathcal{B}(x),$$

so that $b(\lambda) := \ln \psi(\lambda) = b_0(\lambda) + b_1(\lambda)$. It is easy to see that $b_0(\lambda) < \infty$ for all $\lambda \in \mathbb{R}$,

$$\sup\{\lambda : b_1(\lambda) < \infty\} = \lambda_+, \quad \inf\{\lambda : b_1(\lambda) < \infty\} = \lambda_-$$

and that **[C]** is equivalent to the inequality

$$\lambda_+ - \lambda_- > 0.$$

If $S(t)$ has a finite number of jumps on $[0, T]$ with probability 1, i.e., function $r(v)$ can be represented as

$$r(v) = iqv - \frac{v^2\sigma^2}{2} + \mu \int_{-\infty}^{\infty} (e^{ivx} - 1)\, dF_\zeta(x), \tag{4.9.1}$$

where F_ζ is a cumulative distribution function of jumps $\zeta_1, \zeta_2, \ldots$ of a compound Poisson process with intensity $\mu > 0$, then it is also possible to obtain for $\psi_\zeta(\lambda) := \mathbf{E}e^{\lambda\zeta}$, $\zeta \underset{d}{=} \zeta_1$, the result

$$\sup\{\lambda : \psi_\zeta(\lambda) < \infty\} = \lambda_+, \quad \inf\{\lambda : \psi_\zeta(\lambda) < \infty\} = \lambda_-,$$

so that Cramér's conditions **[C]** for $S(1)$ and ζ are equivalent.

If the following condition

[C$_\infty$] *$\psi_\zeta(\lambda) < \infty$ is met for all $\lambda \in \mathbb{R}$*
holds then in [19] the l.d.p. could be established for trajectories of processes

$$s_T = s_T(t) := \frac{S(tT)}{T}, \qquad t \in [0, 1],$$

constructed for a process $S(t)$ of the form (4.1.1) with $\sigma = q = 0, T \to \infty$ (below in Theorem 4.9.5 this assertion will be extended to a more general case).

In [124] processes $S(t)$ with trajectories from the space $\mathbb{V}[0, T]$ of functions on segment $[0, T]$ with bounded variations were considered, i.e. processes for which the function $r(v)$ can be expressed in the form (compare with (4.9.1))

$$r(v) = iqv - \int_{-\infty}^{\infty} (e^{ivx} - 1)\nu(dx),$$

where ν is an arbitrary measure (not necessarily finite) such that $\int |x|\nu(dx) < \infty$. When Cramér's condition

[C$_0$] *$\psi(\lambda) < \infty$ for some neighbourhood of the point* $\lambda = 0$,
is met, the l.d.p. is established (Theorem 5.1 in [124]) for a sequence $\{s_T\}$ in the space $\mathbb{V}[0, 1]$ equipped with the following weak convergence topology: $f_n \to f$, if $\int g(t)df_n(t) \to \int g(t)df(t)$ for any function $g = g(t)$ that is continuous on the segment $[0, 1]$. It should be noted that the maximum of the trajectory $\overline{f} = \sup_{0 \leqslant t \leqslant 1} f(t)$ (which is an important functional in bounded problems and applications) has discontinuities in this topology, so that Theorem 5.1 from [124] cannot be used to estimate $\ln \mathbf{P}(\overline{s}_T \geqslant v)$. In order to find asymptotics for $(1/T) \ln \mathbf{P}(\overline{s}_T \geqslant v)$ as $T \to \infty$ with the l.d.p., a stronger topology should be used (for instance, a topology induced by the uniform metric, or by the metric ρ defined in section 4.2, or by the Skorokhod metric).

We should also mention [85], where under assumptions similar to those considered in [124], the processes

$$\{s_T(t) := \frac{1}{T} S(tT); \ \ 0 \leqslant t < \infty\}$$

on an infinite time interval are studied. For this family, the l.d.p. is also established for a very weak (and hard-to-define) topology.

The purposes of the present section are: (a) to obtain inequalities similar to the inequalities in section 4.3; (b) to prove the local large deviation principle for trajectories $s_T(t)$, $0 \leqslant t \leqslant 1$, without any moment conditions; (c) to prove under the assumption of condition **[C$_\infty$]** the 'usual' l.d.p. in the space $\mathbb{D}$ (of functions without discontinuities of the second kind) with uniform metric $\rho_\mathbb{C}$.

4.9.1 Inequalities

In what follows, the notation for the spaces $\mathbb{C}$, $\mathbb{C}_a$, $\mathbb{D}$, the metrics $\rho_\mathbb{C}$, $\rho_\mathbb{D}$, ρ and the integrals $I(f)$, $J(f)$ used earlier remains valid; moreover,

$$I(B) = \inf_{f \in B} I(f), \qquad J(B) = \inf_{f \in B} J(f).$$

Theorem 4.9.1. *For any convex set B that is open in $(\mathbb{D}, \rho_{\mathbb{D}})$,*

$$\mathbf{P}(s_T \in B) \leqslant e^{-TJ(B\mathbb{C}_a)} \leqslant e^{-TJ(B)}. \qquad (4.9.2)$$

Let $f \in \mathbb{C}$ and $(f)_{\mathbb{C},\varepsilon}$ be a set of functions $g \in \mathbb{D}$ such that $\rho_{\mathbb{C}}(f,g) < \varepsilon$ (it is an ε-neighbourhood in $\mathbb{D}$ of the function $f \in \mathbb{C}$ with respect to the uniform metric). It is not difficult to see that if $g \in (f)_{\mathbb{C},\varepsilon}$ for a given continuous function f, and $\varepsilon > 0$, then:

(a) there exists a δ such that $(g)_{\mathbb{C},\delta} \subset (f)_{\mathbb{C},\varepsilon}$;
(b) there exist a non-decreasing function $q \in \mathcal{Q}$ and a real δ such that

$$\rho_{\mathbb{C}}(q, \mathrm{e}) = \sup_{t\in[0,1]} |q(t) - \mathrm{e}(t)| < \delta, \qquad g\big(q(t)\big) \in (f)_{\mathbb{C},\varepsilon}.$$

In other words, rather small 'up-and-down and sideways shifts' of the function $g \in (f)_{\mathbb{C},\varepsilon}$ preserve g in $(f)_{\mathbb{C},\varepsilon}$. But this means that the set $(f)_{\mathbb{C},\varepsilon}$, being open with respect to $\rho_{\mathbb{C}}$, is also open with respect to the metric $\rho_{\mathbb{D}}$. Since the set $(f)_{\mathbb{C},\varepsilon}$ is in addition convex, this implies, according to Theorem 4.9.1, the following:

Corollary 4.9.2. *For any function $f \in \mathbb{C}$ and arbitrary $\varepsilon > 0$,*

$$\mathbf{P}(s_T \in (f)_{\mathbb{C},\varepsilon}) \leqslant e^{-TJ((f)_{\mathbb{C},\varepsilon})}.$$

The inequalities below would be useful as well for a random variable

$$\overline{S}(T) = \sup_{0\leqslant t\leqslant T} S(t)$$

(analogous to that in Theorem 1.1.1).

Theorem 4.9.3. (i) *For all $T > 0$, $x \geqslant 0$, $\lambda \geqslant 0$,*

$$\mathbf{P}(\overline{S}(T) \geqslant x) \leqslant e^{-\lambda x} \max\left\{1, \psi^T(\lambda)\right\}. \qquad (4.9.3)$$

(ii) *Let $\mathbf{E}S(1) < 0$, $\lambda_1 := \max\{\lambda : \psi(\lambda) \leqslant 1\}$. Then, for all $T > 0$, $x \geqslant 0$,*

$$\mathbf{P}(\overline{S}(T) \geqslant x) \leqslant e^{-\lambda_1 x}. \qquad (4.9.4)$$

If $\lambda_+ > \lambda_1$ then $\psi(\lambda_1) = 1$, $\Lambda(\alpha) \geqslant \lambda_1\alpha$ for all α, $\Lambda(\alpha_1) = \lambda_1\alpha_1$, where

$$\alpha_1 := \arg\left\{\lambda(\alpha) = \lambda_1\right\} = \frac{\psi'(\lambda_1)}{\psi(\lambda_1)}, \qquad (4.9.5)$$

so that a line $y = \lambda_1\alpha$ is tangent at the point $(\alpha_1, \lambda_1\alpha_1)$ to the convex function $y = \Lambda(\alpha)$. Along with (4.9.4) for $\alpha := x/T$ the next inequality holds:

$$\mathbf{P}(\overline{S}(T) \geqslant x) \leqslant e^{-T\Lambda_1(\alpha)}, \qquad (4.9.6)$$

where

$$\Lambda_1(\alpha) := \begin{cases} \lambda_1\alpha, & \text{for } \alpha \leqslant \alpha_1, \\ \Lambda(\alpha), & \text{for } \alpha > \alpha_1. \end{cases}$$

If $\alpha \leqslant \alpha_1$ *then inequality* (4.9.6) *coincides with* (4.9.4); *in the case* $\alpha > \alpha_1$ *it is stronger than* (4.9.4).

(iii) *Suppose that* $\mathbf{E}S(1) \geqslant 0$ *and* $\alpha = x/T \geqslant \mathbf{E}S(1)$. *Then for all* $T > 0$ *one has*

$$\mathbf{P}\big(\overline{S}(T) \geqslant x\big) \leqslant e^{-T\Lambda(\alpha)}. \tag{4.9.7}$$

Theorems 4.9.1 and 4.9.3 extend inequalities established for random walks generated by sums of random variables, to random processes with independent increments.

Theorem 4.9.3 distinguishes three disjoint possibilities:

(a) $\mathbf{E}S(1) < 0$, $\lambda_+ = \lambda_1$,
(b) $\mathbf{E}S(1) < 0$, $\lambda_+ > \lambda_1$;
(c) $\mathbf{E}S(1) \geqslant 0$,

where $\mathbf{P}\big(\overline{S}(T) \geqslant x\big)$ is bounded by the right-hand sides of inequalities (4.9.4), (4.9.6) and (4.9.7) respectively. If, however, some natural conventions are accepted then all three proposed inequalities might be written in the unified form (4.9.6). Indeed, let us turn to definition (4.9.5) of α_1. As noted above, $\lambda(\alpha)$ is a solution of the equation $\psi'(\lambda)/\psi(\lambda) = \alpha$, which has a unique solution when

$$\alpha \in [\alpha_-, \alpha_+], \quad \alpha_+ := \lim_{\lambda \uparrow \lambda_+} \frac{\psi'(\lambda)}{\psi(\lambda)}, \quad \alpha_- := \lim_{\lambda \downarrow \lambda_-} \frac{\psi'(\lambda)}{\psi(\lambda)}.$$

When $\alpha \geqslant \alpha_+$, the function $\lambda(\alpha)$ is defined as a constant λ_+. This means that if $\lambda_1 = \lambda_+$ then α_1 is not uniquely defined and could take an arbitrary value from α_+ to ∞, so that by setting $\alpha_1 = \max\{\alpha : \lambda(\alpha) = \lambda_1 = \lambda_+\} = \infty$, we turn inequality (4.9.6) in the case $\lambda_1 = \lambda_+$ (i.e. in case (a)) into the inequality

$$\mathbf{P}\big(\overline{S}(T) \geqslant x\big) \leqslant e^{-T\lambda_1\alpha} = e^{-\lambda_1 x},$$

i.e. into inequality (4.9.4).

If $\mathbf{E}S(1) \geqslant 0$ then $\lambda_1 = 0$. If $\lambda_+ = 0$ then $\lambda_+ = \lambda_1$, and we have the same situation as before but now $\mathbf{P}\big(\overline{S}(T) \geqslant x\big)$ allows only the trivial bound 1. If $\lambda_+ > 0$ then $\alpha_1 = \mathbf{E}S(1)$; if $\alpha \leqslant \alpha_1$ then the bound (4.9.6) is again trivial, and if $\alpha > \alpha_1$, then it coincides with (4.9.7).

Corollary 4.9.4. *If we set*

$$\alpha_1 := \max\{\alpha : \lambda(\alpha) = \lambda_1\} = \begin{cases} \dfrac{\psi'(\lambda_1)}{\psi(\lambda_1)}, & \text{if } \lambda_+ > \lambda_1, \\ \infty, & \text{if } \lambda_+ = \lambda_1, \end{cases}$$

then the inequality (4.9.6) *is correct without any conditions on* $\mathbf{E}S(1)$ *and* λ_+, *and involves inequalities* (4.9.4) *and* (4.9.7).

It is easy to deduce from the large deviation principle that the exponents in inequality (4.9.6) are asymptotically unimprovable:

$$\lim_{T\to\infty} \frac{1}{T} \ln \mathbf{P}\big(\overline{S}(T) \geqslant x\big) = -\Lambda_1(\alpha) \quad \text{with} \quad \frac{x}{T} = \alpha.$$

Proof of Theorem 4.9.3. (i) The random variable

$$\eta(x) := \inf\{t > 0 : S(t) \geqslant x\}$$

is a stopping time. Therefore, the event $\{\eta(t) \in dt\}$ and the random variable $S(T)-S(t)$ are independent, and

$$\begin{aligned}\psi^T(\lambda) = \mathbf{E}e^{\lambda S(T)} &\geqslant \int_0^T \mathbf{E}\big(e^{\lambda S(T)};\ \eta(x) \in dt\big) \\ &\geqslant \int_0^T \mathbf{E}\big(e^{\lambda(x+S(T)-S(t))};\ \eta(x) \in dt\big) \\ &= e^{\lambda x} \int_0^T \psi^{T-t}(\lambda)\mathbf{P}\big(\eta(x) \in dt\big) \geqslant e^{\lambda x} \min\{1, \psi^T(\lambda)\}\mathbf{P}\big(\eta(x) \leqslant T\big).\end{aligned}$$

Hence we obtain (i).

(ii) Inequality (4.9.4) immediately follows from (4.9.3), assuming $\lambda = \lambda_1$. Now let $\lambda_+ > \lambda_1$. Then, obviously, $\psi(\lambda_1) = 1$, and it follows from the definition of the function $\Lambda(\alpha)$ that

$$\Lambda(\alpha) \geqslant \lambda_1\alpha - \ln\psi(\lambda(\alpha_1)) = \lambda_1\alpha.$$

Furthermore,

$$\Lambda(\alpha_1) = \lambda_1\alpha_1 - \ln\psi(\lambda(\alpha_1)) = \lambda_1\alpha,$$

so that the curves $y = \lambda_1\alpha$ and $y = \Lambda(\alpha)$ are tangent to each other at the point $(\alpha_1, \lambda_1\alpha_1)$.

Next, it is clear that $\psi(\lambda(\alpha)) \geqslant 1$ if $\alpha \geqslant \alpha_1$. For such an $\alpha = x/T$, an optimal choice for λ in (4.9.3) is $\lambda = \lambda(\alpha)$. With such a $\lambda(\alpha)$, $\alpha = x/T$, we obtain

$$\mathbf{P}\big(\overline{S}(T) \geqslant x\big) \leqslant e^{-T\Lambda(\alpha)}.$$

Together with (4.9.4) this proves (4.9.6). It is also clear that $\Lambda(\alpha) > \lambda_1\alpha$ if $\alpha > \lambda_1$, which proves the last statement of (ii).

(iii) Since $\lambda\big(\mathbf{E}S(1)\big) = 0$ and $\lambda(\alpha)$ is non-decreasing, $\lambda(\alpha) \geqslant 0$ if $\alpha \geqslant \mathbf{E}S(1)$. In the case $\mathbf{E}S(1) \geqslant 0$, the inequality $\psi(\lambda) \geqslant 1$ holds for $\lambda \geqslant 0$. Therefore, $\psi(\lambda(\alpha)) \geqslant 1$ if $\alpha \geqslant \mathbf{E}S(1)$. By substituting into (4.9.3) $\lambda = \lambda(\alpha)$ if $\alpha \geqslant \mathbf{E}S(1)$, we obtain (4.9.7). The theorem is proved. □

Proof of Theorem 4.9.1. The proof will be split into several stages.

(i) Construct a polygonal line $s_{T,n} = s_{T,n}(t)$, $t \in [0,1]$, with nodal points $(k/n, s_T(k/n))$, $k = 0, \ldots, n$, and note that the distributions of the processes $s_{T,n}$ weakly converge as $n \to \infty$ to the distribution of the process s_T in the space $\mathbb{D}$ of functions on $[0,1]$ without discontinuities of the second kind, equipped with the metric $\rho_{\mathbb{D}}$. To prove this statement one can use the following criterion of weak convergence in $(\mathbb{D}, \rho_{\mathbb{D}})$ (see e.g. [94], §§1–3).

The following conditions are necessary and sufficient for the distributions of processes Z_n to weakly converge to the distribution of process Z:

(1) *There exists a set $\mathcal{S}$ that is countable and everywhere dense in* $[0, 1]$, *which is such that finite-dimensional distributions of* $\{Z_n(t) : t \in \mathcal{S}\}$ *weakly converge to finite-dimensional distributions of* $\{Z(t) : t \in \mathcal{S}\}$.

(2) *For any* $\varepsilon > 0$,

$$\lim_{\Delta\to 0}\varlimsup_{n\to\infty} \mathbf{P}\big(\omega_\Delta^{\mathbb{D}}(Z_n) > \varepsilon\big) = 0, \tag{4.9.8}$$

where

$$\omega_\Delta^{\mathbb{D}}(f) := \sup_{t\in[0,1]} \min\{\omega^+(t,\Delta),\ \omega^-(t,\Delta)\},$$

$$\omega^{\pm}(t,\Delta) := \sup_{u\in(0,\Delta),\ t\pm u\in[0,1]} \big|f(t) - f(t\pm u)\big|.$$

Fulfilment of the first condition for the processes $s_{T,n}$ and s_T is evident if one takes the set of rational numbers as $\mathcal{S}$. The second condition is satisfied as well, since: (a) owing to the necessity of condition (4.9.8), it still holds if one substitutes $s_{T,n}$ by s_T; (b) the following trivial inequality holds:

$$\omega_\Delta^{\mathbb{D}}(s_{T,n}) \leqslant \omega_\Delta^{\mathbb{D}}(s_T).$$

Thus, the desired convergence in $(\mathbb{D}, \rho_{\mathbb{D}})$ is proved. It follows from this that, for any open set B in $(\mathbb{D}, \rho_{\mathbb{D}})$ the next inequality holds:

$$\varliminf_{n\to\infty} \mathbf{P}(s_{T,n} \in B) \geqslant \mathbf{P}(s_T \in B).$$

Since the function $s_{T,n}$ is continuous, on the left-hand side of the above inequality instead of B one can write $B\mathbb{C}$, so that

$$\mathbf{P}(s_T \in B) \leqslant \varliminf_{n\to\infty} \mathbf{P}(s_{T,n} \in B\mathbb{C}). \tag{4.9.9}$$

Note that the topologies generated by the metrics $\rho_{\mathbb{C}}$ and $\rho_{\mathbb{D}}$ in the space $\mathbb{C}$ coincide; therefore an open set $B\mathbb{C}$ in $(\mathbb{C}, \rho_{\mathbb{D}})$ is also open in the space $(\mathbb{C}, \rho_{\mathbb{C}})$.

(ii) Now bound the probability on the right-hand side of (4.9.9) if set B (and therefore $B\mathbb{C}$) is convex. In that case set $B\mathbb{C}$ is convex and open and, according to Theorem 4.3.1,

$$\mathbf{P}(s_{T,n} \in B\mathbb{C}) \leqslant \exp\big\{-nJ_{(\xi)}(B\mathbb{C}_a)\big\},$$

where

$$\xi := \xi^{T,n} := \frac{n}{T} S\left(\frac{T}{n}\right), J_{(\xi)}(f)$$

is a deviation integral for the random variable ξ. Since $J_{(\xi)}(B\mathbb{C}_a) \geqslant J_{(\xi)}(B)$, we obtain that

$$\mathbf{P}(s_{T,n} \in B\mathbb{C}) \leqslant \exp\big\{-nJ_{(\xi)}(B)\big\}. \tag{4.9.10}$$

(iii) We can now find the value of $J_{(\xi)}(B)$. As long as for $\xi = \xi^{T,n}$ the equation

$$\ln \mathbf{E}e^{\lambda\xi} = \frac{T}{n}\ln\psi\left(\frac{n}{T}\lambda\right)$$

holds, where $\psi(\lambda) = \mathbf{E}e^{\lambda S(1)}$, then

$$\Lambda_{(\xi)}(\alpha) = \sup_{\lambda}\left\{\lambda\alpha - \frac{T}{n}\ln\psi\left(\frac{n}{T}\lambda\right)\right\} = \frac{T}{n}\Lambda_{(S(1))}(\alpha).$$

Hence, we obtain

$$nI_{(\xi)}(f) = TI(f)$$

for $f \in \mathbb{C}_a$, where $I(f)$ corresponds to the random variable $S(1)$. Furthermore, under obvious notational conventions we have

$$\lambda_{\pm}^{(\xi)} = \frac{T}{n}\lambda_{\pm}^{(S(1))}.$$

Therefore

$$nJ_{(\xi)}(f) = TJ(f), \quad nJ_{(\xi)}(B) = TJ(B),$$

where $J(f), J(B)$ correspond to the random variable $S(1)$. Using (4.9.10) we obtain

$$\mathbf{P}(s_{T,n} \in B\mathbb{C}) \leqslant \exp\big\{-TJ(B)\big\}.$$

Returning to (4.9.9), because of (4.9.10) we obtain inequality (4.9.2). The theorem is proved. □

4.9.2 Large deviation principles

Theorem 4.9.5. *For any function $f \in \mathbb{C}$, $f(0) = 0$, the following statement holds:*

$$\lim_{\varepsilon\to 0}\overline{\lim_{T\to\infty}}\frac{1}{T}\ln\mathbf{P}\big(s_T \in (f)_{\mathbb{C},\varepsilon}\big) = \lim_{\varepsilon\to 0}\underline{\lim_{T\to\infty}}\frac{1}{T}\ln\mathbf{P}(s_T \in (f)_{\mathbb{C},\varepsilon}) = -J(f), \tag{4.9.11}$$

where $(f)_{\mathbb{C},\varepsilon}$ is an ε-neighbourhood in $\mathbb{D}$ of f in the uniform metric $\rho_{\mathbb{C}}$.

Theorem 4.9.6. *For any convex set B that is open in $(\mathbb{D}, \rho_{\mathbb{D}})$ there exists the limit*

$$\lim_{T\to\infty}\frac{1}{T}\ln\mathbf{P}(s_T \in B) = -J(B).$$

Theorem 4.9.7. *Under the assumption of condition* $[\mathbf{C}_\infty]$, *for any measurable set B the following inequalities hold:*

$$\overline{\lim_{T\to\infty}}\frac{1}{T}\ln\mathbf{P}(s_T \in B) \leqslant -J\big([B]\big) = -I\big([B]\big),$$

$$\underline{\lim_{T\to\infty}}\frac{1}{T}\ln\mathbf{P}(s_T \in B) \geqslant -J\big((B)\big) = -I\big((B)\big),$$

where $[B]$, (B) *are correspondingly the closure and the open interior of set* B *in the uniform metric* $\rho_{\mathbb{C}}$.

According to Definitions 4.1.1–4.1.3, we may say that Theorem 4.9.5 indicates the fulfilment of the local large deviation principle (local l.d.p.) in the space $(\mathbb{D}, \rho_{\mathbb{C}})$ with parameters (T, J) in the subclass of functions from $\mathbb{C}$.

Analogously, Theorem 4.9.6 indicates the fulfilment of the l.d.p. in $(\mathbb{D}, \rho_{\mathbb{C}})$ with parameters (T, J) in the subclass of convex sets. Theorem 4.9.7 indicates the fulfilment of the 'usual' l.d.p. in $(\mathbb{D}, \rho_{\mathbb{C}})$ with parameters (T, J).

Note also that, along with (4.9.11), other forms of l.l.d.p. notation are available, according to the equivalent definitions 4.1.1–4.1.3.

The conclusion of Theorem 4.9.7 for compound Poisson processes was established in [19]. In the general case Theorems 4.9.5–4.9.7 were proved in [65].

Proof of Theorem 4.9.5. According to the results of section 4.1, to prove the l.l.d.p. (4.9.11) it is sufficient to show that the inequality

$$\overline{\lim_{T\to\infty}} \frac{1}{T} \ln \mathbf{P}\big(s_T \in (f)_{\mathbb{C},\varepsilon}\big) \leqslant -J\big((f)_{\mathbb{C},\delta(\varepsilon)}\big) \tag{4.9.12}$$

holds for any function $f \in \mathbb{C}$, where $\delta(\varepsilon) \to 0$ as $\varepsilon \to 0$, and that the inequality

$$\underline{\lim}_{T\to\infty} \frac{1}{T} \ln \mathbf{P}\big(s_T \in (f)_{\mathbb{C},\varepsilon}\big) \geqslant -I(f) \tag{4.9.13}$$

holds for any function $f \in \mathbb{C}_a$ and real $\varepsilon > 0$ (see Definition 4.1.3). Corollary 4.9.2 implies that the uniform upper bound (4.9.12) holds if $\delta(\varepsilon) = \varepsilon$. It remains to obtain the lower bound (4.9.13).

We will split the proof of inequality (4.9.13) into four stages.

(i) Assume first that condition $[\mathbf{C}_\infty]$ holds, and that the bound for $T = n$ is

$$P_n := \mathbf{P}\big(\rho_{\mathbb{C}}(s_T, s_{T,n}) \geqslant \delta\big) \leqslant n\mathbf{P}\Big(\sup_{t\in[0,1]} \big|S(t) - tS(1)\big| \geqslant \delta n\Big).$$

Denote

$$A_n = \Big\{\max_{t\in[0,1]} \big(S(t) - t(S(1)\big) \geqslant \delta n\Big\}.$$

It is not difficult to see that

$$\begin{aligned}\mathbf{P}(A_n) &= \mathbf{P}\big(A_n;\ S(1) \geqslant 0\big) + \mathbf{P}\big(A_n;\ S(1) < 0\big)\\ &\leqslant \mathbf{P}\Big(\max_{t\in[0,1]} S(t) \geqslant \delta n\Big) + \mathbf{P}\Big(\min_{t\in[0,1]} S(t) \leqslant -\delta n\Big).\end{aligned}$$

Similar inequalities are correct for an event

$$\Big\{\min_{t\in[0,1]} \big(S(t) - tS(1)\big) \leqslant -\delta n\Big\}.$$

Therefore,

$$P_n \leqslant 2n\bigg[\mathbf{P}\Big(\max_{t\in[0,1]} S(t) \geqslant \delta n\Big) + \mathbf{P}\Big(\min_{t\in[0,1]} S(t) \leqslant -\delta n\Big)\bigg].$$

While obtaining bounds for P_n, one can assume without loss of generality that $\mathbf{E}S(1) = 0$ (this requirement is inessential; see also Lemma 4.4.4). Hence, by Theorem 4.9.3,

$$P_n \leqslant 2n\Big[e^{-\Lambda(\delta n)} + e^{-\Lambda(-\delta n)}\Big].$$

Since

$$\frac{\Lambda(\pm\delta n)}{\delta n} \to \infty$$

as $n \to \infty$ when the condition $[\mathbf{C}_\infty]$ is met, for any fixed $N > 0$ and $n \to \infty$ we have

$$P_n = o(e^{-nN}). \tag{4.9.14}$$

(ii) Now we obtain the lower bound $\big(\delta \in (0, \varepsilon)\big)$:

$$\begin{aligned}\mathbf{P}\big(s_T \in (f)_{\mathbb{C},\varepsilon}\big) &\geqslant \mathbf{P}\big(s_T \in (f)_{\mathbb{C},\varepsilon};\ \rho_{\mathbb{C}}(s_T, s_{T,n}) < \delta\big)\\ &\geqslant \mathbf{P}\big(s_{T,n} \in (f)_{\mathbb{C},\varepsilon-\delta}, \rho_{\mathbb{C}}(s_T, s_{T,n}) < \delta\big)\\ &\geqslant \mathbf{P}\big(s_{T,n} \in (f)_{\mathbb{C},\varepsilon-\delta}\big) - \mathbf{P}\big(\rho(s_T, s_{T,n}) \geqslant \delta\big).\end{aligned} \tag{4.9.15}$$

By virtue of the l.l.d.p. for $s_{T,n}$ (see Theorem 4.4.2),

$$\varliminf_{n\to\infty} \frac{1}{n} \ln \mathbf{P}\big(s_{T,n} \in (f)_{\mathbb{C},\varepsilon-\delta}\big) \geqslant -I(f).$$

Consequently, (4.9.15) and (4.9.14) imply the fulfilment of (4.9.13) (recall that we let $T = n$).

(iii) We now show that condition $[\mathbf{C}_\infty]$ is redundant for the fulfilment of (4.9.13). This is done via standard arguments (see e.g. section 4.4) by using truncations for the jumps ζ_k. Let ${}^{(N)}s_n$ be a random process of the same type as s_n but with jumps ${}^{(N)}\zeta_k$, $|{}^{(N)}\zeta_k| \leqslant N$, having distributions

$$\mathbf{P}({}^{(N)}\zeta \in B) = \mathbf{P}\big(\zeta \in B \,|\, |\zeta| \leqslant N\big).$$

Let Q_n be a the number of jumps of the process s_n on segment $[0, n]$ that are larger than 1 in magnitude, and let

$$B_n := \{Q_n \leqslant Rn\}.$$

Then, for a given $M > 0$, we can find an R such that, for all sufficiently large values of n,

$$\mathbf{P}(\overline{B}_n) \leqslant e^{-Mn}. \tag{4.9.16}$$

Next, for $C_n := \big\{|\zeta_j| \leqslant N \text{ for all } j \leqslant Rn\big\}$ we have

$$\begin{aligned}\mathbf{P}\big(s_T \in (f)_{\mathbb{C},\varepsilon}\big) &\geqslant \mathbf{P}\big(s_T \in (f)_{\mathbb{C},\varepsilon};\ B_nC_n\big)\\ &= \mathbf{P}\big(s_T \in (f)_{\mathbb{C},\varepsilon};\ B_n \,|\, C_n\big)\mathbf{P}(C_n)\\ &= \mathbf{P}\big({}^{(N)}s_T \in (f)_{\mathbb{C},\varepsilon};\ B_n\big)\mathbf{P}(C_n)\\ &\geqslant \Big[\mathbf{P}\big({}^{(N)}s_T \in (f)_{\mathbb{C},\varepsilon}\big) - \mathbf{P}(\overline{B}_n)\Big]\Big(1 - \mathbf{P}\big(|\zeta| > N\big)\Big)^{Rn}.\end{aligned}$$

Since ${}^{(N)}\zeta$ meets the condition $[\mathbf{C}_\infty]$, by virtue of part (ii) of the proof and (4.9.16) we obtain, for sufficiently large values of M,

$$\begin{aligned}
&\underline{\lim}\ \frac{1}{n}\ln \mathbf{P}\big(s_n \in (f)_{\mathbb{C},\varepsilon}\big)\\
&\quad\geqslant \underline{\lim}\ \frac{1}{n}\ln\Big[\mathbf{P}\big({}^{(N)}s_n \in (f)_{\mathbb{C},\varepsilon}\big) - \mathbf{P}(\overline{B}_n)\Big] + R\ln\Big(1-\mathbf{P}\big(|\zeta|>N\big)\Big)\\
&\quad\geqslant -{}^{(N)}I(f) + R\ln\Big(1-\mathbf{P}\big(|\zeta|>N\big)\Big),
\end{aligned}\tag{4.9.17}$$

where ${}^{(N)}I(f)$ is a deviation integral constructed from the deviation function ${}^{(N)}\Lambda$ for random variables ${}^{(N)}S(1)$ corresponding to the jumps ${}^{(N)}\zeta$. Since ${}^{(N)}\Lambda(\alpha) \to \Lambda(\alpha)$ and ${}^{(N)}I(f) \to I(f)$, it follows that $\ln\Big(1-\mathbf{P}\big(|\zeta|>N\big)\Big) \to 0$ as $N \to \infty$, and the left-hand side of (4.9.17) does not depend on N. The inequality (4.9.13) for $T = n$ is proved.

(iv) Now we prove (4.9.13) without the constraint $T = n$. Let

$$\overline{f}(t) := \begin{cases} f(t), & \text{if } \ t \in [0,1]\\ f(1), & \text{if } \ t > 1.\end{cases}$$

Taking into account that the process $S(t)$ is defined for all $t \geqslant 0$, for any function $f \in \mathbb{C}_a$ and $\varepsilon > 0$, $n := [T]$, we have

$$\begin{aligned}
\big\{s_T \in (f)_{\mathbb{C},\varepsilon}\big\} &= \left\{\sup_{0\leqslant t\leqslant T}\left|\frac{1}{T}S(t) - f\Big(\frac{t}{T}\Big)\right| < \varepsilon\right\}\\
&\supset \left\{\sup_{0\leqslant t\leqslant n+1}\left|\frac{1}{T}S(t) - \overline{f}\Big(\frac{t}{T}\Big)\right| < \varepsilon\right\}\\
&= \left\{\sup_{0\leqslant t\leqslant n+1}\left|\frac{1}{n+1}S(t) - \frac{T}{n+1}\overline{f}\Big(\frac{t}{T}\Big)\right| < \frac{T}{n+1}\varepsilon\right\}.
\end{aligned}$$

Since

$$\lim_{T\to\infty}\sup_{0\leqslant t\leqslant n+1}\left|\frac{T}{n+1}\overline{f}\Big(\frac{t}{T}\Big) - f\Big(\frac{t}{n+1}\Big)\right| = 0,$$

for all sufficiently large values of T we have

$$\begin{aligned}
&\left\{\sup_{0\leqslant t\leqslant n+1}\left|\frac{1}{n+1}S(t) - \frac{T}{n+1}\overline{f}\Big(\frac{t}{T}\Big)\right| < \frac{T}{n+1}\varepsilon\right\}\\
&\qquad\supset \left\{\sup_{0\leqslant t\leqslant n+1}\left|\frac{1}{n+1}S(t) - f\Big(\frac{t}{n+1}\Big)\right| < \frac{1}{2}\varepsilon\right\} = \big\{s_{n+1} \in (f)_{\mathbb{C},\varepsilon/2}\big\}.
\end{aligned}$$

Therefore, by part (iii) we get

$$\underline{\lim}_{T\to\infty}\ \frac{1}{T}\ln\mathbf{P}\big(s_T \in (f)_{\mathbb{C},\varepsilon}\big) \geqslant \underline{\lim}_{n+1\to\infty}\ \frac{1}{n+1}\ln\mathbf{P}\big(s_{n+1} \in (f)_{\mathbb{C},\varepsilon/2}\big) \geqslant -I(f),$$

and inequality (4.9.13) is proved. At the same time Theorem 4.9.5 is proved. □

Proof of Theorem 4.9.6. It follows from Theorem 4.9.1 that

$$\overline{\lim_{T\to\infty}} \frac{1}{T} \ln \mathbf{P}(s_T \in B) \leqslant -J(B). \tag{4.9.18}$$

Now we derive the lower bound. From the second equality in (4.9.11) (Theorem 4.9.5) we obtain that, for all $\varepsilon > 0, f \in \mathbb{C}$,

$$\underline{\lim_{T\to\infty}} \frac{1}{T} \ln \mathbf{P}\big(s_T \in (f)_{\mathbb{C},\varepsilon}\big) \geqslant -J(f),$$

and, hence,

$$\underline{\lim_{T\to\infty}} \frac{1}{T} \ln \mathbf{P}\big(s_T \in (f)_{\mathbb{D},\varepsilon}\big) \geqslant -J(f). \tag{4.9.19}$$

For any set B that is open in $(\mathbb{D}, \rho_{\mathbb{D}}), f \in B$ and small enough $\varepsilon > 0$,

$$\mathbf{P}(s_T \in B) \geqslant \mathbf{P}\big(s_T \in (f)_{\mathbb{D},\varepsilon}\big).$$

Therefore, by (4.9.19),

$$\underline{\lim_{T\to\infty}} \frac{1}{T} \ln \mathbf{P}(s_T \in B) \geqslant -J\big((B\mathbb{C})\big).$$

From statement (vi) in Theorem 4.9.3 it follows that $J\big((B\mathbb{C})\big) = J\big((B)\big)$. Together with (4.9.18) this proves Theorem 4.9.6. □

Proof of Theorem 4.9.7. When condition $[\mathbf{C}_\infty]$ is met,

$$J(f) := \begin{cases} I(f), & \text{if } f \in \mathbb{C}_a, \\ \infty, & \text{in other cases,} \end{cases}$$

so that

$$J(B) = I(B\mathbb{C}_a).$$

Therefore, the set

$$K_v := \{f \in \mathbb{D} : J(f) \leqslant v\}$$

lies in $\mathbb{C}_a$ and is compact in $(\mathbb{C}_a, \rho_{\mathbb{C}})$ and in $(\mathbb{D}, \rho_{\mathbb{C}})$. We use Lemma 4.4.8, by virtue of which for $s_{T,n}$ the condition $[\mathbf{K}]_0$ is fulfilled with $T = n$; this means that for any N there exists a compact K in $(\mathbb{C}, \rho_{\mathbb{C}})$ such that

$$\overline{\lim_{n\to\infty}} \frac{1}{n} \ln \mathbf{P}(s_{T,n} \notin K) \leqslant -N.$$

Next, using the bounds (4.9.14), we obtain that for s_T the condition $[\mathbf{K}]$ is fulfilled, which means that for any N there exists a compact K in $(\mathbb{C}, \rho_{\mathbb{C}})$ such that, for any $\varepsilon > 0$,

$$\overline{\lim_{n\to\infty}} \frac{1}{T} \ln \mathbf{P}\big(s_T \notin (K)_{\mathbb{C},\varepsilon}\big) \leqslant -N.$$

By Theorem 4.1.6, the l.l.d.p. implies the e.l.d.p., i.e. the statement of Theorem 4.9.7 where

$$J(B+) := \lim_{\varepsilon\to 0} J\big((B)_{\mathbb{C},\varepsilon}\big) = J\big([B]\big).$$

Theorem 4.9.7 is proved. □

4.9.3 Conditional large deviation principles

We saw in the previous subsection (see Theorem 4.9.7) that when the condition $[\mathbf{C}_\infty]$ is met for $S(1)$, the l.d.p. for $\overline{s}_n$ with $\xi_k = S(k)$ carries over completely to the processes

$$s_T = s_T(t) = \frac{S(tT)}{T}, \qquad t \in [0,1],$$

where $S(t)$ is a process with independent increments on $[0,T]$. Here, the l.d.p. has the same form as for a *continuous* version s_n of the random walk $\{S_k = S(k)\}$, i.e. it appears to be a 'usual' l.d.p. in the space $(\mathbb{D}, \rho_{\mathbb{C}})$ with *uniform* metric and parameters (T, I). This arises from the fact that under condition $[\mathbf{C}_\infty]$ the trajectories of all three processes, $\overline{s}_n$, a continuous version of s_n for the random walk $\{S_k = S(k)\}$ and s_T (with $T = n$ for processes with independent increments), behave in the same manner 'in terms of the l.d.p.', since in this case the relations (4.5.12) and (4.5.14) hold (see Lemmas 4.1.11 and 4.5.5).

Moreover, integro-local theorems for $\mathbf{P}\big(S(T) \in \Delta[x)\big)$ with $T \to \infty$ would have the same form as for $\mathbf{P}\big(S_n \in \Delta[x)\big)$ (on replacement of n by T). The reason is that in the proofs of the integro-local theorems 1.5.1, 1.5.3 and 2.2.2 the fact that n is integer does not play a significant role (see e.g. the right-hand side of relation (1.5.6) in the proof of the key theorem, Theorem 1.5.1). This right-hand side determines the desired asymptotics for $\mathbf{P}\big(S(T) \in \Delta[x)\big)$ on replacement of n by T, as $\mathbf{E}e^{i\lambda S(T)} = \varphi(\lambda)^T$ where $\varphi(\lambda) = \mathbf{E}e^{-\lambda S(1)}$.

All the above is also related to the conditional l.d.p. studied in section 4.8.

Thus, let $S(1)$ satisfy condition $[\mathbf{C}_\infty]$ and for simplicity let $x = n = T$. Then the statements of Corollary 4.8.1 and Theorems 4.8.2–4.8.8 still hold if one replaces $\overline{s}_n$ with s_T (for $T = n$); here one should consider Λ as a deviation function for $S(1)$ and assume by analogy with the preceding discussion that $b \in (s_-, s_+)$, where $s_\pm$ are support bounds of the random variable $S(1)$. Non-lattice conditions in the analogues of Theorems 4.8.5 and 4.8.6 and arithmetic conditions in the analogues of Theorems 4.8.7 and 4.8.8 should be imposed on the random variable $S(1) - q$, where q is a drift of the process. With this, under condition $[\mathbf{C}_\infty]$ the sets $\{G : J(G) \leqslant v\}$, $v \geqslant 0$, turn out to be compact in $(\mathbb{C}, \rho_{\mathbb{C}})$. Hence, as noted above, it is not difficult to obtain that $J\big(B^{(b)}+\big), J\big(B^{(b)}\big)$ in statements (i) and (ii) of Theorems 4.8.6 and 4.8.8 can be replaced by $I\big([B]_{\mathbb{C}}^{(b)}\big)$ and $I\big((B)_{\mathbb{C}}^{(b)}\big)$ respectively, and the condition of continuity (iii) in Theorems 4.8.6 and 4.8.8 will have the form

$$I\big([B]^{(b)}_{\mathbb{C}}\big) = I\big((B)^{(b)}_{\mathbb{C}}\big). \tag{4.9.20}$$

In particular, the analogue of Theorem 4.8.8 given below holds true; we will need this later.

We call the process $S(t)$, $t \in [0, T]$, *arithmetic* if it is a compound Poisson processes with integer-valued jumps and some drift q. It is clear that for arithmetic processes the random variable $S(t) - qt$ is arithmetic. In order to simplify the required assertions, we assume that the drift q and the value b are such that $b/q = r/m$ is rational (both r and m are integer numbers; if $q = 0$, then we assume the rationality of b). Then for $T_1 := m/q$ we obtain that the numbers $T_1 q = m$ and $T_1 b = r$ are integer, so that $T(b - q)$ is also integer if T is divisible by T_1 and, therefore,

$$\mathbf{P}\big(s_T(1) = b\big) = \mathbf{P}\big(s_T(1) - q = b - q\big) > 0,$$

if $\mathbf{P}\big(S(T) - qT = T(b - q)\big) > 0$.

Theorem 4.9.8. *Let $S(t)$ be an arithmetic process and let $S(1)$ meet the condition* $[\mathbf{C}_\infty]$.

(i) *Then, for any function $f \in \mathbb{D}$, $f(0) = 0$, $f(1) = b$, the following relation holds:*

$$\lim_{\varepsilon\to 0}\lim_{T\to\infty} \frac{1}{T} \ln \mathbf{P}\big(s_T \in (f)_\varepsilon \,|\, s_T(1) = b\big)\big) = -I(f) + \Lambda(b),$$

where $T \to \infty$ by values divisible by T_1.

(ii) *If, moreover, the condition* $[\mathbf{B}, \mathbf{b}]$ *holds* (see (4.8.5)) *then*

$$\begin{aligned} \overline{\lim_{T\to\infty}} \frac{1}{T} \ln \mathbf{P}\big(s_T \in B \,|\, s_T(1) = b\big) &\leqslant -I\big([B]^{(b)}_{\mathbb{C}}\big) + \Lambda(b), \\ \underline{\lim_{T\to\infty}} \frac{1}{T} \ln \mathbf{P}\big(s_T \in B \,|\, s_T(1) = b\big) &\geqslant -I\big((B)^{(b)}_{\mathbb{C}}\big) + \Lambda(b), \end{aligned} \tag{4.9.21}$$

where $T \to \infty$ by values divisible by T_1.

If b/q is not rational then, instead of the condition $\{s_T(1) = b\}$ in (4.9.21), one can examine the condition

$$\left\{s_T(1) = \frac{[T(b - q)] + Tq}{T}\right\},$$

which has positive probability. It is evident that in this case the right-hand side of (4.9.21) remains under the condition $[\mathbf{B}, \mathbf{b}]$.

4.9.4 Versions of Sanov's theorem

Let $F^*_n(t)$ be an empirical distribution function constructed from a sample of size n from a continuous distribution $\mathbf{F}$. Having in mind the following problems, the distribution **F** can be assumed, without loss of generality, to be uniform on $[0, 1]$ (with distribution function $F(t) = t$, $t \in [0, 1]$). Further, let $S(t)$ be the standard Poisson process on $[0, n]$ with rate 1. Then it is known that the distribution of $F_n^*(t)$ coincides with the distribution of the process $s_T(t) = \frac{1}{T}S(Tt)$ with $T = n$ under the condition that $S(n) = n$ ($s_T(1) = 1$). Here, as already noted, condition $[\mathbf{C}_\infty]$ is met and deviation function $\Lambda(\alpha)$ for $S(1)$ is equal to $\Lambda(\alpha) = \alpha \ln \alpha - \alpha + 1$ (see Example 1.1.8).

Let $\mathbb{F}$ be a class of distribution functions on $[0, 1]$. On the class $\mathbb{F}$, the metric ρ is equivalent to Lévy's metric. For any function $G \in \mathbb{F}$ which is absolutely continuous with respect to F, and with $T = n$, one has

$$\mathbf{P}\big(F_n^* \in (G)_\varepsilon\big) = \mathbf{P}\big(s_T \in (G)_\varepsilon \mid s_T(1) = 1\big),$$

$$J(G) = \int_0^1 \big[G'(t) \ln G'(t) - G'(t) + 1\big]dt = \int_0^1 \ln \frac{dG(t)}{dF(t)} dG(t),$$

so that $J(G)$ coincides with the Kullback–Leibler distance between the distributions G and F. Moreover, here $\Lambda(1) = 0$. As a result, we obtain from Theorem 4.9.8 the following versions of Sanov's theorem (see [164]).

Theorem 4.9.9. (i) *For any $G \in \mathbb{F}$,*

$$\lim_{\varepsilon \to 0} \overline{\lim_{n \to \infty}} \frac{1}{n} \ln \mathbf{P}\big(F_n^* \in (G)_\varepsilon\big) = \lim_{\varepsilon \to 0} \underline{\lim_{n \to \infty}} \frac{1}{n} \ln \mathbf{P}\big(F_n^* \in (G)_\varepsilon\big) = -I(G).$$

(ii) *For any measurable set B satisfying the condition* $[\mathbf{B}, \mathbf{b}]$ *with $b = 1$,*

$$\overline{\lim_{n \to \infty}} \frac{1}{n} \ln \mathbf{P}(F_n^* \in B) \leqslant -I\big([B]_{\mathbb{C}}^{(1)}\big). \tag{4.9.22}$$

$$\underline{\lim_{n \to \infty}} \frac{1}{n} \ln \mathbf{P}(F_n^* \in B) \geqslant -I\big((B)_{\mathbb{C}}^{(1)}\big). \tag{4.9.23}$$

As we have already observed, the assumption that $F(t)$ is uniform means no loss of generality here: $F(t)$ can be any other continuous distribution function, $\mathbb{F}$ then being the class of distribution functions absolutely continuous with respect to F. If the set $B^{(b)}$ satisfies the continuity condition $I\big([B]_{\mathbb{C}}^{(1)}\big) = I\big((B)_{\mathbb{C}}^{(1)}\big)$ then (4.9.22), (4.9.23) imply that there exists the limit

$$\lim_{n \to \infty} \frac{1}{n} \ln \mathbf{P}(F_n^* \in B) = -I(B^{(b)}).$$

That assertion was proved by other arguments in [19].

Note also that the proof of Sanov's theorem can be also obtained with the l.d.p. for sums of random elements in Banach spaces (see e.g. [1], [6], [13]).

Let us explore the problem of finding the logarithmic asymptotics of $\mathbf{P}(F_n^* \in B)$ in an explicit form for some specific sets B. In order to do this, consider the centred empirical cumulative distribution function

$$F_n^{*0}(t) := F_n^*(t) - t, \qquad t \in [0, 1]$$

and, correspondingly, a centred Poisson process $S(t)$ on $[0, T]$ with rate 1 and shift $q = -1$. Let $b = 0$, $T = n$ be an integer, and let set B have the form

$$B = \Big\{f : \sup_{t\in[0,1]} \big(f(t) - g(t)\big) > 0\Big\} \tag{4.9.24}$$

or the form

$$B = \Big\{f : \inf_{t\in[0,1]} \big(f(t) + g(t)\big) < 0\Big\}, \tag{4.9.25}$$

where $g_0 := \inf_{t\in[0,1]} g(t) > 0$. As follows from the commentaries on the example given in (4.8.8), for such sets the condition $[\mathbf{B}, \mathbf{b}]$ for $b = 0$ is always met.

For the process $F_n^{*0}(t)$ introduced above and the centred process $s_T(t) = S(tT)/T$ we will evidently have that, for $T = n$,

$$\mathbf{P}(F_n^{*0} \in B) = \mathbf{P}\big(s_T \in B \,|\, s_T(1) = 0\big). \tag{4.9.26}$$

Let us consider the continuity condition (4.9.20) for sets of the form (4.9.24). Condition $[\mathbf{C}_\infty]$ holds for the proposed centred Poisson process, and

$$\Lambda(\alpha) = (\alpha + 1)\ln(\alpha + 1) - \alpha \tag{4.9.27}$$

(see Example 1.1.4), so the function $\Lambda(\alpha)$ is analytic on $(-1, \infty)$: $\Lambda(0) = 0$, $\Lambda(-1) = 1$, $\Lambda(\alpha)$ increases on $\alpha \geqslant 0$ and decreases on $\alpha \leqslant 0$ and $\Lambda(\alpha) = \infty$ for $\alpha < -1$. Note also that $\mathbf{P}(s_T(1) = 0) > 0$ for integer $T = n$.

Next, observe that in order to find the value of

$$I(B^{(b)}) = \inf_{f\in B^{(b)}} \int_0^1 \Lambda\big(f'(t)\big)dt \tag{4.9.28}$$

we have to find the most probable trajectory from the space $B^{(b)}$. As already noted, the most probable trajectory connecting two given points (t_1, g_1) and (t_2, g_2) is a segment connecting these points. Next, in (4.9.24), for $f \in B^{(b)}$ with $b = 0$ there necessarily exists a point $t \in (0, 1)$ such that

$$f(t) > g(t).$$

Since $f(1) = 0$, the smallest value of the integral in (4.9.28) for such a function f is

$$I(t, g) := t\Lambda\left(\frac{g(t)}{t}\right) + (1 - t)\Lambda\left(-\frac{g(t)}{1 - t}\right). \tag{4.9.29}$$

Therefore, for the sets (4.9.24),

$$I(B^{(0)}) = \inf_{t\in[0,1]} I(t, g). \tag{4.9.30}$$

Since $t\Lambda(g_0/t) \to \infty$, as $t \to 0$ and $\Lambda(-g_0/(1-t)) \to \infty$ as $t \to 1$, the infimum in (4.9.30) cannot be reached in the vicinities of the points $t = 0$ and $t = 1$, as $I(t, g) \geqslant I(t, g_0) \to \infty$ for $t \to 0$ or $t \to 1$. It means that there exists $\theta > 0$ such that

$$I(B^{(0)}) = \inf_{t\in[\theta,1-\theta]} I(t,g).$$

Let there be a point $t_0 \in [\theta, 1-\theta]$ at which infimum in (4.9.30) is reached (such a point always exists if g is continuous on $[\theta, 1-\theta]$). In this case, if

$$g(t_0) < 1 - t_0 \tag{4.9.31}$$

then

$$I(B^{(0)}) = I(t_0, g)$$

and this functional of g would evidently be continuous over g with respect to the uniform metric, as the values $g(t_0)/t_0$ and $-g(t_0)/(1-t_0)$ (see (4.9.29)) lie in the domain of analyticity of the function Λ. This means that under (4.9.31) the continuity condition (4.9.20) for sets (4.9.24) with $b = 0$ is met. Thus, we obtain that Theorem 4.9.8 implies the following.

Corollary 4.9.10. *Let $S(t)$ be a centred Poisson process with rate* 1, *and let there exist a point t_0 at which the infimum in* (4.9.30) *is reached and for which let $g(t_0) < 1 - t_0$ (this implies $t_0 \in [\theta, 1-\theta]$, $\theta > 0$). Then, for sets B of the form* (4.9.24),

$$\lim_{T\to\infty} \frac{1}{T} \ln \mathbf{P}\big(s_T \in B \,|\, s_T(1) = 0\big) = -I(t_0, g) \tag{4.9.32}$$

(*see* (4.9.29), (4.9.27)), *where $T \to \infty$ by integer values.*

By virtue of (4.9.26), we can apply Corollary 4.9.10 to find logarithmic asymptotics of $\mathbf{P}(F_n^{*0} \in B)$.

Remark 4.9.11. If the condition in Corollary 4.9.10 regarding the existence of the point t_0 such that $g(t_0) < 1 - t_0$ is violated then the statement of the corollary would have the form

$$\overline{\lim_{T\to\infty}} \frac{1}{T} \ln \mathbf{P}\big(s_T \in B \,|\, s_T(1) = 0\big) \leqslant -J\big([B]_{\mathbb{C}}^{(0)}\big),$$

$$\underline{\lim_{T\to\infty}} \frac{1}{T} \ln \mathbf{P}\big(s_T \in B \,|\, s_T(1) = 0\big) \geqslant -J\big((B)_{\mathbb{C}}^{(0)}\big).$$

Analogously one can consider the case when the set B has the form (4.9.25). In this case, on the right-hand side of (4.9.32) there will be $-I(1-t_0, g)$, where t_0 has the same meaning. As a result we can find the logarithmic asymptotic behaviour of the probability that the trajectory F_n^{*0} will cross at least one of the two borders $g_1(t) > 0$ and $-g_2(t) < 0$. This logarithm behaviour will be determined by the value of $\min\big[I(t_0, g_1), I(1-t_0, g_2)\big]$.

4.10 On large deviation principles for compound renewal processes

In addition to the processes considered above, one can identify a number of random processes for which the l.d.p. has the same form as in Theorems 4.6.1 and 4.6.2 (or 4.4.2 and 4.4.7 for continuous versions of the processes). Among them there are compound renewal processes and sums of random variables defined on states (or on transitions) of a finite ergodic Markov chain. Increments of such processes on large non-overlapping time intervals are 'asymptotically independent' (as for the processes considered above in this chapter), so that the deviation functional for functions $f \in \mathbb{C}_0$ has the form of the integral

$$J(f) = \int_0^1 \Lambda\big(f'(t)\big)dt, \tag{4.10.1}$$

but the deviation function Λ has a different form.

In this section we will prove l.d.p. for values of compound renewal processes at increasing times and will find the form of the corresponding deviation function. The of results for the trajectories and proofs appear to be rather difficult and cumbersome; they goes beyond the present monograph. Therefore we will just concentrate on an easier problem.

Suppose that (τ_0, ζ_0) is a random vector that is independent of a given sequence of independent identically distributed random vectors $(\tau, \zeta), (\tau_1, \zeta_1), (\tau_2, \zeta_2), \dots,$ where $\tau_0 > 0$, $\tau > 0$. Set

$$T_0 = 0, \qquad T_n = \sum_{j=1}^{n} \tau_j \quad \text{for} \quad n \geqslant 1, \qquad T_{0,n} = \tau_0 + T_n \quad \text{for} \quad n \geqslant 0,$$

$$Z_0 = 0, \qquad Z_n = \sum_{j=1}^{n} \zeta_j \quad \text{for} \quad n \geqslant 1, \qquad Z_{0,n} = \zeta_0 + Z_n \quad \text{for} \quad n \geqslant 0,$$

$$T_{0,-1} = Z_{0,-1} = 0.$$

For $t > 0$, let

$$\nu(t) = \max\{k \geqslant -1 : T_{0,k} < t\} = \eta(t) - 1,$$

where

$$\eta(t) = \min\{k \geqslant 0 : T_{0,k} \geqslant t\}.$$

The compound renewal process $Z(t)$, $t \geqslant 0$, is defined as

$$Z(0) = 0, \qquad Z(t) = Z_{0,\nu(t)} \quad \text{for} \quad t > 0.$$

In order to find the deviation function $\Lambda^{(Z)}(\alpha)$ in (4.10.1) corresponding to the process, we have to get the asymptotic relation

$$\frac{1}{T} \ln \mathbf{P}\left(\frac{Z(T)}{T} \in (\alpha)_{\varepsilon_T}\right) \sim -\Lambda^{(Z)}(\alpha) \quad \text{as} \quad T \to \infty, \quad \varepsilon_T \to 0, \tag{4.10.2}$$

x i.e. to prove the local l.d.p. for $Z(T)/T$. Here we will assume that $\xi = (\tau, \zeta)$ satisfies condition $[\mathbf{C}_0]$. Set

$$\xi_{(1)} = \tau, \quad \xi_{(2)} = \zeta, \quad \xi = (\xi_{(1)}, \xi_{(2)}) \equiv (\tau, \zeta), \quad \xi_0 = (\tau_0, \zeta_0),$$
$$\psi(\lambda) = \mathbf{E}e^{\langle \lambda, \xi \rangle}, \quad \psi_0(\lambda) = \mathbf{E}e^{\langle \lambda, \xi_0 \rangle},$$
$$\mathcal{A} = \{\lambda = (\lambda_{(1)}, \lambda_{(2)}) : \psi(\lambda) < \infty\}, \qquad \mathcal{A}_0 = \{\lambda : \psi_0(\lambda) < \infty\}.$$

According to condition **[C₀]**, the set $(\mathcal{A})$ is not empty and contains the point $\lambda = 0$. If ξ_0 has the same distribution as ξ, the renewal process is called *homogeneous*. If

$$\psi_0(\lambda) = \frac{\psi(\lambda) - \psi(0, \lambda_{(2)})}{\lambda_{(1)} \mathbf{E}\tau} \tag{4.10.3}$$

then the process $Z(t)$ has *stationary increments*, i.e. the distribution of the differences $Z(t+h) - Z(t)$ is independent of t. Such a process on $[0, \infty)$ can be represented as the limit (by distribution) on $[0, \infty)$ as $N \to \infty$ of a sequence of homogeneous processes $Z_N(t)$ that start not at the point $t = 0$ but at the point $t = -N$. Then the limiting joint distribution of the first positive time among the times $-N + T_k$, $k = 1, 2, \ldots$, and the jump of the process $Z_N(t)$ at this point in time will have, by virtue of the local renewal theorem, the following form for non-arithmetic τ:

$$\begin{aligned}
&\mathbf{P}(\tau_0 > u, \zeta_0 \in B) \\
&= \lim_{N\to\infty} \left[\mathbf{P}(\tau > N + u, \zeta \in B) + \int_0^N dH(t) \mathbf{P}(\tau > N - t + u, \zeta \in B) \right] \\
&= \frac{1}{\mathbf{E}\tau} \int_0^\infty \mathbf{P}(\tau > t + u, \zeta \in B) du.
\end{aligned} \tag{4.10.4}$$

The same result holds for arithmetic τ. It is not difficult to see that the Laplace transform of this distribution is equal to (4.10.3). The independence of the distribution of $Z(t+h) - Z(t)$ from $t > 0$ for such a process follows directly from its construction.

Hereafter, when using the results of section 2.9, the following condition plays a significant role:

$$\mathcal{A}^{\leqslant 0} \subset [\mathcal{A}_0], \quad \text{where} \quad \mathcal{A}^{\leqslant 0} = \{\lambda : A(\lambda) \leqslant 0\} = \{\lambda : \psi(\lambda) \leqslant 1\} \tag{4.10.5}$$

(see Theorem 2.9.10). If τ and ζ are independent then, under a natural convention about notation we have $\psi(\lambda) = \psi^{(\tau)}(\lambda_{(1)}) \psi^{(\zeta)}(\lambda_{(2)})$ and, in (4.10.3),

$$\psi_0(\lambda) = \frac{\psi^{(\zeta)}(\lambda_{(2)})(\psi^{(\tau)}(\lambda_{(1)}) - 1)}{\lambda_{(1)} \mathbf{E}\tau}.$$

Therefore, $\mathcal{A}_0 = \mathcal{A}$ and condition (4.10.5) is met.

Condition (4.10.5) is always met when ζ satisfies condition **[C$_\infty$]**. Then $\psi(0, \lambda_{(2)})$ in (4.10.3) is finite for any $\lambda_{(2)}$ and $\mathcal{A}_0 = \mathcal{A}$.

In the case of dependent τ and ζ, it can happen that for $\lambda_{(1)} < 0$ there exists a point $\lambda = (\lambda_{(1)}, \lambda_{(2)}) \in \mathcal{A}$ such that $\psi\big(0, \lambda_{(2)}\big) = \infty$ and condition $\mathcal{A} \subset [\mathcal{A}_0]$ is not met. In this case additional arguments are required to clarify which points λ will be included in further assertions.

In the same way as in sections 2.8 and 2.9, we set

$$S_{0,n} = (T_{0,n}, Z_{0,n}) = \xi_0 + S_n, \qquad S_n = \sum_{k=1}^n \xi_k,$$

where, ξ_k are independent and have the same distributions as $\xi = (\tau, \zeta)$. In section 2.9, integro-local theorems and large deviation principles for $S_{0,n}$ were established.

In order to simplify computation while we are searching for the function $\Lambda^{(Z)}$ in (4.10.2), instead of the asymptotics of $\ln \mathbf{P}\big(Z(T)/T \in (\alpha)_{\varepsilon_T}\big)$, $\varepsilon_T \to 0$, we will consider the asymptotics of $\ln \mathbf{P}\big(Z(T)/T \in \Delta[\alpha)\big)$, where $\Delta = \Delta_T \to 0$ slowly enough as $T \to \infty$. First, let the half-open interval $\Delta[\alpha)$ not contain the point 0. Then

$$\Lambda^{(Z)}(\alpha) = -\lim_{T\to\infty} \frac{1}{T} \ln \mathbf{P}\left(\frac{Z(T)}{T} \in \Delta[\alpha)\right), \tag{4.10.6}$$

where

$$\begin{aligned}\mathbf{P}\big(Z(T) \in T\Delta[\alpha)\big) &= \int_0^T \sum_{n=0}^{\infty} \mathbf{P}\big(T_{0,n} \in dt,\, Z_{0,n} \in T\Delta[\alpha),\, \tau_{n+1} \geqslant T - t\big) \\ &= \int_0^T H_0\big(dt, T\Delta[\alpha)\big)\mathbf{P}(\tau \geqslant T - t) \end{aligned} \tag{4.10.7}$$

and the function

$$H_0(B) = \sum_{n=0}^{\infty} \mathbf{P}(S_{0,n} \in B), \qquad B \subset \mathbb{R}^2,$$

is a renewal function for the sequence of $S_{0,n}$. The integral on the right-hand side of (4.10.7) can be roughly approximated (bearing in mind that the asymptotics are the logarithmic) with a sum over k of values

$$H_0\big(T\Delta[k\Delta),\, T\Delta[\alpha)\big)\mathbf{P}\big(\tau \geqslant T(1 - k\Delta)\big) \quad \text{for} \quad k = 0, 1, \ldots, N-1,$$

assuming for simplicity that $N = 1/\Delta$ is integer. When the condition $\mathcal{A}^{\leqslant 0} \subset [\mathcal{A}_0]$ is met, the exponential part of $H_0\big(T\Delta[k\Delta),\, T\Delta[\alpha)\big)$ as $T \to \infty$ will have the form, according to Theorem 2.9.10 (concerning the l.l.d.p. for H_0), will have the form

$$\exp\{-TD(k\Delta, \alpha)\}, \tag{4.10.8}$$

where $D(v, \alpha)$ is a second deviation function corresponding to the vector $\xi = (\tau, \zeta)$ defined in section 2.9. Let

$$D_\Lambda(v, \alpha) = \inf_{\theta > 0} \theta \Lambda\left(\frac{v}{\theta}, \frac{\alpha}{\theta}\right)$$

and

$$\mathcal{D} := \big\{(v, \alpha) : D_\Lambda(v, \alpha) < \infty\big\}.$$

If $(v, \alpha) \notin \partial\mathcal{D}$ then

$$D(v, \alpha) = D_\Lambda(v, \alpha)$$

(see section 2.9).

Set

$$k\Delta = v \leqslant 1.$$

Then, by virtue of (4.10.7) and (4.10.8), in order to find the asymptotics of $-\frac{1}{T}\ln \mathbf{P}\big(Z(T) \in T\Delta[\alpha]\big)$ as $T \to \infty$, one needs to find for large T the minimal value over v of the function

$$D(v,\alpha) - \frac{1}{T}\ln \mathbf{P}\big(\tau > (1-v)T\big),$$

or, equivalently, the minimal value over $\theta \in (0,\infty)$ and $v \in [0,1]$ of function

$$\theta\Lambda\left(\frac{v}{\theta},\frac{\alpha}{\theta}\right) - \frac{1}{T}\ln \mathbf{P}\big(\tau > (1-v)T\big). \tag{4.10.9}$$

Denote

$$L(T,v) = -\frac{1}{T}\ln \mathbf{P}\big(\tau > T(1-v)\big). \tag{4.10.10}$$

Then

$$-\frac{1}{T}\ln \mathbf{P}\left(\frac{Z(T)}{T} \in \Delta[\alpha]\right) \sim \inf_{v\in[0,1]}\big[D(v,\alpha) + L(T,v)\big] \quad \text{as} \quad T \to \infty. \tag{4.10.11}$$

The limit of the right-hand side of (4.10.11) as $T \to \infty$ (if it exists) gives the desired value $\Lambda^{(Z)}(\alpha)$.

Set

$$\lambda_+^{(\tau)} := \sup\big\{\lambda_{(1)} : \psi^{(\tau)}(\lambda_{(1)}) < \infty\big\},$$
$$\lambda_{+,0} := \sup\big\{\lambda_{(1)} : \lambda \in \mathcal{A}^{\leqslant 0}\big\} \equiv \sup\big\{\lambda_{(1)} : \psi(\lambda) \leqslant 1\big\}.$$

Let v_α be a point where the infimum is reached in the definition of function $\widehat{D}(\alpha)$:

$$\widehat{D}(\alpha) := \inf_v\big[D(v,\alpha) + (1-v)\lambda_+^{(\tau)}\big]. \tag{4.10.12}$$

Therefore, if $(v_\alpha,\alpha) \notin \partial$ then

$$\widehat{D}(\alpha) = \inf_{v,\theta}\left[\theta\Lambda\left(\frac{v}{\theta},\frac{\alpha}{\theta}\right) + (1-v)\lambda_+^{(\tau)}\right],$$

where the infimum is taken over the domain $v \in [0,1]$, $\theta \in (0,\infty)$.

Further, the following assertion will be useful.

Lemma 4.10.1. *If* $\lambda_{+,0} \leqslant \lambda_+^{(\tau)}$ *then, for all* α, *one has*

$$\widehat{D}(\alpha) = D(1,\alpha). \tag{4.10.13}$$

The relation (4.10.13) could be written in the form $v_\alpha \equiv 1$. It follows from the lemma that under $\lambda_+^{(\tau)} = \infty$ the equalities (4.10.13) and $v_\alpha \equiv 1$ always hold.

Proof of Lemma 4.10.1[3] By virtue of Theorem 2.9.2, for any $\lambda \in \mathcal{A}^{\leqslant 0}$ and $v \geqslant 0$ we have

$$D(v,\alpha) \geqslant v\lambda_{(1)} + \alpha\lambda_{(2)} \geqslant \lambda_{(1)} + \alpha\lambda_{(2)} - \lambda_{(1)}(1-v). \tag{4.10.14}$$

For given $\varepsilon > 0$ and $N < \infty$ choose a point $\widehat{\lambda} = \widehat{\lambda}(\varepsilon, N) \in \mathcal{A}^{\leqslant 0}$ such that

$$\widehat{\lambda}_{(1)} + \alpha\widehat{\lambda}_{(2)} \geqslant \min\big\{D(1,\alpha), N\big\} - \varepsilon$$

(if $D(1,\alpha) < \infty$ then we can assume that $N = \infty$). Owing to (4.10.14), we have

$$\begin{aligned} D(v,\alpha) + \lambda_+^{(\tau)}(1-v) &\geqslant \widehat{\lambda}_{(1)} + \alpha\widehat{\lambda}_{(2)} + (1-v)(\lambda_+^{(\tau)} - \widehat{\lambda}_{(1)}) \\ &\geqslant \min\big\{D(1,\alpha), N\big\} - \varepsilon + (1-v)(\lambda_+^{(\tau)} - \widehat{\lambda}_{(1)}). \end{aligned}$$

Since $\lambda_+^{(\tau)} \geqslant \lambda_{+,0}$, for all $v \in [0,1]$ we have

$$(1-v)(\lambda_+^{(\tau)} - \widehat{\lambda}_{(1)}) \geqslant 0$$

and hence

$$D(v,\alpha) + \lambda_+^{(\tau)}(1-v) \geqslant \min\big\{D(1,\alpha), N\big\} - \varepsilon.$$

Therefore,

$$\widehat{D}(\alpha) \geqslant \min\big\{D(1,\alpha), N\big\} - \varepsilon.$$

Since N and ε are arbitrary, we have

$$\widehat{D}(\alpha) \geqslant D(1,\alpha).$$

The inverse inequality $\widehat{D}(\alpha) \leqslant D(1,\alpha)$ is obvious. The lemma is proved. □

The main assertion of the section is the following theorem.

Theorem 4.10.2. *Assume that for* (τ,ζ), (τ_0,ζ_0) *conditions* $[\mathbf{C}_0]$, $\mathcal{A}^{\leqslant 0} \subset [\mathcal{A}_0]$ *and* $(v_\alpha,\alpha) \notin \partial\mathcal{D}$ *are met.*

In the following statements (ii)–(v) *in the case* $\alpha = 0$ *it is additionally assumed that* $\lambda_+^{(\tau_0)} := \sup\big\{t : \psi^{(\tau_0)}(t) < \infty\big\} \geqslant \lambda_+^{(\tau)}$.

(i) *Let* $\lambda_+^{(\tau)} = \infty$. *Then* (4.10.6) *holds and*

$$\Lambda^{(Z)}(\alpha) = D(1,\alpha). \tag{4.10.15}$$

(ii) *Let* $\lambda_+^{(\tau)} < \infty$ *and a 'rough' smoothness condition for the distribution of* τ *be met:*

$$\ln \mathbf{P}(\tau \geqslant T) \sim -\lambda_+^{(\tau)} T \quad as \quad T \to \infty. \tag{4.10.16}$$

Then (4.10.6) *holds and*

$$\Lambda^{(Z)}(\alpha) = \widehat{D}(\alpha). \tag{4.10.17}$$

[3] The proof of Lemma 4.10.1, representations (2.9.10) in Theorem 2.9.2 and the upper bounds in Theorem 2.9.10 were proposed by A.A. Mogul'skii.

(iii) *If* $\lambda_+^{(\tau)} < \infty$ *and for all* (t, α) *the inequality*

$$\frac{\partial}{\partial t} \Lambda(t, \alpha) \leqslant \lambda_+^{(\tau)} \tag{4.10.18}$$

or the inequality

$$\lambda_{+,0} \leqslant \lambda_+^{(\tau)} \tag{4.10.19}$$

is valid then $\widehat{D}(\alpha) = D(1, \alpha)$ *and* (4.10.6) *holds together with* (4.10.15).

(iv) *If* τ *and* ζ *are independent then* (4.10.6) *also holds together with* (4.10.15).

(v) *In the vicinity of* $\alpha_0 := \mathbf{E}\zeta/\mathbf{E}\tau$ *the condition* (4.10.6) *holds, where, as* $\alpha \to \alpha_0$,

$$\Lambda^{(Z)}(\alpha) = D(1, \alpha) = \frac{(\alpha - \alpha_0)^2}{2} D''_{\alpha_0} + O\big(|\alpha - \alpha_0|^3\big). \tag{4.10.20}$$

The coefficient

$$D''_{\alpha_0} := \frac{\partial^2}{\partial \alpha^2} D(1, \alpha)\big|_{\alpha=\alpha_0} > 0 \tag{4.10.21}$$

can be found in an explicit form.

Note that if τ and ζ are independent then the condition $(v_\alpha, \alpha) \notin \partial\mathcal{D}$ takes the form:

$$\alpha \neq 0, \quad \text{if } \xi \textit{ takes values having the same sign.}$$

If the possible values of ξ have different signs then there are no restrictions on the domain of (v_α, α).

Thus, under wide assumptions, the deviation function $\Lambda^{(Z)}(\alpha)$ for the sequence $Z(T)$ coincides with the second deviation function $D(v, \alpha)$ (see section 2.9) at the point $v = 1$. Hence, the asymptotics for $\ln \mathbf{P}\big(Z(T) \in T\Delta[\alpha)\big)$ coincides with the asymptotics for $\ln H\big(T\Delta[(1, \alpha))\big)$ and $\ln P\big(T\Delta[(1, \alpha))\big)$ (see section 2.9), where the renewal function H and probability P are relative to the bivariate random walk $\{S_n\}$.

Remark 4.10.3. Note that the deviation function $\Lambda^{(Z)}(\alpha)$, by virtue of (4.10.6), is lower semicontinuous (see section 4.1). Since the function $D(1, \alpha)$ is lower continuous with respect to α (see Theorem 2.9.2), the statement of the theorem about the fact that $\Lambda^{(Z)}(\alpha) = D(1, \alpha)$ will be correct for all α, and the condition $(v_\alpha, \alpha) \notin \partial\mathcal{D}$ will be redundant. Clearly, for similar reasons, it will also be redundant for the statement $\Lambda^{(Z)}(\alpha) = \widehat{D}(\alpha)$ (see (4.10.17)).

The 'rough' smoothness condition (4.10.16) means that the distribution of τ meets the l.d.p. with deviation function $\alpha\lambda_+$ (see Remark 4.1.13).

Proof of Theorem 4.10.2. First, let us assume that $\alpha \neq 0$.

(i) Since $L(T, 1) = 0$, the right-hand side of (4.10.11) is not greater than $D(1, \alpha) < \infty$.

Since $\lambda_+^{(\tau)} = \infty$, we have $\Lambda(T) \gg T$ as $T \to \infty$ and, for all $v < 1 - \varepsilon$, $\varepsilon > 0$, by virtue of Chebyshev's inequality,

$$L(T, v) \geqslant \frac{1}{T} \Lambda\big(T(1 - v)\big) \to \infty \tag{4.10.22}$$

as $T \to \infty$. It means, that the infimum on the right-hand side of (4.10.11) can only be reached at points $v_{(T)} \to 1$ as $T \to \infty$. However, that right-hand side is greater than $D(v_{(T)}, \alpha)$. But $(1, \alpha) \notin \partial\mathcal{D}$ and the second deviation function $D(v, \alpha)$ of the vector ξ is continuous in the vicinity of the point $(1, \alpha)$ and hence by virtue of Theorem 2.9.2 it is continuous with respect to v in the vicinity of $v = 1$. It follows that, when $T \to \infty$, the right-hand side of (4.10.11) is greater than $D(1, \alpha) + o(1)$. It means that there exists a limit of the right-hand side of (4.10.11) as $T \to \infty$, which is equal to $D(1, \alpha)$. Relation (4.10.15) is proved.

(ii) By the 'rough' smoothness condition (4.10.16) of the distribution of τ we have

$$-\lim_{t \to \infty} \frac{1}{T} \ln \mathbf{P}\big(\tau \geqslant (1 - v)T\big) = (1 - v)\lambda_+^{(\tau)},$$

and the limit of the right-hand side of (4.10.11) is equal to

$$\inf_v \big\{D(v, \alpha) + (1 - v)\lambda_+^{(\tau)}\big\} = \inf_{v,\theta} \left\{\theta\Lambda\left(\frac{v}{\theta}, \frac{\alpha}{\theta}\right) + (1 - v)\lambda_+^{(\tau)}\right\} = \widehat{D}(\alpha). \tag{4.10.23}$$

(iii) If conditions (4.10.16) and (4.10.18) are met then the derivative with respect to v of the function under the infimum sign on the right-hand side of (4.10.23) is non-positive and, therefore, the infimum is reached when $v = v_\alpha = 1$; the equality $\Lambda^{(Z)}(\alpha) = D(1, \alpha)$ is proved.

If the smoothness condition (4.10.16) is not met then the lower bound of (4.10.9) with respect to θ for large T and $v = 1$ still equals $D(1, \alpha)$, and the value of (4.10.9) for $v < 1$, by virtue of Chebyshev's exponential inequality, is not less than

$$\theta\Lambda\left(\frac{v}{\theta}, \frac{\alpha}{\theta}\right) + (1 - v)\lambda_+^{(\tau)},$$

which by the preceding discussion is not less then $D(1, \alpha)$. Again, this proves (4.10.15).

If condition (4.10.19) is met then the equality $\widehat{D}(\alpha) = D(1, \alpha)$ follows from Lemma 4.10.1.

(iv) We now show that for independent τ and ζ the condition (4.10.18) is always met. Indeed, in this case under a natural convention about notation we have

$$\Lambda(t, \beta) = \Lambda^{(\tau)}(t) + \Lambda^{(\zeta)}(\beta),$$

$$\frac{\partial}{\partial t} \Lambda(t, \beta) = \lambda^{(\tau)}(t) \leqslant \lambda_+^{(\tau)},$$

i.e. condition (4.10.18) is met.

It is obvious that for independent τ and ζ we have $\lambda_{+,0} = \lambda_+^{(\tau)}$ and, therefore, condition (4.10.19) is also met.

(v) Let $a_\tau = \mathbf{E}\tau$, $a_\zeta = \mathbf{E}\zeta$ and $\alpha = \alpha_0 + \gamma a_\tau$. Then

$$D(1,\alpha) = \frac{1}{a_\tau} D(a_\tau, a_\zeta + \gamma).$$

The behaviour of the function $D(\cdot)$ in the vicinity of the point $a = (a_\tau, a_\zeta)$ was studied in section 2.9 (see Theorem 2.9.4). In that theorem, α denoted a *vector* (this discrepancy in notation is insurmountable; in what follows it should not cause ambiguity). In relation to our case, the coordinates $\alpha_{(1)}, \alpha_{(2)}$ of that vector should be taken as $\alpha_{(1)} = a_\tau, \alpha_{(2)} = a_\zeta + \gamma$, so that the vector δ in Theorem 2.9.4 has the form of $\delta = (0,\gamma)$ and cannot be collinear to the vector $a = (a_\tau, a_\zeta)$ since $a_\tau > 0$. This means that the inequality (4.10.21) and the second equality in (4.10.20) are immediate corollaries of Theorem 2.9.4. The coefficient D''_α can be easily found with the help of the representation (2.9.21), where one should take $\delta_{(1)} = 0$ and note that the desired coefficient coincides up to a constant factor $1/a_\tau$ with the second derivative of quadratic form in (2.9.21) with respect to $\delta_{(2)}$ at the point $\delta_{(2)} = 0$.

Since $L(T,v) \geqslant \lambda_+^{(\tau)}(1-v)$ (with an obvious interpretation of this inequality in the case $\lambda_+^{(\tau)} = \infty$), then (4.10.10), (4.10.11) and the fact that $D(v,\alpha) \geqslant c(\varepsilon) > 0$ for $|(v,\alpha) - (1,\alpha_0)| \geqslant \varepsilon$ imply the first equality in (4.10.20).

The theorem is proved in the case $\alpha \neq 0$.

Consider now the case $\alpha = 0$. In this case the term $\mathbf{P}(\tau_0 \geqslant T)$ is added to the integral (4.10.7). Under conditions of (i), $\lambda_+^{(\tau)} = \infty$, $\lambda_+^{(\tau_0)} = \infty$, we have

$$-\frac{1}{T} \ln \mathbf{P}(\tau_0 \geqslant T) \to \infty$$

as $T \to \infty$, and the appearance of a new summand does not change the foregoing argument.

Now let $\alpha = 0$, $\lambda_+^{(\tau)} < \infty$. Then, by the statement of the theorem, $\lambda_+^{(\tau_0)} \geqslant \lambda_+^{(\tau)} < \infty$ and

$$-\frac{1}{T} \ln \mathbf{P}(\tau_0 \geqslant T) \geqslant \lambda_+^{(\tau)} + o(1).$$

Therefore, in consideration of (ii)–(v), the minimum of the previously found value of $\Lambda^{(Z)}(\alpha)$ with $\alpha = 0$ and some new value that is not lower than $\lambda_+^{(\tau)}$ should be taken as $\Lambda^{(Z)}(0)$. But, according to the definition of the function $\widehat{D}(\alpha)$ (see part (ii) and (4.10.12)),

$$\widehat{D}(\alpha) \leqslant \lim_{\theta \to 0} \left(\theta \Lambda(0,0) + \lambda_+^{(\tau)}\right) = \lambda_+^{(\tau)}$$

for $\alpha = 0$. Thus, statement (ii) remains true. In a similar (but simpler) way, one can verify that statements (iii)–(v) also remain true. The theorem is proved. □

Remarks

(1) If random variables τ and ζ are dependent then the infimum in (4.10.23) might be reached when $v < 1$. Consider, for example, the case $\zeta \equiv \tau$. Then $\Lambda(t,\alpha) = \Lambda^{(\tau)}(\alpha)$ for $t = \alpha$ and $\Lambda(t,\alpha) = \infty$ for $t \neq \alpha$. Therefore, if $\alpha < 1$ then the infimum in (4.10.23) might only be reached when $v = \alpha < 1$. It is evident that, for non-degenerate distributions of (τ, ζ), when $\zeta = \tau + \gamma$ and the value of γ does not depend on τ, $|\gamma| < \delta$ for small δ, then $\Lambda(t,\alpha) = \infty$ for $|t - \alpha| > \delta$ and the infimum in (4.10.23) might also be reached when $v < 1$.

(2) If $D(1,\alpha) = \Lambda^{(Z)}(\alpha) = \infty$ (e. g. that is the case when $\zeta = \tau$, $\lambda_+^{(\tau)} = \infty$ and $\alpha < 1$) then this does not imply (as in the case of the l.d.p. for sums S_n) that the corresponding prelimit probabilities $\mathbf{P}\big(Z(T) \in T\Delta[\alpha)\big)$ are equal to 0. If, say, $\tau \geqslant 1$, $\mathbf{P}(\tau \geqslant t) \sim ce^{-t^2}$, then in the example mentioned above, where $\zeta = \tau$, we will have

$$-\frac{1}{T^2} \ln \mathbf{P}\big(Z(T) \in t\Delta[\alpha)\big) \sim (1-\alpha)^2.$$

(3) If $\lambda_+^{(\tau)} < \infty$, the infimum in the definition of $D(\alpha)$ is reached when $v < 1$ but the 'rough' smoothness condition for the distribution of τ fails, then the limit as $T \to \infty$ of the right-hand side of (4.10.11) might not exist and the l.d.p. for processes $Z(t)$ would fail. This can be seen by the following example: $\tau_0 \overset{=}{\underset{d}{}} \tau$ and τ takes values 2^k, $k = 1, 2, \ldots$, with probabilities $ck^{-2}e^{-2^k}$. Then

$$\psi^{(\tau)}(1) < \infty,\ \lambda_+ = 1, \text{ but } \lim_{T\to\infty} \frac{1}{T} \ln \mathbf{P}\big(\tau \geqslant T(1-v)\big) \text{ does not exist.}$$

The assertions that we have obtained show that in any case the form of the deviation function under the integral sign of $J(f)$ in (4.10.1) becomes significantly complicated here. However, the function $\Lambda^{(Z)}$ still possesses all the basic properties of the deviation function mentioned in section 1.1.

Theorem 4.10.2 implies that if $\lambda_+^{(\tau)} = \infty$ or τ and ζ are independent then $\Lambda^{(Z)}(\alpha) = D(1,\alpha)$. Otherwise, to find $\Lambda^{(Z)}(\alpha)$ one needs to use additional conditions. Let us bring in another assertion with upper and lower bounds for the considered probabilities. This assertion complements Theorem 4.10.2 and in a number of cases makes it possible to deduce the l.l.d.p. for $Z(T)$ in a stronger form – the for probability of hitting a domain that is narrower than $T\Delta[\alpha)$, e.g. a half-open interval $\Delta_{(2)}[T\alpha)$ for fixed $\Delta_{(2)} > 0$ (the notation $\Delta_{(1)}$ is used for the first coordinate of the vector (v, α)). In what follows, we will assume that $\Delta_{(2)}$ is an arbitrary fixed positive real, if ζ is non-lattice, and $\Delta_{(2)} \geqslant 1$ is an arbitrary fixed integer if ζ is arithmetic. In the latter case ζ_0 is assumed to be an integer-valued random variable.

Theorem 4.10.4. *Let (τ, ζ) meet the condition* $[\mathbf{C}_0]$.

(i) *If $\lambda_+^{(\tau)} < \infty$ then*

$$\overline{\lim_{T\to\infty}} \frac{1}{T} \ln \mathbf{P}\big(Z(T) \in T\Delta_T[\alpha)\big) \leqslant -\widehat{D}(\alpha), \tag{4.10.24}$$

where $\Delta_T \to 0$ slowly enough as $T \to \infty$.

(ii) *Let* $\lambda_+^{(\tau)} < \infty$ *and conditions* (4.10.16), $\mathcal{A} \subset [\mathcal{A}_0]$ *be met. Then*

$$\varliminf_{T\to\infty} \frac{1}{T} \ln \mathbf{P}\big(Z(T) \in \Delta_{(2)}[T\alpha)\big) \geqslant -\widehat{D}(\alpha). \tag{4.10.25}$$

(iii) *In any case,*

$$\varliminf_{T\to\infty} \frac{1}{T} \ln \mathbf{P}\big(Z(T) \in \Delta_{(2)}[T\alpha)\big) \geqslant -D(1,\alpha). \tag{4.10.26}$$

Clearly, condition $\mathcal{A} \subset [\mathcal{A}_0]$ in part (ii) could be weakened.

Proof. (i) Since $\Lambda^{(\tau)}(T) \sim \lambda_+^{(\tau)} T$ as $T \to \infty$, the first statement of the theorem follows from (4.10.11) and Chebyshev's inequality:

$$\varlimsup_{T\to\infty} \frac{1}{T} \ln \mathbf{P}\big(\tau \geqslant (1-v)T\big) \leqslant (1-v)\lambda_+^{(\tau)}.$$

(ii) Let v_α, θ_α be such that $v_\alpha < 1$,

$$\widehat{D}(\alpha) = \theta_\alpha \Lambda\left(\frac{v_\alpha}{\theta_\alpha}, \frac{\alpha}{\theta_\alpha}\right) + (1 - v_\alpha)\lambda_+^{(\tau)}.$$

Set $n = [T\theta_\alpha]$ and consider a rectangular half-open parallelepiped $\Delta[x) = \Delta_{(1)}[x_{(1)}) \times \Delta_{(2)}[x_{(2)})$ with a vertex at the point $x = (x_{(1)}, x_{(2)})$ and sides of length $\Delta_{(1)}$ and $\Delta_{(2)}$, where the same convention as for $\Delta_{(2)}$, stated before Theorem 4.10.4, acts for $\Delta_{(1)}$ but with respect to the random variable τ. Then

$$P_T := \mathbf{P}\big(Z(T) \in \Delta_{(2)}[T\alpha)\big) \geqslant \mathbf{P}\big(T_{0,n} \in \Delta_{(1)}[Tv_\alpha),\, Z_{0,n} \in \Delta_{(2)}[T\alpha)\big) \\ \times \mathbf{P}\big(\tau > (1 - v_\alpha)T\big).$$

According to the l.l.d.p. in Theorem 2.8.13 and condition (4.10.16), we have, as $T \to \infty$,

$$\frac{1}{T} \ln P_T \sim -\frac{n}{T}\Lambda\left(\frac{v_\alpha T}{n}, \frac{\alpha T}{n}\right) - (1 - v_\alpha)\lambda_+^{(\tau)} \sim -\widehat{D}(\alpha).$$

(iii). Let θ_α be such that

$$\theta_\alpha \Lambda\left(\frac{1}{\theta_\alpha}, \frac{\alpha}{\theta_\alpha}\right) = D(1,\alpha), \qquad n = [T\theta_\alpha].$$

Then

$$P_T \geqslant \mathbf{P}\left(T_{0,n} \in \Delta_{(1)}[T - \Delta_{(1)}),\, Z_{0,n} \in \Delta_{(2)}[T\alpha)\right) \mathbf{P}(\tau \geqslant \Delta_{(1)}).$$

By virtue of Theorem 2.8.13,

$$\frac{1}{T} \ln P_T \geqslant -\frac{n}{T}\Lambda\left(\frac{T}{n}, \frac{\alpha T}{n}\right) + \frac{\ln \mathbf{P}(\tau \geqslant \Delta_{(1)})}{T} + o(1),$$

where we choose $\Delta_{(1)}$ such that $\mathbf{P}(\tau \geqslant \Delta_{(1)}) > 0$. This implies (4.10.26).

The theorem is proved. □

From parts (i) and (iii) of Theorem 4.10.4 we obtain

Corollary 4.10.5. *If the conditions* $[\mathbf{C}_0]$ *are satisfied,* $\lambda_+^{(\tau)} < \infty$ *and* $\widehat{D}(\alpha) = D(1,\alpha)$*, then there exists the limit*

$$\lim_{T\to\infty} \frac{1}{T} \ln \mathbf{P}\big(Z(T) \in \Delta_{(2)}[T\alpha)\big) = -D(1,\alpha).$$

As we saw in section 4.9, the form of the function $\Lambda^{(Z)}(\alpha)$ becomes significantly simplified if $Z(t)$ is a compound Poisson process, i.e. if τ has an exponential distribution.

4.11 On large deviation principles for sums of random variables defined on a finite Markov chain

Let X_n, $n = 0, 1, \ldots$, be an ergodic Markov chain with a finite number d of states $1, \ldots, d$ and a transition matrix $\|p_{kj}\|_{k,j=1}^{d}$. Let $\xi^{(k,j)}$, $\xi_n^{(k,j)}$, $n = 1, 2, \ldots$, for all (k,j) be independent (among themselves) sequences of independent identically distributed random variables, satisfying Cramér's condition $[\mathbf{C}]$: for all $k,j = 1, \ldots, d$

$$\psi^{(k,j)}(\lambda) := \mathbf{E}e^{\lambda\xi^{(k,j)}} < \infty$$

is fulfilled for $\lambda \in (\lambda_-, \lambda_+)$ and some $\lambda_- < \lambda_+$.

Next, let

$$S_n := \sum_{m=1}^{n} \xi_m^{(X_{m-1},X_m)},$$

$$p_{kj}(n,B) := \mathbf{P}\big(S_n \in B,\ X_n = j \,|\, X_0 = k\big),$$

$$P_{kj}(n,\lambda) := \int e^{i\lambda t} p_{kj}(n,dt), \qquad P(n,\lambda) := \|P_{kj}(n,\lambda)\|,$$

$$P(\lambda) := \|p_{kj}\psi^{(k,j)}(\lambda)\| = P(1,\lambda).$$

Then

$$P(n,\lambda) = P^n(\lambda).$$

If $\psi(\lambda)$ is a maximal eigenfunction of the matrix $P(\lambda)$ (the solution of the equation $\big|P(\lambda) - zE\big| = 0$, where E is an identity matrix) then, according to Perron's theorem,

$$P_{kj}(n,\lambda) = \psi^n(\lambda)\pi_j(\lambda) + O\big(q^n\psi^n(\lambda)\big) \tag{4.11.1}$$

as $n \to \infty$, where $\psi(0) = 1$, $q < 1$, $\pi_j(\lambda)$ are functions that are analytic on (λ_-, λ_+) and continuous at the point $\lambda = 0$ and $\{\pi_j(0)\}$ is a stationary distribution of the chain $\{X_n\}$. For simplicity, let the distribution of the sum $\sum_{k,j} \xi^{(k,j)}$ have an absolutely continuous component. Then, by applying an inversion formula to find $p_{k,j}\big(n, \Delta[x)\big)$, we get *as a principal part* n integral of the form (1.5.6) (up to the factor $\pi_j(\lambda)$ under the integral sign), if $\varphi(\lambda)$ is taken as $\varphi(\lambda) = \psi(i\lambda)$. Recall, that we had this integral (1.5.6) in the inversion formula for the distribution of sums S_n of independent identically distributed random variables. The only difference is that now the role of the Laplace transform $\psi(\lambda) = \mathbf{E}e^{\lambda\xi}$ in formula of the type (1.5.6) is played by the maximal eigenfunction of the matrix $P(\lambda)$. In addition, as is easy to deduce, for instance from the results of [130] or [106], the function $A(\lambda) = \ln \psi(\lambda)$ is convex. Therefore, for all α the deviation function (the Legendre transformation of $A(\lambda)$)

$$\Lambda(\alpha) = \sup_{\lambda} \big(\lambda\alpha - A(\lambda)\big), \tag{4.11.2}$$

is defined, which in its main properties (analyticity on (λ_-, λ_+), convexity, the relation $\lambda(\alpha) = \Lambda'(\alpha)$, where $\lambda(\alpha)$ is the point where the supremum in (4.11.2) is reached and so on) is not different from the deviation function $\Lambda(\alpha)$ studied in Chapter 1. By changing in (1.5.6) the path of integration in (1.5.6) in such a way that it passes the point $-i\lambda(\alpha)$, $\alpha = x/n$ (i.e. by applying the method of steepest descent; in other words, a Cramér transformation with consequent application of the Laplace method), we will obtain for $\lambda(\alpha) \in (\lambda_-, \lambda_+)$ integro-local theorems for the probabilities $p_{kj}\big(n, \Delta[x)\big)$ for $x = \alpha n$, wherein the exponential factor will have the form $\exp\big\{-n\Lambda(\alpha)\big\}$.

With the same arguments as were used in the proof of Theorem 1.1.2, we obtain the next assertion (the l.l.d.p. for the sums S_n)

Theorem 4.11.1. *For all α and $\varepsilon > 0$ there exist the limits*

$$\lim_{n\to\infty} \frac{1}{n} \ln \mathbf{P}\bigg(\frac{S_n}{n} \in (\alpha)_\varepsilon,\ X_n = j \,|\, X_0 = k\bigg) = -\Lambda\big((\alpha)_\varepsilon\big),$$

$$\lim_{\varepsilon\to 0}\lim_{n\to\infty} \frac{1}{n} \ln \mathbf{P}\bigg(\frac{S_n}{n} \in (\alpha)_\varepsilon,\ X_n = j \,|\, X_0 = k\bigg) = -\Lambda(\alpha). \tag{4.11.3}$$

Proof. As before, this consists of obtaining upper and lower bounds.

Upper bounds can be obtained with Chebyshev's inequalities: for $\alpha = x/n \geqslant 0$, $\lambda \geqslant 0$,

$$\mathbf{P}\bigg(\frac{S_n}{n} \geqslant \alpha,\ X_n = j \,|\, X_0 = k\bigg) \leqslant e^{-\lambda x} P_{kj}(n, \lambda).$$

If $\alpha \geqslant \sum_{k,j} \pi_k(0) p_{kj} \mathbf{E}\xi^{(k,j)}$ then $\lambda(\alpha) \geqslant 0$ and, by setting $\lambda = \lambda(\alpha)$, we will obtain, by virtue of (4.11.1), the upper bound for the probability under consideration,

$$\pi_j\big(\lambda(\alpha)\big) e^{-n\Lambda(\alpha)} + O\big(q^n e^{-n\Lambda(\alpha)}\big).$$

We now repeat the arguments in the proof of Theorem 1.1.2 to obtain the upper bound, which has the form of relation (4.11.3) when lim is replaced by $\overline{\lim}$ and the sign = is replaced by the sign $\leqslant$.

The lower bound is obtained in the same manner as in the proof of Theorem 1.1.2, with the help of truncations for the variables $\xi_n^{(k,j)}$ at level N and the use of the aforementioned integro-local theorems for $p_{kj}\big(n, \Delta[x)\big)$ applied to sums of the truncated variables.

It is clear that the l.d.p. for trajectories of $S_{[nt]}$, $t \in [0, 1]$, will also have an integral $J(f)$ of the form (4.10.1) as a deviation functional. For more details see [130]. □

5

Moderately large deviation principles for the trajectories of random walks and processes with independent increments

As before, let $\xi, \xi_1, \xi_2, \ldots$ be a sequence of identically distributed random vectors of dimension d and let

$$S_0 := 0, \qquad S_n := \sum_{i=1}^{n} \xi_i \quad \text{for} \quad n \geqslant 1.$$

In what follows, we assume that $\mathbf{E}|\xi|^2 < \infty$. We can suppose, without loss of generality, that in the problems to be considered in sections 5.1–5.4 one has $\mathbf{E}\xi = 0$, and that the covariance matrix of the vector ξ is a unit matrix.

As in Chapter 4, we will study the asymptotic behaviour of

$$\ln \mathbf{P}(s_n \in B), \quad \text{as} \quad n \to \infty,$$

for continuous random polygons $s_n = s_n(t)$, $0 \leqslant t \leqslant 1$, with nodes at the points

$$\left(\frac{k}{n}, \frac{1}{x} S_k\right), \qquad k = 0, \ldots, n,$$

and where B is a measurable set of continuous functions on $[0, 1]$. However, now we will consider 'moderately large' deviations $x = x(n)$ with the properties

$$\lim_{n\to\infty} \frac{x}{\sqrt{n}} = \infty, \quad \lim_{n\to\infty} \frac{x}{n} = 0. \tag{5.0.1}$$

5.1 Moderately large deviation principles for sums S_n

5.1.1 Moment conditions and corresponding deviation zones

Along with the condition $[\mathbf{C}_0]$ we will define a more general moment condition describing distributions with heavier tails. To this end, introduce a class $\mathcal{L}_\beta$, $\beta \in (0, 1)$, of functions $l(t)$ such that:

(1) *$l(t) = t^\beta L(t)$ is a regularly varying function (r.v.f.) ($L(t)$ is a slowly varying function (s.v.f.) as $t \to \infty$);*

(2) *$l(t)$ satisfies the relation*

$$l(t+v) - l(t) = \beta v \frac{l(t)}{t}\big(1+o(1)\big) + o(1)$$

as $t \to \infty$ with $v = o(t)$.

The above-mentioned condition, more general than **[C$_0$]**, has the form:

[C(β)]

$$\mathbf{P}\big(|\xi| \geqslant t\big) \leqslant e^{-l(t)} \quad \textit{for} \quad t > 0,$$

where $l(t) \in \mathcal{L}_\beta$, $\beta \in (0,1)$.

In the one-dimensional case the class of distributions $\mathcal{S}e$ such that $\mathbf{P}(\xi \geqslant t) = e^{-l(t)}$, $l \in \mathcal{L}_\beta$, is called semi-exponential (see e.g. [42], Chapter 5). We have made use already of this class in subsection 2.3.4. For that class, the probabilities of large deviations were studied in detail for the sums S_n (see e.g. [42], Chapter 5, and [55]). There exist several other distribution classes that are close to $\mathcal{S}e$ for which the large deviation probabilities for S_n have also been studied (see e.g. [102], [142], [144], [150], [162], [165], [127]).

When condition **[C$_0$]** is met, we will be dealing with deviations $x = x(n)$ defined by (5.0.1). If condition **[C$_0$]** is not satisfied but condition **[C(β)]** is met, we will consider a narrower class of moderately large deviations $x = x(n)$ with properties

$$\lim_{n\to\infty} \frac{x}{\sqrt{n}} = \infty, \quad \lim_{n\to\infty} \frac{x}{\widehat{x}(n)} = 0,$$

where the regularly varying sequence $\widehat{x}(n)$ grows more slowly than n as $n \to \infty$. The sequence is defined as follows. For the function $l \in \mathcal{L}_\beta$ from condition **[C(β)]** consider the function $\omega(t) := l(t)t^{-2} = t^{\beta-2}L(t)$ and define $\widehat{x}(n)$ as the value

$$\widehat{x}(n) = \omega^{(-1)}(1/n)$$

of the generalised function

$$\omega^{(-1)}(t) := \sup\big\{u \geqslant 0 : \ \omega(u) \geqslant t\big\} \tag{5.1.1}$$

inverse to $\omega(t)$ at the point $t = 1/n$. It is known (see e.g. [42], p. 238) that the function $\omega^{(-1)}(1/n)$ has the form

$$\omega^{(-1)}\left(\frac{1}{n}\right) = n^{1/(2-\beta)}L_1(n) = o(n),$$

where $L_1(n)$ is an s.v.f. as $n \to \infty$. In the special case where the s.v.f. $L(t)$ satisfies the additional condition

$$L(tL^{1/(2-\beta)}(t)) \sim L(t) \quad \text{as} \quad t \to \infty,$$

the s.v.f. $L_1(t)$ has the form

$$L_1(t) \sim L^{1/(2-\beta)}(t^{1/(2-\beta)}) \quad \text{as} \quad t \to \infty,$$

so in that case one has

$$\widehat{x}(n) = n^{1/(2-\beta)} L^{1/(2-\beta)}(n^{1/(2-\beta)}) \quad \text{as} \quad n \to \infty.$$

5.1.2 Moderately large deviation principles for sums S_n

First, we will establish a *local* m.l.d.p. Put

$$\Lambda_0(\alpha) := \frac{1}{2}|\alpha|, \quad \alpha \in \mathbb{R}^d, \qquad \Lambda_0(B) = \inf_{\alpha \in B} \Lambda_0(\alpha).$$

Later we need the following condition: *either condition* **[C$_0$]** *or condition* **[C(β)]** *is met, and the deviations $x = x(n)$ are such that as $n \to \infty$ one has*

$$\frac{x}{\sqrt{n}} \to \infty, \quad x(n) = \begin{cases} o(n), & \textit{if condition } [\mathbf{C}_0] \textit{ is met}, \\ o\big(\widehat{x}(n)\big), & \textit{if condition } [\mathbf{C}(\beta)] \textit{ is met}, \end{cases} \tag{5.1.2}$$

Theorem 5.1.1 (the local m.l.d.p. for S_n). (i) *If $x = o(n)$ and $x/\sqrt{n} \to \infty$ as $n \to \infty$ then, for any fixed $\alpha \in \mathbb{R}^d$, $\varepsilon > 0$,*

$$\varliminf_{n\to\infty} \frac{n}{x^2} \ln \mathbf{P}\left(\frac{S_n}{x} \in (\alpha)_\varepsilon\right) \geqslant -\Lambda_0(\alpha). \tag{5.1.3}$$

(ii) *If condition* (5.1.2) *is satisfied then, for any fixed $\alpha \in \mathbb{R}^d$ and $\varepsilon > 0$, one has*

$$\lim_{n\to\infty} \frac{n}{x^2} \ln \mathbf{P}\left(\frac{S_n}{x} \in (\alpha)_\varepsilon\right) = -\Lambda_0\big((\alpha)_\varepsilon\big), \tag{5.1.4}$$

$$\lim_{\varepsilon\to 0} \lim_{n\to\infty} \frac{n}{x^2} \ln \mathbf{P}\left(\frac{S_n}{x} \in (\alpha)_\varepsilon\right) = -\Lambda_0(\alpha). \tag{5.1.5}$$

Thus, for deviations $x = x(n)$ which satisfy conditions (5.1.2), there exist the limits

$$\lim_{\varepsilon\to 0} \varliminf_{n\to\infty} \frac{n}{x^2} \ln \mathbf{P}\left(\frac{S_n}{x} \in (\alpha)_\varepsilon\right) = \lim_{\varepsilon\to 0} \varlimsup_{n\to\infty} \frac{n}{x^2} \ln \mathbf{P}\left(\frac{S_n}{x} \in (\alpha)_\varepsilon\right) = -\Lambda_0(\alpha).$$

Therefore, according to the terminology from section 4.1, the sequence $\{S_n/x\}$ satisfies the local large deviation principle in the space $\big(\mathbb{R}^d, |\cdot|\big)$ with parameters $(x^2/n, \Lambda_0)$, where $|\cdot|$ denotes the Euclidean norm.

Now turn to the 'integral' m.l.d.p. for the sums S_n. Recall, that for a Borel set $B \subset \mathbb{R}^d$,

$$\Lambda_0(B) := \inf_{\alpha \in B} \Lambda_0(\alpha).$$

Denote by (B) and $[B]$, as before, the interior and closure of the Borel set $B \subset \mathbb{R}^d$.

Theorem 5.1.2 (The m.l.d.p. for S_n). (i) *If $x = o(n)$ and $x/\sqrt{n} \to \infty$ as $n \to \infty$ then, for any Borel set $B \subset \mathbb{R}^d$, one has*

$$\varliminf_{n\to\infty} \frac{n}{x^2} \ln \mathbf{P}\left(\frac{S_n}{x} \in B\right) \geqslant -\Lambda_0\big((B)\big).$$

(ii) *Under condition* (5.1.2), *for any Borel set* $B \subset \mathbb{R}^d$ *one has*

$$\overline{\lim_{n\to\infty}} \frac{n}{x^2} \ln \mathbf{P}\left(\frac{S_n}{x} \in B\right) \leqslant -\Lambda_0([B]). \tag{5.1.6}$$

So, in the deviation regions (5.1.2), the sequence $\{S_n/x\}$ satisfies the large deviation principle in the space $(\mathbb{R}^d, |\cdot|)$ with parameters $(x^2/n, \Lambda_0)$ (see section 4.1).

All the proofs in the present discussion will be given for the one-dimensional case $d = 1$. Transition to the general case $d \geqslant 1$ should cause no difficulties and either is obvious or can be done by considering the projections of the random walk S_k onto the coordinate axes. When the computations in the multivariate case are no more cumbersome than in the one-dimensional case, we will consider the multivariate case.

5.1.3 Proof of Theorem 5.1.1

I. Assertion (i) of Theorem 5.1.1 follows from a stronger assertion, which we will formulate as a lemma and will need in the next subsection (when proving Theorem 5.2.1). That lemma will allow us to avoid certain repetitions in the proofs of Theorems 5.1.1 and 5.2.1. Introduce the event A_ε by the formula

$$A_\varepsilon = A_\varepsilon(n, \alpha) := \left\{\max_{j\leqslant n}\left|\frac{S_j}{x} - \frac{j}{n}\alpha\right| < \varepsilon\right\}. \tag{5.1.7}$$

Lemma 5.1.3. *For any* $\varepsilon > 0$ *one has*

$$\underline{\lim_{n\to\infty}} \frac{n}{x^2} \ln \mathbf{P}(A_\varepsilon) \geqslant -\Lambda_0(\alpha). \tag{5.1.8}$$

Clearly (5.1.8) implies (5.1.3).

Proof. The proof is based on bounding from below the probability $\mathbf{P}(A_\varepsilon)$ by that of a bundle of trajectories that is narrower and simpler than A_ε (in the sense of its probabilistic structure). To construct that bundle in the one-dimensional case $d = 1$, consider in the (t, u)-plane 'strips' (sets of continuous functions $g(\cdot)$)

$$\mathcal{B}(b, \alpha, \varepsilon) := \left\{g(\cdot) : \left|g(t) - \frac{\alpha x t}{n}\right| \leqslant \frac{\varepsilon x}{2} \text{ for all } t \in [0, b],\ \left|g(b) - \frac{\alpha x b}{n}\right| \leqslant \frac{\varepsilon x b}{2n}\right\},$$

where $b > 0$ is an integer. We will say that these strips have length b (along the abscissa) and width εx (along the ordinate). Any such strip runs along the ray $(\alpha x t)/n$, $t > 0$, and has at its end a 'window' of width $(\varepsilon x b)/n$, whose central point $(\alpha x b)/n$ is located on the ray; the value $g(b)$ must be inside that window. Further, let $S_{(t)}$ be the continuous piecewise linear function on the half-line $t \geqslant 0$ that has its nodes at the points

$$(k, S_k), \quad k = 0, \ldots, \tag{5.1.9}$$

and let B_0 be the event

$$B_0 := \{S_{(\cdot)} \in \mathcal{B}(b, \alpha, \varepsilon)\}, \tag{5.1.10}$$

where $S_{(\cdot)}$ is the restriction of $S_{(t)}$ to the time interval $[0,b]$. Events B_j for $j \geqslant 1$ are defined in a similar way, but in (5.1.10) instead of $S_{(\cdot)}$ we use the projection of the increment $S_{(jb+t)} - S_{(jb)}$, $t \geqslant 0$, onto the interval $[0,b]$. In that way, we obtain

$$m+1 := \left[\frac{n}{b}\right]+1$$

events $B_0,\dots,B_m$ (note that to get them one may need the sums S_k with $k > n$). These events have the following two important properties:

(1) The events B_j are jointly independent and, on the intersection

$$B := \bigcap_{j=0}^{m} B_j,$$

the trajectory $S_{(t)}$ for $t \in [0,n]$ lies in an expanding strip (it expands at the time points jb by steps of size $(\varepsilon xb)/n$) such that the trajectory does not intersect the 'maximal' upper boundary

$$\frac{\alpha xt}{n} + \frac{\varepsilon xb}{n}m \leqslant \frac{\alpha xt}{n} + \varepsilon x.$$

A similar statement holds for the lower boundary. That means that the intersection B implies the event A_ε in (5.1.7), so that

$$\mathbf{P}(A_\varepsilon) \geqslant \mathbf{P}(B) = \prod_{j=0}^{m} \mathbf{P}(B_j) = \mathbf{P}^{m+1}(B_0). \tag{5.1.11}$$

(2) We have at our disposal the 'free' parameter b, which we choose such that the probability $\mathbf{P}(B_0)$ can be evaluated using the central limit theorem. The event B_0 can be written as

$$B_0 = \left\{\left|S_b - \frac{\alpha xb}{n}\right| < \frac{\varepsilon xb}{2n},\ \max_{k\leqslant b}\left|S_k - \frac{\alpha xk}{n}\right| < \frac{\varepsilon x}{2}\right\} = B_{0,1} \bigcap B_{0,2},$$

where

$$B_{0,1} := \left\{\left|\frac{S_b}{\sqrt{b}} - \frac{\alpha x\sqrt{b}}{n}\right| < \frac{\varepsilon x\sqrt{b}}{2n}\right\},$$

$$B_{0,2} := \left\{\max_{k\leqslant b}\left|S_k - \frac{\alpha xk}{n}\right| < \frac{\varepsilon x}{2}\right\}. \tag{5.1.12}$$

Consider the event $B_{0,1}$. For a fixed $N > 0$, set

$$b = \left[\frac{N^2n^2}{x^2}\right],$$

so that $b \to \infty$, and

$$m = \left[\frac{n}{b}\right] \sim \frac{x^2}{nN^2} \quad \text{as } n \to \infty. \tag{5.1.13}$$

Note also that

$$\frac{\alpha x\sqrt{b}}{n} \to \alpha N, \quad \frac{\varepsilon x\sqrt{b}}{2n} \to \frac{\varepsilon N}{2} \quad \text{as} \quad n \to \infty.$$

Therefore, by virtue of the central limit theorem,

$$\mathbf{P}(B_{0,1}) \to \Phi\big(N(\alpha - \varepsilon/2)\big) - \Phi\big(N(\alpha + \varepsilon/2)\big), \tag{5.1.14}$$

where

$$\Phi(t) = \frac{1}{\sqrt{2\pi}} \int_t^\infty e^{-v^2/2} dv.$$

Now consider the event $B_{0,2}$. Suppose for definiteness that $\alpha > 0$. The maximal value of $(\alpha xk)/n$ in (5.1.12) is attained at $k = b$ and is equal to

$$\frac{\alpha xb}{n} = o\left(\frac{\varepsilon x}{2}\right) \quad \text{as} \quad n \to \infty.$$

Therefore, for all large enough n, one has

$$B^*_{0,2} := \left\{\max_{k \leqslant b} S_k \leqslant \frac{\varepsilon x}{3}\right\} \subset B_{0,2}.$$

Since $b/x^2 \sim Nn^2/x^4 \to 0$ as $n \to \infty$, we find from Kolmogorov's inequality (see e.g. Lemma 4.1.5), that

$$\mathbf{P}(B_{0,2}) \geqslant \mathbf{P}(B^*_{0,2}) \geqslant 1 - \frac{9b}{(\varepsilon x)^2} \to 1 \quad \text{as} \quad n \to \infty.$$

It follows from this and (5.1.14) that, for any fixed $\varepsilon > 0$, as $n \to \infty$ one has

$$\mathbf{P}(B_0) \to \Phi\big(N(\alpha - \varepsilon/2)\big) - \Phi\big(N(\alpha + \varepsilon/2)\big).$$

Since

$$\Phi(t) \sim \frac{1}{t\sqrt{2\pi}} e^{-t^2/2} \text{ as } t \to \infty,$$

there exists an $N_0 = N_0(\alpha, \varepsilon) < \infty$ such that for $N \geqslant N_0$ one has the inequality $\Phi(N(\alpha - \varepsilon/2)) - \Phi(N(\alpha + \varepsilon/2)) \geqslant \Phi(N\alpha)$, and therefore

$$\lim_{n\to\infty} \mathbf{P}(B_0) \geqslant \Phi(N\alpha).$$

Hence, owing to (5.1.11), for $N \geqslant N_0$ one has

$$\varliminf_{n\to\infty} \frac{n}{x^2} \ln \mathbf{P}(A_\varepsilon) \geqslant \varliminf_{n\to\infty} \frac{n(m+1)}{x^2} \ln \Phi(N\alpha),$$

where, owing to (5.1.13),

$$\frac{n(m+1)}{x^2} \sim \frac{1}{N^2} \text{ as } n \to \infty.$$

Thus, for any fixed $\varepsilon > 0$ and $N \geqslant N_0$,

$$\varliminf_{n\to\infty} \frac{n}{x^2} \ln \mathbf{P}(A_\varepsilon) \geqslant \frac{1}{N^2} \ln \Phi(N\alpha), \tag{5.1.15}$$

where the right-hand side can be made arbitrarily close to $-\alpha^2/2$ by choosing N large enough. This means that the left-hand side of (5.1.15), which does not depend on N, is greater than or equal to $-\alpha^2/2$, and (5.1.8) holds true.

The same argument works in the case of negative α and also in the multivariate case. For $\alpha = 0$ the assertion of the lemma is trivial. Lemma 5.1.3 is proved. □

Note that, together with Lemma 5.1.3, we have proved assertion (i) of Theorem 5.1.1.

Similarly to the proof of Theorem 1.1.4 (see section 2.7) we obtain that along with (5.1.3) the following inequality holds:

$$\underline{\lim}_{n\to\infty} \frac{n}{x^2} \ln \mathbf{P}\Big(\frac{S_n}{x} \in (\alpha)_\varepsilon\Big) \geqslant -\Lambda_0\big((\alpha)_\varepsilon\big). \tag{5.1.16}$$

(ii) Now we will prove the second assertion of Theorem 5.1.1. First, we make use of the observation that, under the condition $[\mathbf{C}_0]$, for any $\alpha \in \mathbb{R}^d$ one has

$$\overline{\lim}_{n\to\infty} \frac{n}{x^2} \ln \mathbf{P}\Big(\frac{1}{x} S_n \in (\alpha)_\varepsilon\Big) \leqslant -\Lambda_0\big((\alpha)_\varepsilon\big). \tag{5.1.17}$$

The inequality follows from the exponential inequality of Chebyshev type for the multivariate case (see Corollary 1.3.7):

$$\mathbf{P}\big(S_n \in x(\alpha)_\varepsilon\big) \leqslant \exp\Big\{e^{-n\Lambda(\frac{x}{n}(\alpha)_\varepsilon)}\Big\}, \tag{5.1.18}$$

where $\Lambda(\alpha)$ is the deviation function for ξ and $\Lambda(B) := \inf_{\alpha\in B}\Lambda(\alpha)$. The inequalities (5.1.16) and (5.1.17) imply (5.1.4).

Further, as the function Λ is convex, there exists a point α_ε on the boundary of the zone $(\alpha)_\varepsilon$ such that

$$\Lambda\Big(\frac{x}{n}(\alpha)_\varepsilon\Big) = \Lambda\Big(\frac{x}{n}\alpha_\varepsilon\Big).$$

Since $x\alpha_\varepsilon/n \to 0$ as $n \to \infty$, one has

$$\Lambda\Big(\frac{x}{n}\alpha_\varepsilon\Big) = \frac{x^2}{2n^2}|\alpha_\varepsilon|^2\big(1+o(1)\big), \quad \overline{\lim}_{n\to\infty} \frac{n}{x^2} \ln \mathbf{P}\big(S_n \in x(\alpha)_\varepsilon\big) \leqslant -\frac{1}{2}|\alpha_\varepsilon|^2. \tag{5.1.19}$$

Moreover, $\alpha_\varepsilon \to \alpha$ as $\varepsilon \to 0$. Therefore, (5.1.3) and (5.1.19) imply (5.1.5).

Now assume that condition $[\mathbf{C}(\beta)]$, $\beta \in (0,1)$, is satisfied. In the one-dimensional case we obtain from Corollary 5.2.1(ii) in [42] that if $tx = o\big(\widehat{x}(n)\big)$ then for any $\delta > 0$ and $n \to \infty$ one has

$$\mathbf{P}(S_n \geqslant tx) \leqslant \exp\Big\{e^{-\frac{(tx)^2}{2n(1+\delta)}}\Big\}. \tag{5.1.20}$$

Since for $\alpha > 0$ the inequality

$$\mathbf{P}\big(S_n \in x(\alpha)_\varepsilon\big) \leqslant \mathbf{P}\big(S_n \geqslant x(\alpha-\varepsilon)\big),$$

holds, then

$$\overline{\lim_{n\to\infty}} \frac{n}{x^2} \ln \mathbf{P}\big(S_n \in x(\alpha)_\varepsilon\big) \leqslant -\frac{1}{2}(\alpha-\varepsilon)^2.$$

This, together with (5.1.16), proves the inequality (5.1.4). Letting ε go to 0, we obtain (5.1.5) for arbitrary $\alpha > 0$. For $\alpha < 0$ one can prove equalities (5.1.4), (5.1.5) in the same way; for $\alpha = 0$ they are obvious. Theorem 5.1.1 is proved.

5.1.4 Proof of Theorem 5.1.2

(i) By virtue of Theorem 4.1.1, assertion (i) of Theorem 5.1.2 follows from assertion (i) of Theorem 5.1.1.

(ii) By virtue of the same Theorem 4.1.1 and Lemma 5.1.1 assertion (ii) of Theorem 5.1.2 follows from assertion (ii) of Theorem 5.1.1 since the condition $[\mathbf{K}_0]$ is satisfied for the sequence S_n/x in $\mathbb{R}^d$. In our case, condition $[\mathbf{K}_0]$ is as follows: *for any $N < \infty$ there exists $v = v(N) < \infty$ such that*

$$\overline{\lim_{n\to\infty}} \frac{n}{x^2} \ln \mathbf{P}\left(\frac{1}{x}|S_n| > v\right) \leqslant -N. \tag{5.1.21}$$

Since (5.1.21) is a consequence of (5.1.20), assertion (ii) is proved. Theorem 5.1.2 is proved.

5.2 Moderately large deviation principles for trajectories s_n

5.2.1 Statement of results

Put

$$I_0(f) := \begin{cases} \frac{1}{2}\int_0^1 |f'(t)|^2 dt = \int_0^1 \Lambda_0\big(f'(t)\big)dt, & \text{if } \ f(0)=0, \quad f \in \mathbb{C}_a, \\ \infty, & \text{otherwise.} \end{cases}$$

The functional $I_0(f)$ has the following properties: it is *convex and lower semicontinuous*, and, *for any $v \geqslant 0$ the sets $\{f : I(f) \leqslant v\}$ are compact in* $(\mathbb{C}, \rho_{\mathbb{C}})$ (for more details, see section 4.2).

Theorem 5.2.1 (The local m.l.d.p. for the trajectories s_n). (i) *If $x = o(n)$ and $x/\sqrt{n} \to \infty$ as $n \to \infty$ then, for any function $f \in \mathbb{C}$, $f(0) = 0$, and any $\varepsilon > 0$ one has*

$$\underline{\lim_{n\to\infty}} \frac{n}{x^2} \ln \mathbf{P}\big(s_n \in (f)_{\mathbb{C},\varepsilon}\big) \geqslant -I_0(f). \tag{5.2.1}$$

(ii) *If condition* (5.1.2) *is met then for any function $f \in \mathbb{C}$ one has the inequality*

$$\lim_{\varepsilon\to 0} \overline{\lim_{n\to\infty}} \frac{n}{x^2} \ln \mathbf{P}\big(s_n \in (f)_{\mathbb{C},\varepsilon}\big) \leqslant -I_0(f). \tag{5.2.2}$$

Now turn to the 'integral' m.l.d.p. for s_n. Denote the Borel σ-algebra of subsets of a metric space $(\mathbb{C}, \rho_{\mathbb{C}})$ by $\mathfrak{B}(\mathbb{C}, \rho_{\mathbb{C}})$, as before. For a set $B \in \mathfrak{B}(\mathbb{C}, \rho_{\mathbb{C}})$

denote by $[B]$ and (B) the closure and interior of that set (in the uniform metric), respectively. Put

$$I_0(B) := \inf_{f \in B} I_0(f).$$

Theorem 5.2.2 (M.l.d.p. for the trajectories s_n). (i) *If $x = o(n)$ and $x/\sqrt{n} \to \infty$ as $n \to \infty$ then for any $B \in \mathfrak{B}(\mathbb{C}, \rho_{\mathbb{C}})$,*

$$\varliminf_{n\to\infty} \frac{n}{x^2} \ln \mathbf{P}(s_n \in B) \geqslant -I_0((B)). \tag{5.2.3}$$

(ii) *If condition* (5.1.2) *is met then, for any $B \in \mathfrak{B}(\mathbb{C}, \rho_{\mathbb{C}})$, one has the inequality*

$$\varlimsup_{n\to\infty} \frac{n}{x^2} \ln \mathbf{P}(s_n \in B) \leqslant -I_0\big([B]\big). \tag{5.2.4}$$

Thus, in the deviation zones (5.1.2, the trajectory sequence $\{s_n\}$ satisfies the local large deviation principle (Theorem 5.2.1) and the 'integral' large deviation principle (Theorem 5.2.2) in the space $(\mathbb{C}, \rho_{\mathbb{C}})$ with parameters $(x^2/n, I_0)$ (see section 4.1).

Note that the lower bounds in the m.l.d.p. (assertions (i) in Theorems 5.2.1, 5.2.2) were obtained under the broad conditions

$$\mathbf{E}\xi = 0, \qquad \mathbf{E}|\xi|^2 < \infty,$$

and that they hold true for the broadest class (5.0.1) of moderately large deviations.

The m.l.d.p.s for random walks (in both local and integral forms) under the condition $[\mathbf{C}_0]$ were obtained in [19] in the one-dimensional case $d=1$ for deviations $x \gg \sqrt{n \ln n}$; in [131] they were established in the case $d \geqslant 1$. Moreover, m.l.d.p.s for some classes of processes with independent increments (namely, for compound Poisson processes and the Wiener process) and empirical distribution functions were established in [19].

Consider along with the space $\big(\mathbb{C}, \mathfrak{B}(\mathbb{C}, \rho_{\mathbb{C}})\big)$ the measurable space $\big(\mathbb{D}, \mathfrak{B}(\mathbb{D}, \rho_{\mathbb{C}})\big)$ of functions without discontinuities of the second kind, endowed with the uniform metric and Borel σ-algebra. It is known that the σ-algebras $\mathfrak{B}(\mathbb{C}, \rho_{\mathbb{C}})$ and $\mathfrak{B}(\mathbb{D}, \rho_{\mathbb{C}})$ coincide with the σ-algebras generated by the cylinders in the spaces $\mathbb{C}$ and $\mathbb{D}$, respectively.

It would be interesting to extend the assertions of Theorems 5.2.1 and 5.2.2 to processes $\overline{s}_n(t)$ from the space $\mathbb{D}$ that are close to $s_n(t)$ but can have discontinuities; for example

$$\overline{s}_n(t) := \frac{1}{x} S_{[nt]}, \quad t \in [0,1]. \tag{5.2.5}$$

Other versions of processes $\overline{s}_n(t)$ that are of interest will be considered in section 5.3. With regard to the processes $\overline{s}_n$, we will assume that, in the deviation zones (5.1.2), for any $h > 0$ one has

$$\varlimsup_{n\to\infty} \frac{n}{x^2} \ln \mathbf{P}\big(\rho_{\mathbb{C}}(s_n, \overline{s}_n) \geqslant h\big) = -\infty. \tag{5.2.6}$$

For example, the process (5.2.5) satisfies this condition for deviations (5.1.2). Indeed, if the condition **[C(β)]** is met then

$$\mathbf{P}\big(\rho_{\mathbb{C}}(s_n, \overline{s}_n) \geqslant h\big) \leqslant \mathbf{P}\big(\max_{k\leqslant n} |\xi_k| \geqslant xh\big) \leqslant n\mathbf{P}\big(|\xi| \geqslant xh\big) \leqslant ne^{-r(xh)},$$

so that

$$\ln \mathbf{P}\big(\rho_{\mathbb{C}}(s_n, \overline{s}_n) \geqslant h\big) \leqslant \ln n - l(xh),$$

where

$$\ln n = o\big(l(xh)\big) \quad \text{for} \quad x > \sqrt{n},\ n \to \infty,$$

and

$$\frac{l(xh)n}{x^2} \to \infty \quad \text{for} \quad x = o(\widehat{x}(n)),\ n \to \infty.$$

Similarly, under condition **[C$_0$]** one has

$$\frac{n}{x^2} \ln \mathbf{P}\big(\rho_{\mathbb{C}}(s_n, \overline{s}_n) \geqslant h\big) \leqslant (\ln n - cxh)\,\frac{n}{x^2} = -ch\frac{n}{x}\,\big(1 + o(1)\big),$$

where $n/x \to \infty$ as $n \to \infty$.

The functional $I_0(f)$ can be extended to the space $\mathbb{D}$ by putting $I_0(f) = \infty$ for $f \in \mathbb{D} \setminus \mathbb{C}$. Thus the extended functional $I_0(f)$ is still convex and lower semicontinuous (see section 4.2), while the set $\{f : I_0(f) \leqslant v\}$ is compact in $(\mathbb{D}, \rho_{\mathbb{C}})$.

The following assertion holds true for the processes $\overline{s}_n$.

Corollary 5.2.3. *Let $\overline{s}_n$ be a random process in $(\mathbb{D}, \rho_{\mathbb{C}})$ that satisfies condition* (5.2.6).

(i) *If the process s_n satisfies the conditions of Theorem 5.2.1 then assertions* (i), (ii) *of Theorem 5.2.1 (i.e., the relations* (5.2.1), (5.2.2)), *hold true for the process $\overline{s}_n$ as well. In* (5.2.1), (5.2.2) *one needs to replace $\mathbb{C}$ by $\mathbb{D}$, and by $(f)_{\mathbb{C},\varepsilon}$, $f \in \mathbb{D}$, one should understand the ε-neighbourhood in the metric $\rho_{\mathbb{C}}$ in the space $\mathbb{D}$.*

(ii) *If the process s_n satisfies the conditions of Theorem 5.2.2 then assertions* (i), (ii) *of Theorem 5.2.2 (i.e., the relations* (5.2.3), (5.2.4)), *in which $\mathfrak{B}(\mathbb{C}, \rho_{\mathbb{C}})$ should be replaced with $\mathfrak{B}(\mathbb{D}, \rho_{\mathbb{C}})$, hold true for $\overline{s}_n$ as well.*

In section section 5.3 we will use Corollary 5.2.3 to establish the m.l.d.p. for homogeneous processes with independent increments.

In this chapter we consider as the metric ρ in the spaces $\mathbb{C}$ and $\mathbb{D}$ only the uniform metric $\rho = \rho_{\mathbb{C}}$. Therefore, in what follows we shall, in general, omit the index $\mathbb{C}$ in the notations $\rho_{\mathbb{C}}$, $[B]_{\mathbb{C}}$, $(B)_{\mathbb{C}}$, $(f)_{\mathbb{C},\varepsilon}$. This should not lead to any misunderstanding.

Proof of Corollary 5.2.3. (i) Put $A_\varepsilon := \{\rho(\overline{s}_n, s_n) < \varepsilon/2\}$. Then

$$\mathbf{P}\big(\overline{s}_n \in (f)_\varepsilon\big) = \mathbf{P}\big(\overline{s}_n \in (f)_\varepsilon,\ A_\varepsilon\big) + \mathbf{P}\big(\overline{s}_n \in (f)_\varepsilon,\ \overline{A}_\varepsilon\big), \qquad (5.2.7)$$

where the second term on the right-hand side in (5.2.7) does not exceed $\mathbf{P}(\overline{A}_\varepsilon)$. The first term lies between the following bounds:

$$\mathbf{P}(s_n \in (f)_{\varepsilon/2}) - \mathbf{P}(\overline{A}_\varepsilon) \leqslant \mathbf{P}\big(s_n \in (f)_{\varepsilon/2},\ A_\varepsilon\big)$$
$$\leqslant \mathbf{P}(\overset{\neg}{s}_n \in (f)_\varepsilon,\ A_\varepsilon) \leqslant \mathbf{P}(s_n \in (f)_{2\varepsilon}). \qquad (5.2.8)$$

If $I_0(f) = \infty$ then assertion (i) follows in an obvious way from Theorem 5.2.1 and (5.2.7), (5.2.8). If $I_0(f) < \infty$ then, by virtue of (5.2.6) and Theorem 5.2.1, one has

$$\mathbf{P}(\overline{A}_\varepsilon) = o\Big(\mathbf{P}\big(s_n \in (f)_{\varepsilon/2}\big)\Big)$$

as $n \to \infty$, so that the required assertion again follows from (5.2.8).

(ii) The lower bound (5.2.3) for $\overset{\neg}{s}_n$ follows from the lower bound (5.2.1) already established for $\overset{\neg}{s}_n$ and from Theorem 4.1.2.

In order to prove the upper bound (5.2.4) for $\overset{\neg}{s}_n$ we make use of the following inequality, which holds true for any $\varepsilon > 0$ (see (5.2.7))

$$\mathbf{P}(\overset{\neg}{s}_n \in B) \leqslant \mathbf{P}\big(s_n \in (B)_\varepsilon\big) + \mathbf{P}(\overline{A}_\varepsilon),$$

where $(B)_\varepsilon$ is the ε-neighbourhood of the set B. From that inequality, by virtue of the inequality (5.2.4) for s_n and relations (5.2.6), one obtains the upper bound

$$\overline{\lim_{n\to\infty}} \frac{n}{x^2} \ln \mathbf{P}(\overset{\neg}{s}_n \in B) \leqslant -I_0(B+)$$

where $I_0(B+) := \lim_{\varepsilon\to 0} I_0\big((B)_\varepsilon\big)$. Since for any $v \geqslant 0$ the set $\{f : I(f) \leqslant v\}$ is compact in $(\mathbb{D}, \rho_\mathbb{C})$, it follows from Lemma 4.1.5 that

$$I_0(B+) = I_0\big([B]\big).$$

This establishes the inequality (5.2.4) for $\overset{\neg}{s}_n$. Corollary 5.2.3 is proved. □

5.2.2 Proof of Theorem 5.2.1

(i) To prove the lower bound (5.2.2) for functions $f \in \mathbb{C}$ that start at zero, consider the class $\mathbb{L}$ of piecewise linear functions $f = f(t) \in \mathbb{C}$ such that $f(0) = 0$. We will need the following assertion.

Lemma 5.2.4. *For any function $f \in \mathbb{L}$ one has the inequality* (5.2.1).

Proof. In agreement with the aforesaid, we will consider only the one-dimensional case. Let $f \in \mathbb{L}$ be a piecewise linear function with nodes at the points $0 = t_0 < t_1 < \cdots < t_K = 1$. Assume for simplicity that all the values t_j are multiples of $1/n$, so that the $k_j := t_j n$ are integers, $j = 0, 1, \ldots, K$.

Let the trajectory $s_n(\cdot)$ start at time t_j from a point $\big(t_j, s_n(t_j)\big)$ such that

$$\big|s_n(t_j) - f(t_j)\big| < \frac{j\varepsilon}{K}, \qquad j \geqslant 1, \qquad (5.2.9)$$

and let

$$f_j' := f'(t) \quad \text{for} \quad t \in (t_{j-1}, t_j), \quad j = 1, \ldots, K.$$

Then, if the deviation of the trajectory $s_n(\cdot)$ on (t_j, t_{j+1}) from the straight line

$$u = s_n(t_j) + (t - t_j)f_{j+1}'$$

does not exceed ε/K, $s_n(t_{j+1})$ has the property

$$\left|s_n(t_{j+1}) - f(t_{j+1})\right| < \frac{(j+1)\varepsilon}{K},$$

i.e. a condition of the form (5.2.9) holds at the time point t_{j+1} as well. Taking into account that $s_n(0) = 0$ and applying the above argument at all the points $t_1, \ldots, t_{K-1}$, we come to the conclusion that $\sup_{0 \leqslant t \leqslant 1} \left|s_n(t) - f(t)\right| < \varepsilon$. More formally, this can be described as follows. Set

$$\Delta_j = (t_j - t_{j-1}), \qquad b_j = \Delta_j n, \qquad L_j = \sum_{i=1}^{j} b_i, \quad j = 1, \ldots, K,$$

and let $\mathcal{A}(b, f', \varepsilon)$ be the set of continuous functions $g(\cdot)$ defined by

$$\mathcal{A}(b, f', \varepsilon) := \left\{ g(\cdot) : \left| g(t) - \frac{tf'x}{n} \right| < \varepsilon x \quad \text{for all} \quad t \in [0, b] \right\}.$$

Further, let $A(1, \varepsilon)$ be the event

$$A(1, \varepsilon) := \left\{ S_{(t)} \in \mathcal{A}\left(b_1, f_1', \frac{\varepsilon}{K}\right);\ t \in [0, b_1] \right\}, \tag{5.2.10}$$

where $S_{(t)}$ denotes the continuous piecewise linear function defined in (5.1.9). The events $A(j, \varepsilon)$ for $j = 2, \ldots, K$ are defined in a similar way (replacing b_1, f_1' by b_j, f_j', respectively), but using in (5.2.10), instead of $S_{(t)}$ the increments

$$S_{(L_{j-1}+t)} - S_{(L_{j-1})}, \quad t \in [0, b_j].$$

These events have the following two properties.

(1) The events $A(j, \varepsilon)$ are independent and, on the intersection $A := \cap_{j=1}^{K} A(j, \varepsilon)$, the trajectory $S_{(t)}$ lies in an expanding, step-like strip (the steps occur at times t_j and are equal to $(\varepsilon x)/K$) centred around the function $f_n(t) := xf(t/n)$; the trajectory does not leave the polygonal strip $(f_n)_{\varepsilon x}$. Therefore

$$A \subset \left\{ s_n \in (f)_\varepsilon \right\}, \quad \mathbf{P}\big(s_n \in (f)_\varepsilon\big) \geqslant \prod_{j=1}^{K} \mathbf{P}\big(A(j, \varepsilon)\big). \tag{5.2.11}$$

(2) By Lemma 5.1.3, the probabilities $\mathbf{P}\big(A(j, \varepsilon)\big)$ for $j = 1, \ldots, K$, $\alpha_j = f_j'\Delta_j$, admit the lower bound

$$\varliminf_{n \to \infty} \frac{n}{x^2} \ln \mathbf{P}\big(A(j, \varepsilon)\big) = \frac{1}{\Delta_j} \varliminf_{n \to \infty} \frac{b_j}{x^2} \ln \mathbf{P}\big(A(j, \varepsilon)\big) \geqslant -\frac{1}{\Delta_j} \Lambda_0(\alpha_j).$$

But

$$\frac{1}{\Delta_j}\Lambda_0(f_j'\Delta_j) = \Delta_j\Lambda_0(f_j'),$$

and so, by virtue of (5.2.11) one has

$$\underline{\lim}_{n\to\infty} \frac{n}{x^2} \ln \mathbf{P}\big(s_n \in (f)_\varepsilon\big) \geqslant -\sum_{j=1}^{K} \Delta_j\Lambda_0(f_j') = -I_0(f).$$

If t_j is not a multiple of $1/n$ then instead of a piecewise linear function f one should take the piecewise linear function $f_{[n]}$ which is close to f and has nodes at the points $\big(t_j^{(n)}, f(t_j)\big)$, where $t_j^{(n)} = [t_j n]/n$, $|t_j^{(n)} - t_j| \to 0$, $\rho_{\mathbb{C}}(f_{(n)}, f) \to 0$, $I_0\big(f_{(n)}\big) \to I_0(f)$ as $n \to \infty$. The relation (5.2.1) then follows from the assertion already proved. The lemma is proved. □

Now we will turn to proving the lower bound (5.2.1) in Theorem 5.2.1. For any function $f \in \mathbb{C}$, $f(0) = 0$, there exists a sequence of piecewise linear functions $f_k \in \mathbb{L}$ such that (see section 4.2)

$$\rho_{\mathbb{C}}(f_k, f) \to 0, \quad I_0(f_k) \to I_0(f) \quad \text{as} \quad k \to \infty.$$

Therefore, by virtue of Lemma 5.2.4, for any $\varepsilon > 0$ and all large enough k,

$$\underline{\lim}_{n\to\infty} \frac{n}{x^2} \ln \mathbf{P}\big(s_n \in (f)_{\mathbb{C},\varepsilon}\big) \geqslant \underline{\lim}_{n\to\infty} \frac{n}{x^2} \ln \mathbf{P}\big(s_n \in (f_k)_{\mathbb{C},\varepsilon/2}\big) \geqslant -I_0(f_k).$$

By choosing a large enough k, the right-hand side of the above inequality can be made arbitrarily close to $-I_0(f)$. The left-hand side of the inequality does not depend on k. That demonstrates (5.2.1). Assertion (i) is proved.

(ii) *Proof of inequality* (5.2.2). As in the previous section, assume first that $f \in \mathbb{L}$ is a continuous piecewise function such that the values t_k, $k = 0, \ldots, K$, of its nodes are multiples of $1/n$. One has

$$\begin{aligned}
\mathbf{P}\big(s_n \in (f)_\varepsilon\big) &\leqslant \mathbf{P}\Big(\big|s_n(t_k) - f(t_k)\big| < \varepsilon, \ k = 1, \ldots, K\Big) \\
&= \mathbf{P}\left(\left|\frac{S_{nt_k}}{x} - f(t_k)\right| < \varepsilon, \ k = 1, \ldots, K\right) \\
&\leqslant \mathbf{P}\Big(\bigcap_{k=1}^{K}\Big\{|S_{nt_k} - S_{nt_{k-1}}| > x\big[|f(t_k) - f(t_{k-1})| - 2\varepsilon\big]\Big\}\Big).
\end{aligned} \tag{5.2.12}$$

Let A_k denote the event within the curly brackets in (5.2.12), and let

$$\Delta_k := t_k - t_{k-1}, \quad b_k := n\Delta_k, \quad \mathrm{D}_k f := f(t_k) - f(t_{k-1}), \quad k = 1, \ldots, K.$$

Then the right-hand side of (5.2.12) does not exceed

$$\prod_{k=1}^{K} \mathbf{P}(A_k) = \prod_{k=1}^{K} \mathbf{P}\Big(|S_{b_k}| > x\big[|\mathrm{D}_k f| - 2\varepsilon\big]\Big), \tag{5.2.13}$$

where, for the deviations (5.1.2) one has the inequality (see (5.1.20), (5.1.17))

$$\mathbf{P}\Big(|S_{b_k}| > x\big[|\mathrm{D}_k f| - 2\varepsilon\big]\Big) \leqslant 2\exp\left\{-\frac{x^2[|\mathrm{D}_k f| - 2\varepsilon]^2}{2b_k(1+\delta_n)}\right\}, \tag{5.2.14}$$

for $\delta_n \to 0$ as $n \to \infty$. It follows from this that

$$\overline{\lim_{n\to\infty}} \frac{n}{x^2} \ln \mathbf{P}(A_k) \leqslant -\frac{[|f_k'\Delta_k| - 2\varepsilon]^2}{2\Delta_k},$$

where $f_k' = \mathrm{D}_k f/\Delta_k$ is the derivative $f'(t)$ on the interval (t_{k-1}, t_k). By virtue of (5.2.12)–(5.2.14) we find that

$$\overline{\lim_{n\to\infty}} \frac{n}{x^2} \ln \mathbf{P}\big(s_n \in (f)_\varepsilon\big) \leqslant -\frac{1}{2}\sum_{k=1}^{K} \frac{1}{\Delta_k}\big[\,|f_k'\Delta_k| - 2\varepsilon\big]^2,$$

$$\lim_{\varepsilon\to 0}\overline{\lim_{n\to\infty}} \frac{n}{x^2} \ln \mathbf{P}\big(s_n \in (f)_\varepsilon\big) \leqslant -\frac{1}{2}\sum_{k=1}^{K} \Delta_k |f_k'|^2 = -I_0(f).$$

The transition to a partition $\{t_k\}$ not consisting of points that are multiples of $1/n$ and an arbitrary function $f \in \mathbb{C}$ is made in the standard way, as above. Assertion (ii) of Theorem 5.2.1 is established. Theorem 5.2.1 is proved.

5.2.3 Proof of Theorem 5.2.2

We will need the following condition, which was used in section 4.1 to obtain an 'integral' l.d.p. from the local l.d.p..

[K] *There exists a family of compact sets K_v in $\mathbb{C}$, $v > 0$, such that for any $N < \infty$ there is a compact K_v, $v = v(N)$, such that, for any $\varepsilon > 0$,*

$$\overline{\lim_{n\to\infty}} \frac{n}{x^2} \ln \mathbf{P}\big(s_n \notin (K_v)_\varepsilon\big) \leqslant -N. \tag{5.2.15}$$

To prove Theorem 5.2.2, we will need the following lemma.

Lemma 5.2.5. *Let condition* **[C$_0$]** *or condition* **[C(β)]** *be met. Then a sequence of the random piecewise linear functions s_n satisfies condition* **[K]**.

Proof. Let

$$\omega_\Delta(f) := \sup_{t,u\in[0,1],\ |t-u|\leqslant\Delta} \big|f(t) - f(u)\big|$$

be the continuity modulus of the functions $f \in \mathbb{C}$. Further, let $\theta(\Delta)$, $\Delta \in [0,1]$, be a positive function such that $\theta(\Delta) \downarrow 0$ as $\Delta \downarrow 0$. The function sets

$$K_v := \big\{f : f(0) = 0,\ \omega_\Delta(f) \leqslant v\theta(\Delta) \quad \text{for all} \quad \Delta \in [0,1]\big\}$$

are compact in $(\mathbb{C}, \rho_{\mathbb{C}})$ for any fixed $v > 0$.

For any fixed integer $M \geqslant 2$ construct a 'merged' piecewise linear function $s_n^{(M)} = s_n^{(M)}(t)$, $t \geqslant 0$, using the node points

$$\left(\frac{im}{n}, \frac{S_{im}}{x}\right), \quad i = 0, \dots, \quad m = m(n) := \left[\frac{n}{M}\right], \tag{5.2.16}$$

and set

$$B_n(M, v) := \{s_n^{(M)} \in K_v\}, \quad C_n(M, \varepsilon) := \{\rho(s_n^{(M)}, s_n) < \varepsilon\}.$$

For the piecewise linear functions s_n one has

$$A(2\varepsilon) := \{s_n \in (K_v)_{\mathbb{C},2\varepsilon}\} \supset B_n(M, v) \bigcap C_n(M, \varepsilon).$$

Therefore

$$\mathbf{P}\big(s_n \notin (K_v)_{\mathbb{C},2\varepsilon}\big) \leqslant \mathbf{P}\big(\overline{B}_n(M, v)\big) + \mathbf{P}\big(\overline{C}_n(M, \varepsilon)\big), \tag{5.2.17}$$

and so it suffices to obtain an upper bound for $\mathbf{P}\big(\overline{B}_n(M, v)\big)$ and $\mathbf{P}\big(\overline{C}_n(M, \varepsilon)\big)$.

To avoid formal complications that are not relevant to the problem *per se* but arise from the facts that, generally speaking, n is not a multiple of $\Delta_0 = 1/M$ and the products im in (5.2.16) can assume values greater than n, we will assume here and in what follows that the increments $S_{i+k} - S_i$ are defined for $i + k > n$ in a natural way by adding new independent summands $\xi_{n+1}, \xi_{n+2}, \dots$

(I) We will obtain a bound for $\mathbf{P}\big(\overline{B}_n(M, v)\big)$. Put $\theta(\Delta) = \sqrt{\Delta}$. Then, for

$$B_{i,j} := \left\{\frac{|S_{im} - S_{jm}|}{x} \leqslant v\theta\left(\frac{|i-j|m}{n}\right)\right\}$$

one has

$$\mathbf{P}\big(\overline{B}_n(M, v)\big) \leqslant \mathbf{P}\big(\cup_{0 \leqslant i,j \leqslant M+1} \overline{B}_{i,j}\big) \leqslant (M+2)^2 \max_{0 \leqslant i \leqslant M+1} \mathbf{P}(\overline{B}_{i,0}).$$

By virtue of the m.l.d.p. for the sums S_n (Theorem 5.1.2),

$$\overline{\lim_{n\to\infty}} \frac{n}{x^2} \ln \mathbf{P}(\overline{B}_{i,0}) \leqslant -\frac{x^2 v^2 (\theta(ik/n))^2}{2ik} = -\frac{x^2}{2n} v^2.$$

Hence, for $v := 2\sqrt{N}$ and any fixed $M \geqslant 2$, one has

$$\overline{\lim_{n\to\infty}} \frac{n}{x^2} \ln \mathbf{P}\big(\overline{B}_n(M, v)\big) \leqslant -N. \tag{5.2.18}$$

(II) For the chosen v and arbitrary $\varepsilon > 0$, we will obtain a bound for $\mathbf{P}\big(\overline{C}_n(M, \varepsilon)\big)$. Clearly,

$$\begin{aligned}
\mathbf{P}\big(\overline{C}_n(M, \varepsilon)\big) &\leqslant \mathbf{P}\Big(\cup_{j=0}^{M} \big\{\max_{jm/n \leqslant t \leqslant (j+1)m/n} \big|s_n^{(M)}(t) - s_n(t)\big| \geqslant \varepsilon\big\}\Big) \\
&\leqslant (M+1)\mathbf{P}\Big(\big\{\max_{0 \leqslant t \leqslant m/n} \big|s_n^{(M)}(t) - s_n(t)\big| \geqslant \varepsilon\big\}\Big) \\
&\leqslant (M+1)\mathbf{P}\big(\max_{1 \leqslant i \leqslant m} |S_i| \geqslant x\varepsilon\big),
\end{aligned}$$

and so, by Kolmogorov's inequality ([39], p. 295),

$$\mathbf{P}\big(\overline{C}_n(M,\varepsilon)\big) \leqslant 2(M+1)\mathbf{P}\big(|S_m| \geqslant x\varepsilon - \sqrt{2m}\big).$$

Next, applying the m.l.d.p. to the sums S_m (Theorem 5.1.2) in order to bound the right-hand side of the last inequality, we obtain

$$\overline{\lim_{n\to\infty}} \frac{n}{x^2} \ln \mathbf{P}(\overline{C}_n(M,\varepsilon)) \leqslant -\frac{1}{2}\varepsilon^2 M.$$

Therefore, for $M = 2N/\varepsilon^2$,

$$\overline{\lim_{n\to\infty}} \frac{n}{x^2} \ln \mathbf{P}\big(\overline{C}_n(M,\varepsilon)\big) \leqslant -N. \tag{5.2.19}$$

The relation (5.2.15) follows from (5.2.17)–(5.2.19). The lemma is proved.

Now we turn to the proof of Theorem 5.2.2. By virtue of Theorem 4.1.1, assertion (i) of Theorem 5.2.2 follows from assertion (i) of Theorem 5.2.1. Further, by Theorem 5.1.1 and Lemma 5.2.5, condition **[K]** holds for s_n in the deviation zones $x = x(n)$ specified by conditions **[C$_0$]** and **[C(β)]**. Since Theorem 5.2.1 establishes in these zones the local m.l.d.p. for the trajectories s_n, the integral m.l.d.p. for s_n, i.e. assertion (ii) of Theorem 5.2.2, follows, again by virtue of Theorem 4.1.6 and Lemma 4.1.5, from the m.l.d.p. and condition **[K]** for s_n (cf. the proof of Theorem 5.1.2). Theorem 5.2.2 is proved. □

5.3 Moderately large deviation principles for processes with independent increments

5.3.1 The main results

Consider a homogeneous process with independent increments $S(t) \in \mathbb{R}$, $t \geqslant 0$. As was noted in section 4.9, the characteristic function of $S(t)$ admits the Lèvy–Khinchin representation

$$\mathbf{E}e^{ivS(t)} = e^{tr(v)}, \quad v \in \mathbb{R}, \quad t \geqslant 0,$$

where

$$r(v) = \gamma(v; q, \sigma, \mathcal{B}) := iqv - \frac{v^2\sigma^2}{2} + \int_{-\infty}^{\infty} \left(e^{ivx} - 1 - \frac{ivx}{1+x^2}\right)\frac{1+x^2}{x^2}\,\mathcal{B}(dx); \tag{5.3.1}$$

$\mathcal{B}$ is a bounded measure on $(-\infty,\infty)$ and $\mathcal{B}(\{0\}) = 0$. Without loss of generality, assume that

$$\mathbf{E}S(1) = 0, \quad \mathbf{E}S^2(1) = 1.$$

Let $R_1 = (-\infty,\infty) \setminus [-1,1]$ and ζ be a random variable with distribution

$$\mathbf{P}(\zeta \in B) := \frac{\mathcal{B}(B \cap R_1)}{\mathcal{B}(R_1)}.$$

This is a conditional distribution for a jump in the process $S(t)$ given that the absolute value of that jump does not exceed 1. Further, let $\psi(\lambda) := \mathbf{E}e^{\lambda S(1)} = e^{r(-i\lambda)}$ and let $\psi_\zeta(\lambda)$ be the Laplace transforms of $S(1)$ and ζ, respectively. Since the integral in (5.3.1) is, when we restrict the integration to the interval $[-1, 1]$, an entire function of, it is clear that the distribution of the jump ζ satisfies condition $[\mathbf{C_0}]$ (or condition $[\mathbf{C}(\beta)]$) if and only if that condition is met for the random variable $S(1)$.

As is well known, the trajectories of processes with independent increments will almost surely belong to the space $\mathbb{D}$ of functions without discontinuities of the second kind. To make the assertions to be stated below as strong as possible (see Remark 5.3.3 below), we endow the space $\mathbb{D}$ with the uniform metric $\rho_\mathbb{C}$.

Consider the processes

$$s_T = s_T(t) := \frac{1}{x}S(tT), \quad t \in [0,1],$$

where, as before, $x = x(T) \to \infty$ as $T \to \infty$. By $(f)_\varepsilon$ we mean the ε-neighbourhood of the function $f \in \mathbb{D}$ in the metric $\rho_\mathbb{C}$, and by $\widehat{x}(T)$ the function defined in (5.1.1) in which the argument n is replaced with T.

An analogue of condition (5.1.2) has the following form: let $T \to \infty$ and

$$\frac{x}{\sqrt{T}} \to \infty, \quad x(T) = \begin{cases} o(T), & \text{if for } S(1) \text{ condition } [\mathbf{C_0}] \text{ is met,} \\ o(\widehat{x}(T)), & \text{if for } S(1) \text{ condition } [\mathbf{C}(\beta)] \text{ is met,} \end{cases} \tag{5.3.2}$$

Theorem 5.3.1 (The local m.l.d.p. for processes with independent increments).

(i) *If $x = o(T)$ and $x/\sqrt{T} \to \infty$ as $T \to \infty$ then, for any function $f \in \mathbb{D}$, $f(0) = 0$, and any $\varepsilon > 0$,*

$$\varliminf_{T\to\infty} \frac{T}{x^2} \ln \mathbf{P}(s_T \in (f)_\varepsilon) \geqslant -I_0(f).$$

(ii) *If condition (5.3.2) is met then, for any function $f \in \mathbb{D}$, $f(0) = 0$,*

$$\lim_{\varepsilon\to 0}\varlimsup_{T\to\infty} \frac{T}{x^2} \ln \mathbf{P}\big(s_T \in (f)_\varepsilon\big) \leqslant -I_0(f). \tag{5.3.3}$$

The integral m.l.d.p. for the processes $\{s_T\}$ can be stated in a similar way. Recall that by $\mathfrak{B}(\mathbb{D}, \rho_\mathbb{C})$ we denote the σ-algebra of Borel subsets of $(\mathbb{D}, \rho_\mathbb{C})$.

Theorem 5.3.2 (The m.l.d.p. for processes with independent increments).

(i) *If $x = o(T)$ and $x/\sqrt{T} \to \infty$ as $T \to \infty$ then, for any $B \in \mathfrak{B}(\mathbb{D}, \rho_\mathbb{C})$,*

$$\varliminf_{T\to\infty} \frac{T}{x^2} \ln \mathbf{P}(s_T \in B) \geqslant -I_0((B)).$$

(ii) *If condition (5.3.2) is met then, for any $B \in \mathfrak{B}(\mathbb{D}, \rho_\mathbb{C})$,*

$$\varlimsup_{T\to\infty} \frac{T}{x^2} \ln \mathbf{P}(s_T \in B) \leqslant -I_0\big([B]\big). \tag{5.3.4}$$

Thus, in the terminology from section 4.1, for the deviations $x = x(T)$ that satisfy condition (5.3.2), the family of processes $\{s_T\}$ satisfies the local l.d.p. (Theorem 5.3.1) and also the usual l.d.p. (Theorem 5.3.2) in the space $(\mathbb{D}, \rho_{\mathbb{C}})$ with parameters $(x^2/T, I_0)$.

5.3.2 Proof of Theorems 5.3.1 and 5.3.2

We will make use of Corollary 5.2.3. First, assume that $T = n$ is integer. In that case, construct the process $s_n = s_n(t)$ as a continuous piecewise linear process with nodes at $(k/n,\ (1/x)S(k))$, $k = 0, 1, \dots, n$, where $S(k) = S_k = \sum_{j=1}^{k} \xi_j$ and $\xi_j := S(j) - S(j-1)$ are independent identically distributed random variables with characteristic function $\varphi(\mu) = e^{r(\mu)}$.

Set $\overline{s}_n = \overline{s}_n(t) := (1/x)S(tn) = s_T(t)$, $t \in [0, 1]$, and show that relation (5.2.6) holds under the conditions of Theorems 5.3.1 and 5.3.2.

(I) Let condition $[\mathbf{C}_0]$ be met and $x = o(n)$ as $n \to \infty$. Then $\overline{s}_n(k) = s_n(k)$ for $k = 0, 1, \dots, n$, and the following inequalities hold true:

$$\begin{aligned}
\mathbf{P}(\rho_{\mathbb{C}}(\overline{s}_n, s_n) > 2h) &\leqslant n\mathbf{P}\Big(\sup_{0\leqslant t\leqslant 1/n} \big|\overline{s}_n(t) - s_n(t)\big| > 2h\Big)\\
&= n\mathbf{P}\Big(\sup_{0\leqslant t\leqslant 1} \big|S(t) - tS(1)\big| > 2xh\Big)\\
&\leqslant n\left[\mathbf{P}\Big(\sup_{0\leqslant t\leqslant 1} \big|S(t)\big| > xh\Big) + \mathbf{P}\Big(\big|S(1)\big| > xh\Big)\right]\\
&\leqslant 2n\mathbf{P}\Big(\sup_{0\leqslant t\leqslant 1} \big|S(t)\big| > xh\Big) \leqslant 2n\left[e^{-\Lambda(xh)} + e^{-\Lambda(-xh)}\right].
\end{aligned} \tag{5.3.5}$$

For the last inequality we made use of Corollary 4.9.2, where $\Lambda(\alpha)$ is the deviation function for the random variable $S(1)$. Since under condition $[\mathbf{C}_0]$ one has $\Lambda(\alpha) \geqslant c_1|\alpha| - c_2$ for some $c_1 > 0$, $c_2 > 0$, we obtain

$$\mathbf{P}\big(\rho_{\mathbb{C}}(\overline{s}_n, s_n) > 2h\big) \leqslant c_3 n e^{-c_1 xh},$$

where $c_3 = 4e^{-c_2}$. Therefore,

$$\frac{n}{x^2} \ln \mathbf{P}\big(\rho_{\mathbb{C}}(\overline{s}_n, s_n) > 2h\big) \leqslant \frac{n}{x^2} [\ln nc_3 - c_1 xh] \to -\infty$$

provided that $x = o(n)$, $x \gg \sqrt{n}$.

(II) Let condition $[\mathbf{C}(\beta)]$ be met and $x = o(\widehat{x}(n))$ as $n \to \infty$. Then, instead of (5.3.5) one should use the Kolmogorov inequality

$$\begin{aligned}
\mathbf{P}\Big(\sup_{0\leqslant t\leqslant 1} \big|S(t)\big| > xh\Big) &\leqslant \mathbf{P}\Big(\big|S(1)\big| > xh - 2\sqrt{\mathbf{D}S(1)}\Big)\\
&= \mathbf{P}\Big(\big|S(1)\big| > xh - 2\Big),
\end{aligned}$$

and the inequality

$$\mathbf{P}\Big(\big|S(1)\big| > xh - 2\Big) \leqslant e^{-l(xh-2)},$$

where $l(x-2) \sim l(x)''x^{\beta/2}$ as $x \to \infty$. Therefore, (5.2.6) also holds true. Thus we have proved Theorems 5.3.1 and 5.3.2 under the additional assumption that $T = n$.

Now assume that T is fractional. Set $n = [T]$ and construct a continuous piecewise linear function $s_n^* = s_n^*(t)$ with nodes at the points $(k/T, S(k)/x)$, $k = 0, 1, \dots, n$, and $(1, S(n)/x)$. Employing our previous argument, it is not hard to show that the probability $\mathbf{P}\big(\rho_{\mathbb{C}}(s_n^*, s_T) > h\big)$ satisfies (5.2.6). Further, we change the time by 'stretching' it T/n times and consider the process

$$s_n^{**}(t) := s_n^*\left(t\,\frac{n}{T}\right) \quad \text{on the interval} \quad [0,1],$$

so that $s_n^{**}(1) = S(n)/x$. Clearly,

$$\max_{t\in[0,1]} \big|s_n^{**}(t) - s_n^*(t)\big| \leqslant \frac{1}{x} \max_{k\leqslant n} |\xi_k|,$$

and so $\mathbf{P}\big(\rho_{\mathbb{C}}(s_n^{**}, s_n^*) > h\big)$ also satisfies (5.2.6). From here it follows that, for any $h > 0$, one has

$$\overline{\lim_{n\to\infty}}\, \frac{n}{x^2} \ln \mathbf{P}\big(\rho_{\mathbb{C}}(s_n^{**}, s_T) > h\big) = -\infty.$$

However, the distribution of s_n^{**} coincides with that of the process s_n considered in the previous sections. By virtue of Corollary 5.2.3, Theorems 5.3.1 and 5.3.2 are proved.

Remark 5.3.3. One can prove the assertions of Theorems 5.3.1 and 5.3.2 for the metric space $(\mathbb{D}, \rho_{\mathbb{D}})$ as well, where $\rho_{\mathbb{D}}$ is the Skorokhod metric. However, these assertions will, in a certain sense, be weaker than ours, since $I([B]_{\mathbb{D}}) - I((B)_{\mathbb{D}}) \geqslant I([B]_{\mathbb{C}}) - I((B)_{\mathbb{C}})$, where $[B]_{\mathbb{D}}$, $(B)_{\mathbb{D}}$ are the closure and interior of the set B in $(\mathbb{D}, \rho_{\mathbb{D}})$, respectively.

Remark 5.3.4. In Theorems 5.3.1 and 5.3.2 we dealt only with one-dimensional processes with independent increments. That restriction was intended to simplify the notation and statements of the results. As we have already pointed out, the assertions of Theorems 5.3.1 and 5.3.2 remain valid for multivariate processes with independent increments. The proofs in the multivariate case are essentially the same.

5.4 Moderately large deviation principle as an extension of the invariance principle to the large deviation zone

It is easy to see that in all the previous considerations in this chapter, the right-hand sides in the limiting relations in the m.l.d.p. do not depend on the distribution of the original process (i.e. the distribution of the random vector ξ, in the situation dealt with in sections 5.1, 5.2), provided that the first two moments of $S(1)$ are fixed.

Let $w = w(t)$; $t \in [0, 1]$, be the standard Wiener process and let $\mathcal{W}$ be the measure in $\mathfrak{B}(\mathbb{C}, \rho_{\mathbb{C}})$, corresponding to that process (the Wiener measure). Further, let $\{S_k\}_0^\infty$ be the random walk considered in Section 5.2 and $\tilde{s}_n$ be a continuous piecewise linear process with nodes at the points $(k/n, S_k/\sqrt{n})$, $k = 0, \dots, n$. One can obtain the following assertion from Theorem 5.3.2 and the invariance principle.

Corollary 5.4.1. *Let the conditions of Theorem 5.3.2 be met.*

(i) *If $B \in \mathfrak{B}(\mathbb{C}, \rho_{\mathbb{C}})$ is such that*

$$I_0\big((B)_{\mathbb{C}}\big) = I_0\big([B]_{\mathbb{C}}\big), \tag{5.4.1}$$

then, as $v \to \infty$,

$$\ln \mathcal{W}(vB) \sim -v^2 I(B). \tag{5.4.2}$$

(ii) *If, for a fixed value $v > 0$, one has*

$$\mathcal{W}((vB)_{\mathbb{C}}) = \mathcal{W}([vB]_{\mathbb{C}}), \tag{5.4.3}$$

then, as $n \to \infty$,

$$\ln \mathbf{P}(\tilde{s}_n \in vB) \sim \ln \mathcal{W}(vB) \tag{5.4.4}$$

(the invariance principle).

(iii) *If* (5.4.1) *holds, $v = x/\sqrt{n} \to \infty$ and the deviations $x = v\sqrt{n}$ satisfy condition* (5.1.2) *as $n \to \infty$, relation* (5.4.4) *still holds (the m.l.d.p.).*

Assertion (5.4.4) remains valid for $\ln \mathbf{P}(\overline{s}_n \in B)$ and $\ln \mathbf{P}(s_T \in B)$ with $B \in \mathfrak{B}(\mathbb{D}, \rho_{\mathbb{C}})$, provided that we replace $\mathcal{W}(vB)$ on the right-hand side with $\mathcal{W}(vB \cap \mathbb{C})$.

The first assertion of Corollary 5.4.1 indicates the explicit form of the asymptotic behaviour of the Wiener measure of remote sets vB (as $v \to \infty$). Assertion (iii) can be viewed as an extension (in terms of the logarithmic asymptotics) of the invariance principle to the large deviation zone. Relation (5.4.4), both for a fixed v and $v \to \infty$, could be called the invariance principle extended to the large deviation zone.

With regard to (5.4.2), one can observe that the measure $\mathcal{W}$ is concentrated on the set of functions of unbounded variation, whereas the value of the functional $I_0(f)$ on such functions is ∞. This, however, does not lead to a contradiction owing to the continuity condition (5.4.1).

Proof of Corollary 5.4.1. (i) It follows from Theorem 5.3.2 that, given the continuity condition (5.4.1) is met for a set $B \in \mathfrak{B}(\mathbb{C}, \rho_{\mathbb{C}})$, then, for the process

$$w_T(t) := \frac{1}{x} w(tT), \quad t \in [0, 1],$$

with $x \gg \sqrt{T}$, $x = o(T)$, $T \to \infty$, one has the following relation:

$$\lim_{T\to\infty} \frac{T}{x^2} \ln \mathbf{P}(w_T \in B) = -I_0(B). \tag{5.4.5}$$

However, the distribution of $w_T(t) = (1/x)w(tT)$ coincides with that of $1/vw(t)$ with $v = x/\sqrt{T}$, while

$$\ln \mathbf{P}\left(\frac{1}{v}w(\cdot) \in B\right) = \ln \mathbf{P}\big(w(\cdot) \in vB\big) = \ln \mathcal{W}(vB).$$

At the same time, it follows from (5.4.5) that, as $v \to \infty$, $v = o(\sqrt{T})$, one has

$$\frac{1}{v^2} \ln \mathbf{P}\left(\frac{1}{v}w(\cdot) \in B\right) \sim -I_0(B).$$

The above implies (i).

(ii) The second assertion is an obvious consequence of the invariance principle.

(iii) According to Theorem 5.2.2, for deviations $x \gg \sqrt{n}$ of the form (5.1.2), there exists the limit

$$\lim_{n\to\infty} \frac{n}{x^2} \ln \mathbf{P}(s_n \in B) = -I_0(B).$$

Setting $v = x/\sqrt{n} \to \infty$, we derive from this that

$$\frac{1}{v^2} \ln \mathbf{P}(s_n \in B) \sim -I_0(B) \sim \frac{1}{v^2} \ln \mathcal{W}(B).$$

Since $\ln \mathbf{P}(s_n \in B) = \ln \mathbf{P}(\tilde{s}_n \in vB)$, the relation (5.4.4), along with Corollary 5.4.1, is proved. □

If one prefers assertions having the form of inequalities (upper and lower bounds; cf. Theorems 5.1.1–5.3.2) then an analogue of Corollary 5.4.1 can be obtained without using the continuity conditions (5.4.1) and (5.4.3).

5.5 Conditional moderately large deviation principles for the trajectories of random walks

This section is similar to section 4.8 but here we consider moderately large deviations.

5.5.1 Conditional m.l.d.p. in the space $(\mathbb{C}, \rho_{\mathbb{C}})$

Note first that here we have a full analogue of Corollary 4.8.1 concerning the description of the logarithmic asymptotics for the conditional probability $\mathbf{P}(s_n \in B_1 \mid s_n \in B_2)$ for arbitrary measurable sets B_1, B_2.

In the following we will consider the one-dimensional case $d = 1$ only and will study the logarithmic asymptotics for the conditional probabilities

$$\mathbf{P}\big(s_n \in (f)_{\mathbb{C},\varepsilon} \mid s_n(1) \in (b)_\varepsilon\big), \qquad \mathbf{P}\big(s_n \in B \mid s_n(1) \in (b)_\varepsilon\big)$$

as $n \to \infty$, $\varepsilon \to 0$, given that the end point of the trajectory s_n is localised in the ε-neighbourhood $(b)_\varepsilon$ of the point b. Here f and B are an arbitrary function and a measurable set from the space $\mathbb{C}$ with uniform metric, respectively. Moreover we assume that $f(0) = 0, f(1) = b$ and that the set B satisfies some consistency condition involving fixing the end point of the trajectory. In this section we will assume, as before, that

$$\mathbf{E}\xi = 0, \qquad \sigma^2 := \mathbf{E}\xi^2 = 1. \tag{5.5.1}$$

We start with the local conditional l.l.d.p.

Since for any $b \in \mathbb{R}$ one has $\Lambda_0(b) < \infty$, by virtue of the l.l.d.p. for the sums $s_n(1)$ (see Theorem 5.1.1 below), the probability

$$\mathbf{P}\big(s_n(1) \in (b)_\varepsilon\big)$$

is positive for all $\varepsilon > 0$ and all large enough n.

The analogue of the conditional l.l.d.p. has the following form.

Theorem 5.5.1. *If condition* (5.1.2) *is met then every function* $f \in \mathbb{C}$, $f(0) = 0$, $f(1) = b$, *satisfies the equalities*

$$\begin{aligned} &\lim_{\varepsilon\to 0}\overline{\lim_{n\to\infty}} \frac{n}{x^2} \ln \mathbf{P}\big(s_n \in (f)_{\mathbb{C},\varepsilon} \mid s_n(1) \in (b)_\varepsilon\big) \\ &\quad = \lim_{\varepsilon\to 0}\underline{\lim_{n\to\infty}} \frac{n}{x^2} \ln \mathbf{P}\big(s_n \in (f)_{\mathbb{C},\varepsilon} \mid s_n(1) \in (b)_\varepsilon\big) = -I_0(f) + \Lambda_0(b). \end{aligned} \tag{5.5.2}$$

The relations (5.5.2) could be called the *conditional l.m.l.d.p.* for the trajectory s_n with localised end point.

Proof. The upper bound. For $f \in \mathbb{C}, f(1) = b$, one has

$$\mathbf{P}\big(s_n \in (f)_{\mathbb{C},\varepsilon} \mid s_n(1) \in (b)_\varepsilon\big) = \frac{\mathbf{P}(s_n \in (f)_{\mathbb{C},\varepsilon})}{\mathbf{P}(s_n(1) \in (b)_\varepsilon)}. \tag{5.5.3}$$

Moreover, by virtue of the l.m.l.d.p. for $s_n(1) = S_n/x$ one has (see Theorem 5.1.1)

$$\lim_{\varepsilon\to 0}\underline{\lim_{n\to\infty}} \frac{n}{x^2} \ln \mathbf{P}\big(s_n(1) \in (b)_\varepsilon\big) \geqslant -\Lambda_0(b). \tag{5.5.4}$$

Therefore, the upper bound in Theorem 5.5.1 follows straightforwardly from the l.m.l.d.p. for s_n (Theorem 5.2.1):

$$\begin{aligned} &\lim_{\varepsilon\to 0}\overline{\lim_{n\to\infty}} \frac{n}{x^2} \ln \mathbf{P}(s_n \in (f)_{\mathbb{C},\varepsilon} \mid s_n(1) \in (b)_\varepsilon) \\ &\quad \leqslant \lim_{\varepsilon\to 0}\overline{\lim_{n\to\infty}} \frac{n}{x^2} \ln \mathbf{P}(s_n \in (f)_{\mathbb{C},\varepsilon}) - \lim_{\varepsilon\to 0}\underline{\lim_{n\to\infty}} \frac{n}{x^2} \ln \mathbf{P}(s_n(1) \in (b)_\varepsilon) \\ &\quad \leqslant -I_0(f) + \Lambda_0(b). \end{aligned}$$

The lower bound. By virtue of the upper bounds in the l.m.l.d.p. for $s_n(1) = S_n/x$ one has (see Theorem 5.1.1)

$$\lim_{\varepsilon\to 0}\overline{\lim_{n\to\infty}} \frac{n}{x^2} \ln \mathbf{P}(s_n(1) \in (b)_\varepsilon) \leqslant -\Lambda_0(b). \tag{5.5.5}$$

Therefore, by virtue of (5.5.3) and the l.m.l.d.p. for s_n (see Theorem 5.2.1), we get the desired lower bounds:

$$\begin{aligned}
&\lim_{\varepsilon\to 0}\varliminf_{n\to\infty}\frac{n}{x^2}\ln \mathbf{P}(s_n\in(f)_{\mathbb{C},\varepsilon}|\, s_n(1)\in(b)_\varepsilon)\\
&\quad\geqslant \lim_{\varepsilon\to 0}\varliminf_{n\to\infty}\frac{n}{x^2}\ln \mathbf{P}(s_n\in(f)_{\mathbb{C},\varepsilon}) - \lim_{\varepsilon\to 0}\varlimsup_{n\to\infty}\frac{n}{x^2}\ln \mathbf{P}(s_n(1)\in(b)_\varepsilon)\\
&\quad\geqslant -I_0(f)+\Lambda_0(b).
\end{aligned}$$

The theorem is proved. □

As in section 4.8, we need the condition **[B, b]** (see (4.8.6) and (4.8.5)) in order to formulate and prove the conditional m.l.d.p. All the comments about this condition made in section 4.8 remain valid.

Theorem 5.5.2. *Let B be a measurable set, the condition* **[B, b]** *be met and the set $B^{(b)}$ be defined as in* (4.8.5).

If condition (5.1.2) *is met, one has the inequalities*

$$\lim_{\varepsilon\to 0}\varlimsup_{n\to\infty}\frac{n}{x^2}\ln \mathbf{P}(s_n\in B|\, s_n(1)\in(b)_\varepsilon)\leqslant -I_0([B^{(b)}])+\Lambda_0(b), \tag{5.5.6}$$

$$\lim_{\varepsilon\to 0}\varliminf_{n\to\infty}\frac{n}{x^2}\ln \mathbf{P}(s_n\in B|\, s_n(1)\in(b)_\varepsilon)\geqslant -I_0((B)^{(b)})+\Lambda_0(b). \tag{5.5.7}$$

For the set $B^{(b)}$ that satisfies the continuity condition

$$I([B^{(b)}])=I_0((B)^{(b)})=I_0(B^{(b)}),$$

the following relation holds:

$$\begin{aligned}
&\lim_{\varepsilon\to 0}\varliminf_{n\to\infty}\frac{1}{n}\ln \mathbf{P}(s_n\in B|\, s_n(1)\in(b)_\varepsilon)\\
&\qquad=\lim_{\varepsilon\to 0}\varlimsup_{n\to\infty}\frac{1}{n}\ln \mathbf{P}(s_n\in B|\, s_n(1)\in(b)_\varepsilon)=-I_0(B^{(b)})+\Lambda_0(b).
\end{aligned}$$

Observe that if $\mathbf{E}\xi=b=0$, then, referring back to the example given in (4.8.8) the left-hand side of inequality (5.5.6) equals zero by the invariance principle. Since $B^{(0)}=B_2$ in that example presented in equations, the right-hand side of (5.5.6) is equal to $-I([B_2])<0$ for all small enough $\delta>0$. That means that the assertion (5.5.6) fails in that example, so that condition **[B, b]** is essential for the assertion of Theorem 5.5.2 to hold true.

If condition **[B, b]** is not satisfied, one may be able to obtain other, more cumbersome, upper bounds for the left-hand side of (5.5.6).

Proof of Theorem 5.5.2. The upper bound. Since (4.8.6) is met, one has

$$\mathbf{P}\Big(s_n\in B\cap\big\{f(1)\in(b)_\varepsilon\big\}\Big)\leqslant \mathbf{P}\Big(s_n\in B_-\times\big\{f(1)\in(b)_\varepsilon\big\}\Big).$$

Since the sets B_- and $\{f(1) \in (b)_\varepsilon\}$ are specified by 'independent' coordinates, one has the following relations:

$$B_- \times \{f(1) \in (b)_\varepsilon\} \subset (B_-)_\varepsilon \times \{f(1) \in (b)_\varepsilon\} = \Big(B_- \times \{f(1) = b\}\Big)_\varepsilon.$$

By virtue of (4.8.5) the right-hand side of this relation is a subset of $(B^{(b)})_\varepsilon$. Hence,

$$\mathbf{P}\Big(s_n \in B \cap \{f(1) \in (b)_\varepsilon\}\Big) \leqslant \mathbf{P}\big(s_n \in (B^{(b)})_\varepsilon\big).$$

Regarding the right-hand side of that inequality, all that was said above with regard to $\mathbf{P}(s_n \in B)$ is applicable, so that, by virtue of Theorem 5.2.2 (concerning the upper bounds in the m.l.d.p.), one has

$$\overline{\lim_{n\to\infty}} \frac{1}{n} \ln \mathbf{P}\big(s_n \in (B^{(b)})_\varepsilon\big) \leqslant -I_0\big([(B^{(b)})_\varepsilon]\big) \leqslant -I_0\big((B^{(b)})_{2\varepsilon}\big),$$

$$\lim_{\varepsilon\to 0} \overline{\lim_{n\to\infty}} \frac{1}{n} \ln \mathbf{P}(s_n \in (B^{(b)})_\varepsilon) \leqslant -I_0(B^{(b)}+),$$

where

$$I_0(B+) := \lim_{\varepsilon\to 0} I_0\big((B)_\varepsilon\big).$$

The functional $I_0(f)$ has the following property: *for any* $u \geqslant 0$, *the set* $\big\{f \in \mathbb{C} : I_0(f) \leqslant u\big\}$ *is compact in the metric space* $(\mathbb{C}, \rho)$. Therefore, by Lemma 4.1.5 one has

$$I_0(B^{(b)}+) = I_0([B^{(b)}])$$

and we get the inequality

$$\lim_{\varepsilon\to 0} \overline{\lim_{n\to\infty}} \frac{n}{x^2} \ln \mathbf{P}\Big(s_n \in B \cap \{f(1) \in (b)_\varepsilon\}\Big) \leqslant -I_0\big([B^{(b)}]\big).$$

Since

$$\mathbf{P}(s_n \in B|\, s_n(1) \in (b)_\varepsilon) = \frac{\mathbf{P}(s_n \in B \cap \{f(1) \in (b)_\varepsilon\})}{\mathbf{P}(s_n(1) \in (b)_\varepsilon)},$$

this, together with (5.5.5), establishes (5.5.6).

The lower bound. The set $(B) \cap \{g \in \mathbb{C} :\ g(1) \in (b)_\varepsilon\}$ is open in $(\mathbb{C}, \rho_{\mathbb{C}})$ and is a subset of $B \cap \{g \in \mathbb{C} :\ g(1) \in (b)_\varepsilon\}$. Hence, by Theorem 5.2.2,

$$\begin{aligned} &\underline{\lim_{n\to\infty}} \frac{n}{x^2} \ln \mathbf{P}\big(s_n \in B,\ s_n(1) \in (b)_\varepsilon\big) \\ &\qquad \geqslant \underline{\lim_{n\to\infty}} \frac{n}{x^2} \ln \mathbf{P}\big(s_n \in (B),\ s_n(1) \in (b)_\varepsilon\big) \\ &\qquad \geqslant -I_0((B) \cap \{g \in \mathbb{C} :\ g(1) \in (b)_\varepsilon\}) \geqslant -I_0((B)^{(b)}). \end{aligned}$$

The theorem is proved. □

5.5.2 Conditional m.l.d.p.s with the trajectory end point localised in a narrower zone

In a number of problems (say, when studying empirical distribution functions), it is of interest to obtain conditional m.l.d.p.s whose trajectory end point is localised in a narrower zone than $(b)_\varepsilon$, for instance, for probabilities of the form

$$\mathbf{P}\big(s_n \in B \mid s_n(1) \in \Delta[b)\big), \tag{5.5.8}$$

where $\Delta[b) = [b, b+\Delta)$ and Δ is of order $1/x$. For a more precise problem formulation, we need to distinguish here between the non-lattice and arithmetic cases. In general, in the arithmetic case we cannot assume that the condition (5.5.1) is met (see the discussion below)

In the *non-lattice* case we will understand by Δ any value θ/x with arbitrary fixed $\theta > 0$.

In the *arithmetic* case the condition $\mathbf{E}\xi = 0$ can restrict the generality of considerations (if the arithmetic random variable has a non-zero mean value then the centering procedure can lead to the loss of the arithmetic property. The case with normalising condition $\sigma^2 = \mathbf{D}\xi = 1$ is similar. In the arithmetic case it is more convenient for us to construct the process $s_n(t)$, $t \in [0,1]$, using the nodes

$$\left(\frac{k}{n}, \frac{S_k - ak}{x}\right), \qquad k = 0, 1, \dots, n, \tag{5.5.9}$$

where $a = \mathbf{E}\xi$. Then the values $xs_n(1) + an = S_n$ are integers and in the conditional probabilities we take the events

$$\big\{S_n - an \in \Delta[bx)\big\}$$

with $\Delta \geqslant 1$ as the conditions, or, equivalently, the events

$$\big\{s_n(1) \in \Delta[b)\big\} \quad \text{for} \quad \Delta \geqslant \frac{1}{x}.$$

Therefore, only one point from the lattice for $s_n(1)$ hits the semi-interval $\Delta[b)$ with $\Delta = 1/x$. In that case the limiting values for the logarithms of the probabilities (5.5.8) have the same form in the non-lattice and arithmetic cases.

Since we do not assume that the normalising condition $\sigma = 1$ is met, everywhere in the following discussion, by the deviation function $\Lambda_0(\alpha)$ we will mean the function

$$\Lambda_0(\alpha) = \frac{\alpha^2}{2\sigma^2},$$

so that

$$I_0(f) = \frac{1}{2\sigma^2} \int_0^1 \big(f'(t)\big)^2 dt.$$

In the following we need a strong versioner of the condition (5.1.2).

We say that *the class $\mathcal{S}e$ of semi-exponential distributions is compatible with the condition* $[\mathbf{C}(\beta)]$ *if for $t > 0$, $\beta_+ \geqslant \beta$,*

$$F_+(t) := \mathbf{P}(\xi \geqslant t) = e^{-l_+(t)}, \qquad l_+ \in \mathcal{L}_{\beta_+},$$

and

$$l(t) = O\big(l_+(t)\big) \quad \textit{for} \quad t \to \infty,$$

where $l(t)$ is the function of the condition $[\mathbf{C}(\beta)]$.

What has been said means that, obviously, $F_+(t)$ can decrease as $t \to \infty$ faster than the dominant $e^{-l(t)}$ for $\mathbf{P}\big(|\xi| \geqslant t\big)$.

Now, consider condition $[\mathbf{C}(\boldsymbol{\beta})]^+$, which is stronger than $[\mathbf{C}(\beta)]$. It has the following form:

$[\mathbf{C}(\boldsymbol{\beta})]^+$ *The condition* $[\mathbf{C}(\beta)]$ *is met and $F_+(t)$ belongs to the class $\mathcal{S}e$, which is compatible with* $[\mathbf{C}(\beta)]$.

The counterpart of condition (5.1.2) is now as follows

Either $[\mathbf{C}_0]$ *or* $[\mathbf{C}(\boldsymbol{\beta})]^+$ *is fulfilled and*

$$\frac{x}{\sqrt{n\ln n}} \to \infty, \quad x(n) = \begin{cases} o(n), & \textit{if } [\mathbf{C}_0] \textit{ is met}, \\ o(\widehat{x}(n)), & \textit{if } [\mathbf{C}(\boldsymbol{\beta})]^+ \textit{ is met}, \end{cases} \tag{5.5.10}$$

as $n \to \infty$, where $\widehat{x}(n)$ is defined as above (see subsection 5.1.1).

If $[\mathbf{C}_0]$ fails then $s_n(1)$ is fixed only on the positive half-axis in the semi-interval $\Delta[b)$, $b \geqslant 0$.

We can now formulate the following analogue of the conditional local m.l.d.p. with narrow localisation of the trajectory end-point.

Theorem 5.5.3. *Let the condition* (5.5.10) *be met.*

(i) *If a random variable ξ is non-lattice then, for every function $f \in \mathbb{C}$, $f(1) = b \geqslant 0$, the equalities*

$$\begin{aligned} &\lim_{\varepsilon\to 0}\overline{\lim_{n\to\infty}}\frac{n}{x^2}\ln \mathbf{P}(s_n \in (f)_\varepsilon |\, s_n(1) \in \Delta[b)) \\ &= \lim_{\varepsilon\to 0}\underline{\lim_{n\to\infty}}\frac{n}{x^2}\ln \mathbf{P}(s_n \in (f)_\varepsilon |\, s_n(1) \in \Delta[b)) \\ &= -I_0(f) + \Lambda_0(b), \end{aligned} \tag{5.5.11}$$

hold, where $\Delta = \theta/x$ for any fixed $\theta > 0$.

(ii) *If ξ is arithmetic then, for the process $s_n(t)$ constructed from the node points* (5.5.9), *the assertions of part (i) of the present theorem are true with $\Delta = \theta/x$ and any $\theta \geqslant 1$.*

If condition $[\mathbf{C}_0]$ *is met then the condition $b \geqslant 0$ is unnecessary.*

The analogue of the conditional 'integral' m.l.d.p. with localisation of the trajectory end-point in the half-interval $\Delta[b)$ has the following form:

Theorem 5.5.4. *Let condition* (5.5.10) *be met.*

(i) If a random variable ξ is non-lattice, B is a measurable set and the condition **[B, b]**, *$b \geqslant 0$, is met then, for $\Delta = \theta/x$ with any fixed $\theta > 0$, one has*

$$\overline{\lim_{n\to\infty}} \frac{n}{x^2} \ln \mathbf{P}\big(s_n \in B |\ s_n(1) \in \Delta[b]\big) \leqslant -I_0\big([B^{(b)}]\big) + \Lambda_0(b), \tag{5.5.12}$$

$$\underline{\lim_{n\to\infty}} \frac{n}{x^2} \ln \mathbf{P}(s_n \in B |\ s_n(1) \in \Delta[b]) \geqslant -I_0((B)^{(b)}) + \Lambda_0(b). \tag{5.5.13}$$

For a set $B^{(b)}$ satisfying the continuity condition

$$I_0\big([B^{(b)}]\big) = I_0\big((B)^{(b)}\big) = I_0(B^{(b)}),$$

the following relation holds:

$$\lim_{n\to\infty} \frac{n}{x^2} \ln \mathbf{P}\big(s_n \in B |\ s_n(1) \in \Delta[b]\big) = -I_0(B^{(b)}) + \Lambda_0(b).$$

(ii) If ξ is arithmetic then, for the process $s_n(t)$ constructed from the node points (5.5.9), *the statements of part (i) of the present theorem are true for $b \geqslant 0$, $\Delta = \theta/x$ and any fixed $\theta \geqslant 1$.*

If condition $[\mathbf{C}_0]$ *is met then the assumption $b \geqslant 0$ is not necessary.*

To prove Theorems 5.5.3 and 5.5.4 we need the following addition to the lower bound in the l.l.d.p. (5.1.3) for the sums S_n.

Lemma 5.5.5. *Let condition* (5.5.10) *be met.*

(i) *If a random variable ξ is non-lattice then, for any $b \geqslant 0$,*

$$\lim_{n\to\infty} \frac{n}{x^2} \ln \mathbf{P}\big(s_n(1) \in \Delta[b]\big) = -\Lambda_0(b), \tag{5.5.14}$$

where $\Delta = \theta/x$ with arbitrary fixed $\theta > 0$.

(ii) *If a random variable ξ is arithmetic then, for $s_n(1) = (S_n - an)/x$, under the conditions of part* I, *one has* (5.5.14) *with $\Delta = \theta/x$ and any fixed $\theta \geqslant 1$.*

If condition $[\mathbf{C}_0]$ *is met then the assumption $b \geqslant 0$ is not necessary.*

Proof. Since

$$\mathbf{P}(s_n(1) \in \Delta[b]) = \mathbf{P}(S_n \in [xb,\ xb + \theta)),$$

the relation (5.5.14) follows in the non-lattice case from the integro-local theorem in [175] (see also [39], Chapter 8) provided that the condition $[\mathbf{C}_0]$ is met, while if condition $[\mathbf{C}(\boldsymbol{\beta})]^+$ is met then this relation follows from the integro-local theorem in [55] or in § 5.8 in [42]. We need the condition $x \gg \sqrt{n \ln n}$ to make the contribution of the power factor $cn^{-1/2}$ in the integro-local theorems negligibly small. In the arithmetic case, we have to use the corresponding analogues of the above-mentioned theorems. The lemma is proved. □

Proof of Theorem 5.5.3. (i) *The upper bound.* Since

$$\begin{aligned}\mathbf{P}\big(s_n \in (f)_{\mathbb{C},\varepsilon},\ s_n(1) \in \Delta[b]\big) &\leqslant \mathbf{P}\big(s_n \in (f)_{\mathbb{C},\varepsilon},\ s_n(1) \in (b)_\varepsilon\big)\\ &= \mathbf{P}\big(s_n \in (f)_{\mathbb{C},\varepsilon}\big),\end{aligned}$$

it follows that (5.5.14) and the l.m.l.d.p. for s_n imply the desired upper bound.

The lower bound for $\mathbf{P}\big(s_n \in (f)_{\mathbb{C},\varepsilon},\ s_n(1) \in \Delta[b)\big)$ will be established first for the case when $f(t) = tb$ for $0 \leqslant t \leqslant 1$. In this case,

$$\mathbf{P}\big(s_n \in (f)_{\mathbb{C},\varepsilon},\ s_n(1) \in \Delta[b)\big) \geqslant \mathbf{P}\big(s_n(1) \in \Delta[b)\big) - P_n, \qquad (5.5.15)$$

where

$$P_n := \mathbf{P}\big(s_n \notin (f)_{\mathbb{C},\varepsilon},\ s_n(1) \in (b)_{\varepsilon/2}\big).$$

Applying Theorem 5.2.2 to estimate from above the probability P_n, we verify that for every $\varepsilon > 0$ there exists a $\delta = \delta_\varepsilon > 0$ such that

$$\overline{\lim_{n\to\infty}} \frac{n}{x^2} \ln P_n \leqslant -\big(\Lambda_0(b) - \delta\big) = -\big(I_0(f) - \delta\big).$$

Since by Lemma 5.1.3 the first summand on the right-hand side of (5.5.15) satisfies the relation

$$\lim_{n\to\infty} \frac{n}{x^2} \ln \mathbf{P}\big(s_n(1) \in \Delta[b)\big) = -\Lambda_0(b) = -I_0(f),$$

for every $\varepsilon > 0$ we get

$$\underline{\lim_{n\to\infty}} \frac{n}{x^2} \ln \mathbf{P}\big(s_n \in (f)_\varepsilon,\ s_n(1) \in \Delta[b)\big) \geqslant -I_0(f). \qquad (5.5.16)$$

Let us now show that, for every polygon $f \in \mathbb{C}$ consisting of a finite number K of linear segments and such that $f(1) = b$, the relation (5.5.16) is preserved. Suppose that the function is linear on each interval $(t_{k-1},\ t_k)$, for $k = 1, \dots, K$, where $0 = t_0 < t_1 < \cdots < t_K = 1$. Assume first that all points t_k have the form $t_k = j_k/n$, where the numbers j_k are integer. Then

$$\ln \mathbf{P}\big(s_n \in (f)_{\mathbb{C},\varepsilon},\ s_n(1) \in \Delta[b)\big) \geqslant \sum_{k=1}^{K} L_k,$$

where, for $k = 1, \dots, K$, $b_k := f(t_k) - f(t_{k-1})$ (so that, $b = b_1 + \cdots + b_K$),

$$L_k := \ln \mathbf{P}\Big(\max_{t_{k-1}\leqslant t\leqslant t_k} \big|s_n(t) - s_n(t_{k-1}) - (f(t) - f(t_{k-1}))\big| < \varepsilon/2,$$
$$s_n(t_k) - s_n(t_{k-1}) \in \Delta^*[b_k)\Big), \qquad \Delta^* = \frac{\Delta}{2K}.$$

Since $f(t)$ is linear on each interval $(t_{k-1},\ t_k)$ we can use the relation (5.5.16) already obtained for a linear function:

$$\underline{\lim_{n\to\infty}} \frac{n}{x^2} L_k \geqslant -(t_k - t_{k-1})\Lambda_0\Big(\frac{b_k}{t_k - t_{k-1}}\Big) = -\int_{t_{k-1}}^{t_k} \Lambda_0(f'(t))dt.$$

Summing the left- and right-hand sides of the last inequality over k, we get (5.5.16).

If the assumption on the form of the points t_k fails then, along with f, we have to consider the polygon $f^{(n)}(t)$ constructed from the nodes $(t_k^{(n)},\ f(t_k^{(n)}))$,

$k = 0, 1, \ldots, K$, where the points $t_k^{(n)}$ have the desired form and converge to the points t_k as $n \to \infty$. For these polygons $f^{(n)}$ we repeat the previous arguments, which will lead to (5.5.16).

Let us now prove that (5.5.16) holds for every $f \in \mathbb{C}$, with $f(1) = b$. To this end, denote by f_k the polygon with nodes at the points $(j/k, f(j/k))$ such that $\rho_{\mathbb{C}}(f_k, f) < \varepsilon/3$. Obviously,

$$\mathbf{P}\big(s_n \in (f)_{\mathbb{C},\varepsilon},\ s_n(1) \in \Delta[b)\big) \geqslant \mathbf{P}\big(s_n \in (f_k)_{\mathbb{C},\varepsilon/3},\ s_n(1) \in \Delta[b)\big).$$

Applying (5.5.16) to estimate the logarithm of the right-hand side of the last inequality, we get

$$\varliminf_{n\to\infty} \frac{n}{x^2} \ln \mathbf{P}\big(s_n \in (f)_{\mathbb{C},\varepsilon},\ s_n(1) \in \Delta[b)\big) \geqslant -I_0(f_k).$$

Therefore, since $I(f_k) \leqslant I(f)$, we obtain (5.5.16). It remains to observe that (5.5.16) and Lemma 5.5.5 prove the lower bound in Theorem 5.5.3. Theorem 5.5.3 is proved. □

Proof of Theorem 5.5.4. The upper bound. Since for every $\varepsilon > 0$ and all sufficiently large n one has

$$\mathbf{P}\big(s_n \in B,\ s_n(1) \in \Delta[b)\big) \leqslant \mathbf{P}\big(s_n \in B,\ s_n(1) \in (b)_\varepsilon\big),$$

the proof of the upper bound repeats the proof of the desired estimate in Theorem 5.5.2.

The lower bound. For every $f \in (B), f(1) = b$ and sufficiently small $\varepsilon > 0$ one has

$$\mathbf{P}(s_n \in B,\ s_n(1) \in \Delta[b)) \geqslant \mathbf{P}\big(s_n \in (f)_{\mathbb{C},\varepsilon},\ s_n(1) \in \Delta[b)\big).$$

Using the lower bound of Theorem 5.5.3, we get

$$\varliminf_{n\to\infty} \frac{n}{x^2} \ln \mathbf{P}\big(s_n \in B,\ s_n(1) \in \Delta[b)\big) \geqslant -I_0(f) + \Lambda_0(b).$$

Repeating the arguments given in the proof of Theorem 5.5.2, and thus making it possible to obtain the lower bound in the m.l.d.p. from the lower bound in the l.m.l.d.p., we conclude that

$$\varliminf_{n\to\infty} \frac{n}{x^2} \ln \mathbf{P}\big(s_n \in B,\ s_n(1) \in \Delta[b)\big) \geqslant -I_0\big((B)^{(b)}\big) + \Lambda(b).$$

The theorem is proved. □

5.5.3 Extension of Theorems 5.5.1–5.5.4 to discontinuous trajectories, including trajectories of processes with independent increments

As above, let $\overline{s}_n(t)$ be the trajectory of a random walk in $\mathbb{D}$ satisfying the condition (see (5.2.6)) that, for any $h > 0$,

$$\varlimsup_{n\to\infty} \frac{n}{x^2} \ln \mathbf{P}\big(\rho_{\mathbb{C}}(\overline{s}_n, s_n) \geqslant h\big) = -\infty. \tag{5.5.17}$$

As already observed in section 5.2, the process

$$\overset{\neg}{s}_n(t) = \frac{1}{x} S_{[nt]}, \qquad t \in [0, 1], \tag{5.5.18}$$

satisfies (5.5.17) for the deviations (5.1.2) provided either condition $[\mathbf{C_0}]$ or condition $[\mathbf{C}(\beta)]$ is met.

Suppose further that $S(t)$, $t \in [0, T]$, is a process with independent increments, as described in Section 5.3.1, such that the random variable $S(1)$ satisfies $[\mathbf{C_0}]$ or $[\mathbf{C}(\beta)]$. As above, put

$$s_T = s_T(t) := \frac{1}{x} S(tT), \qquad t \in [0, 1].$$

Then, as was demonstrated in Section 5.3.2, the process $\overset{\neg}{s}_n = s_T$ with $T = n$ also satisfies (5.5.17) for the deviations (5.1.2), where the process s_n is the continuous polygon with nodes at the points $(k/n, S(k)/x)$, $k = 0, 1, \ldots, n$.

As in the proofs of Theorems 5.3.1 and 5.3.2, we make use of Corollary 5.2.3, which establishes that the processes s_n and $\overset{\neg}{s}_n$ satisfying (5.5.17) are indistinguishable from the standpoint of the m.l.d.p. Then, repeating the arguments of the previous sections, with obvious changes, we will prove that for s_T there are full analogues of Theorems 5.5.1–5.5.4. In Theorems 5.5.3 and 5.5.4 we must additionally require that $\overset{\neg}{s}_n(1) = s_n(1)$ (which is always the case for the processes (5.5.18) and s_T with $T = n$).

Thus, we have proved the following statement.

Theorem 5.5.6. *Theorems 5.5.1–5.5.4 remain fully valid if in them*

(a) *the process s_n is replaced by a process $\overset{\neg}{s}_n$ satisfying condition* (5.5.17),
(b) *the space $(\mathbb{C}, \rho_{\mathbb{C}})$ is replaced by the space $(\mathbb{D}, \rho_{\mathbb{C}})$,*
(c) *in Theorems 5.5.3, 5.5.4 it is additionally assumed that $\overset{\neg}{s}_n(1) = s_n(1)$.*

This theorem immediately implies the desired assertions for s_T when $T = n$ is an integer.

If T is not an integer then one should apply the arguments of the second parts of the proofs of Theorems 5.3.1 and 5.3.2 in Section 5.3.2.

Moreover, for $\mathbf{P}\big(S(T) \in \Delta[x)\big)$ as $T \to \infty$, the integro-local theorems used for proving Theorems 5.5.3 and 5.5.4 have the same form as for $\mathbf{P}\big(S_n \in \Delta[x)\big)$ (with n replaced by T). This follows from the fact that, in the proofs of the integro-local theorems 1.5.1, 1.5.3 and 2.2.2 the fact that n is an integer is not crucial (see, for example, the right-hand side of the relation (1.5.6) in the proof of the key Theorem 1.5.1. It is this right-hand side that defines the desired asymptotics of $\mathbf{P}\big(S(T) \in \Delta[x)\big)$; then, with n replaced by T, we obtain $\mathbf{E}e^{i\lambda S(T)} = \varphi(\lambda)^T$ where $\varphi(\lambda) = \mathbf{E}e^{i\lambda S(1)}$. Thus, we get

Corollary 5.5.7. *Theorems 5.5.1–5.5.4 are preserved if in them*

(a) *s_n is replaced by s_T,*
(b) *$(\mathbb{C}, \rho_{\mathbb{C}})$ is replaced by $(\mathbb{D}, \rho_{\mathbb{C}})$.*

5.5.4 Analogues of Sanov's theorem in the moderately large deviation zone

In this subsection, in addition to the results of subsection 4.9.4, we find the logarithmic asymptotics of the probability of the hitting of a small $(\varepsilon x/n)$-neighbourhood of a distribution function $F + (x/n)g$ close to F (here, $x = o(n)$, $x \gg \sqrt{n}$ as $n \to \infty$ and g is a continuous bounded function such that $g(t) = 0$ for $t \notin (s_-, s_+)$ and (s_-, s_+) is the support of the distribution $\mathbf{F}$) by an empirical distribution function $F_n^*(t)$ corresponding to a sample of size n from the distribution $\mathbf{F}$ with continuous distribution function $F(t)$. If the function $F^{(-1)}(u)$ inverse to $F(t)$ is also continuous then, performing the change of variables $t = F^{(-1)}(u)$, we may assume without loss of generality that the distribution $\mathbf{F}$ is uniform on $[0, 1]$: i.e. $F(t) = t$, $t \in [0, 1]$, $g(t) = 0$ outside $[0, 1]$.

Theorem 5.5.8. *Let $F(t) = t$ for $t \in [0, 1]$ and $x = o(n)$, $x \gg \sqrt{n \ln n}$ as $n \to \infty$. Then one has*

(i)

$$\lim_{\varepsilon\to 0}\overline{\lim_{n\to\infty}} \frac{n}{x^2} \ln \mathbf{P}\left(\frac{n}{x}(F_n^* - F) \in (g)_{\mathbb{C},\varepsilon}\right)$$
$$= \lim_{\varepsilon\to 0}\underline{\lim_{n\to\infty}} \frac{n}{x^2} \ln \mathbf{P}\left(\frac{n}{x}(F_n^* - F) \in (g)_{\mathbb{C},\varepsilon}\right) = -I_0(g),$$

where

$$I_0(g) = \frac{1}{2}\int_0^1 \left(g'(t)\right)^2 dt.$$

(ii) *For any measurable set $B \subset \mathbb{D}$ satisfying condition* **[B, b]** *with $b = 0$, the inequalities*

$$\overline{\lim_{n\to\infty}} \frac{n}{x^2} \ln \mathbf{P}\left(\frac{n}{x}(F_n^* - F) \in B\right) \leqslant -I_0\left([B]^{(0)}\right),$$
$$\underline{\lim_{n\to\infty}} \frac{n}{x^2} \ln \mathbf{P}\left(\frac{n}{x}(F_n^* - F) \in B\right) \geqslant -I_0\left((B)^{(0)}\right)$$

hold.

As already observed, the condition **[B, b]** is met when the inclusion $f \in B$ does not fix the trajectory end point, i.e. the value $f(1)$, so that the set $B^{(b)}$ is not empty.

Proof. (i) As already observed in subsection 4.9.4, the distribution of the process $F_n^*(t) - F(t)$ coincides with the conditional distribution of the process $S(tT)/T$ with $T = n$, given that $S(T) = 0$ where $S(t)$ is a centred Poisson process with parameter 1, so that $\mathbf{E}S^2(1) = 1$. In this case condition $[\mathbf{C}_\infty]$ is met and, for $T = n$, one has

$$\mathbf{P}\left(\frac{n}{x}(F_n^* - F) \in (g)_{\mathbb{C},\varepsilon}\right) = \mathbf{P}\left(s_T \in (g)_\varepsilon \mid s_T(1) = 0\right),$$

where

$$s_T(t) = \frac{S(tT)}{x}.$$

Therefore, we can use Corollary 5.5.7 and Theorem 5.5.3 part (ii), by virtue of which one has

$$\lim_{\varepsilon \to 0} \overline{\lim_{T \to \infty}} \frac{n}{x^2} \ln \mathbf{P}\big(s_T \in (g)_\varepsilon \mid s_T(1) = 0\big)$$
$$= \lim_{\varepsilon \to 0} \underline{\lim_{T \to \infty}} \frac{n}{x^2} \ln \mathbf{P}\big(s_T \in (g)_\varepsilon \mid s_T(1) = 0\big) = -I_0(g),$$

where $T \to \infty$ takes integer values. This proves the first assertion of the theorem.

(ii) Since the set B satisfies the condition **[B, b]** with $b = 0$, the second assertion of the theorem similarly follows from Corollary 5.5.7 and Theorem 5.5.4 part (ii). Theorem 5.5.8 is proved. □

The above theorem was proved in [19]. The results of Section 5.5 can be found in [68].

6

Some applications to problems in mathematical statistics

6.1 Tests for two simple hypotheses. Parameters of the most powerful tests

Let $(\mathcal{X}, \mathfrak{B}_{\mathcal{X}}, \mathbf{P})$ be a given sample probability space, and suppose that the distribution $\mathbf{P}$ can assume two values, $\mathbf{P}_1$ and $\mathbf{P}_2$. Let

$$\mathrm{X} = \mathrm{X}_n = (\mathrm{x}_1, \ldots, \mathrm{x}_n), \qquad \mathrm{x}_k \in \mathcal{X},$$

be a given sample of size n from a population with distribution $\mathbf{P}$. The notation

$$\mathrm{x} \Subset \mathbf{P} \qquad (\text{or} \quad \mathrm{X} \Subset \mathbf{P})$$

means that x_k have the distribution $\mathbf{P}$. The problem is how, given a sample X, to determine which distribution it has, $\mathbf{P}_1$ or $\mathbf{P}_2$. By $H_j, j = 1, 2$, we will denote the hypotheses

$$H_j = \{\mathrm{X} \Subset \mathbf{P}_j\}.$$

By definition, a *statistical test* $\pi(\boldsymbol{x})$, $\boldsymbol{x} = (x_1, \ldots, x_n) \in \mathcal{X}^n$, is a measurable function on $(\mathcal{X}, \mathfrak{B}_{\mathcal{X}})$ which is equal to the probability of acceptance of hypothesis H_2 if $\mathrm{X} = \boldsymbol{x}$. If π is a non-randomised test then $\pi(\boldsymbol{x})$ takes on only the values 0 and 1.

The probability of a type I error $\varepsilon_1(\pi)$ of a test π (or the size of a test π) and its *probability of a type II error* $\varepsilon_2(\pi)$ are, by definition, respectively

$$\varepsilon_1(\pi) = \mathbf{E}_1\pi(\mathrm{X}), \qquad \varepsilon_2(\pi) = 1 - \mathbf{E}_2\pi(\mathrm{X}),$$

where $\mathbf{E}_j$ denotes the expectation with respect to the distribution $\mathbf{P}_j$. The value $1 - \varepsilon_2(\pi) = \mathbf{E}_2\pi(\mathrm{X})$ is called the *power* of a test π.

We will assume that there is a σ-additive measure $\boldsymbol{\mu}(dx)$ on $(\mathcal{X}, \mathfrak{B}_{\mathcal{X}})$ such that the distributions $\mathbf{P}_j$ have densities $p_j(x)$ with respect to this measure:

$$p_j(x) = \frac{\mathbf{P}_j(dx)}{\boldsymbol{\mu}(dx)}.$$

One can take, for example, the measure $\boldsymbol{\mu}(dx) = \mathbf{P}_1(dx) + \mathbf{P}_2(dx)$. Denote

$$\xi_k = \ln \frac{p_2(\mathrm{x}_k)}{p_1(\mathrm{x}_k)}, \qquad S_n = \sum_{k=1}^{n} \xi_k.$$

It is well known (see e.g. [30], §§ 42, 43) that the test

$$\pi_\nu(\mathrm{X}) = \begin{cases} 1, & \text{if} \quad S_n \geqslant \nu n, \\ 0, & \text{if} \quad S_n < \nu n \end{cases} \tag{6.1.1}$$

for each ν is the optimal test, i.e. it has the maximal power (the minimal probability of a type II error $\varepsilon_2(\pi)$) in the class of all tests π of size $\varepsilon_1 = \varepsilon_1(\pi_\nu) = \mathbf{P}_1(S_n \geqslant \nu n)$. That is, for any test π from this class

$$\varepsilon_2(\pi) \geqslant \varepsilon_2(\pi_\nu) = \mathbf{P}_2(S_n < \nu n).$$

If the distribution of S_n is continuous then, for any given $\varepsilon_1 > 0$, it is always possible to find ν_1 such that $\varepsilon_1(\pi_{\nu_1}) = \varepsilon_1$. Moreover, the test π_ν is Bayesian (for corresponding prior probabilities) and minimax (for appropriate ν).

The problem consists in finding approximate values of the parameters of the test π_ν for large n, i.e. the values of the probabilities of the errors $\varepsilon_j(\pi_\nu)$. There are two asymptotic approaches to the computation of tests (i.e. to finding the asymptotics of the $\varepsilon_j(\pi)$; typically, it is impossible to find precise values of the $\varepsilon_j(\pi)$). The first approach is associated with Le Cam. This approach assumes a scheme of series (the distributions $\mathbf{P}_1, \mathbf{P}_2$ depend on n in such a way that

$$\mathbf{E}_2\xi_1 - \mathbf{E}_1\xi_1 = O\left(\frac{1}{\sqrt{n}}\right);$$

hereafter ξ denotes the quantity ξ_1 for brevity). The hypotheses H_1 and H_2 in this case are called *close*. In this case the values $\varepsilon_j(\pi_\nu)$ under a suitable choice of ν and some other additional conditions will converge to proper limits, which can be found using the central limit theorem (or the invariance principle for more complex tests; see e.g. [30], § 43 for details).

The second approach assumes that the hypotheses H_j (the distributions $\mathbf{P}_j$) are fixed (do not depend on n). In this case, as we will see below, $\min_j \varepsilon_j(\pi_\nu)$ will necessarily decrease with exponential speed. The asymptotics of $\varepsilon_j(\pi_\nu)$ can be found in an explicit form.

We will consider the second approach. Since

$$a_1 := \mathbf{E}_1\xi = \int \mathbf{P}_1(dx) \ln \frac{\mathbf{P}_2(dx)}{\mathbf{P}_1(dx)} =: -r(\mathbf{P}_1, \mathbf{P}_2) < 0,$$
$$a_2 := \mathbf{E}_2\xi = \int \mathbf{P}_2(dx) \ln \frac{\mathbf{P}_2(dx)}{\mathbf{P}_1(dx)} =: r(\mathbf{P}_2, \mathbf{P}_1) > 0,$$

where

$$r(\mathbf{P}, \mathbf{Q}) = \int \mathbf{P}(dx) \ln \frac{\mathbf{P}(dx)}{\mathbf{Q}(dx)}$$

is the Kullback–Leibler divergence between the distributions **P** and **Q**, this approach, as one can easily see, is related to the probabilities of large deviations of the sums S_n.

Everywhere we will assume that the distributions $\mathbf{F}_j$ of the random variables ξ under the hypotheses H_j are *non-lattice*.

We will need the following characteristics of the distributions:

$$\psi_j(\lambda) := \mathbf{E}_j e^{\lambda\xi} = \mathbf{E}_j\left(\frac{p_2(x_1)}{p_1(x_1)}\right)^{\lambda},$$

$$\lambda_{j,+} := \sup\{\lambda : \psi_j(\lambda) < \infty\}, \qquad \alpha_{j,+} := \frac{\psi'(\lambda_{j,+})}{\psi(\lambda_{j,+})}$$

(the derivatives are left-sided). In a similar way, we introduce the quantities $\lambda_{j,-}$ and $\alpha_{j,-}$. Also, let

$$\Lambda_j(v) := \sup_{\lambda}\left(\lambda v - \ln \psi_j(\lambda)\right), \qquad \lambda_j(v) = \Lambda_j'(v), \quad j = 1,2.$$

The functions Λ_j are deviation functions that correspond to the distributions $\mathbf{F}_j$. The characteristics $\lambda_{j,\pm}$, $\alpha_{j,\pm}$ are the characteristics $\lambda_\pm$, $\alpha_\pm$ which played an important role in Chapters 1 and 2. The above-mentioned characteristics have the following properties.

Lemma 6.1.1. (i)

$$\psi_2(\lambda) = \psi_1(\lambda+1), \tag{6.1.2}$$

$$\Lambda_2(v) = \Lambda_1(v) - v, \qquad \lambda_2(v) = \lambda_1(v) - 1. \tag{6.1.3}$$

(ii) $\lambda_{1,+} \geqslant 1$ *and, hence,* $\mathbf{F}_1$ *satisfies Cramér's condition* $[\mathbf{C}_+]$*;*
$\lambda_{2,-} \leqslant -1$ *and, hence,* $\mathbf{F}_2$ *satisfies Cramér's condition* $[\mathbf{C}_-]$.

(iii)

$$\alpha_{1,\pm} = \alpha_{2,\pm} =: \alpha_\pm; \tag{6.1.4}$$

$$\alpha_+ \geqslant a_2 = \mathbf{E}_2\xi = r(\mathbf{P}_2,\mathbf{P}_1); \ \alpha_- \leqslant a_1 = \mathbf{E}_1\xi = -r(\mathbf{P}_1,\mathbf{P}_2); \tag{6.1.5}$$

$$\Lambda_1(a_2) = a_2, \qquad \Lambda_2(a_1) = -a_1. \tag{6.1.6}$$

(iv) *Let*

$$\sigma_j^2(v) := \frac{\psi_j''\big(\lambda_j(v)\big)}{\psi_j\big(\lambda_j(v)\big)} - v^2$$

(by $\psi''\big(\lambda_j(v)\big)$ *we mean the value of* $\psi''(\lambda)$ *at the point* $\lambda = \lambda_j(v)$*). Then*

$$\sigma_1(v) = \sigma_2(v) =: \sigma(v). \tag{6.1.7}$$

Moreover,

$$\sigma^2(a_1) = \mathbf{E}_1\xi^2 - a_1^2, \qquad \sigma^2(a_2) = \mathbf{E}_2\xi^2 - a_2^2. \tag{6.1.8}$$

Proof. (i) We have

$$\psi_2(\lambda) = \mathbf{E}_2\left(\frac{p_2(\mathrm{x}_1)}{p_1(\mathrm{x}_1)}\right)^{\lambda} = \mathbf{E}_1\left(\frac{p_2(\mathrm{x}_1)}{p_1(\mathrm{x}_1)}\right)^{\lambda+1} = \psi_1(\lambda+1).$$

Then

$$\begin{aligned}\Lambda_2(v) &= \sup_{\lambda}\big(\lambda v - \ln \psi_2(\lambda)\big)\\ &= -v + \sup_{\lambda}\big((\lambda+1)v - \ln \psi_1(\lambda+1)\big) = -v + \Lambda_1(v).\end{aligned}$$

This implies (6.1.3).

(ii) For $\lambda = 1$ we have

$$\psi_1(\lambda) = \mathbf{E}_1\left(\frac{p_2(\mathrm{x}_1)}{p_1(\mathrm{x}_1)}\right) = \int p_2(x)\mu(dx) = 1,$$

so that $\lambda_{1,+} \geqslant 1$. Owing to (6.1.2), we get $\lambda_{2,-} \leqslant -1$.

(iii) Since the functions $\psi_j(\lambda)$ differ only by a shift in the argument, identities (6.1.4) are obvious. Since the functions $\psi_j'(\lambda)/\psi_j(\lambda)$ are monotonically increasing, we have

$$\begin{aligned}\alpha_+ &\equiv \frac{\psi_1'(\lambda_{1,+})}{\psi_1(\lambda_{1,+})} \geqslant \frac{\psi_1'(1)}{\psi_1(1)} = \mathbf{E}_1\xi e^{\xi} = \mathbf{E}_2\xi = r(\mathbf{P}_2, \mathbf{P}_1) = a_2,\\ \alpha_- &= \frac{\psi_1'(\lambda_{1,-})}{\psi_1(\lambda_{1,-})} \leqslant \psi_1'(0) = a_1 = -r(\mathbf{P}_1, \mathbf{P}_2).\end{aligned}$$

This proves (6.1.5). Then, because

$$\Lambda_2(a_2) = 0,$$

we have

$$\Lambda_1(a_2) = \Lambda_2(a_2) + a_2 = a_2.$$

In a similar way, the second equality in (6.1.6) can be proved.

(iv) Owing to (6.1.2), (6.1.3), we have

$$\psi_2\big(\lambda_2(v)\big) = \psi_2\big(\lambda_1(v) - 1\big) = \psi_1\big(\lambda_1(v)\big).$$

Similarly, $\psi_2''\big(\lambda_2(v)\big) = \psi_1''\big(\lambda_1(v)\big)$. This implies (6.1.7). Since $\lambda_1(v) = 0$ for $v = a_1$, we have

$$\sigma^2(a_1) = \mathbf{E}_1\xi^2 - a_1^2.$$

The second relation in (6.1.8) is proved in a similar way. The lemma is proved. □

Now we find the asymptotics of the errors $\varepsilon_j(\pi_v)$, $j = 1, 2$. We consider the following two cases.

(1) Both the probabilities of the errors $\varepsilon_j(\pi_v)$ are exponentially decreasing as $n \to \infty$. This will be the case if the parameter v is chosen inside the interval (a_1, a_2) and is separated from its ends.

(2) The probability of a type I error $\varepsilon_1(\pi_v)$ is asymptotically fixed, i.e. $\varepsilon_1(\pi_v) \to \varepsilon_1$ for $n \to \infty$ and given $\varepsilon_1 > 0$. This is the case if the value of v is chosen in a special way, such that it is located close to the point $a_1 = \mathbf{E}_1\xi$ (see below). In this case we will need the following additional assumption.

[σ] *There exists*

$$\psi_1''(0) = \mathbf{E}_1\xi^2 = \mathbf{E}_1\left(\ln \frac{p_2(\mathrm{x}_1)}{p_1(\mathrm{x}_1)}\right)^2 = \int p_1(x)\left(\ln \frac{p_2(x)}{p_1(x)}\right)^2 \mu(dx) < \infty.$$

The same condition on $\psi_2''(0) = \mathbf{E}_2\xi^2$ will be needed if we fix the probability of a type II error $\varepsilon_2(\pi_v)$.

Clearly, condition **[σ]** is always satisfied if $\lambda_{1,-} < 0$ (i.e. if $\mathbf{F}_1$ satisfies condition **[$\mathbf{C}_0$]**).

Everywhere below we assume that $|a_j| < \infty, j = 1, 2$.

Theorem 6.1.2. (i) *Let $v \in [a_1 + \delta, a_2 - \delta]$ for some $\delta > 0$. Then, as $n \to \infty$,*

$$\varepsilon_j(\pi_v) = \frac{e^{-n\Lambda_j(v)}}{\sigma(v)|\lambda_j(v)|\sqrt{2\pi n}}\big(1 + o(1)\big). \tag{6.1.9}$$

(ii) *Let $\varepsilon_1 > 0$ be given. Assume that*

$$v = a_1 + \frac{h\sigma(a_1)}{\sqrt{n}},$$

where h is the solution of the equation

$$1 - \Phi(h) = \varepsilon_1 \tag{6.1.10}$$

(the $(1-\varepsilon_1)$th quantile of the normal distribution). Then, as $n \to \infty$,

$$\varepsilon_1(\pi_v) \to \varepsilon_1,$$

$$\varepsilon_2(\pi_v) = \frac{1}{\sigma(a_1)\sqrt{2\pi n}} \exp\left\{na_1 + \sigma(a_1)h\sqrt{n} + \frac{h^2}{2}\right\}\big(1 + o(1)\big). \tag{6.1.11}$$

Recall that $a_1 = -r(\mathbf{P}_1, \mathbf{P}_2)$.

Proof. (i) For $v \in [a_1 + \delta, a_2 - \delta]$, Theorem 2.2.3 implies that, for $n \to \infty$,

$$\varepsilon_1(\pi_v) = \mathbf{P}_1(S_n \geqslant vn) = \frac{e^{-n\Lambda_1(v)}}{\sqrt{2\pi n}\,\sigma(v)\lambda_1(v)}\big(1 + o(1)\big).$$

In a similar way, using the integro-local theorem 2.2.3 with subsequent integration towards smaller values of v, we get for $n \to \infty$

$$\varepsilon_2(\pi_v) = \mathbf{P}_2(S_n < vn) = \frac{e^{-n\Lambda_2(v)}}{\sqrt{2\pi n}\,\sigma(v)|\lambda_2(v)|}\big(1 + o(1)\big).$$

The first claim of the theorem is proved.

(ii) If $v = a_1 + h\sigma(a_1)/\sqrt{n}$ then the convergence (6.1.10) follows from the central limit theorem:

$$\varepsilon_1(\pi_v) = \mathbf{P}_1\big(S_n \geqslant a_1 n + h\sigma(a_1)\sqrt{n}\big) \to 1 - \Phi(h) = \varepsilon_1.$$

Let us now find the asymptotics of $\alpha_2(\pi_v)$. If $\lambda_{2,-} = -1$ then one should use the integro-local theorem 2.4.1 and Corollary 2.4.2 applied to the sums $-S_n$ and again integrate towards smaller values of the argument. This yields (6.1.11).

Now consider the case $\lambda_{2,-} < -1$. It is not difficult to see that the proof, and hence the claim of Theorem 2.4.1 is preserved if, instead of $x = \alpha_+ n + y$, one takes the values $x = \alpha n + y$, where α is an interior point of (α_-, α_+) (respectively, $\lambda(\alpha)$ is an interior point of (λ_-, λ_+)). Therefore, in the case $\lambda_{2,-} < -1$ all the above arguments remain the same. The theorem is proved. □

As for the second claim of the theorem, it is possible to consider the case when a type II error is fixed.

Theorem 6.1.2 and Lemma 6.1.1 imply the following.

Corollary 6.1.3. *If* $v \in [a_1 + \delta,\ a_2 - \delta]$, $\delta > 0$, *then, as* $n \to \infty$,

$$\varepsilon_2(\pi_v) = \varepsilon_1(\pi_v)\,\frac{e^{nv}(1 - \lambda_1(v))}{\lambda_1(v)}\big(1 + o(1)\big). \tag{6.1.12}$$

The relation (6.1.12) means that in order to find $\varepsilon_2(\pi_v)$ it is sufficient to compute $\varepsilon_1(\pi_v)$ (and the value $\lambda_1(v)$, which appears in the right-hand side of (6.1.12), will also be found).

From (6.1.12), it also follows that for

$$v = \frac{1}{n}\ln\frac{\lambda_1(0)}{1 - \lambda_1(0)}$$

we have $\varepsilon_1(\pi_v) \sim \varepsilon_2(\pi_v)$ as $n \to \infty$ and hence the test π_v will be asymptotically minimax; $\lambda_1(0)$ is the value at which $\min_\lambda \psi_1(\lambda)$ is attained. The test π_0 can be considered as close to minimax.

Methods of finding numerical values of the functions $\Lambda_j(v)$, $\lambda_j(v)$, $\sigma(v)$ in particular problems were considered in section 3.10.

6.2 Sequential analysis

6.2.1 Unbounded sample size

In this section we will consider the same problem as in section 6.1. We will use the same notation but will assume that an unlimited number of experiments can be performed.

In the previous paragraph we established that the likelihood test $\pi_v(\mathrm{X})$, defined in (6.1.1), is the most powerful test, so one cannot improve upon it. However, if the execution of each experiment (to obtain values of the elements x_j of a sample X) is costly, it is natural not to assume that the sample size n is fixed but to make it random and also to depend on results already obtained. Doing that, it is possible to improve upon the test $\pi_\nu(\mathrm{X})$; this can be explained using the following example. Suppose that the distributions $\mathbf{P}_1$ and $\mathbf{P}_2$ are not mutually absolutely continuous and that there exist sets B_1 and B_2 from $\mathfrak{B}_{\mathcal{X}}$ such that $p_1(x) > 0$, $p_2(x) = 0$ for $x \in B_1$, and $p_1(x) = 0$, $p_2(x) > 0$ for $x \in B_2$. Then it is clear that if $\mathrm{x}_1 \in B_1$ ($\mathrm{x}_1 \in B_2$), we can immediately claim that the hypothesis H_1 (H_2) takes place and there is no need to make further observations.

Thus, if one does not perform n experiments all at once but sequentially analyses the result of each new series of observations, it is possible to reduce the number of observations.

For this end, A. Wald introduced *sequential tests* for testing hypothesis H_1 against H_2, which assumes the possibility of performing an unlimited number of experiments (see e.g. [30], [180], [168]).

A sequential test is a vector function (ν, π) defined on the space $(\mathcal{X}^\infty, \mathfrak{B}^\infty_{\mathcal{X}})$, in which the first coordinate is a stopping time (a Markov time: $\{\nu \geqslant n\} \in \mathfrak{B}^n_{\mathcal{X}}$), and the second coordinate π is a function on $(\mathcal{X}^\nu, \mathfrak{B}^\nu_{\mathcal{X}})$ taking the values 0 and 1. If $\pi(\mathrm{X}_\nu) = 1$ then the hypothesis H_2 is accepted; if $\pi(\mathrm{X}_\nu) = 0$ then H_1 is accepted.

Consider the following sequential test (ν_Γ, π_Γ), where $\Gamma = (\Gamma_1, \Gamma_2)$ for given values $\Gamma_1 < 0$ and $\Gamma_2 > 0$. One stops performing experiments as soon as the inequalities $\Gamma_1 < S_k < \Gamma_2$ fail to hold. In other words,

$$\nu_\Gamma = \min\{k \geqslant 1 : S_k \notin \Gamma\}$$

is the first exit time of a trajectory $\{S_k\}_{k=1}^\infty$ from the strip Γ. Define the function $\pi_\Gamma(\mathrm{X}_{\nu_\Gamma})$ by

$$\pi_\Gamma(\mathrm{X}_{\nu_\Gamma}) = \begin{cases} 1, & \text{if} \quad S_{\nu_\Gamma} \geqslant \Gamma_2, \\ 0, & \text{if} \quad S_{\nu_\Gamma} \leqslant \Gamma_1. \end{cases}$$

Obviously, (ν_Γ, π_Γ) is a sequential test.

In a manner similar to the above, denote by $\varepsilon_1(\nu, \pi)$ and $\varepsilon_2(\nu, \pi)$ the probabilities of type I and type II errors in a test (ν, π), respectively:

$$\varepsilon_1(\nu, \pi) := \mathbf{E}_1 \pi(\mathrm{X}_\nu), \qquad \varepsilon_2(\nu, \pi) := \mathbf{E}_2\big(1 - \pi(\mathrm{X}_\nu)\big).$$

Then the following results are true (see e.g. [30], Theorem 51.2).

Theorem 6.2.1. *Let the numbers* Γ_1, Γ_2, $\varepsilon_j(\nu_\Gamma, \pi_\Gamma) = \varepsilon_j$, $j = 1, 2$, *be given. Among all sequential tests* (ν, π) *for which*

$$\varepsilon_j(\nu, \pi) \leqslant \varepsilon_j, \quad j = 1, 2, \tag{6.2.1}$$

the test (ν_Γ, π_Γ) *has the smallest average number of experiments* $\mathbf{E}_1 \nu$ *and* $\mathbf{E}_2 \nu$.

This result means, in particular, that if π is a test constructed using a sample X_n, with fixed size n ($\nu \equiv n$), for which $\varepsilon_j(\pi) \leqslant \varepsilon_j$ then

$$\mathbf{E}_j \nu_\Gamma \leqslant n, \quad j = 1, 2.$$

Thus, the test (ν_Γ, π_Γ) minimizes the average number of experiments for fixed probabilities of type I and type II errors. In this sense, the test (ν_Γ, π_Γ) is *optimal*.

The goal of this section is to find approximations of the parameters of this test (i.e. of the values $\varepsilon_j(\nu_\Gamma, \pi_\Gamma)$ and $\mathbf{E}_j\nu_\Gamma,\ j = 1, 2$) for large values of $|\Gamma_j|$. In other words, we will assume that $|\Gamma_j| \to \infty$, $j = 1, 2$ (or, which is the same, $\varepsilon_j(\nu_\Gamma, \pi_\Gamma) \to 0$). As before, denote by $\eta(v)$ the first passage time over the level v by the random walk $\{S_k\}$:

$$\eta(v) = \begin{cases} \min(k \geqslant 1 : S_k \geqslant v), & \text{if} \quad v > 0, \\ \min(k \geqslant 1 : S_k \leqslant v), & \text{if} \quad v < 0. \end{cases}$$

We assume that $\eta(v) = \infty$ if $S := \sup_{k \geqslant 0} S_k < v$ for $v > 0$. A similar agreement is used when $v < 0$.

We have

$$\begin{aligned} \varepsilon_1(\nu_\Gamma, \pi_\Gamma) &= \mathbf{P}_1\big(\eta(\Gamma_2) < \infty,\ \eta(\Gamma_2) < \eta(\Gamma_1)\big) \\ &= \mathbf{P}_1\big(\eta(\Gamma_2) < \infty\big) - \mathbf{P}_1\big(\eta(\Gamma_2) < \infty,\ \eta(\Gamma_2) > \eta(\Gamma_1)\big). \end{aligned}$$

Since $\eta(v)$ is a Markov time,

$$\begin{aligned} &\mathbf{P}_1\big(\eta(\Gamma_2) < \infty,\ \eta(\Gamma_2) > \eta(\Gamma_1)\big) \\ &= \sum_{n=1}^{\infty} \mathbf{P}\big(\eta(\Gamma_2) < \infty;\ \eta(\Gamma_2) > \eta(\Gamma_1) = n\big) \\ &\leqslant \sum_{n=1}^{\infty} \mathbf{P}_1\big(\eta(\Gamma_1) = n\big)\mathbf{P}_1\big(\eta(\Gamma_2 - \Gamma_1) < \infty\big) = \mathbf{P}_1\big(\eta(\Gamma_2 - \Gamma_1) < \infty\big). \end{aligned}$$

Thus,

$$\mathbf{P}_1\big(\eta(\Gamma_2) < \infty\big) - \mathbf{P}_1\big(\eta(\Gamma_2 - \Gamma_1) < \infty\big) \leqslant \varepsilon_1(\nu_\Gamma, \pi_\Gamma) \leqslant \mathbf{P}_1\big(\eta(\Gamma_2) < \infty\big). \tag{6.2.2}$$

The asymptotics of the probability $\mathbf{P}_1\big(\eta(v) < \infty\big) = \mathbf{P}_1(S \geqslant v)$ as $v \to \infty$ was studied in Theorems 3.4.6, and 3.9.6; see also [39], Theorem 12.7.4. Since in our case $\psi_1(1) = 1$, according to those theorems we have

$$\mathbf{P}_1(S \geqslant v) \sim q_1 e^{-v}\big(1 + o(1)\big) \tag{6.2.3}$$

as $v \to \infty$, where q_1 is found in the above-mentioned theorems in an explicit form in terms of the Cramér transform of the distribution $\mathbf{F}_1$. From (6.2.2) and (6.2.3) there follows an asymptotic representation for $\varepsilon_1(\nu_\Gamma, \pi_\Gamma)$. However, a representation for q_1 in (6.2.3) has a rather complicated form, and the proof of (6.2.3) has some related conditions.

At the same time, for the special form of jumps ξ_k that we are dealing with, it is possible to obtain a version of representation (6.2.3) in which the coefficient q_1 has a simpler form. Denote, as before,

$$\chi(v) = S_{\eta(v)} - v.$$

Also, denote by $\chi(\infty)$ the size of the overshoot over an infinitely distant barrier:

$$\mathbf{P}\big(\chi(\infty) \geqslant t\big) := \lim_{v\to\infty} \mathbf{P}\big(\chi(v) \geqslant t\big), \qquad t \geqslant 0. \tag{6.2.4}$$

It is known (see e.g. [39], Chapter 10) that if $\mathbf{E}\xi > 0$ is finite then the limit in (6.2.4) exists. In a similar way, one introduces the random variable $\chi(-\infty)$ in the case $\mathbf{E}\xi < 0$:

$$\mathbf{P}\big(\chi(-\infty) < t\big) := \lim_{v\to\infty} \mathbf{P}\big(\chi(-v) < t\big), \qquad t \geqslant 0.$$

The distributions of the variables $\chi(\pm\infty)$ can be found using explicit methods given in Chapters 10 and 12 in [39]. In the relations provided there, the distributions $\mathbf{P}_j$, $j = 1, 2$, play the role of the distribution $\mathbf{P}$.

For brevity, let $\varepsilon_j(\nu_\Gamma, \pi_\Gamma) = \varepsilon_j$, $j = 1, 2$. The following result is true.

Theorem 6.2.2. *Suppose that* $|a_j| < \infty$, $j = 1, 2$. *Then, for* $\Gamma_j \to \infty$,

(i)

$$\varepsilon_1 \sim e^{-\Gamma_2}\mathbf{E}_2 e^{-\chi(\infty)}, \tag{6.2.5}$$

$$\varepsilon_2 \sim e^{\Gamma_1}\mathbf{E}_1 e^{\chi(-\infty)}. \tag{6.2.6}$$

(ii)

$$\mathbf{E}_j\nu_\Gamma \leqslant \mathbf{E}_j\eta(\Gamma_j), \qquad j = 1, 2, \tag{6.2.7}$$

$$\mathbf{E}_j\nu_\Gamma \sim \frac{\Gamma_j}{|a_j|} \quad \text{as} \quad \Gamma_j \to \infty, \quad j = 1, 2. \tag{6.2.8}$$

Thus, in our case, in (6.2.3) we have

$$q_1 = \mathbf{E}_2 e^{-\chi(\infty)}. \tag{6.2.9}$$

The analogous coefficient in (6.2.6) is equal to

$$q_2 = \mathbf{E}_1 e^{\chi(-\infty)}.$$

Proof. (i) It is known (see e.g. [30], § 51) that the following relations are valid:

$$\varepsilon_1 = e^{-\Gamma_2}\mathbf{E}_2(e^{-\chi(\Gamma_2)};\ \pi_\Gamma = 1) \leqslant e^{-\Gamma_2}(1-\varepsilon_2), \tag{6.2.10}$$

$$\varepsilon_2 = e^{\Gamma_1}\mathbf{E}_1(e^{\chi(\Gamma_1)};\ \pi_\Gamma = 0) \leqslant e^{\Gamma_1}(1-\varepsilon_1), \tag{6.2.11}$$

In (6.2.10), we have

$$\mathbf{E}_2(e^{-\chi(\Gamma_2)};\ \pi_\Gamma = 1) = \mathbf{E}_2 e^{-\chi(\Gamma_2)} - \mathbf{E}_2(e^{-\chi(\Gamma_2)};\ \pi_1 = 0),$$

where the last term on the right-hand side is not greater than $\varepsilon_2 \to 0$ as $\Gamma_2 \to \infty$ (see (6.2.10)). Hence,

$$\mathbf{E}_2(e^{-\chi(\Gamma_2)};\ \pi_\Gamma = 1) \to \mathbf{E}_2 e^{-\chi(\infty)}$$

as $\Gamma_2 \to \infty$. This implies (6.2.5). Then (6.2.6) is proved in a similar way.

The role of the Cramér transform is played here by the transition from the distribution $\mathbf{P}_1$ in (6.2.10) to the distribution $\mathbf{P}_2$ (and, vice versa, the transition from $\mathbf{P}_2$ to $\mathbf{P}_1$ in (6.2.11)).

(ii) The bounds (6.2.7) for $\mathbf{E}_j\nu_\Gamma$ are essentially obvious, since

$$\nu_\Gamma = \min\big(\eta(\Gamma_1), \eta(\Gamma_2)\big).$$

Let us now obtain a lower bound for $\mathbf{E}_1\nu_\Gamma$. We have

$$\mathbf{P}_1(\nu_\Gamma = n) \geqslant \mathbf{P}_1\big(\eta(\Gamma_1) = n,\ \eta(\Gamma_2) > n\big).$$

For any $\gamma \in (0,1)$, define

$$n_- = \frac{\Gamma_1}{a_1}(1-\gamma), \qquad n_+ = \frac{\Gamma_1}{a_1}(1+\gamma), \qquad M = [n_-, n_+].$$

Then

$$\mathbf{E}_1\nu_\Gamma \geqslant \sum_{n\in M} n\mathbf{P}_1\big(\eta(\Gamma_1) = n,\ \eta(\Gamma_2) > n\big) = \sum_{n\in M} n\mathbf{P}_1\big(\eta(\Gamma_1) = n\big) - \Sigma_M, \tag{6.2.12}$$

where

$$\Sigma_M = \sum_{n\in M} n\mathbf{P}_1\big(\eta(\Gamma_1) = n,\ \eta(\Gamma_2) < n\big).$$

Here, according to the renewal theorem and the law of large numbers for $\eta(v)$, we have, as $|\Gamma_1| \to \infty$,

$$\sum_{n\in M} n\mathbf{P}_1\big(\eta(\Gamma_1) = n\big) \sim \frac{\Gamma_1}{a_1}. \tag{6.2.13}$$

Let us estimate from above the sum Σ_M. By virtue of Theorem 1.1.2,

$$\mathbf{P}_1\big(\eta(\Gamma_2) < n\big) = \mathbf{P}_1(\overline{S}_n \geqslant \Gamma_2) \leqslant \begin{cases} e^{-\Gamma_2} & \text{for} \quad n \geqslant \Gamma_2/a_2; \\ e^{-n\Lambda_1(\Gamma_2/n)} & \text{for} \quad n < \Gamma_2/a_2. \end{cases} \tag{6.2.14}$$

(Recall that $\psi_1(1) = 1$, $\psi_1'(1) = \psi_2'(0) = a_2$, $\lambda_1(a_2) = 1$.) Let

$$M_+ = M\bigcap\Big[\Gamma_2/a_2, \infty\Big), \qquad M_- = M\bigcap\Big(0, \Gamma_2/a_2\Big).$$

Then, if we interpret the notation in the natural way,

$$\Sigma_M = \Sigma_{M_-} + \Sigma_{M_+}.$$

One of the sets $M_\pm$ can be empty (not containing integer numbers). Then the corresponding sum $\Sigma_{M_\pm}$ is assumed to be zero. Suppose that M_+ is not empty. Then, according to (6.2.14),

$$\Sigma_{M_+} \leqslant e^{\Gamma_2} \sum_{n\in M_+} n\mathbf{P}_1\big(\eta(\Gamma_1) = n \,|\, \eta(\Gamma_2) < n\big) \leqslant n_+ e^{-\Gamma_2} = o\big(|\Gamma_1|\big)$$

as $\Gamma_2 \to \infty$. Now, again using (6.2.14),

$$\Sigma_{M_-} \leqslant \sum_{n \in M_-} n e^{-n\Lambda_1(\Gamma_2/n)}.$$

Here, for $n \in M_-$ we have $n < \Gamma_2/a_2$ and so

$$\Lambda_1\Big(\Gamma_2/a_2\Big) \geqslant \Lambda_1(a_2) = a_2 > 0,$$

$$\Sigma_{M_-} \leqslant \sum_{n \in M_-} n e^{-na_2} \leqslant c|\Gamma_1| \exp\left\{-\frac{a_2\Gamma_1(1-\gamma)}{a_1}\right\} = o\big(|\Gamma_1|\big)$$

as $|\Gamma_1| \to \infty$. Together with (6.2.13), this yields

$$\mathbf{E}_1 \nu_\Gamma \geqslant \frac{\Gamma_1}{a_1}\big(1 + o(1)\big)$$

as $|\Gamma_j| \to \infty, j = 1, 2$. According to (6.2.9), we get (6.2.8).

A bound for $\mathbf{E}_2\nu_\Gamma$ is obtained in a similar way.

The theorem is proved. □

6.2.2 Truncated sequential analysis (bounded sample size)

It is not always possible to perform an unlimited number of experiments. Therefore, it is desirable to modify the optimal test (ν_Γ, π_Γ) for the case of a finite number of experiments, in such a way that the main properties of the parameters, discussed in Theorem 6.2.2, will remain be preserved.

Let n be the maximal possible number of experiments. If $\nu_\Gamma \leqslant n$, no changes are introduced in the sequential procedure. If $\nu_\Gamma > n$ (then $S_n \in (\Gamma_1, \Gamma_2)$), it is natural to use the likelihood ratio test (see section 6.1), and for given v such that $nv \in (\Gamma_1, \Gamma_2)$ to accept the hypothesis H_2 in the case $S_n \geqslant nv$. If $S_n < nv$, the hypothesis H_1 is accepted. Thus, the modified test along with the pair (ν_Γ, π_Γ) also includes the parameters n and v.

Denote by $\Gamma(n)$ the set of parameters $\Gamma(n) = (\Gamma_1, \Gamma_2, n, v)$ and by $(\nu_{\Gamma(n)}, \pi_{\Gamma(n)})$ the truncated sequential procedure described above, where $\nu_{\Gamma(n)} = \min(n, \nu_\Gamma)$ is a stopping time.

Clearly, if $n < \max(\Gamma_1/a_1, \Gamma_2/a_2)$, the effect of using the sequential procedure will not be significant (it will not 'work' until time n). Therefore, we will assume that, for some $\gamma > 0$,

$$n > \max\left(\frac{\Gamma_1}{a_1}, \frac{\Gamma_2}{a_2}\right)(1 + \gamma). \tag{6.2.15}$$

Moreover, in order to simplify our statements, along with the condition $|\Gamma_j| \to \infty$ we will also assume that

$$\Gamma_2 - nv \to \infty, \qquad nv - \Gamma_1 \to \infty. \tag{6.2.16}$$

This condition is not essential and its absence will only slightly change the statement of the theorem.

In a similar way to that used before, suppose that $\varepsilon_j(\nu_{\Gamma(n)}, \pi_{\Gamma(n)})$ is the probability of a type j error of the test $(\nu_{\Gamma(n)}, \pi_{\Gamma(n)})$.

Theorem 6.2.3. *Suppose that relations* (6.2.15), (6.2.16) *hold true and the distributions* $\mathbf{F}_j$ *satisfy Cramér's condition* $[\mathbf{C}_0]$. *Then, as* $|\Gamma_j| \to \infty$,

$$\varepsilon_j(\nu_{\Gamma(n)}, \pi_{\Gamma(n)}) = q_j e^{-|\Gamma_{j+1}|}\big(1+o(1)\big) + \frac{e^{-n\Lambda_j(v)}\big(1+o(1)\big)}{\sigma(v)\big|\lambda_j(v)\big|\sqrt{2\pi n}}, \quad j=1,2, \tag{6.2.17}$$

where $\sigma(v)$, $\lambda_j(v)$ *are from Theorem 6.1.2,* q_j *are defined in* (6.2.9) *and* $\Gamma_3 = \Gamma_1$.

Proof. We have

$$\begin{aligned}\varepsilon_1(\nu_{\Gamma(n)}, \pi_{\Gamma(n)}) &= \mathbf{P}_1\big(\eta(\Gamma_2) \leqslant n,\ \eta(\Gamma_2) < \eta(\Gamma_1)\big) \\ &\quad + \mathbf{P}_1\big(\eta(\Gamma_2) > n,\ \eta(\Gamma_1) > n,\ S_n \geqslant nv\big).\end{aligned} \tag{6.2.18}$$

The first term on the right-hand side is equal to

$$\mathbf{P}_1\big(\eta(\Gamma_2) \leqslant n\big) - \mathbf{P}_1\big(\eta(\Gamma_1) < \eta(\Gamma_2) \leqslant n\big) = \mathbf{P}_1(\overline{S}_n \geqslant \Gamma_2) + r_n,$$

where $r_n \leqslant e^{\Gamma_1 - \Gamma_2}$ and, according to Theorem 3.4.8, if (6.2.15) is satisfied then

$$\mathbf{P}_1(\overline{S}_n \geqslant \Gamma_2) \sim q_1 e^{-\Gamma_2} \quad \text{as} \quad \Gamma_2 \to \infty.$$

Consider now the second term on the right-hand side of (6.2.18). If $n \gg \max_j |\Gamma_j|$ then

$$\mathbf{P}_1\big(\eta(\Gamma_1) > n\big) \leqslant e^{-n\Lambda_1(\Gamma_1/n)},$$

where $\Lambda_1(\Gamma_1/n)$ is close to $\Lambda_1(0)$, so that the second term in (6.2.18) is $o(e^{-\Gamma_2})$. If $n < c\max_j |\Gamma_j|$, $c = \text{const.}$, then using Theorem 3.2.6 and conditions (6.2.16), one can easily check that

$$\mathbf{P}_1\big(\eta(\Gamma_2) > n,\ \eta(\Gamma_1) > n,\ S_n \geqslant nv\big) \sim \mathbf{P}_1(S_n \geqslant nv) \sim \frac{e^{-n\Lambda_1(v)}}{\sigma(v)\big|\lambda_1(v)\big|\sqrt{2\pi n}}$$

as $|\Gamma_j| \to \infty$.

In a similar way the asymptotics of $\varepsilon_2(\nu_{\Gamma(n)}, \pi_{\Gamma(n)})$ can be found. The theorem is proved. □

If $n\Lambda_1(v) \geqslant |\Gamma_1|$, then the main contribution to $\varepsilon_1(\nu_{\Gamma(n)}, \pi_{\Gamma(n)})$ will be made by the first term in (6.2.17). It is not hard to see that this will always be the case if v is close to Γ_2/n. A similar remark can be made about $\varepsilon_2(\nu_{\Gamma(n)}, \pi_{\Gamma(n)})$.

6.3 Asymptotically optimal non-parametric goodness of fit tests

Assume that we are given a simple sample X of size n from a population with a continuous distribution function $F(t)$. Denote by $F_n^*(t)$ the empirical distribution function constructed from this sample. Suppose that we want to test the hypothesis $H_1 = \{F(t) = F_1(t)\}$ that the sample was obtained from the distribution $\mathbf{F}_1$ against some alternative H_2. As H_2 we will consider the hypothesis that $F(t)$ belongs to some subset of the set $\mathcal{F}$ of continuous distribution functions which are just assumed to be separated from $F_1(t)$ by some fixed positive distance in the sense of the uniform metric.

In the problems we consider below it can be assumed, without loss of generality, that $F_1(t) = t$, $t \in [0, 1]$, is the uniform distribution on $[0, 1]$.

6.3.1 Asymptotically optimal tests for the class of 'upper' alternatives

First consider the subset of alternatives

$$\mathcal{F}_+ = \Big\{F \in \mathcal{F} : \sup_{t\in(0,1)} \big(F(t) - F_1(t)\big) > \gamma_1\Big\} \tag{6.3.1}$$

for some $\gamma_1 > 0$, which will be called *upper* alternatives.

Let θ be an arbitrary number in $(0, 1/2)$ and $g(t)$ an arbitrary function on $(0, 1)$ from the class $\mathcal{G} = \mathcal{G}(\theta)$ of functions such that

$$\inf_{t\in\Theta} g(t) > 0, \quad \text{where} \quad \Theta := [\theta, 1-\theta].$$

As goodness of fit tests for the hypothesis $H_1 = \{\mathbf{F} = \mathbf{F}_1\}$ against the alternative $H_2 = \{F \in \mathcal{F}_+\}$, we will consider tests based on the statistic

$$T_g(X) = \sup_{t\in\Theta} \frac{F_n^*(t) - F_1(t)}{g(F_1(t))}. \tag{6.3.2}$$

Clearly, for sufficiently small θ and $F \in \mathcal{F}_+$ it will also hold that

$$\sup_{t\in\Theta} \big(F(t) - F_1(t)\big) > \gamma \tag{6.3.3}$$

for some $\gamma > 0$. We will also assume that $F(0) = 0$, $F(1) = 1$ for all $F \in \mathcal{F}_+$. (Regarding the later assumption, see Remark 6.3.5 below.)

Thus, we will use the statistic

$$T_g(X) = \sup_{\Theta} \frac{F_n^*(t) - t}{g(t)}$$

to test a hypothesis about the uniformity of $\mathbf{F}_1$ against the class of alternatives

$$\mathcal{F}_+^\theta = \Big\{F \in \mathcal{F} : F(0) = 0,\, F(1) = 1,\, \sup_{t\in\Theta} \big(F(t) - t\big) > \gamma > 0\Big). \tag{6.3.4}$$

The goodness of fit test based on the statistic $T_g(X)$ will have the form

$$\pi_g(X) = \mathbf{I}\big(T_g^\theta(X) \geqslant 1\big),$$

where $\mathbf{I}(B)$ is the indicator of a set B, $\pi_g(X)$ is the probability of acceptance of H_2 and $1 - \pi_g(X)$ is the probability of acceptance of H_1. We will characterise the probability of a type I error of this test by the quantity

$$L(g) := \overline{\lim_{n\to\infty}} \frac{1}{n} \ln \mathbf{P}\big(T_g(X) \geqslant 1 \,|\, \mathbf{F}_1\big).$$

Denote by $\mathcal{G}_v$ the class of functions g such that $L(g) \leqslant -v$, $v > 0$, so that the probability $\varepsilon_1(\pi_g)$ of type I error for $g \in \mathcal{G}_v$ is not greater than

$$e^{-nv+o(n)}$$

as $n \to \infty$ and we are dealing with large deviation probabilities. If one thinks of the function $F \in \mathcal{F}_+^\theta$ as of a strategy of nature and of the function $g \in \mathcal{G}_v$ as a strategy of the player, then it is natural to define the loss function $L(g, F)$ at a point $g \in \mathcal{G}_v$, $F \in \mathcal{F}_+^\theta$ ($L(g, F)$ is the characteristic of the probability of a type II error at the point $F \in \mathcal{F}_+^\theta$) by the equality

$$L(g, F) := \overline{\lim_{n\to\infty}} \frac{1}{n} \ln \mathbf{P}\big(T_g(X) < 1 \,|\, F\big).$$

As we will see below, the operation $\overline{\lim}$ in the definitions of $L(g)$ and $L(g, F)$ can be replaced with the operation lim for all piecewise continuous functions g.

Our goal consists in the optimal choice of a strategy $g \in \mathcal{G}_v$ for the test π_g.

As it has been already noted, the empirical process

$$F_n^*(t) - t \quad \text{for} \quad F(t) = F_1(t) = t, \quad t \in [0, 1],$$

coincides in distribution with the process

$$s_T(t) = \frac{S(Tt)}{n}, \qquad t \in [0, 1],$$

for $T = n$ (see section 4.9), where $S(t)$ is the centred Poisson process with parameter 1 under the condition $S(n) = 0$ $\big(\mathbf{P}(S(n) = 0) > 0\big)$.

Let $\Lambda(\alpha)$ be the deviation function corresponding to the random variable $S(1)$, so that

$$\begin{aligned} \Lambda(\alpha) &= (\alpha + 1)\ln(\alpha + 1) - \alpha \quad \text{for} \quad \alpha \geqslant -1, \\ \Lambda(\alpha) &= \infty \quad \text{for} \quad \alpha < -1 \end{aligned} \tag{6.3.5}$$

(see Example 1.1.8). For each $t \in (0, 1 - \theta]$ consider the equation

$$t\Lambda\left(\frac{g}{t}\right) + (1 - t)\Lambda\left(-\frac{g}{1 - t}\right) = v. \tag{6.3.6}$$

This is the equation for a '*level line*' of a trajectory of the Poisson process $s_n(t)$ connecting the points (0,0), (t, g), (1,0). Since $\Lambda(\alpha) = \infty$ for $\alpha < -1$, the solution $g(t)$ of this equation at a point t may exist only in the zone

$$g(t) \leqslant 1 - t.$$

Since $\Lambda(-1) = 1$, the upper admissible bound $v(t)$ for the bounded left-hand side of (6.3.6), owing to (6.3.5) is equal to

$$v(t) = t\Lambda\left(\frac{1-t}{t}\right) + 1 - t = t\left[\frac{1}{t}\ln\frac{1}{t} - \frac{1-t}{t}\right] + 1 - t = -\ln t,$$

so $v(t) \to \infty$ as $t \to 0$ and $v(t) \to 0$ as $t \to 1$. The left-hand side of (6.3.6) monotonically and continuously increases from 0 to $v(t)$ as g grows from 0 to $1-t$. Thus, the solution of (6.3.6) at a point t exists and is unique, if $v \leqslant -\ln t$, or, equivalently, if $t \leqslant e^{-v}$. This means that for

$$\theta \geqslant 1 - e^{-v}$$

the solution of equation (6.3.6) exists for all $t \in (0, 1-\theta]$. For small v we obtain the bound

$$\theta \geqslant v - \frac{v^2}{2} + O(v^3).$$

Hence, we can choose θ as small as possible; it is only necessary to ensure that v is not greater than

$$v(\theta) := -\ln(1-\theta) \sim \theta \quad \text{as} \quad \theta \to 0.$$

Let $g = \mathfrak{l}_v(t)$ be the solution of (the level line) equation (6.3.6) with respect to g. If v is small then $\mathfrak{l}_v$ should also be small. Since, for small α,

$$\Lambda(\alpha) = \frac{\alpha^2}{2} + O(\alpha^3) \tag{6.3.7}$$

(see (6.3.5)), the left-hand side of (6.3.6) for each fixed $t \in (0,1)$ is approximately equal to

$$\frac{g^2}{2t} + \frac{g^2}{2(1-t)} = \frac{g^2}{2t(1-t)}.$$

Hence, as $v \to 0$, the solution $\mathfrak{l}_v(t)$ is of the form

$$\mathfrak{l}_v(t) = \sqrt{2vt(1-t)}\,\big(1 + o(1)\big). \tag{6.3.8}$$

If $t \to 0$ then, for each fixed $v > 0$, the left-hand side of (6.3.6) is asymptotically equivalent to

$$t\Lambda\left(\frac{g}{t}\right) \sim g\ln\frac{g}{t}.$$

Therefore,

$$\mathfrak{l}_v(t) \sim -\frac{v}{\ln t},$$

so $\mathfrak{l}_g(t) \to 0$ as $t \to 0$ more slowly than any power function t^β, $\beta > 0$.

Theorem 6.3.1. *Let θ be a number in $(0, 1/2)$, $v \leqslant -\ln(1-\theta)$. Then*

(i) *The class $\mathcal{G}_v$ of functions g for which $L(g) \leqslant -v$ consists of the functions*

$$g(t) \geqslant \mathfrak{l}_v(t), \qquad t \in \Theta = [\theta, 1-\theta].$$

(ii) *The function* $g(t) = \mathfrak{l}_v(t)$ *is asymptotically optimal in the class* $\mathcal{G}_v$*, in the following sense: for any* $g \in \mathcal{G}_v$ *and* $F \in \mathcal{F}_+^\theta$ (see (6.3.4)) *it holds that*

$$L(g,F) \geqslant L(\mathfrak{l}_v, F), \tag{6.3.9}$$

where $L(\mathfrak{l}_v, F) > 0$ *for sufficiently small* v *(the upper bound of admissible values of* v *depends on the value of* γ*; see* (6.3.3))*; the function* $L(g,F)$ *can be found in an explicit form* (see below equation (6.3.11)).

Proof. (i) We have

$$P_{g,n} := \mathbf{P}\left(\sup_{t\in\Theta} \frac{F_n^*(t)-t}{g(t)} \geqslant 1 \,|\, F_1\right) = \mathbf{P}\left(\sup_{t\in\Theta} \frac{S(nt)}{ng(t)} \geqslant 1 \,|\, S(n)=0\right).$$

This means that in the notation of section 4.9 (for $T = n$)

$$P_{g,n} = \mathbf{P}\Big(\sup_{t\in\Theta}\big(s_T(t)-g(t)\big) \geqslant 0 \,|\, s_T(1)=0\Big) = \mathbf{P}\big(s_T \in B_g \,|\, s_T(1)=0\big),$$

where

$$B_g := \Big\{f \in \mathbb{D} : \sup_{t\in\Theta}\big(f(t)-g(t)\big) \geqslant 0\Big\}.$$

Let us use Corollary 4.9.4 and assume, according to the notation of section 4.8, that

$$B_g^{(0)} = B_g \bigcap E^{(0)}, \qquad E^{(0)} = \{f \in \mathbb{D} : f(1) = 0\}.$$

In view of Corollary 4.9.4,

$$\varlimsup_{T\to\infty} \frac{1}{T} \ln \mathbf{P}\big(s_T \in B_g \,|\, s_T(1)=0\big) \leqslant -J\big([B_g]^{(0)}\big),$$
$$\varliminf_{T\to\infty} \frac{1}{T} \ln \mathbf{P}\big(s_T \in B_g \,|\, s_T(1)=0\big) \geqslant -J\big((B_g)^{(0)}\big),$$

where $T \to \infty$ along integer values n,

$$J(B) = \inf_{f\in B} J(f), \qquad J(f) = \int_0^1 \Lambda\big(f'(t)\big)dt,$$

and the function Λ is defined in (6.3.5). Choosing the most probable trajectories from $B_g^{(0)}$, which consist of the two line segments connecting the points $(0,0)$, $\big(t, g(t)\big)$ and $(1,0)$, we obtain

$$J(B_g^{(0)}) = \inf_{t\in\Theta}\left(t\Lambda\left(\frac{g(t)}{t}\right) + (1-t)\Lambda\left(\frac{g(t)}{1-t}\right)\right).$$

It is clear that if $g(t) \geqslant \mathfrak{l}_v(t)$ for all $t \in \Theta$ then

$$J(B_g^{(0)}) \geqslant v.$$

On the other hand, if $g(t) = \mathfrak{l}_v(t)$ for at least one point t from Θ then $J(B_g^{(0)}) = J(B_{\mathfrak{l}_v}^{(0)}) = v$. If $g(t) < \mathfrak{l}_v(t)$ for at least at one point t from Θ then obviously $J(B_g^{(0)}) < v$. The first claim of the theorem is proved.

(ii) Now we prove (6.3.9). We have

$$P_{g,n}(F) := \mathbf{P}\left(\sup_{t\in\Theta}\left(\frac{F_n^*(t)-t}{g(t)}\right) < 1 \,\middle|\, F\right). \tag{6.3.10}$$

Let $F^{(-1)}(t)$ be the inverse function of F. Then $F_n^*\big(F^{(-1)}(t)\big)$ is the empirical uniform distribution function on [0, 1]. Let

$$\Theta_F := \big[F(\theta), F(1-\theta)\big], \qquad b(u) = u - F^{(-1)}(u).$$

Then, changing t to $F^{(-1)}(u)$, under the supremum in (6.3.10), we get for $T = n$

$$\begin{aligned} P_{g,n}(F) &= \mathbf{P}\Big(\sup_{u\in\Theta_F}\big(F_n^*(F^{(-1)}(u)) - u + b(u) - g(F^{(-1)}(u))\big) < 0 \,\Big|\, F\Big) \\ &= \mathbf{P}\big(s_T \in B_{g,F} \mid s_T(1) = 0\big), \end{aligned}$$

where

$$B_{g,F} = \left\{f \in \mathbb{D} : \sup_{u\in\Theta_F}\Big(f(u) + b(u) - g\big(F^{(-1)}(u)\big)\Big) < 0\right\}.$$

Using Corollary 4.9.4, we find that

$$L(g,F) = \overline{\lim_{n\to\infty}}\,\frac{1}{n}\ln P_{g,n}(F) \geqslant \underline{\lim_{n\to\infty}}\,\frac{1}{n}\ln P_{g,n}(F) \geqslant -J(B_{g,F}^{(0)}), \tag{6.3.11}$$

where

$$B_{g,F}^{(0)} = \left\{f \in \mathbb{D} : \sup_{u\in\Theta_F}\Big(f(u) + b(u) - g\big(F^{(-1)}(u)\big)\Big) < 0\right\}.$$

But, for $g \in \mathcal{G}_v$ we have $g \geqslant \mathfrak{l}_v$, and so $B_{\mathfrak{l}_v,F} \subset B_{g,F}$,

$$\begin{aligned} J(B_{g,F}^{(0)}) &\leqslant J(B_{\mathfrak{l}_v,F}^{(0)}), \\ L(g,F) &\geqslant L(\mathfrak{l}_v, F). \end{aligned}$$

This proves (6.3.9). Note that if for some $t_0 \in \Theta$ we have $F_0 := F(t_0) > t_0$ then $F^{(-1)}(F_0) = t_0 < F_0$ and therefore for $u = F_0 \in \Theta_F$ it will hold that

$$b(u) = u - F^{(-1)}(u) = F_0 - F^{(-1)}(F_0) > 0,$$

so that $\sup_{u\in\Theta_F} b(u) > 0$. This means that for sufficiently small v we will also have

$$\sup_{u\in\Theta_F}\Big(b(u) - \mathfrak{l}_v\big(F^{(-1)}(u)\big)\Big) > 0.$$

In this case the most probable trajectory $f_{v,F}$ in the set $B_{\mathfrak{l}_v,F}^{(0)}$ has the form of a string stretched between the points $(0,0)$ and $(1,0)$ and lying on the set Θ_F not above the curve $-b(u) + \mathfrak{l}_v\big(F^{(-1)}(u)\big)$. Since this curve assumes negative values on Θ_F,

$$J(B_{\mathfrak{l}_v,F}^{(0)}) = J(f_{v,F}) > 0.$$

The theorem is proved. □

6.3.2 'Lower' alternatives

In a completely similar way one can construct goodness of fit tests $\pi_g^{(-)}(X) = \mathbf{I}\big(T_g^{(-)}(X) \leqslant -1\big)$ based on the statistic

$$T_g^{(-)}(X) = \inf_{t\in\Theta} \frac{F_n^*(t) - t}{g(t)},$$

for testing a hypothesis H_1 about the uniformness of F_1 against the alternative, $H_2 = \{F \in \mathcal{F}_-^\theta\}$, where

$$\mathcal{F}_-^\theta = \Big\{F \in \mathcal{F} : \inf_{t\in\Theta} \big(F(t) - t\big) \leqslant -\gamma\Big\}, \qquad \gamma > 0.$$

Note that the inequality $T_g^{(-)}(X) \leqslant -1$ can be written as $-T_g^{(-)} \geqslant 1$, and the statistic $-T_g^{(-)}$ as

$$-T_g^{(-)}(X) = \sup_{t\in\Theta} \frac{t - F_n^*(t)}{g(t)}.$$

Therefore, all the arguments provided above are preserved if the Poisson process $S(t)$ is replaced by the process $-S(t)$. Since the deviation function $\Lambda^{(-)}(\alpha)$ for $-S(1)$ is equal to $\Lambda(-\alpha)$, where $\Lambda(\alpha)$ was defined in (6.3.5), the equation for a level line $\mathfrak{l}_v^{(-)}(t)$ (see (6.3.6)) will have the form

$$t\Lambda\left(-\frac{g}{t}\right) + (1-t)\Lambda\left(\frac{g}{1-t}\right) = v.$$

Similarly to the previous equation, it will have a unique solution $g \leqslant t$ for $t \in [\theta, 1]$, where $\theta \geqslant 1 - e^{-v}$. It is not difficult to see that the level line $\mathfrak{l}_v^{(-)}(t)$ is equal to

$$\mathfrak{l}_v^{(-)}(t) = \mathfrak{l}_v(1-t),$$

where the function $\mathfrak{l}_v(t)$ was defined as the solution of equation (6.3.6). From (6.3.7) it is clear that the approximation for $\mathfrak{l}_v^{(-)}(t)$ for small v will be of the same form, (6.3.8), as for $\mathfrak{l}_v(t)$.

Furthermore, for each fixed v and $t \uparrow 1$, it holds that

$$\mathfrak{l}_v^{(-)}(t) \sim -\frac{v}{\ln(1-t)}.$$

In a similar way, consider the characteristic

$$L^{(-)}(g) := \overline{\lim_{n\to\infty}} \frac{1}{n} \mathbf{P}\big(T_g^{(-)}(X) \leqslant -1 \,|\, F_1\big)$$

and the loss function

$$L^{(-)}(g, F) = \overline{\lim_{n\to\infty}} \frac{1}{n} \ln \mathbf{P}\big(T_g^{(-)}(X) > -1 \,|\, F\big)$$

for $g \in \mathcal{G}$, $F \in \mathcal{F}_-^\theta$. The following analogue of Theorem 6.3.1 is valid.

Theorem 6.3.2. *Let θ be a number from $(0,1/2)$, $v \leqslant -\ln(1-\theta)$. Then*

(i) *The class $\mathcal{G}_v^{(-)}$ of functions g for which $L_\theta(g) \leqslant -v$ consists of the functions*

$$g(t) \geqslant \mathfrak{l}_v^{(-)}(t) = \mathfrak{l}_v(1-t), \qquad t \in \Theta.$$

(ii) *The function $g(t) = \mathfrak{l}_v(1-t)$ is asymptotically optimal in the class $\mathcal{G}_v^{(-)}$, i.e. for any $g \in \mathcal{G}_v^{(-)}$ and $F \in \mathfrak{F}_-^\theta$ we have*

$$L^{(-)}(g,F) \geqslant L^{(-)}(\mathfrak{l}_v^{(-)},F).$$

One can also obtain additions to the above claims which are similar to those made in Theorem 6.3.1. The proof of Theorem 6.3.2 is very similar (but with the new notation) to the proof of Theorem 6.3.1.

6.3.3 Two-sided alternatives

Theorems 6.3.1 and 6.3.2 allow us to construct an asymptotically optimal non-parametric test for testing the hypothesis H_1 against the 'two-sided' alternative $H_2 = \{F \in \mathcal{F}^\theta\}$, where

$$\mathcal{F}^\theta = \mathcal{F}_+^\theta \bigcup \mathcal{F}_-^\theta = \Big\{F \in \mathcal{F}: \sup_{t\in[\theta,1-\theta]} \big|F(t)-t\big| \geqslant \gamma\Big\}, \quad \gamma > 0.$$

These theorems show that the optimal upper and lower bounds of the critical zone for $F_n^*(t) - t$ will be different. Namely, consider the test

$$\pi(g_-,g_+,X) = \mathbf{I}\Big(\big\{T_{g_+}(X) \geqslant 1\big\} \bigcup \big\{T_{g_-}^{(-)}(X) \leqslant -1\big\}\Big)$$

and denote by $\mathcal{G}_{v\pm}$ the class of pairs of functions (g_-,g_+) such that

$$L(g_-,g_+) := \overline{\lim_{n\to\infty}} \frac{1}{n} \ln \mathbf{P}\big(\pi(g_-,g_+,X) = 1 \,|\, F_1\big) \leqslant -v.$$

For $g_\pm \in \mathcal{G}_{v_\pm}$, $F \in \mathcal{F}^\theta$, let

$$L(g_-,g_+,F) := \overline{\lim_{n\to\infty}} \frac{1}{n} \ln \mathbf{P}\big(\pi(g_-,g_+,X) = 0 \,|\, F_1\big).$$

Theorem 6.3.3. *Let θ be any number from $(0,1/2)$, $v \leqslant -\ln(1-\theta)$. Then*

(i) *The class $\mathcal{G}_{v\pm}$ of functions $g_\pm$ consists of the functions*

$$g_+(t) \geqslant \mathfrak{l}_v(t), \quad g_-(t) \geqslant \mathfrak{l}_v(1-t) \quad \textit{for} \quad t \in [\theta, 1-\theta].$$

(ii) *For testing the hypothesis H_1 against $H_2 = \{F \in \mathcal{F}^\theta\}$, the functions*

$$g_+(t) = \mathfrak{l}_v(t), \quad g_-(t) = \mathfrak{l}_v^{(-)}(1-t)$$

are asymptotically optimal in the class $\mathcal{G}_{v\pm}$; i.e. for any $(g_-,g_+) \in \mathcal{G}_{v\pm}$ and $F \in \mathfrak{F}^\theta$, it holds that

$$L(g_-,g_+,F) \geqslant L(\mathfrak{l}_v,\mathfrak{l}_v^{(-)},F). \tag{6.3.12}$$

If v is sufficiently small then

$$L(\mathfrak{l}_v, \mathfrak{l}_v^{(-)}, F) > 0.$$

Proof. (i) Clearly, for $g_+ \geqslant \mathfrak{l}_v$, $g_- \geqslant \mathfrak{l}_v^{(-)}$, it holds that

$$L(g_-, g_+) = \overline{\lim_{n\to\infty}} \frac{1}{n} \ln \Big[\mathbf{P}\big(T_{g_+}(X) \geqslant 1 \,|\, F_1\big) + \mathbf{P}\big(T_{g_-}^{(-)}(X) \leqslant -1 \,|\, F_1\big)\Big] \leqslant -v. \tag{6.3.13}$$

This follows from Theorems 6.3.1 and 6.3.2 and from that the probability that the process $s_T(t)$ (for this process see the proof of Theorem 6.3.1) crosses both boundaries (upper and lower) is exponentially smaller than each of the terms in the logarithm in (6.3.13). From the proof of Theorem 6.3.1, it follows that the equality in (6.3.13) is attained when at least one of the functions $g_\pm(t)$ ($(g_-, g_+) \in \mathcal{G}_{v\pm}$) touches, in at least one point from $[\theta, 1-\theta]$, its lower boundary (the lower boundaries are $\mathfrak{l}_v$ and $\mathfrak{l}_v^{(-)}$, respectively). If these lower bounds are crossed by the functions $g_\pm(t)$ then the inequality in (6.3.13) will be violated.

(ii) The inequalities (6.3.12), as in Theorems 6.3.1 and 6.3.2, follow from Theorem 4.9.6 and the fact that the zone

$$\big\{T_{g_+}(X) < 1,\ T_{g_-}^{(-)}(X) > -1\big\}$$

(or the zone

$$\big\{f \in \mathbb{D} : -g_-\big(F^{(-1)}(t)\big) < f(t) + b(t) < g_+\big(F^{(-1)}(t)\big)$$
$$\text{for all} \quad t \in \big[F(\theta), F(1-\theta)\big]\big\}\big)$$

only shrinks if one changes $(g_-, g_+) \in \mathcal{G}_{v\pm}$ to $(\mathfrak{l}_v, \mathfrak{l}_v^{(-)})$.

The last claim of the theorem follows from Theorems 6.3.1, 6.3.2 and the equality $\mathcal{F}^\theta = \mathcal{F}_+^\theta \cup \mathcal{F}_-^\theta$.

The theorem is proved. □

Remark 6.3.4. The well-known criterion of A.N. Kolmogorov for testing the hypothesis H_1 against $H_2 = \{F \neq F_1\}$ is based (after a reduction of F_1 to the uniform distribution) on the statistic

$$T_g(X) = \sup_{t\in[0,1]} \frac{|F_n^*(t) - t|}{g(t)} \tag{6.3.14}$$

for $g(t) \equiv 1$. It has the form

$$\pi_g(X) = \mathbf{I}\big(T(X) \geqslant v\big)$$

and is used for small v, of order $1/\sqrt{n}$. In the 1960s, author A.A. Borovkov discussed this criterion with Andrei Nikolaevich, who saw, of course, the drawbacks relating to the small power of the criterion for alternatives F which differ from F_1 in neighbourhoods either of the point $t = 0$ or the point $t = 1$ (for $F_1(t) = t$, $t \in [0, 1]$). Kolmogorov made a claim (and was going to justify it) that the test π_g for $g(t) = g_0(t) := \sqrt{t(1-t)}$ should be in some sense optimal. His arguments were based on the fact that the variance of a Brownian bridge (the limiting process for $\sqrt{n}\big(F_n^*(t) - t\big)$ under the hypothesis H_1 and $n \to \infty$) at a point t is equal to $t(1-t)$, so the function $g_0(t) = \sqrt{t(1-t)}$ is also, in the known sense, a level line. The claim of Theorem 6.3.3, taking into account (6.3.8), for small v shows that the function g_0 is close to optimal in the sense of Theorem 6.3.3 and confirms the hypothesis of A.N. Kolmogorov. This hypothesis is also confirmed by the results of the paper [71], where one-sided upper alternatives and moderately large deviations were considered (more precise results can be obtained using Theorem 5.5.6). However, the boundary $g_0(t)$ in (6.3.14) cannot give the optimal result, since from the law of the iterated logarithm one can easily see that the process $\sqrt{n}\big(F_n^*(t) - t\big)$ under the hypothesis H_1 and $n \to \infty$ crosses the boundary a number of times equal to $v\sqrt{t(1-t)}$ many times in a neighbourhood of the point reduce space $t = 0$ with probability close to 1. Here, in order to preserve some optimality property of the function $g_0(t)$, one should take the supremum in (6.3.14) not over the whole segment $[0, 1]$ but over the set $\Theta = [\theta, 1-\theta]\ \theta > 0$.

In the setting of the problem that we are considering, it is not possible to construct the optimal boundary $g(t)$ in (6.3.14) on the whole segment $[0, 1]$ using results about large deviation probabilities (i.e. with exponentially small probability of error). This has two main causes. First, as we saw, level lines on the whole segment $[0, 1]$ do not exist. Second, it could be possible to consider generalised level lines v, defined under the conditions of Theorem 6.3.1 on the segment $[1-e^{-v}, 1]$, as the function $1-t$ (the left-hand side of (6.3.6) on this boundary will be discontinuous). But at both ends of the segment $[0, 1]$ these generalised level lines will vanish, and for them condition $[\mathbf{B}, \mathbf{b}]$ (for $b = 0$) of Theorem 4.9.6, which we used in the form of Theorem 4.9.1, will not be satisfied and it appears that proofs of Theorems 6.3.1 and 6.3.2 become impossible. Note that inclusion of the linear part $g(t) = 1 - t$ into the boundary on the segment $[1 - e^{-v}, 1]$ also violates the continuity condition in the conditional l.d.p.

Remark 6.3.5. In the case when, after reducing the distribution F_1 to the uniform distribution, there exist alternatives F such that $F(1) - F(0) < 1$, hypothesis testing can be naturally performed in two stages. First, one should check whether the sample X contains observations outside $[0, 1]$. If it does then, with probability 1, the hypothesis H_2 is true. If all the observations are from $[0, 1]$ (the probability of this event is $\big(F(1)-F(0)\big)^n$) then one should use the above procedure of testing H_1 against alternatives with the distribution functions

$$\frac{F(t) - F(0)}{F(1) - F(0)}, \qquad t \in [0, 1].$$

6.4 Appendix. On testing two composite parametric hypotheses

The problem stated in the section heading consists of the following. Suppose that there is a parametric family $\mathbf{P}_\theta$ of distributions on a probability space $(\mathcal{X}, \mathfrak{B}_\mathcal{X}, \mathbf{P}_\theta)$, where $\theta \in \Theta \subset \mathbb{R}^k$ is a set in the k-dimensional Euclidean space $\mathbb{R}^k$. Let Θ_1 and

Θ_2 be two non-intersecting subsets of $\mathbb{R}^k$ that are separated from each other, and let $\Theta = \Theta_1 \cup \Theta_2$. The unknown parameter θ belongs to one of the sets Θ_j, and the hypothesis H_j is of the form

$$H_j = \{\theta \in \Theta_j\}, \qquad j = 1, 2.$$

Given a sample $X \in \mathbf{P}_\theta$, it is necessary to accept one of the hypotheses H_j.

Suppose that the distributions $\mathbf{P}_\theta$, $\theta \in \Theta_1 \cup \Theta_2$, have densities $p_\theta(x)$ with respect to some measure μ in the phase space $\mathcal{X}$, and let

$$p_\theta(X) = \prod_{i=1}^{n} p_\theta(\mathrm{x}_i)$$

be the likelihood function of a sample $X = (\mathrm{x}_1, \ldots, \mathrm{x}_n)$. By the second fundamental theorem of statistical game theory (see e.g. [30], § 75), if some regularity conditions are satisfied for $\mathbf{P}_\theta$, the class of all Bayesian tests for H_1 and H_2 forms a full minimal class of statistical solutions (the minimal class of best tests, among which are, in particular, minimax tests and uniformly most powerful tests, if the latter exist). But the form of Bayesian tests is well known: they are tests $\pi(X)$ (where $\pi(X)$ is the probability of acceptance of the hypothesis H_2), for which $\pi(X) = 1$ if

$$T(X) := \frac{\int_{\Theta_2} p_\theta(X)\mathbf{Q}_2(d\theta)}{\int_{\Theta_1} p_\theta(X)\mathbf{Q}_1(d\theta)} > c, \tag{6.4.1}$$

where $\mathbf{Q}_j$ are prior distributions on $\Theta_j, j = 1, 2$; the number c depends on the prior probabilities $\mathbf{P}(H_j)$. Here, the distributions $\mathbf{P}_\theta$, $\mathbf{Q}_j$ and $\mathbf{P}(H_j)$ are such that

$$\sup_{\theta \in \Theta_j} \mathbf{P}_\theta\big(T(X) \geqslant c\big) \to 0, \qquad \sup_{\theta \in \Theta_2} \mathbf{P}_\theta\big(T(X) < c\big) \to 0 \tag{6.4.2}$$

for $n \to \infty$, and we are dealing with probabilities of large deviations for the statistic $T(X)$. The asymptotics of the probabilities in (6.4.2) when the sets Θ_j and the distributions $\mathbf{Q}_j$ are sufficiently regular was studied in detail in [48]. It turns out that, for the required asymptotics, the behaviour of $\mathbf{Q}_i$ is important only in the neighbourhoods of certain points and that there exist other statistics, in particular, simpler ones, for example, maximum likelihood ratio statistics

$$T^*(X) := \sup_{\theta \in \Theta_2} \ln p_\theta(X) - \sup_{\theta \in \Theta_1} \ln p_\theta(X), \tag{6.4.3}$$

for which the probabilities of large deviations (in the corresponding regions) behave in almost the same way as for the statistic $T(X)$. This allows us to construct, in an explicit form, tests which are asymptotically equivalent to (6.4.1) and, hence, asymptotically optimal. It also turns out to be possible to find the asymptotics of the probabilities of errors for the tests

$$\mathbf{I}\big\{T(X) > c\big\}, \qquad \mathbf{I}\big\{T^*(X) > c\big\}$$

(see [48]).

This is based on a consideration of the probabilities of large deviations in Cramér's zone of the sums of random fields

$$S_n(\theta) = \sum_{i=1}^{n} \ln p_\theta(x_i), \quad \theta \in \Theta$$

(analogues of the integral theorems in section 2.2) and the probabilities of large deviations of the statistics

$$G(S_n) = \sup_{\theta \in \Theta_2} S_n(\theta) - \sup_{\theta \in \Theta_1} S_n(\theta),$$

$$I(S_n) = \ln \int_{\Theta_2} e^{S_n(\theta)} d\theta - \ln \int_{\Theta_1} e^{S_n(\theta)} d\theta.$$

Statements of the results and their proofs are quite cumbersome and so we do not provide them here.

6.5 Appendix. The change point problem

A substantial body of literature has been devoted to the change point problem (see e.g. [36], [82], [109], [167], [7], [44]–[78], [88], [98], [100], [114], [117], [139], [146], [153], [158]; see also the special issue of the journal *Theory of Probability and Its Applications*, volume 53, issue 3 (2008), which is devoted to the change point problem). In this section, we discuss some results from [30] and [45].

Consider the following sequence X of independent observations:

$$X = (x_1, x_2, \ldots, x_{\theta-1}, x_\theta, \ldots),$$

where the first $\theta - 1$ observations have distribution $\mathbf{F}_1$ and the remaining observations have distribution $\mathbf{F}_2 \neq \mathbf{F}_1$ (here we change the notation of Sections 6.1 and 6.2 somewhat, for reasons which will become clear later). Without loss of generality, one can assume that $\mathbf{F}_1$ and $\mathbf{F}_2$ have densities (are absolutely continuous) with respect to some measure μ. For example, one can take as the measure $\mu = \mathbf{F}_1 + \mathbf{F}_2$. The integer parameter $\theta < \infty$ is unknown, and the problem consists in guessing or estimating this parameter based on observations of X.

6.5.1 Estimation of the parameter θ using the whole sample when the distributions $\mathbf{P}_j$ are known

We will consider the asymptotic setting of the problem, when $n \to \infty$, $\theta \to \infty$ and $n - \theta \to \infty$. The case $n = \infty$ is not excluded; in this case only the assumption $\theta \to \infty$ remains.

The goal of this section is to find an estimator θ^*, of the parameter θ, which is close to θ, for example, in the mean-square sense.

Definition 6.5.1. An estimator θ^* is called *asymptotically homogeneous* (in the mean-square sense), if

$$\mathbf{E}_t(\theta^* - t)^2 \to \sigma^2 \leqslant \infty \tag{6.5.1}$$

as $t \to \infty$ (and $n - t \to \infty$, if $n < \infty$). The class of such estimators will be denoted by $K_{(2)}$.

Let

$$f_j(x) = \frac{\mathbf{F}_j(dx)}{\mu(dx)}, \quad j = 1, 2,$$

be the densities of the distributions $\mathbf{F}_j$ with respect to the measure μ and let

$$r_i = \frac{f_1(\mathrm{x}_i)}{f_2(\mathrm{x}_i)}, \qquad R(k) = R(k, X) = \prod_{i=1}^{k-1} r_i.$$

The likelihood function of the parameter θ at a point $\theta = k < n < \infty$ has the form

$$\prod_{i=1}^{k-1} f_1(\mathrm{x}_i) \prod_{i=k}^{n} f_2(\mathrm{x}_i).$$

Dividing it by $\prod_{i=1}^{n} f_2(\mathrm{x}_i)$, we obtain the value $R(k)$. Therefore, the maximum likelihood estimator $\widehat{\theta}^*$ can be given as follows:

$$\widehat{\theta}^* = \min \big\{k : R(k) = \max_{1 \leqslant j \leqslant n} R(j)\big\}.$$

As it was noted above, in this section we change the notation from that used in sections 6.1 and 6.2. By $\mathbf{P}_t$ we will denote the distribution in the space of samples X for $\theta = t$: the random variables

$$\mathrm{y}_i = \ln r_i$$

under different distributions of the elements of a sample will be denoted by different symbols. Let

$$\xi_i = \ln \frac{f_2(\mathrm{x}_i)}{f_1(\mathrm{x}_i)} = -\mathrm{y}_i, \quad \text{if} \quad \mathrm{x}_i \in\!\!\!\!\in \mathbf{F}_1,$$

$$\zeta_i = \ln \frac{f_1(\mathrm{x}_i)}{f_2(\mathrm{x}_i)} = \mathrm{y}_i, \quad \text{if} \quad \mathrm{x}_i \in\!\!\!\!\in \mathbf{F}_2,$$

so that

$$\mathbf{E}\xi_i = -r(\mathbf{F}_1, \mathbf{F}_2) < 0, \qquad \mathbf{E}\zeta_i = -r(\mathbf{F}_2, \mathbf{F}_1) < 0.$$

If one lets $\mathrm{Y}(k) = \sum_{i=1}^{n} \mathrm{y}_i$, so that $\mathrm{Y}(k) = \ln R(k)$, then the point $\widehat{\theta}^*$ will be the maximum point of the non-homogeneous random walk $\{\mathrm{Y}(k)\}$, which is, at first (until a time θ), directed on average upwards, and then after that downwards. It is not difficult to check that the estimator $\widehat{\theta}^*$ is asymptotically homogeneous. For this estimator we will obtain exponential bounds for $\mathbf{P}_\theta\big(|\theta^* - \theta| > k\big)$ in Theorem 6.5.6.

Along with $\widehat{\theta}^*$, we will also need the following estimator $\widetilde{\theta}^*$ which is close to it.

Definition 6.5.2. The estimator $\widetilde{\theta}^*$ defined by

$$\widetilde{\theta}^* = \sum_{k\geqslant 1} kR(k) / \sum_{k\geqslant 1} R(k), \qquad (6.5.2)$$

is called *the average likelihood estimator.*

Clearly, this estimator is not necessarily integer-valued. Along with $\widetilde{\theta}^*$, one can consider an equivalent (in an obvious sense) randomised integer-valued estimator $\widetilde{\theta}^{**}$ assuming the value $[\widetilde{\theta}^*]$ with probability $1 - \{\widetilde{\theta}^*\}$ and the value $[\widetilde{\theta}^*] + 1$ with probability $\{\widetilde{\theta}^*\}$, where $[x]$ and $\{x\}$ are the integer and fractional parts of x respectively.

Further, let

$$S_k = \sum_{i=1}^{k} \xi_i, \qquad Z_k = \sum_{i=1}^{k} \zeta_i.$$

Introduce the random variables

$$E_1 = \sum_{k=1}^{\infty} e^{S_k}, \qquad E_2 = \sum_{k=1}^{\infty} e^{Z_k}, \qquad E_1' = \sum_{k=1}^{\infty} ke^{S_k}, \qquad E_2' = \sum_{k=1}^{\infty} ke^{Z_k}.$$

Let $c > 0$, $c < -\max(\mathbf{E}\xi_1, \mathbf{E}\zeta_1)$. Since, according to the strong law of large numbers, $S_k < -ck < 0$, $Z_k < -ck < 0$ with probability 1 from some k onwards, the series at hand converges and the above random variables are proper (note that $\mathbf{E}e^{S_k} = \mathbf{E}e^{Z_k} = 1$, $\mathbf{E}E_i = \infty$).

Theorem 6.5.3. *The $\mathbf{P}_\theta$-distribution of $\widetilde{\theta}^* - \theta$ weakly converges to the distribution*

$$\frac{E_2' - E_1'}{1 + E_1 + E_2}$$

as $\theta \to \infty$ (and $n - \theta \to \infty$, if $n < \infty$, $n \to \infty$). This convergence holds together with the convergence of the moments at all orders,

$$\widetilde{\sigma}^2 := \mathbf{E}\left(\frac{E_2' - E_1'}{1 + E_1 + E_2}\right)^2 < \infty.$$

The average likelihood estimator $\widetilde{\theta}^*$ is asymptotically optimal in $K_{(2)}$: for any estimator $\theta^* \in K_{(2)}$ with limiting variance σ^2 (see (6.5.1)) we have $\sigma^2 \geqslant \widetilde{\sigma}^2$.

The maximum likelihood estimator $\widehat{\theta}^*$ is close to the estimator $\widetilde{\theta}^*$ and also has some remarkable properties.

Let $S_{(1)} = \sup_{k\geqslant 1} S_k$, $Z_{(1)} = \sup_{k\geqslant 1} Z_k$.

Theorem 6.5.4. *The estimator $\widehat{\theta}^*$ has the properties*

$$\mathbf{P}_t(\widehat{\theta}^* = t) \downarrow \widehat{p} := \mathbf{P}(S_{(1)} < 0)\mathbf{P}(Z_{(1)} \leqslant 0)$$

as $t \to \infty$ (and $n - t \to \infty$, if $n < \infty$), and, for any other estimator θ^* such that $\lim_{t\to\infty} \mathbf{P}_t(\theta^* = t) = p$ exists, it holds that $\widehat{p} \geqslant p$. Moreover,

$$\mathbf{P}_t(\widehat{\theta}^* > t) \to \mathbf{P}(Z_{(1)} > S_{(1)};\, Z_{(1)} > 0),$$
$$\mathbf{P}_t(\widehat{\theta}^* < t) \to \mathbf{P}(Z_{(1)} \leqslant S_{(1)};\, S_{(1)} \geqslant 0),$$
$$\mathbf{P}_t(\widehat{\theta}^* > t+k) \leqslant 2\psi^k, \qquad \mathbf{P}_t(\widehat{\theta}^* < t-k) < 2\psi^k,$$

where $S_{(1)}$ and $Z_{(1)}$ are independent, $\psi = \min_\lambda \mathbf{E}e^{\lambda\xi} < 1$.

Theorems 6.5.3 and 6.5.4 follow from Theorems 72.1 and 72.4 in [30], respectively. See § 72 in [30] for details on the properties of the limiting $\mathbf{P}_\theta$-distributions of the differences $\widehat{\theta}^* - \theta$ and $\widetilde{\theta}^* - \theta$.

6.5.2 Sequential procedures

Sequential procedures are of interest in those cases when it is necessary to make a decision about the presence of change point and localise it in a minimal amount of time (i.e. number of observations). One can reduce the number of observations in a natural way using the following procedure. Denote by $\widehat{\theta}_m^*$ the smallest value k such that $Y(k) \geqslant Y(j)$ for all $j \leqslant m+1$, so that $\widehat{\theta}_n^* = \widehat{\theta}^*$ if we have n observations. Further, we will assume that $n = \infty$ for the sake of specificity. As the time ν when an observation should be stopped, it is natural to take the value of m for which the difference $Y(\widehat{\theta}_m^*) - Y(m)$ becomes a sufficiently large. We will denote this difference by $N + b(m)$. (It is not hard to see that if $b(m) = o(m)$ then, as $m - \theta \to \infty$, this event occurs with probability 1. On the other hand, if $m < \theta$ then $\widehat{\theta}_m^*$ grows together with m and the difference in question stays bounded.) So, we define the stopping time

$$\nu = \min\big\{m : \, Y(\widehat{\theta}_m^*) - Y(m) \geqslant N + b(m)\big\}. \tag{6.5.3}$$

Such a stopping time is said to be obtained by the cumulative sum method (CUSUM). In the case $b(m) \equiv 0$, this method was proposed in [117] and [146]. Clearly, $\mathbf{P}(\nu < \infty) = 1$, since, owing to the strong law of large numbers $Y(\theta) - Y(m) > c(m - \theta)$, for $m > \theta$ we have $c \in (0, -\mathbf{E}\zeta_1)$ for all sufficiently large $m - \theta$.

If $b(m) = \text{const.}$, one can assume that $b(m) = 0$, and for $m < \theta$

$$\big\{Y(\widehat{\theta}_m^*) - Y(m) > N\big\} \supset \Big\{S(m) - \min_{m \geqslant j > 0} S_{m-j} > N\Big\},$$
$$\mathbf{P}(\nu \leqslant m) \geqslant \mathbf{P}\Big(\bigcup_{1 \leqslant j < k \leqslant m} \big\{S(k) - S_{k-j} > N\big\}\Big).$$

If $m \to \infty$ (as $m < \theta \to \infty$) then it is not difficult to see that this probability converges to 1 and the stopping time ν is 'false' with high probability. Hence, in order to avoid this, we should choose increasing sequences $b(m) \uparrow \infty$ (as far as possible, slowly increasing, so that $\nu - \theta$ is not too large). Let $b(m) \uparrow \infty$ in such a way that

$$\sum_{k=1}^{\infty} e^{-b(k)} < \infty.$$

(One can take, for example, $b(k) = \ln k + (1+\beta)\ln\ln k, \ \beta > 0$ for $k \geqslant 3$; $b(1) = b(2) = 0$.)

The properties of the stopping time $\nu = \nu(N)$ (note that the event $\{\nu \leqslant n\}$ belongs to the σ-algebra $\sigma(\mathrm{x}_1, \ldots, \mathrm{x}_n)$ generated by $\mathrm{x}_1, \ldots, \mathrm{x}_n$), defined in (6.5.3), are described by the following statement. In order to simplify arguments, we will assume that the distribution of $-\xi_i$ is continuous on $(0, \infty)$.

Denote by $P(t) := P(t,N) := \mathbf{P}_t\big(\nu(N) < t\big)$ the probability of a false alarm for $\theta = t \leqslant \infty$,

$$S = \sup_{k\geqslant 0} S_k, \qquad Z = \sup_{k\geqslant 0} Z_k, \qquad \nu^+ = \max(0, \nu).$$

Theorem 6.5.5. *Let* $\beta > 0$,

$$b(k) = \ln k + (1+\beta)\ln\ln k, \qquad b = 2 + \sum_{3}^{\infty} k^{-1}\big(\ln(k)\big)^{-(1+\beta)}. \tag{6.5.4}$$

Then, for all t,

$$P(t) \leqslant b e^{-N}. \tag{6.5.5}$$

If the distribution of $-\xi_i$ *is continuous on* $(0,\infty)$ *then* $P(t)$ *and* $P(\infty)$ *depend continuously on N and it is possible to find an* $\varepsilon_0 > 0$ *such that, for any* $\varepsilon < \varepsilon_0$, *there exists an* $N = N(\varepsilon)$ *which solves the equation* $P(\infty, N) = \varepsilon$, *so that* $P(t) \uparrow P(\infty) = \varepsilon$ *for* $t \to \infty$. *Moreover, for non-lattice* ζ_i *as* $t \to \infty$,

$$\mathbf{E}_t(\nu - t \,|\, \nu \geqslant t) = \frac{\mathbf{E}_t(\nu - t)^+}{1 - P(t)}$$
$$= \frac{1}{r}\big(b(t) + N - \mathbf{E}\max(S^+, Z^+) + \mathbf{E}_\chi\big) + o(1), \tag{6.5.6}$$

where $r = r(\mathbf{F}_2, \mathbf{F}_1)$, *S and Z are independent and* χ *is the size of the overshoot over an infinitely distant barrier by the random walk* $\{-Z_k\}$.

The proof and justification of the choice of $b(k)$ in the form (6.5.4) can be found in [30], § 72. The growth rate of $\mathbf{E}_t(\nu - t \,|\, \nu \geqslant t)$ as $t \to \infty$, defined by (6.5.6), is, apparently, the smallest possible up to the term $c \ln\ln t$, $c = \text{const}$.

As an estimator θ^* for the change point time θ, one can take the value $\widehat{\theta_\nu^*}$ or the value $\theta^* = \nu - G$, where G is equal to the right-hand side of (6.5.6).

6.5.3 Estimation of the parameter θ using the whole sample under incomplete information about the distributions $\mathbf{F}_j$

In this section we will not assume that the distributions $\mathbf{F}_j$ are known. Therefore Theorems 6.5.3 and 6.5.5 cannot be applied in this situation.

Different settings of the change point problem when the $\mathbf{F}_j$ are unknown, while θ and n are large, were considered, for example, in the papers [7], [44]–[76], [81], [88], [98], [100]. Those papers and also [77] contain more detailed bibliographies. The estimators θ^* proposed in those papers have a rather difficult form and are either quite too approximate (so that $|\theta^* - \theta| = O_p(\sqrt{n})$) or too difficult, if $|\theta^* - \theta| = O_p(1)$. Moreover, in the second case it is assumed that

$$\underline{\lim}_{n\to\infty} \frac{1}{n} \min\{\theta, n-\theta\} > 0 \tag{6.5.7}$$

and there is no estimate of the proximity of θ^* and θ. At the same time, under rather general conditions on the distributions $\mathbf{F}_j$, it is possible to construct an estimator θ^*, similar to the maximum likelihood estimator $\widehat{\theta}^*$, such that the probabilities

$$\mathbf{P}_\theta\big(|\theta^* - \theta| \geqslant k\big)$$

admit exponential bounds as k grows.

Let us denote by $\mathrm{x}^{(j)}$ a random variable with distribution $\mathbf{F}_j$.

The condition that the distributions $\mathbf{F}_j$ of the observations x_i are different will be often formulated as follows: there exists a *known* measurable function $h : \mathcal{X} \to \mathbb{R}$ such that

[**h**] $$\mathbf{E}h\big(\mathrm{x}^{(1)}\big) \neq \mathbf{E}h\big(\mathrm{x}^{(2)}\big). \tag{6.5.8}$$

If the distributions $\mathbf{F}_j$ are known, then, as we have seen, it is best to take $h(x) = \ln\big(f_1(x)/f_2(x)\big)$ (see subsection 6.5.2 above). If a parametric *class* of distributions, to which the $\mathbf{F}_j$ belong, is known then it usually simplifies the search for appropriate functions h. For example, if $\mathcal{X} = \mathbb{R}$ and the $\mathbf{F}_j$ are normal then it is enough to consider $h(x) = x$ and $h(x) = x^2$. If $\mathcal{X} = \mathbb{R}$ and it is known that only one of the unknown values $a = \mathbf{E}\mathrm{x}_i$ or $\sigma^2 = \mathbf{D}\mathrm{x}_i$ changes (not necessarily for normal populations) then as a 'separating' function it is natural to take $h(x) = x^2$, since $\mathbf{E}\mathrm{x}_i^2 = \sigma^2 + a^2$.

By introducing a new sample $\mathrm{Y} = (\mathrm{y}_1, \mathrm{y}_2, \ldots, \mathrm{y}_n)$, where $\mathrm{y}_i = h(\mathrm{x}_i)$, we can reduce the problem to the consideration of a real-valued sample; thus we will assume, without loss of generality, that we are given a sample; thus we will $(\mathrm{y}_1, \ldots, \mathrm{y}_n)$, $\mathrm{y}_i \in \mathbb{R}$, and that there exists a constant c such that

$$\mathbf{E}\mathrm{y}^{(1)} > c > \mathbf{E}\mathrm{y}^{(2)}.$$

It is clear that without loss of generality we can assume that $c = 0$ (otherwise, it would be enough to consider the observations $\mathrm{y}'_i = \mathrm{y}_i - c$). Then the condition [**h**] takes the form

[$\mathbf{h_0}$] $$\mathbf{E}\mathrm{y}^{(1)} > 0 > \mathbf{E}\mathrm{y}^{(2)}.$$

Let us introduce some notation, which will be used in what follows. Let

$$\mathrm{Y}_k = \sum_{i=1}^{k} \mathrm{y}_i, \quad \overline{\mathrm{Y}}_n = \max_{k\leqslant n} \mathrm{Y}_k.$$

The superscript $^{(j)}$ (we will write, for example, $\overline{\mathrm{Y}}_n^{(j)}$) will mean that the terms of the sums Y_k have the distribution $\mathbf{F}_j$. By the symbol a_j we will denote the expectation of a random variable $\mathrm{y}^{(j)}$ with distribution $\mathbf{F}_j$: $a_j = \mathbf{E}\mathrm{y}^{(j)}$. As before, for the expectation of functionals of a sample $(\mathrm{y}_1, \dots, \mathrm{y}_n)$ for $\theta = t$ we will use the symbol $\mathbf{E}_t$ and by $\mathbf{P}_t$ we will denote the distribution in the space of sequences X in the case $\theta = t$.

In what follows, we will assume that the elements y_i of the new sample, for which condition $[\mathbf{h_0}]$ holds, satisfy Cramér's condition in the following form:

$$\begin{aligned} \psi_1(\lambda) &:= \mathbf{E}e^{\lambda \mathrm{y}^{(1)}} < \infty \quad \text{for some} \quad \lambda < 0, \\ \psi_2(\lambda) &:= \mathbf{E}e^{\lambda \mathrm{y}^{(2)}} < \infty \quad \text{for some} \quad \lambda > 0. \end{aligned} \tag{6.5.9}$$

These conditions do not present an essential restriction, since if there is a function h satisfying condition $[\mathbf{h_0}]$ then it is possible to find a sufficiently large N that the truncations $^{(N)}h(\mathrm{y}^{(j)})$ of the random variables $h(\mathrm{y}^{(j)}), j = 1, 2$, at levels N and $-N$ ($^{(N)}\xi = \max\big[\min(\xi, N), -N\big]$), will still satisfy the condition $[\mathbf{h_0}]$ but will also satisfy condition $[\mathbf{C}_\infty]$, since the variables $^{(N)}h(\mathrm{y}^{(j)})$ are bounded.

Define the estimator θ^* of the change point time θ by the equality

$$\theta^* = \min\{k \leqslant n : \mathrm{Y}_k = \overline{\mathrm{Y}}_n\}, \tag{6.5.10}$$

so that θ^* is the point where the maximum of the random walk $\{Y_k\}_{k=1}^n$ is attained (in this sense, it is a complete analogue of the estimator $\widehat{\theta}^*$).

Let $\psi_j := \min\limits_{\lambda} \psi_j(\lambda) < 1$.

Theorem 6.5.6. *Suppose that the condition* $[\mathbf{h_0}]$ *holds. Then:*

(i) *if the random variable* $\mathrm{y}^{(1)}$ *satisfies Cramér's condition* $[\mathbf{C}_-]$ (see (6.5.9)), *for all k,* $0 < k < \theta$, *we have*

$$\mathbf{P}_\theta(\theta^* - \theta \leqslant -k) \leqslant 2\psi_1^k;$$

(ii) *if the random variable* $\mathrm{y}^{(2)}$ *satisfies Cramér's condition* $[\mathbf{C}_+]$ (see (6.5.9)), *for all k,* $0 \leqslant k < n - \theta$, *we have*

$$\mathbf{P}_\theta(\theta^* - \theta > k) \leqslant 2\psi_2^k. \tag{6.5.11}$$

Since the distributions of $\mathrm{x}^{(j)}$ are, generally, unknown, the statement of Theorem 6.5.6 has a somewhat qualitative nature, stating that $\mathbf{P}_\theta\big(|\theta^* - \theta| > k\big)$ decreases exponentially as k grows. This means that the estimator θ^* has high accuracy.

The proof of Theorem 6.5.6 is rather simple and we will provide it. It is based on the following statement.

Let $\xi, \xi_1, \xi_2, \dots$ be a sequence of independent identically distributed random variables, $\mathbf{E}\xi < 0$. Let

$$\psi(\lambda) := \mathbf{E}e^{\lambda\xi}, \qquad \psi := \min_{\lambda} \psi(\lambda), \qquad S_n = \sum_{i=1}^{n} \xi_i, \qquad \overline{S} = \sup_{k\geqslant 1} S_k, \tag{6.5.12}$$

and introduce the random variable

$$\tau = \min\{k : \; S_k = \overline{S}\}. \tag{6.5.13}$$

The exact asymptotics of the variable τ was obtained in section 3.5 (see (3.5.3) and the subsequent statements; in (3.5.3) the variable τ was denoted by θ). Here we obtain estimates for $\mathbf{P}(\tau > k)$.

The following result holds true.

Lemma 6.5.7. *Suppose that the random variable ξ satisfies Cramér's condition* $[\mathbf{C}_+]$. *Then $\psi < 1$ and, for $k = 1, 2, \dots$, we have*

$$\mathbf{P}(\tau > k) \leqslant \mathbf{P}\big(\max_{j\geqslant k} S_j > 0\big) \leqslant 2\psi^k. \tag{6.5.14}$$

As the results in § 21 of [22] show, the bounds (6.5.14) cannot be improved, the power factor $ck^{-3/2}$.

Proof of Lemma 6.5.7. For any integer k, $k > 0$, we have

$$\{\tau > k\} \subseteq \{\max_{j\geqslant k+1} S_j > \max_{1\leqslant j\leqslant k} S_j\}. \tag{6.5.15}$$

Since

$$\max_{j\geqslant k+1} S_j = \xi_1 + \max_{j\geqslant k+1} (S_j - \xi_1), \qquad \max_{1\leqslant j\leqslant k} S_j \geqslant \xi_1,$$

we have

$$\{\tau > k\} \subset \big\{\max_{j\geqslant k+1} (S_j - \xi_1) > 0\big\},$$

$$\mathbf{P}(\tau > k) \leqslant \mathbf{P}\left(\max_{j\geqslant k} S_j > 0\right) \leqslant \mathbf{P}(S_k > 0) + \int_{-\infty}^{0} \mathbf{P}(S_k \in dt)\mathbf{P}(\overline{S} > -t). \tag{6.5.16}$$

Let $\lambda(0)$ be the point where $\min_{\lambda\geqslant 0} \psi(\lambda)$ is attained (recall that, under the conditions of the lemma, the random variable ξ satisfies Cramér's condition $[\mathbf{C}_+]$ and $\mathbf{E}\xi < 0$). Then by Chebyshev's inequality we obtain

$$\mathbf{P}\,(S_k > 0) \leqslant \psi^k(\lambda(0)) = \psi^k. \tag{6.5.17}$$

Let us estimate the second term in (6.5.16). By the inequality from Theorem 1.1.2, we have

$$\mathbf{P}\left(\overline{S} > u\right) \leqslant e^{-u\lambda_1}, \quad \text{where } \lambda_1 = \sup\{\lambda : \; \psi(\lambda) \leqslant 1\}, \quad u \geqslant 0. \tag{6.5.18}$$

Obviously $\lambda(0) \leqslant \lambda_1$ and hence, for all $u \geqslant 0$, the right-hand side of the inequality in (6.5.18) does not exceed $e^{-\lambda(0)u}$. Therefore,

$$\int_{-\infty}^{0} \mathbf{P}(S_k \in dt)\mathbf{P}(\overline{S} > -t) \leqslant \int_{-\infty}^{0} e^{\lambda(0)t}\mathbf{P}(S_k \in dt) \leqslant \psi^k(\lambda(0)) = \psi^k. \tag{6.5.19}$$

The first inequality in (6.5.14) is established in (6.5.16). Together with (6.5.16) and (6.5.17), the bound (6.5.19) proves the second inequality of the lemma.

Lemma 6.5.7 is proved. □

Proof of Theorem 6.5.6. According to definition (6.5.10), θ^* is the value t which maximizes the sums Y_t. In particular, if $\theta^* > \theta$ then θ^* is the value t which maximizes $\sum_{i=\theta+1}^{t} \mathrm{y}_i$. Therefore, $\theta^* - \theta$ maximizes the sums $\sum_{i=1}^{t} \mathrm{y}_{\theta+i}$ over t, and $\{\mathrm{y}_{\theta+i};\ i \geqslant 1\}$ are independent identically distributed random variables with distribution $\mathbf{F}_2$. Recall that $a_2 = \mathbf{E}\mathrm{y}^{(2)} < 0$ according to the condition $[\mathbf{h_0}]$. Then the claim (i) follows from Lemma 6.5.7.

The bound (ii) is proved in a similar way. Theorem 6.5.6 is proved. □

6.5.4 Sequential procedures under incomplete information about distributions

In this section, as in the previous one, we will construct a procedure which is close to that used for known distributions $\mathbf{F}_j$.

We will again consider the asymptotic setting of the problem, assuming that $n = \infty$, the moment of the change point θ increases unboundedly and condition $[\mathbf{h_0}]$ holds.

As for the stopping time, introduce the same variable as in (6.5.3):

$$\nu = \nu\big(N, b(\cdot)\big) = \min\big\{m : \overline{Y}_m - Y_m \geqslant N + b(m)\big\}, \tag{6.5.20}$$

where N is some large value, and we define the function $b(\cdot)$ below.

It is clear that, as under conditions of subsection 6.3.2, we should choose increasing sequences $b(k) \uparrow \infty$ as $k \uparrow \infty$ (if possible, slowly growing, so that $\nu - \theta$ is not too large; see below). Otherwise, if $N + b(k)$ is bounded but θ is growing, the probability $\mathbf{P}_\theta(\nu < \theta)$ of a 'false alarm' (of crossing a barrier) will be close to 1 (see the remarks regarding (6.5.3)).

Introduce the following notation:

$$\begin{gathered} \lambda_1 := \sup\left\{\lambda :\ \mathbf{E}e^{-\lambda \mathrm{y}^{(1)}} \leqslant 1\right\}, \\ P(t) = P\left(t, N, b(\cdot)\right) := \mathbf{P}_t\left(\nu(N, b(\cdot)) < t\right), \\ \varepsilon_0 = \varepsilon_0(b(\cdot)) = \mathbf{P}_\infty\left(\bigcup_{k=1}^{\infty}\{\overline{Y}_k - Y_k \geqslant b(k)\}\right). \end{gathered} \tag{6.5.21}$$

The following analogue of (6.5.5) is valid. Denote

$$a_2 = -\mathbf{E}\mathrm{y}^{(2)}, \qquad Y^{(1)} = \sup_{k \geqslant 0}\{-Y_k^{(1)}\}, \qquad Y^{(2)} = \sup_{k \geqslant 0} Y_k^{(2)}.$$

Theorem 6.5.8. *Suppose that condition* $[\mathbf{h_0}]$ *is satisfied, the random variable* $\mathrm{y}^{(2)}$ *satisfies Cramér's condition* $[\mathbf{C}_+]$ *and*

$$b(1) = b(2) = 0, \qquad b(k) = \frac{1}{\lambda_1}\big[\ln k + (1+\beta)\ln\ln k\big] \quad \text{for} \quad k \geqslant 3, \quad \beta > 0,$$

$$b = 2 + \sum_{k=3}^{\infty} k^{-1}(\ln k)^{-(1+\beta)}.$$

Then:

(i) *for all t we have*

$$P(t) = P(t, N, b(\cdot)) \leqslant b e^{-\lambda_1 N}; \tag{6.5.22}$$

(ii) *if* $\mathbf{E}\left|\mathrm{y}^{(2)}\right|^2 < \infty$ *then, for a non-lattice distribution* $\mathbf{F}_2$, *as* $t \to \infty$, *the following representation is true:*

$$\begin{aligned} \mathbf{E}_t(\nu - t \mid \nu > t) &= \frac{\mathbf{E}_t(\nu - t)^+}{1 - P(t)} \\ &= \frac{1}{a_2}\Big(b(t) + N - \mathbf{E}\max\big[Y^{(1)}, Y^{(2)}\big] + \mathbf{E}\chi\Big) + o(1), \end{aligned} \tag{6.5.23}$$

where χ *is the magnitude of the overshoot over an infinitely distant barrier by the random walk* $\{-Y_k^{(2)}\}$ *and the random variables* $Y^{(1)}$ *and* $Y^{(2)}$ *are independent and proper.*

Here, as an estimator θ^* of the change point time θ, as in subsection 6.5.2 it is natural to take the value $k = \widehat{\theta_\nu^*}$ which maximizes the sums Y_k for $k \leqslant \nu$, or the difference $\nu - G$ where G is equal to the right-hand side of (6.5.23).

In order to obtain a meaningful corollary from Theorem 6.5.8 when the $\mathbf{F}_j$ are unknown, one should introduce the classes of distributions $\mathbf{F}_j$ which have lower positive bounds for λ_1 and a_2 and an upper bound c for $\mathbf{E}\chi$. Then, for such classes of distributions, estimate (6.5.22) will be valid and, for all sufficiently large t, the inequality

$$\mathbf{E}_t\big(\nu - t \mid \nu > t\big) \leqslant \frac{1}{a_2}\big[b(t) + N + c\big] \tag{6.5.24}$$

will also be valid. Here one should understand the values λ_1 and a_2 as the lower bounds of these quantities.

The proof of Theorem 6.5.8 is provided in [45] (Theorem 9).

Basic notation

This list contains only the notation used systematically throughout the book.

Random variables

$\xi, \xi_1, \xi_2 \dots$ are independent identically distributed random vectors

$\xi^{(\lambda)}$ is the Cramér transform of ξ, i.e. a random vector with distribution

$$\mathbf{P}(\xi^{(\lambda)} \in dt) = \frac{e^{\langle \lambda, t \rangle}\mathbf{P}(\xi \in dt)}{\mathbf{E}e^{\langle \lambda, t \rangle}}$$

${}^{\alpha}\xi = \xi^{(\lambda(\alpha))}$

${}^{(N)}\xi$ is the truncation of a random variable ξ at the level N: ${}^{(N)}\xi = \min\left[\max(\xi, -N), N\right]$

$S_n = \sum_{j=1}^{n} \xi_j$

$S_n^{(\lambda)} = \sum_{j=1}^{n} \xi_j^{(\lambda)}$

${}^{\alpha}S_n = \sum_{j=1}^{n} {}^{\alpha}\xi_j$

$\overline{S}_n = \max_{0 \leqslant k \leqslant n} S_k$

$\underline{S}_n = \min_{o \leqslant k \leqslant n} S_k$

$S = S_\infty$

$S(t)$ is a process with independent increments

$Z(t)$ is a compound renewal process

$s_n(t)$ is the continuous polygonal curve on $[0, 1]$, constructed by node points $\left(\frac{k}{n}, \frac{S_k}{x}\right)$, $k = 0, 1, \dots, n$

$s_T = \frac{S(tT)}{x}$

$\eta(t) = \min\{k \geqslant 1 : S_k \geqslant t\}$ is the first passage time at a level $t > 0$

$\chi(t) = S_{\eta(t)} - t$ is the magnitude of overshoot over a level t

Transforms of random variables and related notation

$\mathbf{F}$ is the distribution of a random variable ξ
$\psi(\lambda)$ is the Laplace transform of the distribution $\mathbf{F}$: $\psi(\lambda) = \mathbf{E}e^{\langle\lambda,\xi\rangle}$
$\psi^{(\zeta)}(\lambda)$ is the Laplace transform of the distribution of a random variable ζ
$\varphi(\lambda) = \psi(i\lambda)$ is the characteristic function of ξ
$A(\lambda) = \ln\psi(\lambda)$
$\mathbf{F}^{(\lambda)}$ is the Cramér transform of a distribution $\mathbf{F}$: $\mathbf{F}^{(\lambda)}(dt) = \dfrac{e^{\langle\lambda,t\rangle}\mathbf{F}(dt)}{\psi(\lambda)}$
$\Lambda(\alpha)$ is the deviation function of a variable ξ or the Legendre transform of the function $A(\lambda)$: $\Lambda(\alpha) = \sup\big(\langle\lambda,\alpha\rangle - A(\lambda)\big)$.
$\lambda(\alpha) = \Lambda'(\alpha)$ is the point where $\sup\big(\langle\lambda,\alpha\rangle - A(\lambda)\big)$ is attained
$\lambda_+ = \sup\big(\lambda : \psi(\lambda) < \infty\big)$
$\lambda_- = \inf\big(\lambda : \psi(\lambda) < \infty\big)$
$\lambda_1 = \sup\big(\lambda : \psi(\lambda) \leqslant 1\big)$
$\alpha_+ = \lim\limits_{\lambda\uparrow\lambda_+} \dfrac{\psi'(\lambda)}{\psi(\lambda)}$
$\alpha_- = \lim\limits_{\lambda\downarrow\lambda_-} \dfrac{\psi'(\lambda)}{\psi(\lambda)}$
$s_\pm$ are the boundaries of the support of a distribution $\mathbf{F}$
$F_+(t) = \mathbf{P}(\xi \geqslant t),\ t > 0$
$F_-(t) = \mathbf{P}(\xi < -t),\ t > 0$
$I(f) = \int_0^1 \Lambda\big(f'(t)\big)dt$
$I_0(f) = \frac{1}{2}\int_0^1 \big(f'(t)\big)^2 dt$
$J(f)$ is the deviation integral (functional), introduced in sections 4.1 and 4.6

Classes of distributions, conditions, regions, and σ-algebras

$\mathbb{R}^d$ is the d-dimensional Euclidean space
$\mathcal{R}$ is the class of functions regularly varying at infinity, i.e. functions that can be represented in the form $t^\beta L(t)$, $t > 0$, where $L(t)$ is a slowly varying function at infinity
$\mathcal{S}e$ is the class of semi-exponential distributions, i.e. distributions for which $F_+(t) = e^{-l(t)}$, $l(\cdot) \in \mathcal{R}$ for $\beta \in (0,1)$
$\mathcal{ER}$ is the class of distributions for which $F_+(t) = e^{-\lambda_+ t}l(t)$, $l(\cdot) \in \mathcal{R}$, $\lambda_+ > 0$
$\mathcal{ES}e$ is the class of distributions for which $F_+(t) = e^{-\lambda_+ t - l(t)}$, $l(\cdot) \in \mathcal{R}$ for $\beta \in (0,1)$
$\mathcal{L}(\beta)$ is the class of distributions for which $F_+(t) \leqslant e^{-l(t)}$, $l(\cdot) \in \mathcal{R}$ for $\beta \in (0,1)$
$\mathcal{A} = \big\{\lambda : \psi(\lambda) < \infty\big\}$
$\mathcal{A}' = \left\{\dfrac{\psi'(\lambda)}{\psi(\lambda)} : \lambda \in \mathcal{A}\right\}$
[C] is Cramér's condition that the set $\mathcal{A}$ is a non-empty set of values λ in $\mathbb{R}^d$

[C$_0$] is Cramér's condition that the point $\lambda = 0$ is an interior point of $\mathcal{A}$
[C$_+$] is Cramér's condition in the one-dimensional case that the intersection $\mathcal{A}\bigcap(0,\infty)$ in non-empty
[C$_-$] is Cramér's condition in the one-dimensional case that the intersection $\mathcal{A}\bigcap(-\infty,0)$ is non-empty
[C$_\infty$] is Cramér's condition $\mathcal{A} = \mathbb{R}^d$
[C(β)] is the condition that the distribution of ξ belongs to the class $\mathcal{L}(\beta)$
$\mathbb{C}$ is the space of continuous functions
$\mathbb{C}_a$ is the space of absolutely continuous functions
$\mathbb{D}$ is the space of functions without discontinuities of the second kind
$\mathbb{V}$ is the space of functions of bounded variation
ρ is the metric in the space $\mathbb{D}$, introduced in sections 4.2 and 4.6
$\rho_{\mathbb{C}}$ is the uniform metric
$\rho_{\mathbb{D}}$ is the Skorokhod metric
$\rho_{\mathbb{V}}$ is the metric in the space $\mathbb{V}$, introduced in subsection 4.7.1
[R] is the condition that ξ is non-lattice
[Z] is the condition that ξ is arithmetic
[R$^\kappa$] is Cramér's condition on the characteristic function of the distribution of ξ, introduced in subsection 2.3.2
[R$_{\text{den}}$] is the condition on the existence of the density of the distribution of ξ, introduced in subsection 2.3.2
[B, b] is the condition introduced in section 4.8
$\mathfrak{B}(Y) = \mathfrak{B}(\mathbb{Y},\rho)$ is the σ-algebra of Borel sets in a metric space $\mathbb{Y}$ with a metric ρ

Various notations

$\sim$ is the asymptotic equivalence symbol: $a_n \sim b_n$ means that $\dfrac{a_n}{b_n} \to 1$
$\underset{d}{=}$: the relation $\xi \underset{d}{=} \zeta$ means that random variables ξ and ζ have the same distribution
$o_p(1)$ is a random variable converging to zero in probability
$\Delta[x)$ is a half-open cube in $\mathbb{R}^d$: $\Delta[x) = \{v \in \mathbb{R}^d : x_{(1)} \leqslant v_{(1)} < x_{(1)} + \Delta, \dots, x_{(d)} \leqslant v_{(d)} < x_{d)} + \Delta\}$
$\mathrm{e}(x) = \dfrac{x}{|x|}$

References

[1] de Acosta, A. On large deviations of sums of independent random vectors. *Probability in Banach spaces*, V (Medford, Mass., 1984), 1–14, *Lecture Notes in Math.*, 1153 (Springer, 1985).

[2] Arndt, K. Asymptotic properties of the distribution of the supremum of a random walk on a Markov chain. *Theory Probab. Appl.*, 25:2 (1981), 309–324.

[3] Arndt, K. On finding the distribution of the supremum of a random walk on a Markov chain in an explicit form. *Trudy Instituta Matematiki SO AN SSSR*, 1 (1982), 139–146. (In Russian.)

[4] Asmussen, S. Subexponential asymptotics for stochastic processes: extremal behaviour, stationary distributions and first passage probabilities. *Ann. Appl. Probab.*, 8:2 (1998), 354–374.

[5] Bahadur, R.R., Ranga Rao, R. On deviations of the sample mean. *Ann. Math. Statist.*, 31:4 (1960), 1015–1027.

[6] Bahadur, R.R., Zabell, S.L. Large deviations of the sample mean in general vector spaces. *Ann. Probab.*, 7:4 (1979), 587–621.

[7] Baron, M. Nonparametric adaptive change-point estimation and on-line detection. *Sequential Anal.*, 19:1–2 (2000), 1–23.

[8] Bertoin, J., Doney, R.A. Some asymptotic results for transient random walks. *Adv. Appl. Probab.*, 28 (1996), 207–227.

[9] Bhattacharya, R.N., Rao, R. *Normal Approximation and Asymptotic Expansions* (Robert E. Krieger Publishing Company, 1986).

[10] Bickel, P.J., Yahav, J.A. Renewal theory in the plane. *Ann. Math. Statist.*, 36 (1965), 946–955.

[11] Billingsley, P. *Convergence of Probability Measures* (Wiley, 1968).

[12] Bingham, N.H., Goldie, C.H., Teugels, J.L. *Regular Variations* (Cambrige University Press, Cambrige, 1987).

[13] Bolthausen, E. On the probability of large deviations in Banach spaces. *Ann. Probab.*, 2:2 (1984) 427–435.

[14] Borisov, I.S. On the rate of convergence in the 'conditional' invariance principle. *Theory Probab. Appl.*, 23:1 (1978), 63–76.

[15] Borovkov, A.A. Limit theorems on the distributions of maxima of sums of bounded lattice random variables. I, II. *Theory Probab. Appl.*, 5:2 (1960), 125–155; 5:4 (1960), 341–355.

[16] Borovkov, A.A. New limit theorems for boundary crossing problems for sums of independent summands. *Sibirskij Mat. J.*, 3:5 (1962), 645–694. (In Russian.)

[17] Borovkov, A.A. Analysis of large deviations in boundary crossing problems with arbitrary boundaries. I. *Sibirskij Mat. J.*, 5:2 (1964), 253–289. (In Russian.)

[18] Borovkov, A.A. Analysis of large deviations in boundary crossing problems with arbitrary boundaries. II. *Sibirskij Mat. J.*, 5:4 (1964), 750–767. (In Russian.)

[19] Borovkov, A.A., Boundary-value problems for random walks and large deviations in function spaces. *Theory Probab. Appl.*, 12:4 (1967), 575–595.

[20] Borovkov, A.A. Some inequalities for sums of multidimensional random variables. *Theory Probab. Appl.*, 13:1 (1968), 156–160.

[21] Borovkov, A.A. The convergence of distributions of functionals on stochastic processes. *Russian Math. Surveys*, 27:1 (1972), 1–42.

[22] Borovkov, A.A. *Stochastic Processes in Queueing Theory* (Springer, 1976).

[23] Borovkov, A.A. Convergence of measures and random processes. *Russian Math. Surveys*, 31:2 (1976), 1–69.

[24] Borovkov, A.A. *Asymptotic Methods in Queueing Theory* (Wiley, 1984).

[25] Borovkov, A.A. On the Cramér transform, large deviations in boundary value problems, and the conditional invariance principle. *Siberian Math. J.*, 36:3 (1995), 417–434.

[26] Borovkov, A.A. On the limit conditional distributions connected with large deviations. *Siberian Math. J.*, 37:4 (1996), 635–646.

[27] Borovkov, A.A. Limit theorems for time and place of the first boundary passage. *Doklady Mathematics* 55:2 (1997), 254–256.

[28] Borovkov, A.A. An asymptotic exit problem for multidimensional Markov chains. *Markov Proc. Rel. Fields*, 3:4 (1997), 547–564.

[29] Borovkov, A.A. *Ergodicity and Stability of Stochastic Processes* (Wiley, 1998).

[30] Borovkov, A.A. *Mathematical Statistics* (Gordon & Breach, 1998).

[31] Borovkov, A.A. Estimates for the distribution of sums and maxima of sums of random variables without the Cramér condition. *Siberian Math. J.*, 41:5 (2000), 811–848.

[32] Borovkov, A.A. Large deviation probabilities for random walks with semi-exponential distributions. *Siberian Math. J.*, 41:6 (2000), 1061–1093.

[33] Borovkov, A.A. On subexponential distributions and asymptotics of the distribution of the maximum of sequential sums. *Siberian Math. J.*, 43:6 (2002), 995–1022.

[34] Borovkov, A.A. On the asymptotic behavior of the distributions of first-passage times, I. *Math. Notes*, 75:1 (2004), 23–37.

[35] Borovkov, A.A. On the asymptotic behavior of distributions of first-passage times, II. *Math. Notes*, 75:3 (2004), 322–330.

[36] Borovkov, A.A. Large sample change-point estimation when distributions are unknown. *Theory Probab. Appl.*, 53:3 (2009), 402–418.

[37] Borovkov, A.A. Integro-local and local theorems on normal and large deviations of the sums of nonidentically distributed random variables in the triangular array scheme. *Theory Probab. Appl.*, 54:4 (2010), 571–587.

[38] Borovkov, A.A. Large deviation principles for random walks with regularly varying distributions of jumps. *Siberian Math. J.*, 52:3 (2011), 402–410.

[39] Borovkov, A.A. *Probability Theory* (Springer, 2013).

[40] Borovkov, A.A., Borovkov, K.A. On probabilities of large deviations for random walks. I. Regularly varying distribution tails. *Theory Probab. Appl.*, 46:2 (2002), 193–213.

[41] Borovkov, A.A., Borovkov, K.A. On probabilities of large deviations for random walks. II. Regular exponentially decaying distributions. *Theory Probab. Appl.*, 49:2 (2005), 189–206.

[42] Borovkov, A.A., Borovkov, K.A. *Asymptotic Analysis of Random Walks. Heavy Tailed Distributions* (Cambridge University Press, 2008).

[43] Borovkov, A.A., Korshunov, D.A. Large-deviation probabilities for one-dimensional Markov chains. Part 2: Prestationary distributions in the exponential case. *Theory Probab. Appl.*, 45:3 (2001), 379–405.

[44] Borovkov, A.A., Linke, Yu.Yu. Asymptotically optimal estimates in the smooth change-point problem. *Math. Methods Stat.*, 13:1 (2004), 1–24.

[45] Borovkov, A.A., Linke, Yu.Yu. Change-point problem for large samples and incomplete information on distribution. *Math. Methods Stat.*, 14:4 (2006), 404–430.

[46] Borovkov, A.A., Mogul'skii, A.A. Probabilities of large deviations in topological spaces, I. *Siberian Math. J.*, 19:5 (1978), 697–709.

[47] Borovkov, A.A., Mogul'skii, A.A. Probabilities of large deviations in topological spaces, II. *Siberian Math. J.*, 21:5 (1980), 653–664.

[48] Borovkov, A.A., Mogul'skii, A.A. *Large Deviations and Testing Statistical Hypotheses* (Nauka, 1992; in Russian).

[49] Borovkov, A.A., Mogul'skii, A.A. Large deviations and testing statistical hypothesis. *Siberian Adv. Math.*, 2 (1993), 3 (1993).

[50] Borovkov, A.A., Mogul'skii, A.A. The second rate function and the asymptotic problems of renewal and hitting the boundary for multidimensional random walks. *Siberian Math. J.*, 37:4 (1996), 647–682.

[51] Borovkov, A.A., Mogulskii, A.A. Itegro-local limit theorems including large deviations for sums of random vectors, I. *Theory Probab. Appl.*, 43:1 (1999), 1–12.

[52] Borovkov, A.A., Mogulskii, A.A. Integro-local limit theorems including large deviations for sums of random vectors, II. *Theory Probab. Appl.*, 45:1 (2001), 3–22.

[53] Borovkov, A.A., Mogul'skii, A.A. Limit theorems in the boundary hitting problem for a multidimensional random walk. *Siberian Math. J.*, 42:2 (2001), 245–270.

[54] Borovkov, A.A., Mogul'skii, A.A. Integro-local theorems for sums of independent random vectors in the series scheme. *Math. Notes*, 79:4 (2006), 468–482.

[55] Borovkov, A.A., Mogul'skii, A.A. Integro-local and integral theorems for sums of random variables with semiexponential distributions. *Siberian Math. J.*, 47:6 (2006), 990–1026.

[56] Borovkov, A.A., Mogul'skii, A.A. On large and superlarge deviations for sums of independent random vectors under Cramér's condition, I. *Theory Probab. Appl.*, 51:2 (2007), 227–255.

[57] Borovkov A.A., Mogul'skii A.A. On large and superlarge deviations of sums of independent random vectors under Cramér's condition, II. *Theory Probab. Appl.*, 51:4 (2007), 567–594.

[58] Borovkov, A.A., Mogul'skii, A.A. On large deviations of sums of independent random vectors on the boundary and outside of the Cramér zone, I. *Theory Probab. Appl.*, 53:2 (2009), 301–311.

[59] Borovkov, A.A., Mogul'skii, A.A. On large deviations of sums of independent random vectors on the boundary and outside of the Cramér zone, II. *Theory Probab. Appl.*, 53:4 (2009), 573–593.

[60] Borovkov, A.A., Mogul'skii, A.A. On large deviation principles in metric spaces. *Siberian Math. J.*, 51:6 (2010), 989–1003.

[61] Borovkov, A.A., Mogul'skii, A.A. Properties of a functional of trajectories which arises in studying the probabilities of large deviations of random walks. *Siberian Math. J.*, 52:4 (2011), 612–627.

[62] Borovkov, A.A., Mogul'skii, A.A. Chebyshev type exponential inequalities for sums of random vectors and random walk trajectories. *Theory Probab. Appl.*, 56:1 (2012), 21–43.

[63] Borovkov, A.A., Mogulskii, A.A. On large deviation principles for random walk trajectories, I. *Theory Probab. Appl.*, 56:4 (2012), 538–561.

[64] Borovkov, A.A., Mogul'skii, A.A. On large deviation principles for random walk trajectories, II. *Theory Probab. Appl.*, 57:1 (2013), 1–27.

[65] Borovkov, A.A., Mogul'skii, A.A. Inequalities and principles of large deviations for the trajectories of processes with independent increments. *Siberian Math. J.*, 54:2 (2013), 217–226.

[66] Borovkov, A.A., Mogulskii, A.A. Large deviation principles for random walk trajectories, III. *Theory Probab. Appl.*, 58:1 (2014), 25–37.

[67] Borovkov, A.A., Mogul'skii, A.A. Moderately large deviation principles for trajectories of random walks and processes with independent increments. *Theory Probab. Appl.*, 58:4 (2014), 562–581.

[68] Borovkov, A.A., Mogul'skii, A.A. Conditional moderately large deviation principles for the trajectories of random walks and processes with independent increments. *Siberian Adv. Math.*, 25:1 (2015), 39–55.

[69] Borovkov, A.A., Foss, S.G. Estimates for overshooting an arbitrary boundary by a random walk and their applications. *Theory Probab. Appl.*, 44:2 (2000), 231–253.

[70] Borovkov, A.A., Rogozin, B.A. Boundary value problems for some two-dimensional random walks. *Theory Probab. Appl.*, 9:3 (1964), 361–388.

[71] Borovkov, A.A., Sycheva, N.M. On asymptotically optimal non-parametric criteria. *Theory Probab. Appl.*, 13:3 (1968), 359–393.

[72] Borovkov, A.A., Mogul'skii, A.A., Sakhanenko, A.A. *Limit Theorems for Random Processes* (VINITI, Moscow, 1995; in Russian).

[73] Borovkov, K.A. Stability theorems and estimates of the rate of convergence of the components of factorizations for walks defined on Markov chains. *Theory Probab. Appl.*, 25:2 (1980), 325–334.

[74] Boukai, B., Zhou, H. Nonparametric estimation in a two change-point model. *J. Nonparametric Stat.*, 3 (1997), 275–292.

[75] Brodskij, B., Darkhovskij, B. *Nonparametric methods in change-point problems* (Kluwer, 1993).

[76] Carlstein, E. Nonparametric change-point estimation. *Ann. Stat.*, 16:1 (1988), 188–197.

[77] Carlstein, E., Muller, H.-G., Sigmund, D., eds. *Change-Point Problems*. Institute of Math. Statist. Lecture Notes, vol. 23 (1994).

[78] Chover, J., Ney, P., Wainger, S. Function of probability measures. *J. Anal. Math.*, 26 (1973), 255–302.

[79] Cohen, J.W. *Analysis of Random Walks* (IOS Press, 1992).

[80] Cramér, H. Sur un nouveau théorème-limite de la théorie des probabilités. In: *Actualités scientifiques et industrielles* (Hermann, 1938; in French).

[81] Darkhovshkii, B.S. A non-parametric method for the a posteriori detection of the disorder time of a sequence of independent random variables. *Theory Probab Appl.*, 21 (1976), 178–183.

[82] Darkhovskii, B.S., Brodskii, B.E. A nonparametric method for fastest detection of a change in the mean of a random sequence. *Theory Probab. Appl.*, 32:4 (1987), 640–648.

[83] Dembo, A., Zeitouni, O. *Large Deviations Techniques and Applications*, 2nd edn. (Springer, 1998).

[84] Dinwoodie, I.H. A note on the upper bound for i.i.d. large deviations. *Ann. Probab.*, 18:4 (1999), 1732–1736.

[85] Dobrushin, R.L., Pecherskii, E.A. Large deviations for random processes with independent increments on an infinite interval. *Probl. Inform. Transmiss.*, 34:4 (1998), 354–382.

[86] Doney, R.A. An analog of the renewal theorem in higher dimensions. *Proc. London Math. Soc.*, 16 (1966), 669–684.

[87] Doney, R.A. On the asymptotic behaviour of first passage times for transient random walk. *Probab. Theory Rel. Fields* 18 (1989), 239–246.

[88] Dumbgen, L. The asymptotic behavior of some nonparametric change-point estimation. *Ann. Stat.*, 19:3 (1991), 1471–1495.

[89] Ellis, R.S. *Entropy, Large Deviations, and Statistical Mechanics* (Springer, 2006).

[90] Emery, D.J. Limiting behaviour of the distribution of the maxima of partial sums of certain random walks. *J. Appl. Probab.*, 9 (1972), 572–579.

[91] Essen, M. Banach algebra methods in renewal theory. *J. Anal. Math.*, 26 (1973) 303–335.

[92] Feller, W. *An Introduction to Probability Theory and its Applications II*, 2nd edn (Wiley, 1968).

[93] Feng, J., Kurtz, T.G. *Large Deviations for Stochastic Processes* (Amer. Math. Soc., Providence, RI, 2006).

[94] Gikhman, I.I., Skorokhod, A.V. *Introduction to the Theory of Random Processes* (Dover, 1996).

[95] Gnedenko, B.V. On a local limit theorem of the theory of probability. *Uspekhi Mat. Nauk*, 3:3 (1948), 187–194. (In Russian.)

[96] Gnedenko, B.V., Kolmogorov, A.N. *Limit Distributions for Sums of Independent Random Variables* (Addison-Wesley, 1954).

[97] Gut, A. *Stopped Random Walks: Limit Theorems and Applications* (Springer, 1988).

[98] Hawkins, D.L. A simple least squares method for estimating a change. *Commun. Statist. Simula.*, 15:3 (1986), 655–679.

[99] Heyde, C.C. Asymptotic renewal results for a natural generalization of classical renewal theory. *J. Roy. Statist. Soc. Ser. B.*, 29 (1967), 141–150.

[100] Huang, W.-T., Chang, Y.-P. Nonparametric estimation in change-point models. *J. Statis. Plan. Inference*, 35 (1993), 335–347.

[101] Hunter, J.J. Renewal theory in two dimensions: asymptotic results. *Adv. Appl. Probab.*, 6 (1974), 546–562.

[102] Ibragimov, I.A., Linnik, Yu.V. *Independent and Stationary Sequences of Random Variables* (Wolters-Noordhoff, 1971).

[103] Ikeda, N., Watanabe, S. *Stochastic Differential Equations and Diffusion Processes* (North-Holland, 1981).

[104] Janson, S. Moments for first-passage and last-exit times. The minimum, and related quantities for random walks with positive drift. *Adv. Appl. Probab.*, 18 (1986), 865–879.

[105] Kantorovich, L.V., Akilov, G.P. *Functional Analysis* (Pergamon Press, 1982).

[106] Keilson, J., Wishart, D.M.G. A central limit theorem for processes defined on a finite Markov chain. *Proc. Camb. Phil. Soc.*, 60 (1964), 547.

[107] Kesten, H., Maller, R.A. Two renewal theorems for general random walks tending to infinity. *Probab. Theory Rel. Fields*, 106:1 (1996), 1–38.

[108] Khodjibaev V.R. Asymptotic analysis of distributions in two-boundaries problems for random walks with continuous time. Limit theorems for sums of random variables. *Trudy Instituta Matematiki SO AN SSSR*, 3 (1984), 77–93. (In Russian.)

[109] Kligene, N., Tel'ksnis, L. Methods of detecting instants of change of random process properties. *Automation and Remote Control*, 44:10 (1983), 1241–1283.

[110] Kolmogorov, A.N., Fomin, S.V. *Elements of the Theory of Functions and Functional Analysis* (Dover, 1968).

[111] Korolyuk, V.S., Borovskikh, Yu.V. *Analytical Problems for Asymptotics of Probability Distributions* (Naukova Dumka, 1981; in Russian).

[112] Krein, S.G., ed. *Functional Analysis* (Nauka, 1964; in Russian).

[113] Kushner, H.J., Clark, D.S. *Stochastic Approximation Methods for Constrained and Unconstrained Systems*. Applied Mathematical Sciences, vol. 26 (Springer, 1978).

[114] Kushner H.J., Dupuis P.,G. *Numerical methods for stochastic control problems in continuous time* (Springer, 1992).

[115] Liggett, T.M. An invariance principle for conditioned sums of independent random variables. *J. Math. Mech.*, 18:6 (1968), 559–570.

[116] Liptser, R.Sh., Shiryaev, A.N. *Statistics of Random Processes* (Springer, 2001).

[117] Lorden, G. Procedures for reacting to a change in distribution. *Ann. Math. Statist.*, 42:6 (1971), 1897–1908.

[118] Lotov, V.I. Asymptotic analysis of distributions in problems with two boundaries, I. *Theory Probab. Appl.*, 24:3 (1979), 480–491.

[119] Lotov, V.I. Asymptotic analysis of distributions in problems with two boundaries, II. *Theory Probab. Appl.*, 24:4 (1979), 869–876.

[120] Lotov, V.I. On the asymptotics of distributions in two-boundaries problems for random walks defined on a Markov chain. Asymptotic analysis of distributions of random processes. *Trudy Instituta Matematiki SO AN SSSR*, 13 (1989), 116–136. (In Russian.)

[121] Lotov, V.I., Khodzhibaev, V.R. Asymptotic expansions in a boundary problem. *Siberian Math. J.*, 25:5 (1984), 758–764.

[122] Lotov, V.I., Khodzhibaev, V.R. On limit theorems for the first exit time from a strip for stochastic processes, I. *Siberian Adv. Math.*, 8:3 (1998), 90–113.

[123] Lotov, V.I., Khodzhibaev, V.R. On limit theorems for the first exit time from a strip for stochastic processes, II. *Siberian Adv. Math.*, 8:4 (1998), 41–59.

[124] Lynch, J., Sethuraman, J. Large deviations for processes with independent increments. *Ann. Prob.*, 15:2 (1987), 610–627.

[125] Malyshev, V.A. *Random Walks. The Wiener–Hopf Equation in a Quadrant of the Plane. Galois Automorphisms* (Moscow University Publishing House, 1972; in Russian).

[126] Markushevich, A.I. *The Theory of Analytical Functions: A Brief Course* (Mir, 1983).

[127] *Mikosh, T., Nagaev, A.V.* Large deviations of heavy-tailed sums with applications in insurance. *Extremes*, 1:1 (1998), 81–110.

[128] Miller, H.D. A matrix factorization problem in the theory of random variables defined on a finite Markov chain. *Proc. Camb. Phil. Soc.*, 58:2 (1962), 265–285.

[129] Mogul'skii, A.A. Absolute estimates for moments of certain boundary functionals. *Theory Probab. Appl.*, 18:2 (1973), 340–347.

[130] Mogul'skii, A.A. Large deviations in the space $C(0, 1)$ for sums given on a finite Markov chain. *Siberian Math. J.*, 15:1 (1974), 43–53.

[131] Mogul'skii, A.A. Large deviations for trajectories of multidimensional random walks. *Theory Probab. Appl.*, 21:2 (1977), 300–315.

[132] Mogul'skii, A.A. Large deviation probabilities of random walks. *Trudy Instituta Matematiki SO AN SSSR*, 3 (1983), 93–124. (In Russian.)

[133] Mogul'skii, A.A. An integro-local theorem applicable on the whole half-axis to the sums of random variables with regularly varying distributions. *Siberian Math. J.*, 49:4 (2008), 669–683.

[134] Mogul'skii, A.A. Local limit theorem for the first crossing time of a fixed level by a random walk. *Siberian Adv. Math.*, 20:3 (2010), 191–200.

[135] Mogul'skii, A.A. The expansion theorem for the deviation integral. *Siberian Adv. Math.*, 23:4 (2010), 250–262.

[136] Mogul'skii, A.A. On the upper bound in the large deviation principle for sums of random vectors. *Siberian Adv. Math.*, 24:2 (2014), 140–152.

[137] Mogul'skii, A.A., Pagma, Ch. Superlarge deviations for sums of random variables with arithmetical super-exponential distributions. *Siberian Adv. Math.*, 18:3 (2008), 185–208.

[138] Mogul'skii, A.A., Rogozin, B.A. Random walks in the positive quadrant, I–III. *Siberian Adv. Math.*, 10:1 (2000), 34–72; 10:2 (2000), 35–103; 11:2 (2001), 35–59.

[139] Moustakides, G.V. Optimal stopping times for detecting changes in distributions. *Ann. Statist.*, 14:4 (1986), 1379–1387.

[140] Nagaev, A.V. Local theorems with an allowance of large deviations. In *Limit Theorems and Random Processes*, 71–88 (Fan, 1967; in Russian).

[141] Nagaev, A.V. Limit theorems for a scheme of series. In *Limit Theorems and Random Processes*, pp. 43–70 (Fan Publishers, 1967; in Russian).

[142] Nagaev, A.V. Integral limit theorems taking into account large deviations when Cramér's condition does not hold, I–II. *Theory Probab. Appl.*, 14:1 (1969), 51–64; 14:2 (1969), 193–208.

[143] Nagaev, A.V. Renewal theorems in $\mathbb{R}^d$. *Theory Probab. Appl.*, 24:3 (1980), 572–581.

[144] Nagaev, S.V. Some limit theorems for large deviations. *Theory Probab. Appl.*, 10:2 (1965), 214–235.

[145] Nagaev, S.V. On the asymptotic behaviour of one-sided large deviation probabilities. *Theory Probab. Appl.*, 26:2 (1982), 362–366.

[146] Page, E.S. Continuous inspection schemes. *Biometrika*, 41:1 (1954), 100–115.

[147] Paulauskas, V.I. Estimates of the remainder term in limit theorems in the case of stable limit law. *Lithuanian Math. J.*, 14:1 (1974), 127–146.

[148] Paulauskas V.I. Uniform and nonuniform estimates of the remainder term in a limit theorem with a stable limit law. *Lithuanian Math. J.*, 14:4 (1974), 661–672.

[149] Petrov, V.V. A generalization of Cramér's limit theorem. *Russian Math. Surveys*, 9:4 (1954), 195–202. (In Russian.)

[150] Petrov, V.V. Limit theorems for large deviations violating Cramér's condition. *Vestnik Leningrad. Univ. Ser. Mat. Meh. Astronom.*, 19 (1963), 49–68. (In Russian.)

[151] Petrov, V.V. *Sums of Independent Random Variables* (Springer, 1975).

[152] Pollak, M. Optimal detection of a change in distribution. *Ann. Statist*, 13:1 (1985), 206–227.

[153] Pollak, M. Average run length of optimal method of detecting a change in distribution. *Ann. Statist.*, 15:2 (1987), 749–779.

[154] Presman, E.L. Factorization methods and boundary problems for sums of random variables given on Markov chains. *Math. USSR Izvestiya*, 3:4 (1969), 815–852.

[155] Pukhalskii, A.A. On the theory of large deviations. *Theory Probab. Appl.*, 38:3 (1993), 490–497.

[156] Puhalskii, A.A. *Large Deviations and Idempotent Probability* (Chapman & Hall/CRC, 2001).

[157] Riesz, F., Szökefalvi-Nagy, B. *Functional Analysis* (Blackie & Son, London, 1956).

[158] Roberts, S.W. A comparison of some control chart procedures. *Technometrics*, 8 (1966), 411–430.

[159] Rockefeller, R.T. *Convex Analysis* (Princeton University Press, 1970).

[160] Rogozin, B.A. On distributions of functionals related to boundary problems for processes with independent increments. *Theory Probab. Appl.*, 11:4 (1966), 580–591.

[161] Rogozin, B.A. Distribution of the maximum of a process with independent increments. *Siberian Math. J.*, 10:6 (1969), 989–1010.

[162] Rozovskii, L.V. Probabilities of large deviations on the whole axis. *Theory Probab. Appl.*, 38:1 (1993), 53–79.

[163] Rvacheva, E.L. On domains of attraction of multi-dimensional distributions. *Select. Transl. Math. Statist. Probab.*, 2 (1962), 183–205.

[164] Sanov, I.N. On the probability of large deviations of random magnitudes. *Sbornik: Mathematics*, 42:1 (1957), 11–44. (In Russian.)

[165] Saulis, L., Statulevicius, V.A. *Limit Theorems for Large Deviations* (Kluwer, Dordrecht, 1991).

[166] Shepp, L.A. A local limit theorem. *Ann. Math. Stat.*, 35 (1964), 419–423.

[167] Shiryaev A.N. Minimax optimality of the method of cumulative sums (CUSUM) in the case of continuous time. *Russian Math. Surveys*, 51:4 (1996), 750–751.

[168] Shiryaev A.N. *Optimal Stopping Rules* (Springer, 2008). (Translated from the 1976 Russian original.)

[169] Skorokhod, A.V. Limit theorems for stochastic processes. *Theory Probab. Appl.*, 1:3 (1956), 261–290.

[170] Slaby, M. On the upper bound for large deviations of sums of i.i.d. random vectors. *Ann. Probab.*, 16:3 (1988), 978–990.

[171] Smith, W.L. Renewal theory and its ramifications. *J. Roy. Statist. Soc. B.*, 20:2 (1961), 95–150.

[172] Stam, A.J. Renewal theory in r-dimensions, I, II. *Composito Math.* 21 (1969), 383–399; 23, 1–13.

[173] Stone, C. On characteristic function and renewal theory. *Trans. Amer. Math. Soc.*, 20:2 (1965), 327–342.

[174] Stone, C. A local limit theorems for nonlattice multidimensional distribution functions. *Ann. Math. Statistics*, 36 (1965), 546–551.

[175] Stone, C. On local and ratio limit theorems. In: *Proc. Fifth Berkeley Symp. Math. Stat. Probab.*, vol. II, part II, 217–224 (University of California, 1966).

[176] Stroock, D.W., Varadhan, S.R.S. *Multidimensional Diffusion Processes* (Springer, 1979).

[177] Varadhan, S.R.S. Asymptotic probabilities and differential equations. *Comm. Pure Appl. Math.* 19:3 (1966), 261–286.

[178] Varadhan, S.R.S. *Large Deviations and Applications* (SIAM, 1984).

[179] Varadhan, S.R.S. Large deviations. *Ann. Probab.* 36:2 (2008), 397–419.

[180] *Wald, A.* Sequential Analysis (Wiley, 1947).

[181] Zaigraev, O. *Large-Deviation Theorems for Sums of Independent and Identically Distributed Random Vectors* (Universytet Mikolaja Kopernika, 2005).

Index

CPSIA information can be obtained
at www.ICGtesting.com
Printed in the USA
LVHW111916030821
694401LV00001B/67

9 781107 074682